T0328910

REDEFINING DIVERSITY AND DYNAMICS OF NATURAL RESOURCES MANAGEMENT IN ASIA

VOLUME 1

Sustainable Natural Resources Management in Dynamic Asia,
Ganesh P. Shivakoti, Ujjwal Pradhan and Helmi; editors

Volume 1 is dedicated to Nobel Laureate Elinor Ostrom who is the source of inspiration in drafting these volumes and all chapter authors of these volumes have benefited from her theoretical framework.

Book Title: Re-defining Diversity and Dynamism of Natural Resource Management in Asia

Book Editors: Ganesh P. Shivakoti, Shubhechchha Sharma and Raza Ullah

Other volumes published:

1) Upland Natural Resources and Social Ecological Systems in Northern Vietnam Volume 2
   Mai Van Thanh, Tran Duc Vien, Stephen J. Leisz and Ganesh P. Shivakoti; editors
2) Natural Resource Dynamics and Social Ecological System In Central Vietnam:
   Development, Resource Changes and Conservation Issues Volume 3
Tran N. Thang, Ngo T. Dung, David Hulse, Shubhechchha Sharma and Ganesh P. Shivakoti; editors
3) Reciprocal Relationship between Governance of Natural Resources and Socio-Ecological
   Systems Dynamics in West Sumatra Indonesia Volume 4

Rudi Febriamansyah, Yonariza, Raza Ullah and Ganesh P. Shivakoti; editors

REDEFINING DIVERSITY AND DYNAMICS OF NATURAL RESOURCES MANAGEMENT IN ASIA

Sustainable Natural Resources Management in Dynamic Asia

VOLUME 1

EDITED BY

GANESH P. SHIVAKOTI, UJJWAL PRADHAN AND HELMI

ELSEVIER

AMSTERDAM • BOSTON • HEIDELBERG • LONDON • NEW YORK • OXFORD
PARIS • SAN DIEGO • SAN FRANCISCO • SINGAPORE • SYDNEY • TOKYO

Elsevier
Radarweg 29, PO Box 211, 1000 AE Amsterdam, Netherlands
The Boulevard, Langford Lane, Kidlington, Oxford OX5 1GB, United Kingdom
50 Hampshire Street, 5th Floor, Cambridge, MA 02139, United States

Notices
Knowledge and best practice in this field are constantly changing. As new research and experience broaden our
understanding, changes in research methods, professional practices, or medical treatment may become necessary.

Practitioners and researchers must always rely on their own experience and knowledge in evaluating and using
any information, methods, compounds, or experiments described herein. In using such information or methods
they should be mindful of their own safety and the safety of others, including parties for whom they have a
professional responsibility.

To the fullest extent of the law, neither the Publisher nor the authors, contributors, or editors, assume any liability
for any injury and/or damage to persons or property as a matter of products liability, negligence or otherwise, or
from any use or operation of any methods, products, instructions, or ideas contained in the material herein.

Library of Congress Cataloging-in-Publication Data
A catalog record for this book is available from the Library of Congress

British Library Cataloguing-in-Publication Data
A catalogue record for this book is available from the British Library

ISBN: 978-0-12-805454-3

For information on all Elsevier publications
visit our website at https://www.elsevier.com/

Working together
to grow libraries in
developing countries

www.elsevier.com • www.bookaid.org

Publisher: Candice G. Janco
Acquisition Editor: Laura S Kelleher
Editorial Project Manager: Emily Thomson
Production Project Manager: Mohanapriyan Rajendran
Cover Designer: Matthew Limbert

Typeset by SPi Global, India

Contents

I

INTRODUCTION AND CONCEPTUAL BACKGROUND

1 Challenges of Sustainable Natural Resources Management in Dynamic Asia

G. SHIVAKOTI, R. ULLAH, U. PRADHAN

2 Theoretical Advances in Community-Based Natural Resources Management: Ostrom and Beyond

K.K. SHRESTHA, H.R. OJHA

3 Governing the Commons Through Understanding of Institutional Diversity: An Agenda for Application of Ostrom's Framework in Managing Natural Resources in Asia

R.C. BASTAKOTI, G. SHIVAKOTI

II

THEORETICAL ISSUES

4 Challenges of Polycentric Water Governance in Southeast Asia: Awkward Facts, Missing Mechanisms, and Working with Institutional Diversity

B. BRUNS

Contributors

E. Achyar Institute of Participatory Approaches, Development & Studies, Padang, West Sumatra, Indonesia

I. Andriyani Jember University, Jawa Timur, Indonesia

R.C. Bastakoti International Water Management Institute, Lalitpur, Nepal

B. Bruns Independent Consulting Sociologist, PO Box 176, Warm Springs, VA, United States

R. Cochard Institute of Integrative Biology, Swiss Federal Institute of Technology, Zurich, Switzerland; Asian Institute of Technology, Klong Luang, Pathumthani, Thailand

Helmi Andalas University, Padang, Indonesia

M. Inoue The University of Tokyo, Tokyo, Japan

D. Jourdain CIRAD, UMR G-EAU, Montpellier, France; Asian Institute of Technology, Klong Luang, Pathumthani, Thailand

B. Kartiwa IAHRI, Indonesian Agriculture Research Institute, Bogor, Indonesia

B. Lidon CIRAD UMR G-EAU, Montpellier, France

Mahdi Andalas University, Padang, Indonesia

H.R. Ojha The University of New South Wales, Sydney, NSW, Australia

Partoyo Universitas Pembangunan Nasional Veteran Yogyakarta, Yogyakarta, Indonesia

S. Perret CIRAD, UMR G-EAU, Montpellier, France

U. Pradhan World Agro-Forestry Center, SE Asian Regional Office; ICRAF-SEA, Bogor, Indonesia

B.A. Rusdi Andalas University, Padang, Indonesia

M.A. Sardjono Mulawarman University, Samarinda, Indonesia

D. Schmidt-Vogt Mountain Societies Research Institute, University of Central Asia, Bishkek, Kyrgyz Republic

S. Sharma WWF-Nepal, Kathmandu, Nepal

G. Shivakoti The University of Tokyo, Tokyo, Japan; Asian Institute of Technology, Bangkok, Thailand

K.K. Shrestha The University of New South Wales, Sydney, NSW, Australia

R.P. Shrestha Asian Institute of Technology, Pathumthani, Thailand

M.T. Sirait ICRAF-SEA, Bogor, Indonesia

P. Soni Asian Institute of Technology, Klong Luang, Pathumthani, Thailand

S. Srisopaporn Asian Institute of Technology, Bangkok, Thailand

A. Surahman Asian Institute of Technology, Klong Luang, Pathumthani, Thailand; Indonesian Agency for Agricultural Research and Development (IAARD), Jakarta Selatan, Indonesia

Ngo T.D. Hue University of Agriculture and Forestry, Hue City, Vietnam

S. Tsuyuki University of Tokyo, Tokyo, Japan

R. Ullah The University of Agriculture, Peshawar, Pakistan

F. von Benda-Beckmann Max Planck Institute for Social Anthropology; Martin Luther University Halle-Wittenberg, Halle, Germany

K. von Benda-Beckmann Max Planck Institute for Social Anthropology; Martin Luther University Halle-Wittenberg, Halle, Germany

B. White Institute of Social Studies (ISS), Den Haag, The Netherlands

W.V.C. Wong University of Tokyo, Tokyo, Japan; University of Malaysia Sabah, Kota Kinabalu, Sabah, Malaysia

Words From Book Editors

CONTEXT

Elinor Ostrom received the Nobel Prize in Economics for showing how the "commons" is vital to the livelihoods of many throughout the world. Her work examined the rhetoric of the "tragedy of the commons," which has been used as the underlying foundation in privatizing property and centralizing its management as a way to protect finite resources from depletion. She worked, along with others, to overturn the "conventional wisdom" of the tragedy of the commons by validating the means and ways that local resources can be effectively managed through common property regimes instead of through the central government or privatization. Ostrom identified eight design principles relating to how common pool resources can be governed sustainably and equitably in a community. Similarly, the Institutional Analysis and Development (IAD) framework summarizes the ways in which institutions function and adjust over time. The framework is a "multi-level conceptual map," which describes a specific hierarchical section of interactions made in a system. The framework seeks to identify and explain interactions between actors and action situations.

As a political scientist, Ostrom has been a source of inspiration for many researchers and social scientists, including this four volumes book. Her theories and approach serve as the foundation for many of the chapters within these volumes. Following in her footsteps, the books is based on information collected during fieldwork that utilized quantitative as well as qualitative data, and on comparative case studies, which were then analyzed to gain an understanding of the situation, rather than starting from a formulated assumption of reality. The case studies in these volumes highlight the issues linked to the management of the environment and natural resources, and seek to bring about an understanding of the mechanisms used in managing the natural resource base in the regions, and how different stakeholders interact with each other in managing these natural resources. The details of the books are as follows:

Volume title		Editors
"Re-defining Diversity and Dynamism of Natural Resources Management in Asia"		Ganesh P. Shivakoti, Shubhechchha Sharma, and Raza Ullah
Volume I	Sustainable Natural Resources Management in Dynamic Asia	Ganesh P. Shivakoti, Ujjwal Pradhan, and Helmi
Volume II	Upland Natural Resources and Social Ecological Systems in Northern Vietnam	Mai Van Thanh, Tran Duc Vien, Stephen J. Leisz, and Ganesh P. Shivakoti
Volume III	Natural Resource Dynamics and Social Ecological Systems in Central Vietnam: Development, Resource Changes and Conservation Issue	Tran Nam Thang, Ngo Tri Dung, David Hulse, Shubhechchha Sharma, and Ganesh P. Shivakoti

Continued

Volume title		Editors
Volume IV	Reciprocal Relationship between Governance of Natural Resources and Socio-Ecological Systems Dynamics in West Sumatra Indonesia	Rudi Febriamansyah, Yonariza, Raza Ullah, and Ganesh P. Shivakoti

These volumes are made possible through the collaboration of diverse stakeholders. The intellectual support provided by Elinor Ostrom and other colleagues through the Ostrom Workshop in Political Theory and Policy Analysis at the Indiana University over the last two and half decades has provided a solid foundation for drafting the book. The colleagues at the Asian Institute of Technology (AIT) have been actively collaborating with the Workshop since the creation of the Nepal Irrigation, Institutions and Systems (NIIS) database; and the later Asian Irrigation, Institutions and Systems (AIIS) database (Ostrom, Benjamin and Shivakoti, 1992; Shivakoti and Ostrom, 2002; Shivakoti et al., 2005; Ostrom, Lam, Pradhan and Shivakoti, 2011). The International Forest Resources and Institutions (IFRI) network carried out research to support policy makers and practitioners in designing evidence based natural resource polices based on the IAD framework at Indiana University, which was further mainstreamed by the University of Michigan. In order to support this, the Ford Foundation (Vietnam, India, and Indonesia) provided grants for capacity building and concerted knowledge sharing mechanisms in integrated natural resources management (INRM) at Indonesia's Andalas University in West Sumatra, Vietnam's National University of Agriculture (VNUA) in Hanoi, and the Hue University of Agriculture and Forestry (HUAF) in Hue, as well as at the AIT for collaboration in curriculum

development and in building capacity through mutual learning in the form of masters and PhD fellowships (Webb and Shivakoti, 2008). Earlier, the MacArthur Foundation explored ways to support natural resource dependent communities through the long term monitoring of biodiversity, the domestication of valuable plant species, and by embarking on long-term training programs to aid communities in managing natural resources.

VOLUME 1

This volume raises issues related to the dependence of local communities on natural resources for their livelihood; their rights, access, and control over natural resources; the current practices being adopted in managing natural resources and socio-ecological systems; and new forms of natural resource governance, including the implementation methodology of REDD+ in three countries in Asia. This volume also links regional issues with those at the local level, and contributes to the process of application of various multimethod and modeling techniques and approaches, which is identified in the current volume in order to build problem solving mechanisms for the management of natural resources at the local level. Earlier, the Ford Foundation Delhi office supported a workshop on Asian Irrigation in Transition, and its subsequent publication (Shivakoti et al., 2005) was followed by Ford Foundation Jakarta office's long term support for expanding the knowledge on integrated natural resources management, as mediated by institutions in the dynamic social ecological systems.

VOLUME 2

From the early 1990s to the present, the Center for Agricultural Research and

Ecological Studies (CARES) of VNUA and the School of Environment, Resources and Development (SERD) of AIT have collaborated in studying and understanding the participatory process that has occurred during the transition from traditional swidden farming to other farming systems promoted as ecologically sustainable, livelihood adaptations by local communities in the northern Vietnamese terrain, with a special note made to the newly emerging context of climate change. This collaborative effort, which is aimed at reconciling the standard concepts of development with conservation, has focused on the small microwatersheds within the larger Red River delta basin. Support for this effort has been provided by the Ford Foundation and the MacArthur Foundation, in close coordination with CARES and VNUA, with the guidance from the Ministry of Agriculture and Rural Development (MARD) and the Ministry of Natural Resources and Environment (MONRE) at the national, regional, and community level. Notable research documentation in this volume includes issues such as local-level land cover and land use transitions, conservation and development related agro-forestry policy outcomes at the local level, and alternative livelihood adaptation and management strategies in the context of climate change. A majority of these studies have examined the outcomes of conservation and development policies on rural communities, which have participated in their implementation through collaborative governance and participatory management in partnership with participatory community institutions. The editors and authors feel that the findings of these rich field-based studies will not only be of interest and use to national policymakers and practitioners and the faculty and students of academic institutions, but can also be equally applicable to guiding conservation and development issues for those scholars interested in understanding a developing country's social ecological systems, and its context-specific adaptation strategies.

VOLUME 3

From the early 2000 to the present, Hue University of Agriculture and Forestry (HUAF) and the School of Environment, Resources and Development (SERD) of AIT supported by MacArthur Foundation and Ford Foundation Jakarta office have collaborated in studying and understanding the participatory process of Social Ecological Systems Dynamics that has occurred during the opening up of Central Highland for infrastructure development. This collaborative effort, which is aimed at reconciling the standard concepts of development with conservation, has focused on the balance between conservation and development in the buffer zone areas as mediated by public resource management institutions such as Ministry of Agriculture and Rural Development (MARD), Ministry of Natural Resources and Environment (MONRE) including National Parks located in the region. Notable research documentation in this volume includes on issues such as local level conservation and development related policy outcomes at the local level, alternative livelihood adaptation and management strategies in the context of climate change. A majority of these studies have examined the outcomes of conservation and development policies on the rural communities which have participated in their implementation through collaborative governance and participatory management in partnership with participatory community institutions.

VOLUME 4

The issues discussed above are pronounced more in Indonesia among the Asian countries and the Western Sumatra is such typical example mainly due to earlier logging concessionaries, recent expansion of State and private plantation of para-rubber and oil palm plantation. These new frontiers have created confrontations among the local community deriving their livelihoods based on inland and coastal natural resources and the outsiders starting mega projects based on local resources be it the plantations or the massive coastal aqua cultural development. To document these dynamic processes Ford Foundation Country Office in Jakarta funded collaborative project between Andalas University and Asian Institute of Technology (AIT) on Capacity building in Integrated Natural Resources Management. The main objective of the project was Andalas faculty participate in understanding theories and diverse policy arenas for understanding and managing common pool resources (CPRs) which have collective action problem and dilemma through masters and doctoral field research on a collaborative mode (AIT, Indiana University and Andalas). This laid foundation for joint graduate program in Integrated Natural Resources Management (INRM). Major activities of the Ford Foundation initiatives involved the faculty from Andalas not only complete their degrees at AIT but also participated in several collaborative training.

1 BACKGROUND

Throughout Asia, degradation of natural resources is happening at a higher rate, and is a primary environmental concern. Recent tragedies associated with climate change have left a clear footprint on them, from deforestation, land degradation, and changing hydrological and precipitation patterns. A significant proportion of land use conversion is undertaken through rural activities, where resource degradation and deforestation is often the result of overexploitation by users who make resource-use decisions based on a complex matrix of options, and potential outcomes.

South and Southeast Asia are among the most dynamic regions in the world. The fundamental political and socioeconomic setting has been altered following decades of political, financial, and economic turmoil in the region. The economic growth, infrastructure development, and industrialization are having concurrent impacts on natural resources in the form of resource degradation, and the result is often social turmoil at different scales. The natural resource base is being degraded at the cost of producing economic output. Some of these impacts have been offset by enhancing natural resource use efficiency, and through appropriate technology extension. However, the net end results are prominent in terms of increasing resource depletion and social unrest. Furthermore, climate change impacts call for further adaptation and mitigation measures in order to address the consequences of erratic precipitation and temperature fluctuations, salt intrusions, and sea level increases which ultimately affect the livelihood of natural resource dependent communities.

Governments, Non-governmental organizations (NGOs), and academics have been searching for appropriate policy recommendations that will mitigate the trend of natural resource degradation. By promoting effective policy and building the capacity of key stakeholders, it is envisioned that sustainable development can be promoted from both the top-down and bottom-up perspectives. Capacity building in the field of natural resource management, and poverty alleviation is, then, an urgent need; and several policy alternatives have been suggested (Inoue and Shivakoti, 2015; Inoue and Isozaki, 2003; Webb and Shivakoti, 2008).

The importance of informed policy guidance in sustainable governance and the management of common pool resources (CPRs), in general, have been recognized due to the conflicting and competing demand for use of these resources in the changing economic context in Asia (Balooni and Inoue, 2007; Nath, Inoue and Chakma, 2005; Pulhin, Inoue and Enters, 2007; Shivakoti and Ostrom, 2008; Viswanathan and Shivakoti, 2008). This is because these resources are unique in respect to their context. The management of these resources are by the public, often by local people, in a partnership between the state and the local community; but on a day-to-day basis, the benefits are at the individual and private level. In the larger environmental context, however, the benefits and costs have global implications. There are several modes of governance and management arrangement possible for these resources in a private-public partnership. Several issues related to governance and management need to be addressed, which can directly feed into the ongoing policy efforts of decentralization and poverty reduction measures in South and South East Asia.

While there has been a large number of studies, and many management prescriptions made, for the management of natural resources, either from the national development point-of-view or from the local-level community perspectives, there are few studies which point toward the interrelationship among other resources and CPRs, as mediated by institutional arrangement, and that have implications for the management of CPRs in an integrated manner, vis-a-vis poverty reduction. In our previous research, we have identified several anomalies and tried to explain these in terms of better management regimes for the CPRs of several Asian countries (Dorji, Webb and Shivakoti, 2006; Gautam, Shivakoti and Webb, 2004; Kitjewachakul, Shivakoti and Webb, 2004;

Mahdi, Shivakoti and Schmidt-Vogt, 2009; Shivakoti et al., 1997; Dung and Webb, 2008; Yonariza and Shivakoti, 2008). However, there are still several issues, such as the failure to comprehend and conceptualize social and ecological systems as coupled systems that adapt, self-organize, and are coevolutionary. The information obtained through these studies tends to be fragmented and scattered, leading to incomplete decision making, as they do not reflect the entire scenario. The shared vision of the diverse complexities, that are the reality of natural resource management, needs to be fed into the governance and management arrangements in order to create appropriate management guidelines for the integrated management of natural resources, and CPR as a whole.

Specifically, the following issues are of interest:

a. How can economic growth be encouraged while holding natural resources intact?
b. How has the decentralization of natural management rights affected the resource conditions, and how has it addressed concerns of the necessity to incorporate gender concerns and social inclusion in the process?
c. How can the sustainability efforts to improve the productive capacity of CPR systems be assessed in the context of the current debate on the effects of climate change, and the implementation of new programs such as Payment for Ecosystem Services (PES) and REDD+?
d. How can multiple methods of information gathering and analysis (eg use of various qualitative and quantitative social science methods in conjunction with methods from the biological sciences, and time series remote sensing data collection methods) on CPRs be integrated into national natural resource

policy guidelines, and the results be used by local managers and users of CPRs, government agencies, and scholars?

e. What are the effective polycentric policy approaches for governance and management of CPRs, which are environmentally sustainable and gender balanced?

2 OBJECTIVES OF THESE VOLUMES

At each level of society, there are stakeholders, both at the public and private level, who are primarily concerned with efforts of management enhancement and policy arrangements. Current theoretical research indicates that this is the case whether it is deforestation, resource degradation, the conservation of biodiversity hotspots, or climate change adaptation. The real struggles of these local-level actors directly affect the management of CPR, as well as the hundreds of people who are dependent upon them for a living. This book is about those decisions as the managers of natural resources. Basically, the authors of these chapters explore outcomes after decentralization and economic reforms, respectively. The volumes of this book scrutinize the variations of management practices with, and between, communities, local administration, and the CPR. Economic growth is every country's desire, but in the context of South and South East Asia, much of the economic growth is enabled by the over use of the natural resource base. The conundrum is that these countries need economic growth to advance, but the models of economic growth that are advanced, negatively affect the environment, which the country, depends upon. Examples of this are seen in such varied contexts as the construction of highways through protected areas, the construction of massive hydropower dams, and the conversion of traditional agricultural fields into rubber and oil palm plantations.

The research also shows that the different levels of communities, administration, and people are sometimes highly interactive and overlapping, for that reason, it is necessary to undertake coordinated activities that lead to information capture and capacity building at the national, district, and local levels. Thus the impacts of earlier intervention efforts (various policies in general and decentralization in particular) for effective outcomes have been limited, due to the unwillingness of higher administrative officials to give up their authority, the lack of trust and confidence of officials in the ability of local communities in managing CPR, local elites capturing the benefits of decentralization in their favor, and high occurrences of conflicts among multiple stakeholders at the local level (IGES, 2007).

In the areas of natural resource management particular to wildlife ecology monitoring and climate change adaptation, the merging of traditional knowledge with science is likely to result in better management results. Within many societies, daily practices and ways of life are constantly changing and adapting to new situations and realities. Information passed through these societies, while not precise and usually of a qualitative nature, is valued for the reason that it is derived from experience over time. Scientific studies can backstop local knowledge, and augment it through the application of rigorous scientific method derived knowledge, examining the best practices in various natural resource management systems over spatial and temporal scales. The amalgamation of scientific studies and local knowledge, which is trusted by locals, may lead to powerful new policies directed toward nature conservation and livelihood improvements.

Ethnic minorities, living in the vicinity to giant infrastructure projects, have unequal access, and control over, resources compared to other more powerful groups. Subsistence agriculture, fishery, swiddening, and a few off-farm options are the livelihood activities for these individuals. But unfortunately, these livelihood options are in areas that will be hit the most by changing climatic scenarios, and these people are the least equipped to cope; a situation that further aggravates the possibility of diversifying their livelihood options. Increasing tree coverage can help to mitigate climate change through the sequestering of carbon in trees. Sustainably planting trees requires technical, social, and political dimensions that are mainly possible through the decentralization of power to local communities to prevent issues of deforestation and degradation. The role of traditional institutions hence becomes crucial to reviving social learning, risk sharing, diversifying options, formulating adaptive plans and their effective implementation, fostering stress tolerance, and capacity building against climate change effects.

Though, the role of institutions in managing common pool resources has been explained in literature, it is also worth noting that institutions play significant roles in climate change adaptation. A study conducted by Gabunda and Barker (1995) and Nyangena (2004) observed that household affiliations in social networks were highly correlated with embracing soil erosion retaining technologies. Likewise, Jagger and Pender (2006) assumed that individuals involved in natural resource management focused programs were likely to implement land management expertise, regardless of their direct involvement in particular organizations. Friis-Hansen (2005) partially verifies that there is a positive relationship among participation in a farmer's institution and the adoption of smart agriculture technology. Dorward et al. (2009) correspondingly notes that institutions are vital in shaping the capability of local agrarians to respond to challenges and opportunities. This study has also shown that institutions are the primary attribute in fostering individuals and households to diversify livelihoods in order to adapt to a changing climate. In the context of REDD+, a system is required that can transcend national boundaries, interconnect different governance levels, and allow both traditional and modern policy actors to cooperate. Such a system emphasizes the integration of both formal and informal rule making mechanisms and actor linkages in every governance stage, which steer toward adapting to and mitigating the effects of local and global environmental change (Corbera and Schroeder, 2010).

Based on the above noted discussions, the volumes in this book bring these issues forward for a global audience and policy makers. Though earlier studies show that the relationship between scientific study and outcomes in decision making are usually complex; we hope that the studies examined and discussed here can have some degree of impact on academics, practitioners, and managers.

G.P. Shivakoti, S. Sharma, and R. Ullah

References

Balooni, K.B., Inoue, M., 2007. Decentralized forest management in South and Southeast Asia. J. Forest. 2007, 414–420.

Corbera, E., Schroeder, H., 2010. Governing and Implementing REDD+. Environ. Sci. Pol. http://dx.doi.org/10.1016/j.envsci.2010.11.002.

Dorji, L., Webb, E., Shivakoti, G.P., 2006. Forest property rights under nationalized forest management in Bhutan. Environ. Conservat. 33 (2), 141–147.

Dorward, A., Kirsten, J., Omamo, S., Poulton, C., Vink, N., 2009. Institutions and the agricultural development challenge in Africa. In: Kirsten, J.F., Dorward, A.R., Poulton, C., Vink, N. (Eds.), Institutional Economics Perspectives on African Agricultural Development. IFPRI, Washington DC.

Dung, N.T., Webb, E., 2008. Incentives of the forest land allocation process: Implications for forest management in Nam Dong District, Central Vietnam. In: Webb, E., Shivakoti, G.P. (Eds.), Decentralization, Forests and Rural Communities: Policy outcome in South and South East Asia. SAGE Publications, New Delhi, pp. 269–291.

Friis-Hansen, E., 2005. Agricultural development among poor farmers in Soroti district, Uganda: Impact Assessment of agricultural technology, farmer empowerment and changes in opportunity structures. Paper presented at Impact Assessment Workshop at CYMMYT, Mexico, 19–21. October. http://citeseerx.ist.psu.edu/viewdoc/download?doi=10.1.1.464.8651&rep=rep1&type=pdf.

Gautam, A., Shivakoti, G.P., Webb, E.L., 2004. A review of forest policies, institutions, and the resource condition in Nepal. Int. Forest. Rev. 6 (2), 136–148.

Gabunda, F., Barker, R., 1995. Adoption of hedgerow technology in Matalom, Leyte Philipines. Mimeo. In: Bluffstone, R., Khlin, G. (Eds.), 2011. Agricultural Investment and Productivity: Building Sustainability in East Africa. RFF Press, Washington, DC/London.

IGES, 2007. Decentralization and State-sponsored Community Forestry in Asia. Institute for Global Environmental Studies, Kanagawa.

Inoue, M., Isozaki, H., 2003. People and Forest-policy and Local Reality in Southeast Asia, the Russian Far East and Japan. Kluwer Academic Publishers, Netherlands.

Inoue, M., Shivakoti, G.P. (Eds.), 2015. Multi-level Forest Governance in Asia: Concepts, Challenges and the Way Forward. Sage Publications, New Delhi/California/London/Singapore.

Jagger, P., Pender, J., 2006. Impacts of Programs and Organizations on the Adoption of Sustainable Land Management Technologies in Uganda. IFPRI, Washington, DC.

Kijtewachakul, N., Shivakoti, G.P., Webb, E., 2004. Forest health, collective behaviors and management. Environ. Manage. 33 (5), 620–636.

Mahdi, Shivakoti, G.P., Schmidt-Vogt, D., 2009. Livelihood change and livelihood sustainability in the uplands of Lembang Subwatershed, West Sumatra, Indonesia, in a changing natural resource management context. Environ. Manage. 43, 84–99.

Nath, T.K., Inoue, M., Chakma, S., 2005. Prevailing shifting cultivation in the Chittagong Hill Tracts, Bangladesh: some thoughts on rural livelihood and policy issues. Int. For. Rev. 7 (5), 327–328.

Nyangena, W., 2004. The effect of social capital on technology adoption: empirical evidence from Kenya. Paper presented at 13th Annual Conference of the European Association of Environmental and Resource Economics, Budapest.

Ostrom, E., Benjamin, P., Shivakoti, G.P., 1992. Institutions, Incentives, and Irrigation in Nepal: June 1992. (Monograph) Workshop in Political Theory and Policy Analysis, Indiana University, Bloomington, Indiana, USA.

Ostrom, E., Lam, W.F., Pradhan, P., Shivakoti, G.P., 2011. Improving Irrigation Performance in Asia: Innovative Intervention in Nepal. Edward Elgar Publishers, Cheltenham, UK.

Pulhin, J.M., Inoue, M., Enters, T., 2007. Three decades of community-based forest management in the Philippines: emerging lessons for sustainable and equitable forest management. Int. For. Rev. 9 (4), 865–883.

Shivakoti, G., Ostrom, E., 2008. Facilitating decentralized policies for sustainable governance and management of forest resources in Asia. In: Webb, E., Shivakoti, G.P. (Eds.), Decentralization, Forests and Rural Communities: Policy Outcomes in South and Southeast Asia. Sage Publications, New Delhi/Thousand Oaks/London/Singapore, pp. 292–310.

Shivakoti, G.P., Ostrom, E. (Eds.), 2002. Improving Irrigation Governance and Management in Nepal. Institute of Contemporary Studies (ICS) Press, California, Oakland.

Shivakoti, G.P., Vermillion, D., Lam, W.F., Ostrom, E., Pradhan, U., Yoder, R., 2005. Asian Irrigation in Transition-Responding to Challenges. Sage Publications, New Delhi/Thousand Oaks/London.

Shivakoti, G., Varughese, G., Ostrom, E., Shukla, A., Thapa, G., 1997. People and participation in sustainable development: understanding the dynamics of natural resource system. In: Proceedings of an International Conference held at Institute of Agriculture and Animal Science, Rampur, Chitwan, Nepal. 17–21 March, 1996. Bloomington, Indiana and Rampur, Chitwan.

Viswanathan, P.K., Shivakoti, G.P., 2008. Adoption of rubber integrated farm livelihood systems: contrasting empirical evidences from Indian context. J. For. Res. 13 (1), 1–14.

Webb, E., Shivakoti, G.P. (Eds.), 2008. Decentralization, Forests and Rural Communities: Policy Outcomes in South and Southeast Asia. Sage Publications, New Delhi/Thousand Oaks/London/Singapore.

Yonariza, Shivakoti, G.P., 2008. Decentralization and co-management of protected areas in Indonesia. J. Legal Plur. 57, 141–165.

Foreword

Natural resource management issues have attracted increasing attention in recent decades, particularly in Asia, partly in response to a sequence of crises in energy, food, water, and other resources. Effective governance and management of resources have always been important, but have become increasingly challenging in the face of changing climate, livelihoods, and market pressures. Many Asian countries have compromised their natural resource base for the sake of development, and are consequently facing various environmental challenges. The pressure on natural resources has potentially been aggravated by the development of infrastructure, advancement in extraction techniques, and expanding product markets that enlarge extraction opportunities for concession holders as well as local populations. Under such circumstances, the quality of land, water, and forest is threatened, and the regenerating capacity of resources is hardly guaranteed. The haphazard use, and the conflicts over use, of natural resources pose serious threats to the viability and sustainability of the natural resources in Asia. These conflicts over natural resources are largely fueled by the dual goals of the government and the community for both the preservation of and utilization of protected zones, and the political and economic value of natural resources that has also increased inter-ethnic tensions and struggles.

What, then, can research contribute, in the face of these multifaceted challenges? The legacy of Elinor Ostrom, exemplified in this volume, provides guidance on many of these issues. The Institutional Analysis and Development (IAD) and Social Ecological System (SES) frameworks she proposed offer entry points in analyzing how different stakeholders interact with each other and their natural environment, and how communities can successfully manage Common Pool Resources (CPRs) through local self-governance. The design principles for effective CPR management, developed by Elinor Ostrom, have wide applicability in analyzing the governance of many resources in Asia. These frameworks and design principles are tested in various resource systems in Asia, in the present volume. The iteration between theory and nuanced field research, reflected in the overall structure of this volume, as well as in its individual chapters, is another hallmark of Ostrom's work.

Following themes explored by Ostrom and colleagues, various studies reported in this volume analyze community participation, nonattenuated property rights, and collective action as potential remedies for these issues and concerns. Examples of such initiatives are the Forest Management Units (FMUs), the collective community forestry stewardship program (*Hutan Kemasyarakatan HKm*), and the village forest stewardship program (Hutan Dsa HD) for forest management in Indonesia. The Zanjera irrigation system in the northern Philippines, the Subaks in Bali, and the communal *muang fai* irrigation schemes in Northern Thailand are also examples of long-enduring, locally-governed farmer managed irrigation systems. However, the presence of a legal framework, or issuing formal rights, might be of little help, unless these rights can be translated into secure access to, and management of, natural resources. The main concern is not the formal presence of policies on legal rights, but the effective enforcement of such policies in resource allocation and dispute resolution.

Though environmental and conservation policies and legislation exist in many countries in Asia, implementation and effective enforcement of these policies is a concern. In particular, many countries in the region have adopted decentralization policies, delegating authority to lower levels and expanding community rights over natural resources. But many of these decentralization policies have not been fully implemented, and the capacity of local groups and institutions to take over state roles, or effectively manage resources, is not guaranteed. Chapters in this volume show how Ostrom's work on institutional diversity and polycentric governance provides guidance on possible institutional arrangements that go beyond relying on state *or* local groups, to developing institutional arrangements governing a range of resources.

The first book published by Elinor Ostrom and colleagues, after she won the Nobel Prize in Economics, was titled *Working together: collective action, the commons, and multiple methods in practice.*[1] This title epitomizes much of her work, and also captures the essence of this present volume in her honor. First, this volume represents extensive collaboration. Many of the chapters' authors have worked together over decades, sharing insights; yet, there are also new voices and perspectives brought in. Second, dealing with different types of commons, including forest, land, water, biodiversity, and ecosystems, helps to identify what is specific to one type of resource, and what are broader principles. Finally, addressing these complex issues requires integrated methodologies considering the environment as well as social conditions, which, in turn, calls for interdisciplinary approaches. This volume reflects these emerging challenges with an emphasis on multistakeholder participatory research approaches to natural resource management that recognizes the competing demands on resource use, and conservation for social, economic, and environmental benefits. There are no panaceas in this area, but the studies in this volume show how careful analysis can lead to improved outcomes in policy and practice.

R. MEINZEN-DICK
International Food Policy Research Institute
Washington, DC

---

[1] Poteete, A.R., Janssen, M.A., Ostrom, E., 2010. Working Together: Collective Action, the Commons, and Multiple Methods in Practice. Princeton University Press, Princeton NJ.

Preface

In such diverse and dynamic settings, Asian natural resource management (NRM) issues cannot only be solved by technology and panacea; it is equally important to understand the nuts and bolts of local institutions and their specific contexts, livelihood dependency, the opportunities of NRM in the context of market pressure, and its global link to local resources. The consideration of effective governance, and the management of such dynamic resources, requires an even greater attention to the context of dramatic climate changes, and their immediate impact on local communities and the surrounding resources. To cope with such issues, several governments are advocating decentralization and devolution of management regimes through good governance, which have both positive and negative impacts on resources. The local community has adapted accordingly. There have been several case studies conducted that evaluate the success and failure of such management regimes at the local level, which can be an important feed-back mechanism for suggesting improvements in policy formulation.

The issues discussed have created confrontations among the local communities, who derive their livelihoods based on inland and coastal natural resources, and the outsiders starting megaprojects based on local resources, be it plantations or massive coastal aqua-cultural development. In order to document these dynamic processes, the Ford Foundation Country Office in Jakarta funded a long-term collaborative project between the Andalas University, located in West Sumatra, Indonesia, and the Asian Institute of Technology (AIT), a regional graduate training and capacity building institution located in Bangkok, Thailand, on capacity building in integrated natural resources management (INRM). Major activities of the Ford Foundation initiatives involved the faculty from Andalas, who not only completed their degrees at AIT, but also participated in several collaborative training programs, such as the International Forestry Resources and Institution's (IFRI) training which was jointly organized by the Ostrom Workshop in Political Theory and Policy Analysis at Indiana University, the Natural Resources Institute at University of Michigan, and at AIT. They also developed an international collaborative network through participating in the biennial International Association for the Study of Commons (IASC). The intellectual contribution of Elinor Ostrom has been a source of inspiration for the authors of these chapters in the present volume, and many have benefited from her theoretical framework. The outcome of such intellectual capital is in this current volume, which serves as an important textbook on INRM, covering theory, its application, and related case studies.

This volume has been made possible through direct, or indirect, contributions from several organizations and individuals. We wish to acknowledge the generous support provided by the Ford Foundation Country Offices in Delhi and Jakarta, which paved the way for academic collaboration between the Asian Institute of Technology and Andalas University, with the intellectual support provided by the Ostrom Workshop in Political Theory and Policy Analysis at Indiana University. We wish to acknowledge the tremendous efforts of our colleagues at the Ostrom Center for the Advanced Study in Natural Resource Governance (OCEAN) at the Asian Institute of Technology, who took the burden upon their shoulders to take this volume to press. We appreciate the endless administrative and editing efforts of Laura Kelleher and Emily Thomson at Elsevier in putting everything into a coherent manner.

W. Kanok-Nukulchai
Asian Institute of Technology, Bangkok, Thailand
T. Husni
Rector Andalas University, Padang, Indonesia

INTRODUCTION AND CONCEPTUAL BACKGROUND

1

Challenges of Sustainable Natural Resources Management in Dynamic Asia

G. Shivakoti[*,§], *R. Ullah*[†], *U. Pradhan*[‡]

[*]The University of Tokyo, Tokyo, Japan [†]The University of Agriculture, Peshawar, Pakistan
[‡]World Agro-Forestry Center, SE Asian Regional Office, Bogor, Indonesia
[§]Asian Institute of Technology, Bangkok, Thailand

1.1 BACKGROUND

Asia has been a very dynamic region in the last several decades. The economic and financial crises in many countries in the region followed by political turmoil have changed the socioeconomic and political setting. Natural resources management (NRM) in such a dynamic setting is not only a technology fix and a panacea but also an understanding of the nuts and bolts of local institutions, their specific contexts, livelihood dependency, and opportunities of NRM in the context of market pressure and global link of local resources. The consideration of effective governance and management of such dynamic resources requires even greater attention in the context of dramatic climate changes and their immediate impact on local communities and surrounding resources. The importance of local communities participating on Reducing Emissions from Deforestation and Forest Degradation (REDD+) and Payment for Environmental Services in addressing local environmental issues cannot be undermined, especially in improving methodology for effective carbon credit accounting and facilitating policies recognized by both national and international policy stakeholders. To cope with such issues, several governments are advocating decentralization and devolution of management regimes through good governance, which have both positive and negative impacts on resources; and local communities have adapted accordingly. There have been several case studies conducted to evaluate the success and failure of such management regimes at the local level that can be important feedback mechanisms to suggest improvement in policy formulation.

In the race to be a developed country, many of the Asian countries have foregone natural resources through haphazard use. As a result, the people are faced with numerous environmental challenges, particularly deforestation and forest degradation, biodiversity loss, ecosystem degradation, reduction in soil quality, and a reduction in available water quantity. This is prominently displayed through the occurrence of haze clouds in Asia, often considered to be caused by fire; whereby, slash and burn is a common technique for converting natural forest to oil palm plantations. The 1997–98 haze incident affected larger areas encircling Indonesia to neighboring countries of Brunei Darussalam, Papua New Guinea, the Philippines, Singapore, and Thailand. This resulted in an estimated loss of US$9 billion that damaged 9 million hectares of land and adversely affected 70 million people (APFED, 2004). While strong environmental and conservation policies and legislation exist in many countries, effective implementation and enforcement is always a concern. To make it worse, corrupt practices do not make it easier.

1.2 IMPACTS OF ECONOMIC AND FINANCIAL CRISES ON NATURAL RESOURCE MANAGEMENT IN ASIA

Asian countries have still not recovered from the Asian financial crisis of 1997–98 and have faced a wide range of challenges spanning their economic, social, and environmental concerns. The financial crises have overturned developments in a poverty situation that was achieved over years of dynamic economic growth. Severe currency depreciation and government fiscal imbalances have forced dramatic adjustments within Asian economies. Owing to the financial crises, the natural resources that were already under heavy pressure due to rapid economic growth, especially in the more dynamic economies of the region, were put under even more severe pressure from troubled manufacturers and displaced workers, both of whom intensified exploitation of natural resources. Moreover, the pollution control measures have been set aside in efforts to reduce costs by manufacturers. The governments are unable to respond to these intensified environmental pressures due to reduced budgetary resources caused by the economic slowdown.

1.3 DECENTRALIZATION AND THE NEED FOR COLLABORATIVE NATURAL RESOURCES MANAGEMENT

The transition in Asia from a centrally controlled and authoritarian model toward a decentralized and more democratic way, and from an elite bias toward a popular and participatory management, presents a strong challenge for capacity building in the transition of NRM directed to alleviate poverty through expansion of community rights over natural resources. The global trends toward decentralizing and devolving NRM responsibilities and benefits to local stakeholders including households, user groups, communities, nongovernmental organizations, as well as the private sector are echoing strongly in Asia. These trends are accelerated by rapid transition toward market economies, increasing commitment to community-based NRM, increasing concern for more equitable sharing of benefits from natural resources, and the realization of the fact that centralized models for NRM have been ineffective in protecting and conserving the natural resources during the past several decades. This necessitates the need for initiating an accelerated effort through a learning process, synergy, and sharing experience among societies in the region with particular focus on NRM as most of the rural communities in the region are directly or indirectly dependent on these natural resources for their livelihoods.

The decentralized model of resource management in Asia requires new methods and capacities to support interdisciplinary action research in the field of integrated natural resource management (INRM) to empower stakeholders. Stakeholders need access to knowledge, understanding of the problem at one hand, skills and tools to facilitate communication and negotiation, and the capability to articulate their differing objectives, negotiate their demands, and finally on the other hand adopt and adapt appropriate interventions at the right scale. The main research challenge is to generate pertinent INRM knowledge and the tools for using it, and to facilitate their free exchange among researchers, policymakers, managers, and resource users (Bousquet et al., 2006). A key challenge in the current backdrop is to foster institution building and policy formulation at the interface between (1) community-based practices and interests and (2) regional regulations for managing resources to support resilient livelihoods, particularly for the most vulnerable sectors of society including women (Bousquet et al., 2006).

1.4 CARBON GOVERNANCE AND GHG EMISSION REDUCTION MECHANISMS: NEED FOR COMMUNITY PARTICIPATION

Community-based forest management is the involvement of local communities in the protection, conservation, and management of public forests (Rath, 2010) to prevent degradation from overuse, ensure sustainable forest management, and respond to the basic social and economic needs of local people. When the people who depend on forest resources for their livelihoods are jointly responsible for managing and protecting them, they tend to do so in a more sustainable manner by focusing on the long-term benefits rather than the immediate short-term gains. However, when tenure rights are weak, unclear, insecure, or offer limited benefits, people are incited to extract more immediate benefits, resulting in suboptimal forest management and the reduction of carbon stocks. The transfer of forest ownership, management, and user rights to local people is therefore expected to lead to improvements in forest protection and conditions as well as improved livelihoods. There is considerable evidence in the literature to suggest that when local people acquire secure tenure and forest management rights and receive adequate benefits from forest resources, this indeed leads to improved forest management, conservation of biodiversity, stronger local livelihoods (FAO 2006, 2011), and more participation in developing and auditing effective indicators for new initiatives for carbon governance and greenhouse gas (GHG) emission reduction mechanisms.

1.5 FORD FOUNDATION INITIATIVES FOR ACADEMIC COLLABORATION

The issues discussed above are more pronounced in Indonesia among the Southeast Asian countries and West Sumatra is a typical example mainly due to earlier logging concessionaries and recent expansion of state and private Para rubber and oil palm plantations. These new frontiers have created confrontations among the local community deriving their livelihoods based on inland and coastal natural resources and the outsiders starting mega projects based on local resources, be it the plantations or the massive coastal aquacultural development. To document these dynamic processes the Ford Foundation country office in Jakarta funded a collaborative project between Andalas University and the Asian Institute of Technology (AIT) on capacity building in INRM. The main objective of the project was that Andalas faculty participate in understanding theories and diverse policy arenas for understanding and managing common pool resources (CPRs), which have a collective action problem and dilemma through masters and doctoral field research in a collaborative mode (AIT, Indiana University, and Andalas). This laid a foundation for a joint graduate program in INRM (AIT faculty participated in teaching at Andalas until its number of faculty finished their higher studies in AIT and other institutions of higher learning in INRM). Major activities of the Ford Foundation initiatives involved the faculty from Andalas not only completing their degrees at AIT but also participating in several collaborative training programs such as the International Forestry Resources and Institutions (IFRI) training program jointly organized by the Workshop in Political Theory and Policy Analysis of Indiana University, the Natural Resources Institute at the University of Michigan, and AIT but also developed by an international collaborative network through participating in the biennial International Association for the Study of Commons (IASC). The outcome of such intellectual capital is the current volume, which serves as an important textbook on INRM covering theory, its application, and related case studies. As a result, Andalas University now has a full-fledged graduate program and the graduates from the program are already involved in influencing the policy and implementing effective local level governance and management projects for INRM in West Sumatra in particular and Indonesia in general.

1.5.1 Lessons Learned From the Program

Several lessons have been learned during the collaboration in understanding the major issues related to NRM including the following:

a. Theoretical

The theoretical issues related to the management of the environment and natural resources are presented in this volume to help understand the mechanisms in managing the natural resource base and analyze how different stakeholders interact with each other in managing the natural resources.

b. Intellectual Contribution of Nobel Laureate Elinor Ostrom in Pursuing Analysis of Policies and Institutions

Ostrom (1990) analyzed how communities had successfully managed irrigation systems, fisheries, forests, and other CPRs by identifying institutional design principles favoring local self-governance. Ostrom (2005) applied the Institutional Analysis and Development framework to analyze autonomous rulemaking by resource users, polycentric governance, and ways of coping with threats to commons governance. The intellectual contribution of Elinor Ostrom has been a source of inspiration for the authors of chapters in the present volume and many have benefited from her theoretical framework. Earlier documentation of the Zanjera irrigation system in the northern Philippines, Subaks in Bali, and communal muang fai schemes in northern Thailand are among the examples of long-enduring, locally governed, farmer managed irrigation systems that form the basis for improving policies on governing the commons.

Ostrom proposed the Social-Ecological System (SES) framework, which emerged from a long process of collaboration with other scholars. The SES framework was originally designed for application to a relatively well-defined domain of CPRs management situations in which resource users extract resource units from a resource system. The resource users also provide for the maintenance of the resource system according to rules and procedures determined by an overarching governance system and in the context of related ecological systems and broader sociopolitical-socioeconomic settings. The processes of extraction and maintenance were identified as among the most important forms of interactions and outcomes that were located in the very center of this framework, as illustrated in slightly different forms in Ostrom's (2007, 2009) initial work. The SES framework allows researchers from diverse disciplinary backgrounds working on different resources in distinct geographic areas, biophysical conditions, and temporal domains to share a common vocabulary for the construction and testing of alternative theories and models that determine which influences on processes and outcomes are especially critical in specific empirical settings (McGinnis and Ostrom, 2014).

c. Methodological

Owing to the number of actors involved and the complexity of social-ecological interaction, large-scale environmental problems, including degradation of the ozone layer, deterioration of migratory fish stocks, and pollution of international watersheds, are arguably the most difficult to address. By nature they also impact the welfare of large numbers of people. While some large-scale environmental problems have been successfully addressed, extensive governance and analytical challenges must still be met to systematically understand and confront these types of problems (Cox, 2014).

Due to these dynamic and complex issues of the natural resources, integrated methodological approaches that consider the environment and social science together are required to be studied together. Recent attempts to deal with this complexity in NRM settings required input from many scientific disciplines and used contextual approaches, whereby the nature of the substantive domain is understood through the framing of questions (Bishop and Browne, 2007). These complexities and issues in natural resources render opportunities for reciprocity between natural resource governance and embedded SES.

Although CPR theory is one of the most conspicuous contemporary theories of environmental governance, previous literature including Keohane and Ostrom (1995), Dietz et al. (2003), Stern (2011), and Araral (2014) has provided conflicting answers on how CPR theory can be applied to study resource systems with large spatial extents and large numbers of users. There have not been systematic tests of its applicability to large-scale forest governance. Specifically, it is unclear which variables and design principles from CPR theory can be applied at these larger scales, or whether the logic of collective action underlying CPR theory can be used to study cases involving large numbers of actors (Fleischman et al., 2014a). Nevertheless, a number of scholars have successfully applied CPR theory to understand water allocation and infrastructure provision in larger river basins (eg, Kerr, 2007; Schlager and Heikkila, 2011; Heikkila and Schlager, 2012) and with some qualification and extension, CPR theory can be used in explaining pollution management in large transboundary river basins

(Villamayor-Tomas et al., 2014). Fleischman et al. (2014b) pointed to the need to reinterpret the meaning of CPR theories for them to be applicable at larger scales.

Various methodological approaches have been found to be effective in dealing with the complex and dynamic issues related to natural resources, including

i. The land-use change model, which allows analysis of the causes and consequences of land-use change to better understand the function of the land-use system and to support land-use planning and policy. The model aims to simulate the function of land-use systems and to conduct spatially explicit simulation of land-use patterns in the future (Verburg et al., 2004).

ii. The dynamic modeling approach is developed based on system dynamic modeling methodology. Dynamic modeling is an abstraction or simplification of a complex system for representing the real condition. Based on this model, a scenario is simulated with logical assumptions (Sterman, 2002).

iii. In the collaborative forest governance approach, Inoue (2011a,b) proposed nine prototype design guidelines as follows: (1) degree of local autonomy; (2) clearly defined resource boundary; (3) graduated membership; (4) commitment principle; (5) fair benefit distribution; (6) two-storied monitoring system; (7) two-storied sanctions; (8) nested conflict management mechanism; and (9) trust building.

iv. Environmental social science is a highly interdisciplinary and frequently participatory area of research that incorporates many scientific approaches, including institutional analysis, ecology, political ecology, geography, and anthropology, as well as the study of complex SES (Cox, 2015).

d. Case studies in Asia

Various case studies are reported in this volume for Asia regarding the management of natural resources in the region. These studies include (i) deforestation and forest degradation issues and forest governance in Asia; (ii) issues related to land rights, land tenure, and social insecurity and its effect on the community's access to natural resources and in turn livelihoods; (iii) peatland agriculture for food security and mitigating GHG emission, provision of information for good agricultural practices (GAP) in Thailand; (iv) issues related to land-use change and land management; and (v) policy and practice analysis with implication for REDD+ implementation in the region.

e. Link between the Asia Region and West Sumatra

Regional NRM issues, including weak forest governance, GHG emission, deforestation, ecosystem degradation, biodiversity losses, land-use and land-cover changes, issues related to agriculture, food, and livelihood insecurity, weak water governance, less community participation, and prospects for REDD+ implementation, are discussed in the current volume. The Regional volume (volume IV in the series Rudi, Yonariza, and Ullah, ed.) deals with local dynamic NRM and governance issues. These issues are raised in various papers from Masters and PhD thesis/dissertation at the Andalas University located at West Sumatra. The Regional volume raises issues related to the dependence of local communities on natural resources for their livelihood, their rights, access, and control over natural resources, current practices being adopted in managing natural resources, and the SES and new forms of natural resource governance including implementation methodology of REDD+ in West Sumatra. These volumes may link the regional issues (Asia) with those at the local level (West Sumatra) and contribute in the process of application of various multimethod and modeling techniques and approaches, identified in the current volume to build problem solving mechanisms for management of natural resources at the local level (West Sumatra). The two volumes combine in supporting, encouraging, and coordinating regional cooperation for solving NRM issues at the regional level based on local context-specific solutions and provide opportunities for learning individual influences in policy formulation at the regional and local levels.

1.6 ISSUES RELATED TO NATURAL RESOURCE AND ITS MANAGEMENT IN ASIA

There are a number of issues identified in the case studies related to natural resources and its management in Asia. These include

1. Rising levels of deforestation and forest degradation in Asia: The major causes of the deforestation include overexploitation, particularly illegal logging and forest conversion for other utilization, especially extensive oil palm and coal mining industries.
2. Agricultural production and food security in the region: Reduced agricultural production is due to climate change and has affected food security. Rapidly increasing population leads to increased demand for food. Marginal land such as peatland and upland, despite having low productivity, are used for food production. Using marginal area for producing food has consequences in terms of technology innovation to improve land productivity and to reduce environmental degradation.
3. Natural ecosystem degradation and biodiversity loss: Aceh province in Sumatra Indonesia is one of the biodiversity rich provinces harboring vast tropical lowland, upland, and mountain forests, and many of the endangered species including Sumatran rhinoceros, tigers, elephants, and orangutan which are classified by the International Union for Conservation of Nature (IUCN) as critically endangered.
4. Rising conflicts over the use, control, and management of natural resources: Because of the contradictory dual goals of government and community for preservation and utilization of protected zones in Indonesia and Thailand, conflicts between governments and communities are arising. There is serious competition and conflict over land allocated for large-scale, commercially oriented concessions and upland peasant farming. There are also serious issues of inequality within upland peasant societies for those engaged in multiple opportunities provided by large-scale agricultural and forestry concessions. These conflicts adversely affect the natural resources base as the basic needs of local communities induce them to break rules and engage in haphazard harvesting of the natural resources.
5. Unsustainable use of natural resources: The natural resources in the region are generally overexploited and utilized in an unsustainable manner, which result in changed quantity, quality, and distribution of the natural capital.
6. Rights and social security: Access of local communities to natural resources is restricted through regulations that do not allow the cutting of trees, hunting, cultivating land, or house construction. In several cases, a local community's access to land and natural resources has been completely terminated by leasing rights over the land and resources to private-sector companies within production forests, or by classifying the area as protected forests.

The concept of social security is associated with rights to natural resources. The general distribution and availability of natural resources and other forms of wealth in a community are important for social security; however, the actual mix of sources of wealth each individual has is equally important for the overall social security. Because of the resource competition, social security conditions pose serious problems at all levels of social organization.

1.7 BRIEF OUTLINE AND SUMMARY OF ISSUES ADDRESSED IN THE VOLUME

This volume comprised of 20 chapters that deal with diverse issues related to natural resources and its management in Asia. The chapters are further categorized under five sections: (I) introduction and theoretical background, (II) theoretical issues, (III) learning from the field: cases/issues, (IV) looking forward, and (V) conclusion.

SECTION I: INTRODUCTION AND CONCEPTUAL BACKGROUND

The applications of the theoretical framework proposed by Elinor Ostrom are discussed in an Asian context along with issues related to community participation in environmental management (Chapter 2) and managing CPRs in Asia (Chapter 3). The main theme of the section is to answer how the theoretical framework proposed by Elinor Ostrom can be used for effective management of CPRs in Asia.

SECTION II: THEORETICAL ISSUES

This section discusses theoretical issues related to the management of natural resources in Asia. Challenges of water governance in Southeast Asia (Chapter 4) and issues of social insecurity and legal complexity are discussed (Chapter 6), which are of crucial importance and interlinked with property rights, economic growth, and sustainable use of natural resources. Moreover, the effect of conservation and livelihood policies on community land use and management is also presented in this section to answer how land-use change reduces cropland and will threaten the food security situation in the region.

SECTION III: LEARNING FROM THE FIELD: CASES/ISSUES

This section presents some case studies concerning NRM and describes diverse issues related to natural resources and its management in Asia. Lack of access to land rights and land reforms issues pose threats to the effective management of natural resources in the region. Similarly, issues related to forest fire management, which has become an emerging issue and environmental concern around the globe, and the potential losses due to forest fires and sustainable forest fires management practices through a multistakeholder forum in West Sumatra along with the intensity and frequency of forest fires and the resultant impacts on the ecosystem are discussed in Chapter 10.

The Ministry of Forestry (MoF) in Indonesia established Forest Management Units (FMUs), also known as Kesatuan Pemangkuan Hutan (KPH) in Indonesia, to respond to the need for an on-site forestry institution for the huge and inaccessible forest in many parts of the country. However, challenges still exist in collaborative governance of forest resources in Indonesia as the rapid sociopolitical dynamics in the last decade has resulted in more complex issues and therefore demanded a better relationship among the main forestry stakeholders, particularly among government levels, to make KPHs more effective. These challenges are discussed in Chapter 11 of this volume. Past and present environmental issues and developments including deforestation, coastal ecosystem degradation, and biodiversity losses and significant environmental changes in the Aceh province in Indonesia since the 2004 Asian tsunami, which shock the Acehnese, bearing considerable political, social, and environmental change are described and discussed in Chapters 12 and 13.

SECTION IV: LOOKING FORWARD

The dynamic modeling approach (Chapter 15) to evaluate how sustainably a reclaiming peatlands and uplands can be used for food production to suppress food insecurity and reduce GHG emissions are discussed in this section. The section also describes information provision for GAP adoption of rice farmers in the Central Plains of Thailand, a comparison of the implementation of REDD+ in three countries, namely, Thailand, Indonesia, and Vietnam. The lessons learned from these case studies may serve as a basis for effective policy intervention in natural resources governance. There are various issues related to NRM discussed in this section, including the conflicts arising among water users due to land-use and land-cover changes, lack of conservation practices that results in erosion in the upstream of the watershed, sedimentation in river and irrigation systems, and water shortage downstream. This section also compares the governance attributes that are prerequisites for successful REDD+ implementation in three countries (Thailand, Indonesia, and Vietnam), and enquires about how the existing policy framework and positions are communicated, translated, and supported into actions and programs.

SECTION V: CONCLUSION

This section concludes the overall discussion on various issues and management practices in managing CPRs. Based on the findings of the studies conducted the section also provides some recommendations/suggestions for the improvement of management practices.

References

Araral, E., 2014. Ostrom, Hardin and the commons: a critical appreciation and a revisionist view. Environ. Sci. Policy 36, 11–23.

Asia-Pacific Forum for Environment and Development (APFED). (2004). Paradigm Shift towards Sustainability for Asia and the Pacific - Turning Challenges into Opportunities. Available online on http://www.apfed.net/pub/apfed1/final_report/pdf/final_report.pdf.

Bishop, B., Browne, A., 2007. Natural resource management methodology: lessons for complex community settings. Aust. Community Psychol. 19 (1), 124–136.

Bousquet, F., Castella, J.C., Ekasingh, B., Thai, H.C., Kam, S.P., Manichon, H., Ni, D.V., Quang, D.D., Trébuil, G., Tuong, T.P., 2006. Ecoregional research for integrated natural resource management in Southeast Asian rice ecosystems. Available online at https://cgspace.cgiar.org/bitstream/handle/10568/17353/17353.pdf?sequence=1.

Cox, M., 2014. Understanding large social-ecological systems: introducing the SESMAD project. Int. J. Commons 8 (2), 265–276.

Cox, M., 2015. A basic guide for empirical environmental social science. Ecol. Soc. 20 (1), 63. http://dx.doi.org/10.5751/ES-07400-200163.

Dietz, T., Ostrom, E., Stern, P.C., 2003. The struggle to govern the commons. Science 302, 1907–1912.

FAO, 2006. Global Forest Resource Assessment 2005: Progress Towards Sustainable Forest Management. FAO, Rome, Italy. FAO Forestry Paper 147.

FAO, 2011. Reforming Forest Tenure Issues, Principles and Process. Food and Agriculture Organization of the United Nations, Rome. FAO forestry paper 165.

Fleischman, F.D., Loken, B., Garcia-Lopez, G.A., Villamayor-Tomas, S., 2014a. Evaluating the utility of common-pool resource theory for understanding forest governance and outcomes in Indonesia between 1965 and 2012. Int. J. Commons 8 (2), 304–336.

Fleischman, F.D., Ban, N.C., Evans, L.S., Epstein, G., Garcia-Lopez, G., Villamayor-Tomas, S., 2014b. Governing large-scale social-ecological systems: lessons from five cases. Int. J. Commons 8 (2), 428–456.

Heikkila, T., Schlager, E.C., 2012. Addressing the issues: the choice of environmental conflict-resolution venues in the United States. Am. J. Polit. Sci. 56 (4), 774–786. http://dx.doi.org/10.1111/j.1540-5907.2012.00588.x.

Inoue, M., 2011a. Prototype design guidelines for 'collaborative governance' of natural resource. In: Presented at 13th Biennial Conference of the International Association for the Study of the Commons, Hyderabad, India, January 12.

Inoue, M., 2011b. Summary of the Design Guidelines: Instruction for the Authors of Multi-level Forest Governance in Asia – Recognizing Diversity. GSALS University of Tokyo, Tokyo.

Keohane, R.O., Ostrom, E., 1995. Local Commons and Global Interdependence: Heterogeneity and Cooperation in Two Domains. Sage Publications, London; Thousand Oaks, CA.

Kerr, J., 2007. Watershed management: lessons from common property theory. Int. J. Commons 1 (1), 89–110.

McGinnis, M.D., Ostrom, E., 2014. Social-ecological system framework: initial changes and continuing Challenges. Ecol. Soc. 19 (2), 30. http://dx.doi.org/10.5751/ES-06387-190230.

Ostrom, E., 1990. Governing the Commons: The Evolution of Institutions for Collective Action. Cambridge University Press, New York.

Ostrom, E., 2005. Understanding Institutional Diversity. Princeton University Press, Princeton, NJ.

Ostrom, E., 2007. A diagnostic approach for going beyond panaceas. Proc. Natl. Acad. Sci. 104 (39), 15181–15187. http://dx.doi.org/10.1073/pnas.0702288104.

Ostrom, E., 2009. A general framework for analyzing sustainability of social-ecological systems. Science 325, 419–422. http://dx.doi.org/10.1126/science.1172133.

Rath, B., 2010. Redefining community forestry: for a better approach and better world. http://www.rcdcindia.org/redefining%20community%forestry.pdf.

Schlager, E., Heikkila, T., 2011. Left high and dry? Climate change, common-pool resource theory, and the adaptability of western water compats. Public Adm. Rev. 71 (3), 461–470. http://dx.doi.org/10.1111/j.1540-6210.2011.02367.x.

Sterman, 2002. All models are wrong: reflections on becoming a systems scientist. On line publication: http://jsterman.scripts.mit.edu/docs/Sterman-2002-AllModelsAreWrong.pdf. Access on January 5th 2014.

Stern, P.C., 2011. Design principles for global commons: natural resources and emerging technologies. Int. J. Commons 5 (2), 213–232.

Verburg, P.H., Schot, P.P., Dijst, M.J., Veldkamp, A., 2004. Land use change modelling: current practice and research priorities. GeoJournal 61 (4), 309–324.

Villamayor-Tomas, S., Fleischman, F.D., Ibarra, I.P., Thiel, A., Laerhoven, F.V., 2014. From Sandoz to Salmon: conceptualizing resource and institutional dynamics in the Rhine watershed through the SES framework. Int. J. Commons 8 (2), 361–395.

2

Theoretical Advances in Community-Based Natural Resources Management: Ostrom and Beyond

K.K. Shrestha, H.R. Ojha

The University of New South Wales, Sydney, NSW, Australia

2.1 INTRODUCTION

Since the publication of Elinor Ostrom's seminal work on common property institutions and natural resources management (NRM) in the early 1990s, there have been considerable advances in theoretical and policy debates relating to community-based NRM. While community institutions continue to remain a strong element in the theory of natural resource governance and public policy reforms, the recent shifts in scholarly and policy debates are varied and diverse. Rapid change in contexts of environmental management, such as the mediatization of society and the emergence of transnational networks, have had a profound influence on the way we understand and enact ideas of institutional diversity, public policy, collective action, and governance. In this context, there is a need for adequately mapping out such evolving and rich theoretical and policy streams of work. Exploring and resituating Ostrom's work is therefore helpful in advancing our theoretical understandings of collective action, governance, and public policy in NRM, in a way that can provide both a critical and pragmatic lens for scholars and policy makers to catalyze debates and social learning on environmental governance.

The aim of this chapter is to situate Elinor Ostrom's work in relation to contemporary advances in natural resource governance theories and practical innovations that are taking place in Asia. Our aim is not to review and analyze the comprehensive work of Ostrom. Acknowledging the rich and diverse theoretical works of Ostrom, our focus here is on collective action and common property aspects of her work, with an objective to explore the link between Ostrom and other contemporary advances in the theory of community-based natural resource governance. In so doing, first we briefly explore how Ostrom's work evolved by herself and others, and how her work advanced in response to critiques during her lifetime

13

and after her death. Second, and in a more depth, we chart out seven domains of work that advances or complements her work, namely, (1) the question of embeddedness, (2) political ecology, (3) decentralization, (4) participation, (5) equity and justice, (6) deliberative governance, and (7) critical action research. We map out how Ostrom's works are linked to these domains and demonstrate the ways in which Ostrom's ideas are being reproduced, discontinued, or transformed. We also discuss how these new concepts can gain insights from Ostrom's work. Finally, we conclude by highlighting the need to work toward revitalization of Ostrom's ideas of common property resource management in the context of unprecedented socioeconomic and environmental change.

2.2 OSTROM'S WORK ON COLLECTIVE ACTION AND GOVERNANCE OF COMMON POOL RESOURCES

Elinor Ostrom (1933–2012) received Nobel Prize in Economic Sciences in 2009. Ostrom was recognized for her analysis of economic governance, especially the common pool resources (CPRs), but the actual work that was recognized involved elements of economics, sociology, and political science. In her earlier work, she analyzed the role of public choice in decision-making processes influencing the production of public goods and services. Her better-known works in this area involve the study of the polycentricity of police functions in California (Ostrom, 2010; Aligica and Tarko, 2012). Ostrom's major work is known as the Institutional Analysis and Development (IAD) framework (Ostrom, 2011). The IAD framework considers institutions as a human-made system within which individual choices take place and which configure consequences of the respective choices. The IAD framework is a multilevel conceptual map with which one could zoom in and out of particular hierarchical parts of the regularized interactions in an established social system. It consists of (1) context including biophysical environmental, socioeconomic context, and rules in use; (2) action situations; (3) interaction and outcome; and (4) evaluation criteria (Ostrom, 2005, 2011; Blomquist and deLeon, 2011). The IAD framework helps to perceive complex social phenomenon by dividing them into smaller pieces of practically understandable function. The important aspect of the IAD framework is that outcome is influenced by the institutional arrangements created by local actors in a given context (Ostrom, 2011). This work emerged from the Workshop in Political Theory and Policy Analysis at Indiana University established by Elinor and Vincent Ostrom in 1973. This work is considered to be a distinct school of thought from public policy (Mitchell, 1988), where she examined the use of collective action, trust, and cooperation in the management of CPRs. She made seminal contributions to the theory of collective action among humans sharing concerns of commons (Araral, 2014).

Ostrom made a major contribution in the theory of CPR and community-based NRM policy and practices by identifying eight "design principles" of stable local CPR management (Ostrom, 1990). They are

a. Clearly defined boundaries (clear definition of the contents of the CPR and effective exclusion of external unentitled parties);

b. Rules regarding the appropriation and provision of common resources that are adapted to local conditions;

c. Collective-choice arrangements that allow most resource appropriators to participate in the decision-making process;

d. Effective monitoring by monitors who are part of or accountable to the appropriators;

e. A scale of graduated sanctions for resource appropriators who violate community rules;

f. Mechanisms of conflict resolution that are cheap and of easy access;

g. Self-determination of the community recognized by higher-level authorities;

h. In the case of larger CPRs, organization in the form of multiple layers of nested enterprises, with small local CPRs at the base level.

Ostrom's design principles have been applied in various contexts around the world (Agrawal, 2007; Agrawal and Benson, 2011; Huntjens et al., 2012; Arielle Levine and Richmond, 2015). These principles have since been expanded to include a number of variables that affect the effectiveness of self-organized governance systems, including communication, trust, reciprocity, context, power, and the nature of the resource system as a whole (Araral, 2009; Poteete et al., 2010; Blomquist and deLeon, 2011; Willis, 2012; Huntjens et al., 2012; Thiel et al., 2015; Barr et al., 2015).

Her now famous work is considered to be the development of social-ecological system (SES) framework (Ostrom, 2009). This framework is based on extensive empirical studies on the local management of pastures in Africa and irrigation systems in Nepal; her seminal works focused on analyzing how people interact with CPRs such as forests, fisheries, grazing lands, and irrigation systems to manage and maintain sustainable yields (Ostrom, 2010, 2011; Ostrom and Cox, 2010; Aligica, 2014; McGinnis and Ostrom, 2014; Kimbrough and Vostroknutov, 2015). She analyzed the multifaceted nature of human-ecosystem interaction and explained how societies have developed diverse institutional arrangements for managing natural resources and avoided the tragedy of the commons. She argues against any single "panacea" for individual SES problems (Ostrom and Cox, 2010). It is argued that Ostrom cautioned against single governmental units at the global level to solve the collective action problem of coordinating work against environmental destruction because of the diversity of actors involved (Ostrom, 2010). She argued for a polycentric approach, where key management decisions should be made as close to the scene of events and the actors involved as possible (Ostrom, 2010). The SES framework is considered to be a still-evolving theory of CPRs and collective self-governance and is being applied to a variety of situations (Aligica, 2014; Cox, 2014; Thiel et al., 2015).

Her work presents robust explanations of socioenvironmental interdependence within the context of common property resource management. Unlike other discipline-focused analysis, her work explains, with rigorous empirical evidence and analyses, why and how people find it worthwhile to act together to pursue goals of sustaining natural resources and improve livelihoods at the local community level; aspects that are critical in the study of the human-environment relationship (Ostrom, 1990, 2009, 2010; Poteete et al., 2010; McGinnis and Ostrom, 2014). To establish the robust foundations of collective action, she challenged the dominant idea that collective action emerges essentially from the pursuit of self-interest by people with similar interests. Traditionally, it was assumed that interdependent individuals are well placed to motivate themselves to act collectively and produce collective outcomes (Pandey and Yadama, 1990). However, Ostrom argues that interdependent individuals do not necessarily produce collective action. Instead, rational individuals pursuing their self-interest may actually harm collective interests. Informed by the exploration of the foundations of collective action, Ostrom offered "design principles" for effective organization of collective action involving local people to regulate the use of resources (Ostrom, 1990; Agrawal, 2014; Aligica, 2014). Despite widespread citations of normatively framed design principles, we see that Ostrom's focus is on understanding the possibility of sustained collective action, which as Ostrom argues, depends on how individuals interact in groups to make rules and decision-making structures to institutionalize and control the access to and use of the resources.

Ostrom's (1990) work has largely emerged in response to Hardin's (1968) famous dictum of the "The Tragedy of the Commons". For Hardin, there is little possibility of a collective institution in a community of self-interested members locked into the unavoidable problem of free riding. Through empirical case studies and comparative analyses, Ostrom challenged Hardin, arguing that collective action can emerge to address the problem of free riding, institutional failures, and overexploitation of resources. Expanding Ostrom's argument, scholars of CPRs brought the institutional version of the community to the policy and practice of environmental management (eg, Agrawal and Gibson, 1999; Agrawal, 2007; Wollenberg et al., 2007; Agrawal and Benson, 2011).

While Ostrom's contributions on collective action are seen as being pivotal to policy reforms around the world to improve sustainable NRM outcomes, it cannot be assumed that individuals are necessarily inclined to cooperate among themselves on matters related to the sustainability of natural resources systems. Several scholars have critiqued Ostrom's work, interrogating, expanding, and enriching her work on common property institutions. Ostrom's work on collective action responds to the zero-sum contribution thesis, which draws principally from the rational choice tradition; the idea that individuals act in their own self-interest. Probably the two most influential works are Olson's, 1965 book, *The Logic of Collective Action: Public Goods and the Theory of Groups*, and Hardin's, 1968 article in *Science*, "The Tragedy of the Commons." Olson (1965) challenged the presumption that the possibility of benefit for a group would be sufficient to generate collective action to achieve that benefit. He argued that unless the number of individuals is quite small, unless there is coercion or some other special device to make individuals act in their common interests, rational self-interested individuals will not act to achieve their common or group interests (Olson, 1965). While Olson asserts that a rational individual pursuing his or her own self-interest will free ride at the expense of the group's welfare, he tends to avoid the social nature of human life, in which people work together for the benefit of others (Petrzelka and Bell, 2000). Some analysts even deny that a free-rider problem significantly affects collective action. Hardin (1968) echoed Olson's logic of collective action by providing an example of an English pasture that is open to all. He examined the situation from the perspective of a rational herder, assuming that herders on a commons will seek to increase their herd size as much as possible until overgrazing sets in. Each herder receives a direct benefit from his own animals and suffers delayed costs from the deterioration of the commons when his and others cattle overgraze. The individual's rational action results in collectively irrational outcomes. Hardin recommends "mutual coercion, mutually agreed upon" as a solution to the cooperation problem. Ophuls (1973), one of Hardin's strong supporters, asserts that because the tragedy of the commons and environmental problems cannot not be solved through cooperation, the use of coercive power by the government to manage resources is justified (Ciriacy-Wantrup and Bishop, 1975; Bromley and Cernea, 1989).

Ostrom responded, rather vigorously, to the zero-contribution thesis championed by Olson and Hardin. She and colleagues argued that Hardin was fundamentally confused between a free-for-all open access situation and a common property resource regime where internally informed rules could discourage overuse (Bromley and Cernea, 1989; Ostrom, 1990, 2005, 2010; Agrawal, 2014; Aligica, 2014). In responding to Hardin, Ostrom and others used examples of failed state-managed access regimes to call for a return to local, collectively organized management settings (Berkes, 1989; Ostrom, 1990, 2003, 2009; Li, 1996; Aligica, 2014). Others oppose Hardin's pessimistic prediction of tragedy and argue in favor of collective action and common property (eg, Dawes, 1980). Ostrom's work on common property theory primarily draws on institutional economics to argue against the assumption that CPRs are inherently inefficient. She argues that the zero-contribution thesis fails to account for factors that encourage collective action and self-regulating capabilities of user groups (Runge, 1986). Geographers, political scientists, and anthropologists argue that many societies have devised, maintained, or adapted collective arrangements to manage CPRs (Blaikie and Brookfield, 1987; Ostrom, 1990, 2002; Agrawal, 2007; Agrawal and Benson, 2011; Barnes-Mauthe et al., 2013; Araral, 2014; Arielle Levine and Richmond, 2015).

Ostrom (1990), in her seminal work *Governing the Commons: The Evolution of Institutions for Collective Action*, strongly criticizes the assumption that self-interested rational individuals will not cooperate. Instead, she argues for the importance of institutional development as a basis for collective action. Her work is focused on how a group of principals who are in an interdependent situation can organize and govern themselves to obtain continuing joint benefits when all face temptations to free ride, shirk, or otherwise act opportunistically (Ostrom, 1990). She analyzed successful and unsuccessful cases to identify the factors that impede or enhance the capabilities of individuals to govern CPRs. She produced a list of eight design principles that are "associated with the establishment of coordinated or organised strategies for managing common-pool resources" (Ostrom, 1992, p. 294). Ostrom and others suggest that collective action is possible among self-interested individuals, but it needs to be formalized so that individuals are constrained to behave collectively. Many theorists share Ostrom's argument of individuals being capable of collective action in NRM (Agrawal, 2007; Agrawal and Benson, 2011; Araral, 2014; Barnes et al., 2016). The list of conditions for successful collective action have been developed by Wade (1988), Ostrom (1990), Baland and Platteau (1996), Agrawal (2007), Poteete et al. (2010), and Araral (2014). Ostrom has further contributed to the theory of CPRs and collective action. She lists the attributes of CPRs and its appropriators, which "are associated with an increased likelihood of self-organization" (Ostrom, 2000, p. 40). Ostrom's design principles and the theory of CPR are popular in the analysis of CPR and being used by many international organizations to enhance environmental and social outcomes.

Ostrom's works are advanced in response to critiques during her lifetime and after her death. Analysts have shown that collective action among self-interested individuals is possible, but only under certain conditions (National Research Council, 1986; Ostrom, 1990; Agrawal, 2007, 2014). Other scholars argue that the behavior and actions of individuals are not exclusively determined by self-interest, but trust, norms, and power influence actions and thereby offset pure self-interest (Petrzelka and Bell, 2000; Granovetter and Swedberg, 2001). Therefore, collective action and resource management are better understood by analyzing them as embedded in social, economic, and political situations (Peters, 1987; Fisher, 1994; Shrestha and McManus, 2007). Ostrom responded to these criticisms by placing an action arena embedded within the wider socioeconomic and political setting (Ostrom, 2009; Araral, 2014). These criticisms and responses, while showing the importance and global nature of her work, instruct us to discuss Ostrom's contributions, critiques, and theoretical advances.

Critics argue that design principles are normative and prescriptive blueprint criteria for success (Steins and Edwards, 1999), and too focused on internal dynamics of resource management, while ignoring the wider political economy (Steins and Edwards, 1999; Barnes et al., 2016). In response to this, Ostrom repeatedly highlighted that design principles are not a blueprint for success and that there is no panacea; hence the need for a context-specific, multitired diagnostic approach to socioecological analysis (Ostrom and Cox, 2010; Cox, 2014; Barnes-Mauthe et al., 2014). Critics also point out that the principles are inattentive to issues of conflict and power (Mosse, 1997; Leach et al., 1999; Tucker, 1999; Rangan and Lane, 2001; Agrawal, 2007; Wollenberg et al., 2007; Mansuri and Rao, 2013), and fail to account for the important role of the state to organize collective action and manage local resources, particularly when the resource is scarce and in high demand (Johnson and Forsyth, 2002; Araral, 2009). The state is not necessarily a disinterested party in NRM (Li, 1996, 2002; Shrestha and McManus, 2007; Araral, 2014). Ostrom's work remains affiliated with the rational choice perspective, failing to account for the historical, social, economic, and political processes within which CPRs are embedded (Cleaver, 2000; Petrzelka and Bell, 2000; Barnes et al., 2016).

A critical appreciation of Ostrom's work has also helped to advance the fundamental ideas that Ostrom championed in her decorated career. Three major critiques are useful for reiteration because of their contributions to advance Ostrom's reach, impact, and legacy, perhaps not even anticipated by herself at the time of her writings. First, critics argue that Ostrom's common property draws mainly from a narrow rational choice tradition. The problem is their underlying assumption that individuals are rational actors who behave in their best self-interest to maximize material economic gain (Petrzelka and Bell, 2000; Granovetter and Swedberg, 2001). Instead, trust, norms, and power all influence actions and thereby offset pure self-interest (Petrzelka and Bell, 2000). Actions always take place in a broader social context that affects the action of individuals and interferes with self-interest (Granovetter and Swedberg, 2001).

Second, Ostrom's assumption that humans can be reduced to autonomous, profit maximizing individuals and that their economic behavior can be extracted from social relations and culture is problematic (Uphoff, 1992; Fisher, 1994). Extracting or separating economic actions from social relations is misleading because common property institutions are social phenomena, not an isolated activity (Petrzelka and Bell, 2000). Responding to Ostrom's point of economic interdependence, Fisher (1994) argues that people are interdependent in many ways; resource management is just part of this interdependence and cooperation is encouraged not just by relatively narrow economic interdependence, but by the overall interdependence between people. The underlying implication is that many scholars have been disappointed by the self-interest driven, instrumental, and reductionist approach of economic analysis and the rational choice tradition.

Third, Ostrom's work is seen as weak in relation to the analysis of context. The president of the International Association for the Study of Common Property (IASCP) admitted that "The tragedy-of-the-commons theory … has come under attack from an 'embeddedness' perspective, which places situation and context as primary" (McCay, 2000, p. 3). There are increasing calls for the recognition of the concept of embeddedness within CPR analysis. For example, Peters (1987, p. 193) argues that "without keener sense of the relations in which individual users are embedded, we cannot penetrate the dynamics of a commons, which is necessarily a social system." Similarly, Fisher (1994, p. 70) stresses that it is important for "looking at embedded social relationships in understanding the commons." Mosse (1997, p. 467) argues that CPR analysis should be based on "more historically and politically grounded understanding of resources, rights and entitlements." Some analysts have criticized CPR analysis for not giving sufficient attention to contextual factors or local and global dynamic forces affecting the choice of the actors (Edwards and Steins, 1999; Cleaver, 2000). These critiques highlight some limitations of Ostrom's work, but Ostrom and associates addressed many of these by the development of the SES framework (Ostrom, 2009, 2010; Ostrom and Cox, 2010; Aligica, 2014; Cox, 2014; Thiel et al., 2015). Yet, there are spaces to explore how different complementary ideas have emerged and benefited from Ostrom's work, which are useful for advancing her ideas in relation to other contemporary theoretical advances in collective action and governance so as to maximize her work's reach and impacts.

2.3 SITUATING OSTROM IN RELATION TO ADVANCES ON COLLECTIVE ACTION THEORY

2.3.1 The Question of Embeddedness

Ostrom's work has been enriched, advanced, and complemented by contributions in such diverse sectors as NRM, geography, anthropology, economics, development studies, and political science. One separate theoretical advance, which complements and enriches Ostrom's work, is in the field of environmental sociology. Through the lens of "embeddedness," conventionally understood as economic actions being socially situated, this body of work highlights the criticality of engaging with social contexts. The notion of embeddedness was a key concept of anthropology (but is not so now). Malinowski (1926) highlighted the importance of social context and discussed why people conform to social rules in the absence of a formal legal system. He explained cooperation in terms of reciprocity that occurs in the context of kinship, religious, and other obligations. The concept of embeddedness was popularized during the 1950s in Economic Sociology by Karl Polanyi who argued that "man's economy, as a rule is enmeshed in his social relationships" (Polanyi, 1957, p. 46). Drawing on a term developed by Geertz (1973), McCay and Jentoft (1998) stress the role of "thick description," indicating an ethnographic and careful specification of the systems of resource use and their embeddedness within historical, social, and political relations and environmental conditions. During the 1980s, Granovetter (1985) revitalized the concept, representing the reenergization of New Economic Sociology as a field of study that advocates that "many economic problems … can be better analysed by taking sociological considerations into account" (Granovetter and Swedberg, 2001, p. 2). Schatzki (1993) classifies embeddedness into two conceptual facets: vertical facet and horizontal facet. Vertical facet relates to "hierarchical linkages of individuals and corporate actors at the local level to the larger society, economy and polity of which they are part" (Schatzki, 1993, p. 740). The larger context constraints local action, but also by providing new opportunities, allows local maneuvering at local levels. On the other hand, horizontal facet refers to "the interpenetration of societal/cultural domains" (Schatzki, 1993, p. 740).

The concept of embeddedness is useful to enrich Ostrom's work as it adds value to understand the sociological failings of neoclassical schemes such as common property institutions. Its implications can be extended to collective action in the analysis of CPR. In this sense, collective action in common property institutions can be a form of social action, which is embedded in social, economic, and political contexts across different scales. The concept facilitates the analysis of natural resources as being situated in the local context (sociocultural, economic, and historical), which is also influenced by the historically changing government policies and government and nongovernment institutions at different levels.

2.3.2 Political Ecology

Another area of complementarity on Ostrom's work can be seen from the burgeoning literature on political ecology. This body of literature corresponds to Ostrom's limited focus on power relations and knowledge. Political ecology provides useful insights into the importance of focusing on an analysis of decision making at different levels and the relationships between them. In a general sense, it is an outgrowth of ecological and social science that combines social and political investigation with environmental processes (Bryant, 1992; Batterbury et al., 1997). It

> … combines the concerns of ecology with a broadly defined political economy. Together this encompasses the constantly shifting dialectic between society and land-based resources, and also within classes and groups within society itself. *(Blaikie and Brookfield, 1987, p. 17)*

Early political ecological analyses were focused on land-based explanations of environmental change, focusing more on political control over natural resources and less on ecological processes (eg, Blaikie and Brookfield, 1987; Bryant, 1992, 1997). The focus has now shifted to discursive relations between humans and their environment and in narratives that support power relations, which in turn maintain hegemony over people and the environment (Stott and Sullivan, 2000). Many writers have highlighted the use of environmental orthodoxies or myths in policymaking, despite the accumulation of evidence to suggest they are flawed (Fairhead and Leach, 1995; Forsyth, 2003). These orthodoxies are translated into policies and, subsequently, these policies impose unnecessary restrictions on the livelihoods of marginalized people (Forsyth, 2003).

A critical issue in political ecology is that state policies are not developed in a political and economic vacuum, but "result from struggle between competing actors seeking to influence policy formulation … [policy content] often facilitating the interests of powerful economic elites" (Bryant, 1992, p. 18). The impacts of policies depend more on the manner in which the policy is implemented than its content because social divisions are fully recognized during the implementation (Bryant, 1992). Responding to the lack of local specificity in third-world political ecology, Batterbury (2001) promoted a "local political ecology" that focused on local decision-making processes and contextualizing these decisions in wider social and political systems. It has been recognized that villagers do not operate in isolation to the wider economy, but their decisions are made in response to the local availability of natural resources (Bryant, 1992; Peet and Watts, 1993).

Political ecological analysis is useful for the study of collective action and decision making at the local level for three reasons. First, there is increasing conflict between the global move toward decentralization and devolution, giving rise to community-based NRM, while the discourses of global environmental change promote global approaches to environmental problems (Dryzek, 1997). Second, the political ecological analysis helps to understand the impacts of decisions taken at different levels, in terms of the environmental change and the livelihoods of local communities. Decisions taken at higher levels are less likely to take into account the social and ecological variations at the local level (Adger et al., 2002). Third, the political analysis helps to distinguish between those who make decisions, those who enforce and monitor them and those who are subjected to decisions (Adger et al., 2002). If the voices and priorities of the recipients are not engaged in the decision-making process, effective implementation is less likely. The best option to address the mismatch between decision makers and decision recipients is the use of inclusive decision-making processes, in which multiple interests are negotiated (Holmes and Scoones, 2000).

2.3.3 Decentralization

Literature on decentralization has had many complementarities with Ostrom's work. In a global movement that promotes democracy, justice, and sustainability, decentralization is now a centerpiece of policy reforms around the world. Decentralization is the process where a central government relinquishes some of its powers and management responsibilities to local governments, local leaders, or community institutions (Ribot, 2004; Dressler et al., 2010). Decentralization has been advocated for various reasons, chief among these are better local participation, equity, efficiency, and sustainability outcomes (Larson and Ribot, 2004), fostering rural development (Uphoff, 1993), increasing public service performance and legitimacy of government (Larson and Ribot, 2004). At least 60 countries claim to be decentralizing some aspect of NRM (Agrawal, 2001) and the number of countries implementing decentralized NRM has grown significantly in the last decade (Dressler et al., 2010).

Decentralization has been employed in NRM in various ways such as the government ceding the power to local governments, to regional or local outposts, and sometimes sharing the power with local users (Ribot, 2004). However, the most popular form of decentralised NRM is community based (Enters et al., 2000; Blaikie, 2006; Dressler et al., 2010), where decision-making and implementation processes, at least in theory, are instigated, controlled, and conducted at the local community level under the auspices of a government agency (Li, 2002). Supporters argue that the benefits include improved plan development and implementation (Gray et al., 2001), better access to local agency knowledge and improved implementation of context-sensitive plans (Li, 2002). Community-based NRM is now a major strategy in many developed and developing countries such as in Nepal, the Philippines, and Mexico. The United Nations is also a strong supporter of decentralization in NRM (Leach et al., 1999). Indeed, community-based NRM is very much seen as the preferred response to top-down management approaches (Li, 2002; Shrestha and McManus, 2007; Pulhin and Dressler, 2009).

While the popularity of community-based NRM is not in dispute, the literature highlights a number of significant issues. The first issue concerns power relations, which links well with the response of political ecologists to Ostrom's work on power. Community-based NRM involves a diverse range of stakeholders usually unequal in wealth, social status, education, and gender. There are two interrelated issues in relation to unequal power. The first is that in theory community-based NRM requires the real transfer of power from government to the community. Practice shows clearly that community-based NRM often maintains the status quo where government often dominates the decisions, with little benefit the environment or local empowerment (Ribot, 2004). In a review of the progress of community forestry (CF) programs in Asia, Fisher (2010) argues that decentralized forestry programs rarely involve real devolution of power. The second issue is that community-based NRM requires the empowerment of all individuals within the relevant communities. Empirical evidence also indicates that any power devolved to the local level can easily be captured for the benefit of local elites (Shrestha and McManus, 2008; Shrestha, 2009). Sometimes, powerful individuals control resources, even against the rules produced by the local council, broader society, or state (Agrawal and Ribot, 1999). A substantial body of literature critical of community-based NRM exists in relation to the problems of unequal power relations and entrenched institutional structures that affect decision-making and implementation processes, blocking the path to improving social equity and environmental sustainability outcomes (eg, Ojha, 2008; Ojha et al., 2008; Dressler, et al., 2010). Another issue is that collective action in community-based NRM assumes that stakeholders work collectively to manage and use natural resources to promote a collective interest of equity and sustainability.

2.3.4 Equity and Social Justice

Ostrom's work is enriched by a broader debate on equity and justice. Equity is a normative concept that has an important place in social thinking and is associated with many social, economic, and environmental issues. Concern over equity is "one of the fundamental principles of community involvement in forest management" (Anon., 2003, p. 1), and is generally considered as the legitimate basis for community-based NRM (Li, 1996). Many assumptions are made about equity without actually defining it. Central to the equity debate is the concern that the poor, women, and minorities are exposed disproportionately to high levels of risk (Ringquist, 1998; ESRC GEC Programme, 2001). Traditionally, equity has been narrowly understood as a concept that focuses on the distributional consequences of decisions. However, its scope has widened to include procedural, geographic, and social equity (Bullard and Johnson, 2000). Some people consider equity in distribution, decision making, and fund allocation (Chhetri and Nurse, 1992; Bosma, 1995). Fisher (1989) argues that equity involves getting a fair share, not necessarily an equal share. Messerschmidt (1981) argues that equity may be defined differently in hierarchical societies, where unequal outcomes are not necessarily seen as inequitable. The important point remains that an equitable system should not further disadvantage the poor (Gilmour and Fisher, 1991).

Two critical dimensions of equity are often highlighted. First distributional equity, which generally refers to "the physical movement of goods to people" (Seymour-Smith, 1986, p. 79). More specifically, it refers to distributional consequences of decisions. People's judgments of distributive fairness are based on the "need" principle (Deutsch, 2000). However, people do not always respond positively to outcomes, in which they have voiced their opinions, especially when the procedures result in repeated unfavorable outcomes. People show "frustration"; the frustration effect (Tyler, 1994; Deutsch, 2000). Once the impression of fairness is established, it is extremely difficult to change (Deutsch, 2000). Second, procedural equity refers to affirmative action, in which the poor, women, and other disadvantaged groups have greater access to, and influence over, the way decisions are made and implemented. Previous studies found that two key factors are influential in procedural fairness. First, and most important is "voice effects," which suggest that a procedure is fair when people are given the opportunity to voice their opinions (Lind et al., 1990). Overall, if procedures are seen to be fair, then it is more likely that the outcomes will also be perceived to be fair (Lind and Earley, 1992; Deutsch, 2000). Procedural justice is a more pervasive concern than fair outcomes (Paavola and Adger, 2002). Second is "dignitary effects," which propose that the processes are fair when people evaluate procedures based on trust, neutrality, and standing of the organization responsible for developing and implementing a decision (Deutsch, 2000). The perceptions of procedural fairness are important as they influence how the community evaluates the government authorities and others and the decisions that they make (Deutsch, 2000).

The debate on equity is situated on the wider literature on justice. Justice is the indifferent (but often equal) treatment to different people. It generally attends to the questions associated with distribution of goods and services. Justice can be interpreted as a rationale for equity. Theories of justice are important to understand equity issues. One of the most influential contributions in the theory of justice is given by Rawls (1971, p. 303):

> All social primary goods – liberty and opportunity, income and wealth, and the bases of self-respect – are to be distributed equally unless an unequal distribution of any or all of these goods is to the advantage of the least favoured.

Following Rawls, inequalities should be tolerated only if people will work for the advantage of the least well off. The need-based distribution was also highlighted by Galston (1986), who argued that basic needs of goods and services are to be fulfilled on the basis of needs, but the opportunities are to be allocated through a competition, in which all have a fair chance to participate. On the other hand, some writers opposed the distribution according to need, but argued that to do justice is to distribute in accordance with ability and hard work (Taylor, 1988). Walzer (1983) asserted that the road to an egalitarian society not only depends on equal opportunity and equality of outcomes, but also on the equality of conditions or circumstances, which influence the opportunities and the outcomes. He argued, "[an] equal start is also important in addition to [an] open road ... today's inequalities of opportunities derive from yesterday's victories and defeats" (Walzer, 1983, p. 144). Justice is also interpreted on the basis of entitlement (see Nozick, 1974; Sen, 1999). It is argued that things come into the world already attached to people having entitlements over them. The justice depends upon where this comes about (Nozick, 1974).

Discussions on equity and justice invite some clarification on the idea of equality. Equality and equity are two different concepts. Equality broadly refers to the same (ie, equal) in size, amount, value, and number of the matter under consideration, while equity refers to fairness. In theoretical studies, theorists suggest that equality and equity carry different ideas and are related differently in different situations. Parfit (1991) claims that equality and priority are distinct ideas; the fundamental difference between them is that equality is relational in terms of how each person's level is compared with the level of other people, whereas priority is not. However, these two concepts are not rivals because they occupy different places in moral thinking (Norman, 1999).

Equity is generally considered as superior to equality. Parfit (1991, p. 19), who formulated the "Priority View," which argues that "Benefiting people matters more, the worse off these people are." He adds "equality is the default: what we should aim for when we cannot justify distributing unequally" (p. 15). While Parfit has not argued for or against priority or equality, he implies that priority is a more useful concept than equality. The explicit emphasis on priority comes from Raz (1986), who stated that "egalitarian principles often lead to waste" (p. 227). He argued in favor of priority based on the argument of "concern":

> ... what makes us care about various inequalities is not the inequality but the concern identified by the underlying principle. It is the hunger of the hungry, the need of the needy, the suffering of the ill, and so on. The fact that they are worse off in the relevant respect than their neighbours is relevant. But it is relevant not as an independent evil of inequality. Its relevance is in showing that their hunger is greater, their need more pressing, their suffering more hurtful, and therefore our concern for the hungry, the needy, the suffering, and not our concern for equality, makes us give them the priority *(Raz, 1986, p. 240)*.

Within the context of NRM, while there is evidence to suggest that community-based NRM can deliver good conservation outcomes (Mahanty et al., 2006), there is a growing body of literature highlighting the problem of equity issues of such outcomes (Malla et al., 2003; Shrestha, 2009; Dressler et al., 2010). Equity is a normative concept and generally refers to "fairness" in processes of decision making (procedural equity) and distribution of outcomes (distributional equity) (Paavola and Adger, 2002). The processes are seen to be fair when people are given the opportunity to voice their opinions ("voice effects") (Lind and Earley, 1992), and when these processes are based on the trust, neutrality, and standing of the decision-making organization ("dignitary effects") (Deutsch, 2000). Distributional equity is achieved when local communities get a fair share, not necessarily an equal share (Fisher, 1989). Equity is usually understood as equality of outcomes and procedural equality, but neither equal distributional outcomes nor procedural equality necessarily lead to equity in the sense of "fair" outcomes. Equity as positive discrimination in favor of the poor and disadvantaged groups (equity as priority) is considered as superior to equality principles (Raz, 1986; Parfit, 1991). If procedures are seen to be fair, it is more likely that the outcomes will also be perceived to be fair (Lind and Earley, 1992; Deutsch, 2000). Some studies claim that the application of formal equality has mostly benefited the elites, but added costs to the poor and minority groups, and worsened their livelihoods (eg, Shrestha, 2009).

2.3.5 Participation

Ostrom's work on collective action has been enriched by the literature and practice of participatory development. Globally, the question of how communities function effectively to deliver socioenvironmental benefits remains an enigma, despite many years of research and practice. Critics have branded participation as a "tyranny," which dominates development discourses and yet delivers little on the ground (Kothari, 2001). "Community participation" paradoxically contributes to the exclusion of women and disadvantaged groups in some settings (Agarwal, 2009). Problems also abound, as promoters tend to focus narrowly at the local level, ignoring politics outside of communities themselves (Mohan and Stokke, 2000). Alongside these failures, others see the project of participation as an extension of neocolonial development ideology and a discursive weapon of Western development agencies (Escobar, 1995; Li, 1996; Blaikie, 2006).

Even the World Bank, which has promoted participation as a key strategy for social capital building (Woolcock and Narayan, 2000), has demonstrated that community participation has remained "supply-driven," usually reinforcing the status quo (Mansuri and Rao, 2004). Thus, despite years of research, policy experiments, and practice, there remains a need to revisit most basic questions such as who is participating, how and in what ways, and for whose benefit (Cornwall, 2008), as well as the need for interrogating the paradoxes underlying the idea of participation (Dill, 2009).

Responding to the above critiques, attempts have been made to revitalize the potential of participation by invoking notions of citizenship (Mohan and Hickey, 2004), "deliberative" practices (Fischer, 2006), and empowerment of communities (Gaventa, 2004). Despite these conceptual innovations, evidence from the field suggests that the practice of participation has not improved as anticipated (Mansuri and Rao, 2004, 2013). This project seeks to reconcile the conceptual debates between participation as "tyranny" (Kothari, 2001) and participation as "transformation" (Mohan and Hickey, 2004) into a more productive framework. Rather than criticizing or celebrating the normative appeals of participation, it examines practice of participation in a specific context, drawing on the rich policy experiments and multilevel dynamics in Nepal's CF governance. Nepal offers an excellent opportunity to explore the pitfalls and potential of community participation given the complex ways in which the Himalayan conservation agenda (Guthman, 1997; Ives, 2004) has been bound up with political upheavals, reflecting the demand for equitable development and inclusive democracy (Blaikie and Brookfield, 1987; Lawoti, 2008).

A crucial element of the above definition is about power, which must be devolved to a body, representing and accountable to the community. The community's decisions may then internalize social and ecological costs or assure equitable decision making and use (Lawoti, 2008). The Brundtland report places participation at the heart of sustainable development as

> … the recognition of traditional rights must go hand in hand with measures to protect the local institutions that enforce responsibility in resource use. And this recognition must also give local communities a decisive voice in the decisions about resource use in their area. *(WCED, 1987, pp. 115–116)*

Sherry Arnstein was one of the first to critique participation in terms of the concept of power. She developed a "ladder of participation" as a model, in which she discussed different degrees of involvement of participants and application of decision-making power (Arnstein, 1969). While Arnstein herself admits that the ladder is a simplified model of participation, it captures the important point that there are different types of participation, but genuine participation requires decision-making power. Many people can be disempowered in decision-making processes if their participation is not emphasized. Nelson and Wright (1995, pp. 7–8) identify two types of participation: "participation as a means" (ie, a process of achieving the aims of a project more efficiently, effectively, or cheaply) and "participation as ends" (ie, a process of giving control of a development agenda to a community or group). In a participatory policy, such as CF, the participation can ideally be considered as an ends. However, in practice, it is generally seen as a means to achieve certain outcomes by focusing on lower levels of the ladder.

One of the major hurdles in collective action is the problem of elite domination. Many well-intended projects fail when local elites misrepresent community interests and seize control of the project (Cernea, 1993). Without effective participation, the participatory policies, such as CF may become a form of "covert privatization" that leads to centralized resource control, yet existing within common property ownership (Anderson, 2000). The question of power to influence resource management decisions is important.

2.4 EXPANDING OSTROM'S APPROACH: DELIBERATIVE GOVERNANCE AND CRITICAL ACTION RESEARCH

Given that Ostrom's seminal work on collective action has been increasingly questioned for its universal applicability in more complex and dynamic contexts of natural resource governance, questions on tackling power and enhancing free and open deliberation among local people and external stakeholders have become even more crucial. We argue that Ostrom-inspired collective action theory can be advanced further through an effective dialogue with ideas of deliberative governance and critical action research (CAR). These two related lenses help in asking questions on the procedural aspect of change, in a way that recognizes the potential of social actors to engage in more productive and meaningful dialogue, while also tackling questions of power. Of these, deliberative governance can provide greater democratic foundations to further collective action and common property institutions. The other lens, CAR, captures insights from practice on how collective action and CPR management can be made more deliberative, democratic, and equitable by challenging power inequality and hegemonic knowledge systems that underpin collective action in natural resources and environmental management. Below we outline these two approaches, showing how they can help expand Ostrom's work in relation to the contemporary challenges to natural resource governance.

2.4.1 Deliberative Governance

Proponents of the deliberative practice of governance have attempted to show various ethical, normative, and epistemological values of deliberation. Deliberation is founded in the Kantian moral tradition that all humans are equal and no one has any right to strategically manipulate the other. This means that decisions affecting multiple groups should be settled through open deliberation—without the powerful manipulating the other strategically. This is precisely what Habermas meant by "communicative rationality." Deliberation involves use of verbal and nonverbal means of communication in relation to social conflict and coordination. Humans have immense potential to symbolically represent and communicate social experiences not only through literal and coded language but also through rhetoric, humor, emotion, testimony, storytelling, and gossip (Dryzek, 2000). Such communicative practices do not just make governance possible but are constitutive of it. Most problems of governance are usually associated with distortions in communication.

Proponents of deliberation make a case that social norms are more legitimate if passed through a test of quality of deliberation. Every moment of decision making implies commitment of the participants to abide by the obligations. One of Habermas's (1990) theses, known as "discourse ethics" is that the legitimacy of any use of restraint (such as "rules" or "penalties") is only justified when the subjects choose it freely. As he argues, "only those norms can claim to be valid that meet (or could meet) with the approval of all affected in their capacity as participants in a practical discourse" (Habermas, 1996). Because for him, private and public dimensions of human existence are co-original (Habermas, 1996, p. 129), one leading to the other, social agents in an ethical community "have to" to engage in communicative interactions to arrive at understanding so that both individual and the collective gains are jointly optimized (Haller, 1994). This means that participants who get opportunities to express their views are more likely to value and appreciate collective understanding or decision better in practical life (Fearon, 1998). Deliberative processes demand that people should be "treated not merely as objects of legislation, as passive subjects to be ruled, but as autonomous agents who take part in the governance of their society, directly or through their representatives" (Gutmann and Thompson, 2004, p. 3). Deliberation is considered a "discussion intended to change the preferences on the basis of which people decide how to act" (Przeworski, 1998).

The object of deliberation may be ultimate ends or beliefs from which preferences are derived (Elster, 1998). Dryzek (2000, p. 1) holds that deliberation is a process of argumentation to arrive at some understanding (though not necessarily a consensus) at moments of collective decision, and not just for the sake of discussion. To him, deliberation is a "social process," which is different from "other kinds of communication in that deliberators are amenable to changing their judgments, preferences, and views during the course of their interactions, which involve persuasion rather than coercion, manipulation or deception." Dryzek (2000) argues all forms of argument—rhetoric, humor, emotion, testimony, storytelling, and gossip—should be admitted so long as they are noncoercive and capable of connecting the particular to the general, contestations to be allowed, no precommitment to process values of deliberation. This view expands the rational choice model of interaction that underpins much of Ostrom's collective action theory.

Forester (1999) argues "rationality to be an interactive and argumentative process," and cautions against considering deliberation as mere "process" and distracting "talks," suggesting deliberative planners (such as public managers) not only manage the physical spaces but also the "dialogic spaces." He expands the scope of deliberative dialogues to include not just reasons and argument but "issues of political membership and identity, memory and hope, confidence and competence, appreciation and respect, acknowledgement and the ability to act together." To him, "the transformations at stake are those not only of knowledge and or class structure, but also of people more or less able to act practically together to better their lives, people we might call citizens" (Forester, 1999, pp. 115–116). Forester proposes a "transformative theory of social learning" to allow analysis of not only how we argue but also how we change through dialogical interaction. He elaborates: "at the heart of this transformative account is a view of" how citizens "not only pursue interests strategically and display themselves expressively, but reproduce and reconstitute their social and political relationships with one another too" (Forester, 1999, p. 130).

For some, apart from the external collective aspect, deliberative democracy involves "internal-reflective" aspects, too (Goodin, 2003, p. 57). From this point of view, a process of judgment takes place during which he or she imagines or empathizes the position of others, without the latter being "conversationally present." They connect their argument with Habermas's idea of "oppositional public sphere" to serve as the source of deliberatively democratic inputs into the ordinary political sphere, which appears like a representative adaptation away from "directly deliberatively democratic" practice.

Young (1997, p. 385) argues that differentiation of citizen groups in various social positions/ classes is actually a resource for democratic communication through which participants can "cooperate, reach understanding and do justice." Without such group-differentiated social perspectives (corresponding to cultural, economic, social, and environmental aspects) there is no possibility of deliberation. The primary resource, which she claims comes from difference is diversity of perspectives on structures, relations, and events of the society. Each differentiated group has its own point of view and perspectives on history and the current situation, which during deliberation (a) enables groups to express their proposals as appeals to justice rather than mere self-interests or preference; (b) teaches individuals the partiality of their experience and reveals their own experience as perspective; and (c) adds synergistically to social knowledge through questioning, challenging, and dialogue. However, if difference coincides with political inequality then the process of deliberation is unduly influenced by the powerful.

Reich (1990, p. 8) (cited in Fisher, 2003, p. 207) relates deliberation to the fundamental process of civic engagement through which public managers must seek to build legitimacy for public decisions. He argues that deliberation is a process of social learning about public problems and possibilities. Christiano (1997, p. 251) argues that "participating in discussing matters of great moral importance is an essential or at least irreducible component of good life" and that "public deliberation is an expression of a kind of mutual respect among citizens in the society." Deliberation thus implies a duty on the part of public managers to respect and seek out opinions from citizens on matters of the latters' concerns, and the process of deliberation must take place in "public" beyond private domains in ways and languages accessible to ordinary citizens (Gutmann and Thompson, 2004, p. 4).

Given these values of deliberation in human interaction, a number of questions remain as to how and when deliberation can be organized to achieve such values. A key question is the location of deliberation, especially in the continuum between citizen-public-sphere-state (Dryzek, 2000, p. 81). Others hold that institutions of deliberation are not well developed (Smith, 2003, p. 80). Media is also increasingly considered an element of the public sphere in modern society (Chambers and Costain, 2000) but is always differentiated (Chambers, 2004). Chambers argues that "the democratic public sphere is a huge and amorphous entity that is not really one public sphere, but many public spheres made up of countless crisscrossing conversations in a multiplicity of venues from talk radio to house committees, newspaper editorials to Greenpeace meetings" (Chambers, 2004, p. 409). Gutmann and Thompson (2003) critiques Habermas's proceduralist version of deliberative democracy, arguing that mere procedural democracy may not bring substantive outcomes of justice. They defend the inclusion of substantive principles (such as basic liberty and fair opportunity) in a theory of deliberative democracy, though often the procedural and substantive principles may not be distinguishable (Thompson, 2003, p. 32).

Feminists criticize that the idea of public sphere as a key venue of deliberation is male-biased and ignores the household level public responsibilities undertaken by women in many societies. The very concept of deliberative democracy is a Western creation, and Foucauldians would claim that it is part of Western-led hegemonic discourse rather than something having any emancipatory promise. Another critique is that deliberation with an emphasis on communicative interaction is likely to reinforce, rather than transform, oppressive political hierarchies. In particular, the kind of communication emphasized in deliberation (dispassionate, reasoned, and logical, including idioms) favor certain citizens and disadvantage others (speech culture of women and minorities) (Sanders, 1997; Young, 2003). Deliberation tends to seek unity among diverse perspectives, which may be better kept differentiated and contested rather than reaching compromise (Bohman, 1995). Young (2003, p. 108) challenges political deliberations as a practice suited to "political elites" who treat each other as equals and may hold deliberative meetings, but when it comes to deliberation between unequal classes, then the power of "respectful argument" is gone. She contends, "Under these circumstances of structural inequality and exclusive power, good citizens should be protesting outside these meetings, calling public attention to the assumption made in them, the control exercised, and the resulting limitations or wrongs of their outcomes" (Young, 2003, p. 108).

2.4.2 Enhancing Deliberative Governance Through Critical Action Research

Given the ubiquity of power inequality, it is not easy to foster deliberative processes in practice. One important way forward is to envision the role of critical intellectuals who can work with citizens and political leaders to expand deliberative processes. Here we outline the CAR approach to enhance the interplay between research and social movement practices. We argue such interplay is crucial to improve the quality of the democratic policy process. Such interplay has the potential to address some of the concerns related to the continued lack of effective deliberation in the policy processes.

CAR can enhance deliberation in a variety of ways. It is not likely that the lay public will be able to resolve all the environmental knowledge problems they face. The substantive depth of information needed before a policy decision is made requires the knowledge of scientists, on top of the ordinary knowledge of citizens (Sabatier, 1991). And it is also not likely that a policy question can be fully relegated to scientific resolution. What is important is that concerned groups of people or citizens deliberate to define the scope of the problem and expert inquiry, and develop judgments on the type of expert they need. Then the role of the expert can be (a) gathering information and analysis, (b) maintaining critical communication with the concerned groups of citizens, and (c) critiquing the doxa of ordinary people that limits moral inquiry.

Absolute dependence on, and romanticization of, local knowledge is also problematic (Sillitoe, 1998), as local knowledge is inscribed within the day-to-day pressures of livelihoods and the larger sociopolitical structure that shapes learning. If discursive knowledge is just a thin tip of a thick doxa (Crossley, 2003; Hayward, 2004) that naturalizes social agents into particular sets of practices, then there is a role for critical social science to trigger self-reflexivity among the ordinary people at all stages of the inquiry. Here the role of science can be broadened to include the praxis of external epistemological critique (using the methods of science to challenge practical doxa) to trigger rethinking of accepted assumptions in the practice of natural resource governance. But the challenges remain as to how an external expert engages critically and at the same time within the democratic control of the people being critiqued. But the Nepal case of CF indicates that those who are taking the position of scientists are themselves the victims of timber-oriented forestry doxa, whereas the land-poor farmers have to some extent responded to changing contexts of livelihoods, and adjusted forest harvesting techniques (harvesting bushes).

Under the CAR approach, the purpose of scientific inquiry should be to "improve the quality of policy argumentation in public deliberation" (Fischer, 1998). As Fischer argues, the shift from positivism involves turning from proof or verification to discursive and contextual inquiry so as to provide a normative framing without rejecting the empirical aspect. And knowledge is augmented through the "dialectical clash of competing interpretations" (Fischer, 1998). Through processes of deliberation, the possibility exists for citizens and experts to reach a consensus concerning what will be taken as a "valid explanation" of the problem (Fischer, 1998). The role of experts in the public organizations is therefore to manage ongoing processes of deliberation and education rather than making and implementing decisions (Reich, 1990). The Nepal case of forest harvesting does not indicate a dialectical clash between local forest users and foresters. It is only the foresters as experts who have prescribed what is best.

Bohman's synthesis of Dewean pragmatism and critical social science provides further insights into how scientists and ordinary people should relate to each other in the process of inquiry. He argues for a social control of scientific practice as the autonomous practice of scientists can lead to technocratic rule (Bohman, 1999). He identifies the role of expert inquiry in the process of democracy, which can be democratic under two conditions: (a) it must establish a free and open interchange between experts and the lay public, and (b) discover ways of resolving recurrent cooperative conflicts about the nature and distribution of social knowledge. For him, two features of science are relevant for democracy: disciplined discussion and experimentation. If inquiry is organized in a cooperative way, then Bohman sees a possibility of knowledge thus being owned socially. This would also create a possibility of combining depth (of scientists) and breadth (of ordinary people) dimensions of knowledge in the governance of natural resources. In the community forest user group (CFUG) case, foresters do not see a need to engage in cooperative inquiry with the villagers as regards how best forest can be managed for the benefits of local users, and local users are also not in a position to force the latter to do so.

Even when spaces are created to ensure critical communication between scientists and ordinary citizens, they are not neutral, and hence not equally accessible to all groups. Structural inequality among local people means that some are more able to invest resources and time to generate "formal" knowledge, which they could articulate in the deliberative processes. As such, when scientific initiatives operate at the local level, when they require investment of time or local resources, they are more likely to draw the involvement of local elites. Both through immediate benefits of participation for the elite, such as status, information, stipends, or other benefits and in terms of long-term policy implications of the research, this association between scientists and local elites may tend to reinforce the existing social inequality among the different groups of local people (Vernooy and McDougall, 2003). It is therefore essential to envision multiple spaces of deliberation, and exploring the cultural politics of deliberative spaces (Fischer, 2006) to identify avenues that suit the disadvantaged people should itself be an issue of first order (before the substantive issue) deliberative practices, which could open onto possibilities for inclusive deliberation in more substantive issues of environmental governance. In Nepal's forestry case, the resource-poor farmers are less likely to articulate their concerns in front of the forest officials, and only a series of small-scale meetings, with forest officials disempowering themselves, may lead to deliberative transformation of forest governance.

We argue that the CAR approach has the potential to go beyond the dominant understanding on democratic environmental governance, such as the one related to participatory research (Gaventa, 2004; Görsdorf, 2006; Mansuri and Rao, 2013). Critical scholars recognize that a crisis has set in the promise of participatory development policy and practice (Kothari, 2001). After a history of over four decades, the participatory policy paradigm faces increasing concerns as citizens are losing control of the policy politics that affect their choices and positions (Sassen, 1996). In many situations, even the heavy investment in participatory projects has failed to foster effective deliberation, empowerment of communities, and enhance equitable resource entitlements (Mosse, 2005).

A long heritage of critical inquiry has unraveled a whole range of interpretive strategies and discourse analysis (Yanow, 2007). But such endeavors have also become confined to what Bourdieu called "scholastic reason" (Bourdieu, 2009). This situation means that critical scholars are great at critiquing why a certain practice does not work, but often fail to offer a nuanced theory of practice through which one can see how critical reasoning can contribute to address practical challenges in the real world. Parallel to this criticism, some have argued for a more engaged role of activist intellectuals to contribute to policy and practice (Hale, 2008). For us, explaining why policies do not work is different from exploring how exactly they can work (Ojha, 2013). As Kurt Lewin argues: "You cannot understand a system, unless you try to change it." Paulo Friere demonstrated processes to conscientize the oppressed through critical pedagogies (Freire, 1970/2000). Fals-Borda and Rahman also emphasized the emancipatory and liberation potential of participatory action research by advancing an approach for critical researchers to work with the exploited groups (Fals-Borda, 1987; Fals-Borda and Rahman, 1991). Yet, there are still limited attempts to blend critical inquiry, action orientation, and participatory research for better understanding and transforming policy practices.

As technoscientific approaches to understanding policy challenges are facing trenchant critique, the whole enterprise of science is under pressure to stimulate constructive policy dialogues (Fischer, 1998) and to facilitate agreement in conflict without aiming for the ultimate truth (Rorty, 2009). A more engaged view of such interactions can be explored through the recourse to "transactional inquiry" (Dewey and Bentley, 1949), wherein researchers and participants cocreate themselves in and through the dialogues or communicative transmissions. In line with the Dewean view, critical action researchers emphasize a deeper appreciation of and commitment to transactional inquiry, working with people in particular contexts not only to generate critical evidence, but also to help people know and do something in action to influence the policy process both as an ethical obligation of good citizens as well as for enhancing the quality of knowledge.

As an important consideration, CAR actors take practice as a source of learning and critical reflections, aiming for research results that build on some form of actions on the ground. The idea is that action-based evidence is more contextually grounded, and also carries the perspectives of the various social groups. Policy research is not just an epistemic project, but also one involving engagement and relationship to prepare grounds for social learning and political negotiations (Hall, 1993). The action aspect has both moral and epistemological dimensions. First, it means genuine effort to tackle or change the problematic situation in context. Second, the action is a mechanism for validation of new knowledge through expanding the definition of evidence (to include action-reflection dynamics), which will otherwise not be adequately robust. Participatory inquiry projects have oscillated between an action focus and a research focus, and CAR has evolved out of a need to develop a more thorough and integrated view of action as a form of critical research, reconciling both moral and epistemological dimensions. In CAR, we highlight maximizing both the immediate moral outcomes and the knowledge that can contribute to justice and development through improved policy decisions.

Multiscalar engagement is also an important aspect of CAR. In an increasingly networked world with multilevel drivers of change, it is hard for critical researchers to influence policy by acting only at the local level, and also within the arbitrarily delineated domains of formal institutions (Hajer, 2003). For example, local forest governance in Nepal is shaped by global processes related to knowledge, power, money, and strategic interests (Ojha, 2014). Several United Nations resolutions have begun to supersede national legislation. Likewise, international development aid has emerged as a strong political force in shaping the policy systems in the developing world. Clearly, the policy process has become a multiscale deliberative system (Mansbridge, 1999; Mansbridge et al., 2012), where not just sovereign citizens but a wide variety of actors have become legitimate players in the national policy field. More importantly, it is not just the number or the type of actors that make a policy system politically complicated, it is, in the language of Bourdieu (1989), the social field of practices with their own history, culture, and power relations, that influence policy in the increasingly networked society. This means that the task of a critical policy inquiry is not complete simply by looking at the interactions among individuals; how the wider political economy of knowledge is produced and legitimated in the policy process needs examination.

The CAR approach also signals some fundamental conceptual shift from participatory and demand-driven research and development approaches. In standard participatory approaches, the association between scientists and local elites may tend to reinforce the existing social inequality among the different groups of local people (Vernooy and McDougall, 2003), and thus reduce the possibility for deliberative governance. CAR projects can rarely start simply by waiting for its "clients" or "beneficiaries" to ask for a "research service" (Ojha, 2013). In all three cases, there is significant space for researchers working more proactively. Again, in the Nepal and Indian case, the level of proactive engagement is much higher, while in the landcare case in Australia, the current practice of research is based mostly on demand. This is perhaps because communities also vary in their capacity to demand—as slum communities in Indian cities where the Spatial Planning and Analysis Research Center (SPARC) worked often lack adequate capacity and knowledge to demand critical research partnerships. In any case, the impact of CAR can be enhanced if they actively find a space to undertake research, even in highly disabling political environments. This is why the notion of "critical" is particularly important here, although it has been left out in many strands of action research and policy inquiry. Being critical also means a disposition to engage, not to dismiss or reject, the hegemonic power (Ojha, 2013).

The technical possibility of the simultaneous production of environmental goods at different scales and limited possibility of fully privatizing resource arenas of governance (due to public good nature) require deliberation to take place between social agents occupying positions in different scales. Here, the role of technical experts should be to stimulate deliberation, rather than prescribe a cross-scale relationship. Forest degradation in Nepal's Himalayas was considered to be one of the factors responsible for floods in Bangladesh. And tropical rain forests are actually the global carbon sink. This multiscale production possibility expands the sphere of legitimate actors in the moral discourse, and complicates the possibility of arriving at understanding. While a significant movement in decentralization and local management of environmental resources has been promoted recently, they are driven more by the instrumental purpose of "conservation" than the genuine goal of political empowerment and equity. While the need for conservation cannot be denied, the way conservation is organized—by shifting burdens to locals while allowing the nonlocals to have a greater share of the benefits—is problematic. The conservation science has advocated centralized models, which provide little space to the local actors regarding how they want to manage such areas (Newmann, 1997). Also, leaving local resources with local people without adequate linkages established to enable them to exchange the resource values does not result in much benefit. Without realizing the value-in-exchange, the emphasis on devolution may actually create an extra burden on the local people to take care of globally significant environmental resources. In this connection, the challenge for a deliberative scientific practitioner is to reveal the conditions and possibilities of more equitable cross-scale governance of environmental resources (Cameron and Ojha, 2006), and thus facilitate equitable negotiation between scales.

2.5 CONCLUSION

In this chapter we have reviewed the work of Ostrom on collective action theory and have also put her work in the context of contemporary advances on theories of governance.

The chapter has briefly explored how Ostrom's work evolved by herself and others, and how her work advanced in response to critiques during her lifetime and after her death. In addition, we have charted out seven domains of work that advances or complements her work, namely, (1) the question of embeddedness, (2) political ecology, (3) decentralization, (4) participation, (5) equity and justice, (6) deliberative governance, and (7) CAR. We map out how Ostrom's works are linked to these domains and demonstrate the ways in which Ostrom's ideas are being reproduced, discontinued, or transformed. We also discussed how these new concepts can gain insights from Ostrom's work. In doing so, we have also identified gaps and issues in Ostrom's framework in the context of contemporary NRM challenges.

We conclude that while Ostrom's work continues to remain very relevant and important, there are important areas where contemporary advances can complement and enrich her work. Such a dialogue between Ostrom and other works is critical to facilitate a robust theorizing on environmental governance that can work in the changing and dynamic context. There is also a pragmatic necessity to advance our understanding on procedural aspects of change—including how power inequality can be tackled and how spaces of negotiation and deliberation among social actors can be enhanced. In this regard, we suggest that connecting Ostrom's work with ideas on deliberative governance and CAR can provide some important epistemological and political grounding for scholars aiming to catalyze progressive change in governance.

Acknowledgments

Authors would like to acknowledge support of Australian Centre for International Agricultural Research (ACIAR) funded research project on enhancing livelihood and food security through community forestry and agroforestry systems in Nepal to offer some insights into this chapter. Krishna Shrestha would like to acknowledge a SoSS Project Grant (SPG) awarded by the School of Social Sciences at the University of New South Wales Australia, and is grateful to Department of Geography at the University of Cambridge and Asia Research Institute (ARI) at Singapore National University for hosting him during sabbatical to finalise this chapter.

References

Adger, N., Brown, K., Fairbrass, J., Jordan, A., Paavola, J., Rosendo, S., Seyfang, G., 2002. Governance for Sustainability: Towards a 'Thick' Understanding of Environmental Decision Making. Centre for Social and Economical Research on the Global Environment (CSERGE), University of East Anglia, Norwich.

Agarwal, B., 2009. Rule making in community forestry institutions: the difference women make. Ecol. Econ. 68 (8–9), 2296–2308.

Agrawal, A., 2001. The regulatory community: decentralization and the environment in the van panchayats (forest councils) of Kumaon. Mt. Res. Dev. 21 (3), 208–211.

Agrawal, A., 2007. Forests, governance, and sustainability: common property theory and its contributions. Int. J. Commons 1 (1), 111–136.

Agrawal, A., 2014. Studying the commons, governing common-pool resource outcomes: some concluding thoughts. Environ. Sci. Pol. 36, 86–91.

Agrawal, A., Benson, C.S., 2011. Common property theory and resource governance institutions: strengthening explanations of multiple outcomes. Environ. Conserv. 38 (2), 199–210.

Agrawal, A., Gibson, C., 1999. Enchantment and disenchantment: the role of community in natural resource conservation. World Dev. 27 (4), 629–649.

Agrawal, A., Ribot, J., 1999. Accountability in decentralisation: a framework with South Asian and West African cases. J. Dev. Areas 33, 473–502.

Aligica, P.D., 2014. Institutional Diversity and Political Economy: The Ostroms and Beyond. Oxford University Press, Oxford.

Aligica, P., Tarko, V., 2012. Polycentricity: from Polanyi to Ostrom, and beyond. Gov.: Int. J. Policy Adm. Inst. 25 (2), 237–262.

Anderson, J., 2000. Four considerations for decentralised forest management: subsidiarity, empowerment, pluralism and social capital. In: Enters, T., Durst, P.B., Victor, M. (Eds.), Decentralisation and Devolution of Forest Management in Asia and the Pacific. RECOFTC, Bangkok. Report No 18.

Anon., 2003. Editorial: community forestry. Mekong Update Dialogue 6 (2), 1.

Araral, E., 2009. What explains collective action in the commons? Theory and evidence from the Philippines. World Dev. 37 (3), 687–697.

Araral, E., 2014. Ostrom, Hardin and the commons: a critical appreciation and a revisionist view. Environ. Sci. Pol. 36, 11–23.

Arielle Levine, A., Richmond, L., 2015. Using common-pool resource design principles to assess the viability of community-based fisheries co-management systems in American Samoa and Hawai. Mar. Policy 62, 9–17.

Arnstein, S., 1969. The ladder of participation. J. Am. Plan. Assoc. 35, 216–224.

Baland, J.M., Platteau, J., 1996. Halting Degradation of Natural Resources: Is There a Role for Rural Communities? FAO and Clarendon Press, Oxford.

Barnes, M., Kalberg, K., Pan, M., Leung, P., 2016. When is brokerage negatively associated with economic benefits? Ethnic diversity, competition, and common-pool resources. Soc. Networks 45, 55–65.

Barnes-Mauthe, M., Arita, S., Allen, S.D., Gray, S.A., Leung, P.S., 2013. The influence of ethnic diversity on social network structure in a common-pool resource system: implications for collaborative management. Ecol. Soc. 18 (1), 23–36.

Barnes-Mauthe, M., Gray, S.A., Arita, S., Lynham, J., Leung, P., 2014. What determines social capital in a social–ecological system? Insights from a network perspective. Environ. Manage. 55, 392–410.

Barr, A., Dekker, M., Fafchamps, M., 2015. The formation of community-based organizations: an analysis of a quasi-experiment in Zimbabwe. World Dev. 66, 131–153.

Batterbury, S., 2001. Landscapes of diversity: a local political ecology of livelihood diversification in South-Western Niger. Ecumene 8 (4), 437–464.

Batterbury, S., Forsyth, T., Thomson, K., 1997. Environmental transformation in developing countries: hybrid research and democratic policy. Roy. Geogr. Soc. 163, 126–131.

Berkes, F., 1989. Cooperation from the perspectives of human ecology. In: Berkes, F. (Ed.), Common Property Resources: Ecology and Community-based Sustainable Development. Belhaven Press, London, pp. 70–88.

Blaikie, P., 2006. Is small really beautiful? community-based natural resource management in Malawi and Botswana. World Dev. 34 (11), 1942–1957.

Blaikie, P., Brookfield, H., 1987. Land Degradation and Society. Methuen, London, New York.

Blomquist, W., deLeon, P., 2011. The design and promise of the institutional analysis and development framework. Policy Stud. J. 39 (1), 1–6.

Bohman, J., 1995. Public reason and cultural pluralism. Polit. Theor. 23 (2), 253–290.

Bohman, J., 1999. Democracy as inquiry, inquiry as democratic: pragmatism, social science and the cognitive division of labor. Am. J. Polit. Sci. 43 (2), 590–607.

Bosma, W., 1995. Benefits from Community Forestry: A Study from Community Forests and Benefit Sharing within Forest User Groups in the Koshi Hills of Nepal. Nepal UK Community Forestry Project, Kathmandu.

Bourdieu, P., 1989. Social space and symbolic power. Sociol Theory 7 (1), 14–25.

Bourdieu, P., 2009. The scholastic point of view. Cult. Anthropol. 5 (4), 380–391.

Bromley, D., Cernea, M., 1989. The Management of Common Property Natural Resources: Some Conceptual and Operational Fallacies. The World Bank, Washington, DC.

Bryant, R., 1992. Political ecology: an emerging research agendas in third-world studies. Political Geogr. 11 (1), 12–36.

Bryant, R., 1997. The Political Ecology of Forestry in Burma 1824–1994. University of Hawaii Press, Honolulu.

Bullard, R., Johnson, G., 2000. Environmental justice: grassroots activism and its impact on public policy decision making. J. Soc. Issues 56 (3), 555–578.

Cameron, J., Ojha, H., 2006. A deliberative ethic for development: a Nepalese journey from Bourdieu through Kant to Dewey and Habermas. Int. J. Soc. Econ. 34 (1), 66–87.

Cernea, M., 1993. The sociologist's approach to sustainable development. Finance Dev. 30 (4), 11–13.

Chambers, S., 2004. Giving up (on) rights? The future of rights and the project of radical democracy. Am. J. Polit. Sci. 48 (2), 185–200.

Chambers, S., Costain, A., 2000. Editorial introduction. In: Chambers, S., Costain, A. (Eds.), Deliberation, Democracy and the Media. Rowman & Littlefield Publishers, Inc, New York and Oxford, pp. xi–xiv.

Chhetri, R., Nurse, M., 1992. Equity in User Group Forestry: Implementation of Community Forestry in Central Nepal. Nepal-Australia Community Forestry Project, Kathmandu.

Christiano, T., 1997. The significance of public deliberation. In: Bohman, J., Rehg, W. (Eds.), Deliberative Democracy: Essays in Reason and Politics. MIT Press, Masachussets and London, pp. 243–277.

Ciriacy-Wantrup, S., Bishop, R., 1975. Common property as a concept in natural resources policy. Nat. Resour. J. 15, 713–727.

Cleaver, F., 2000. Moral ecological rationality, institutions and the management of common property resources. Dev. Change 28 (3), 361–383.

Cornwall, A., 2008. Unpacking 'participation': models, meanings and practices. Community Dev. J. 43 (3), 269–283.

Cox, M., 2014. Applying a social-ecological system framework to the study of the Taos Valley irrigation system. Hum. Ecol. 42 (2), 311–324.

Crossley, N., 2003. From reproduction to transformation: social movement fields and the radical habitus. Theory Cult. Soc. 20 (6), 43–68.

Dawes, R., 1980. Social dilemmas. Annu. Rev. Psychol. 31, 169–193.

Deutsch, M., 2000. Justice and conflict. In: Deutsch, M., Coleman, P.T. (Eds.), The Handbook of Conflict Resolution. Jossey-Bass, San Francisco, pp. 41–64.

Dewey, J., Bentley, A., 1949. Knowing and the Known. Greenwood Press, Westport, CT.

Dill, B., 2009. The paradoxes of community-based participation in Dar es Salaam. Dev. Change 40 (4), 717–743.

Dressler, W., Buscher, B., Schoon, B., Brockington, D., Hayes, T., Kull, C., McCarthy, J., Shrestha, K., Buscher, B., 2010. From hope to crisis and back again? A critical genealogy of the global CBNRM narrative. Environ. Conservat. 37 (1), 5–15.

Dryzek, J., 1997. The Politics of the Earth: Environmental Discourses. Oxford University Press, Oxford.

Dryzek, J., 2000. Deliberative Democracy and Beyond: Liberals, Critics, Contestations. Oxford University Press, Oxford.

Edwards, V., Steins, N., 1999. A framework for analysing contextual factors in common pool resource research. J. Environ. Policy Plann. 1, 205–221.

Elster, J., 1998. Introduction. In: Deliberative Democracy. Cambridge University Press, Cambridge, pp. 1–18.

Enters, T., Durst, P.B., Victor, M., 2000. Decentralization and Devolution of Forest Management in Asia and the Pacific. RECOFTC, Bangkok. Report No. 18.

Escobar, A., 1995. Encountering Development: The Making and Unmaking of the Third World. Princeton University Press, Princeton, NJ.

Esrc GEC Programme, 2001. Environmental Justice. SPRU, University of Sussex, Brighton.

Fairhead, J., Leach, M., 1995. False forest history, complicit social analysis: rethinking some West African environmental narratives. World Dev. 23 (6), 1023–1035.

Fals-Borda, O., 1987. The application of participatory action-research in Latin America. Int. Sociol. 2 (4), 329–347.

Fals-Borda, O., Rahman, M.A., 1991. Action and Knowledge: Breaking the Monopoly with Participatory Action Research. Intermediate Technology Publications, The Apex Press, London.

Fearon, J.D., 1998. Deliberation as discussion. In: Elster, J. (Ed.), Deliberative Democracy. Cambridge University Press, Cambridge, pp. 44–68.

Fischer, F., 1998. Beyond empiricism: policy inquiry in post positivist perspective. Policy Stud. 26 (1), 129–146.

Fischer, F., 2006. Participatory governance as deliberative empowerment. Am. Rev. Public Adm. 36 (1), 19.

Fisher, R., 1989. Indigenous Systems of Common Property Forest Management in Nepal. East-West Centre, Environment and Policy Institute, Honolulu.

Fisher, R., 1994. Indigenous forest management in Nepal: why common property is not a problem. In: Allen, M. (Ed.), Anthropology of Nepal: Peoples, Problems and Processes. Mandala Book Point, Kathmandu, pp. 64–81.

Fisher, R., 2010. Devolution or persistence of state control? In: Wittayapak, C., Vandergeest, P. (Eds.), The Politics of Decentralization. Mekong Press, Bangkok, pp. 21–37.

Forester, J., 1999. The Deliberative Practitioner: Encouraging Participatory Planning Processes. The MIT Press, Massachusetts.

Forsyth, T., 2003. Critical Political Ecology. Routledge, London.

Freire, P., 1970/2000. Pedagogy of the Oppressed. Continuum, New York.

Galston, W., 1986. Equality of opportunity and liberal theory. In: Lucash, F.S. (Ed.), Justice and Equality Here and Now. Cornell University Press, Ithaca, New York.

Gaventa, J., 2004. Towards participatory governance: assessing the transformative possibilities. In: Hickey, S., Mohan, G. (Eds.), Participation – From Tyranny to Transformation? Exploring New Approaches to Participation in Development. Zed Books, London, New York, pp. 25–41.

Geertz, C., 1973. The Interpretations of Cultures. Basic Books, New York.

Gilmour, D.A., Fisher, R.J., 1991. Villagers, Forests and Foresters: The Philosophy, Process and Practice of Community Forestry in Nepal. Sahayogi Press, Kathmandu.

Görsdorf, A., 2006. Inside deliberative experiments: dynamics of subjectivity in science policy deliberations. Policy Soc. 25 (2), 177–206.

Goodin, R.E., 2003. Reflective Democracy. Oxford University Press, Oxford.

Granovetter, M., 1985. Economic action and social structure: the problems of embeddedness. Am. J. Sociol. 91, 481–510.

Granovetter, M., Swedberg, R. (Eds.), 2001. The Sociology of Economic Life. Westview Press, Boulder and Oxford.

Gray, G.J., Enzer, M.J., Kusel, J., 2001. Understanding community-based forest management: an editorial synthesis. In: Gray, G.J., Enzer, M.J., Kusel, J. (Eds.), Understanding Community-Based Forest Ecosystems Management. Hawarth Press Inc., New York, pp. 1–23.

Guthman, J., 1997. Representing crisis: the theory of Himalayan environmental degradation and the project of development in post-rana Nepal. Dev. Change 28 (1), 44–69.

Gutman, A., Thompson, D., 2003. Deliberative democracy beyond process. In: Fishkin, J.S., Laaslett, P. (Eds.), Debating Deliberative Democracy. Blackwell, Oxford, pp. 31–53.

Gutmann, A., Thompson, D., 2004. Why Deliberative Democracy? Princeton University Press, New Jersey.

Habermas, J., 1990. Moral Consciousness and Communicative Action. MIT Press, Cambridge, Massachusetts.

Habermas, J., 1996. Between Facts and Norms: Contributions to a Discourse Theory of Law and Democracy. MIT Press, Massachusetts.

Hajer, M., 2003. Policy without polity? Policy analysis and the institutional void. Policy Sci. 36 (2), 175–195.

Hale, C.R. (Ed.), 2008. Engaging Contradictions: Theory, Politics, and Methods of Activist Scholarship. University of California Press, Berkeley.

Hall, P., 1993. Policy paradigms, social learning, and the state: the case of economic policy making in Britain. Comp. Politics 25 (3), 275–296.

Haller, M., 1994. The Past as Future - Jurgen Habermas Interviewed (M. Pensky, Trans. and Ed.). Polity Press, Cambridge.

Hardin, G., 1968. The tragedy of the commons. Science 162, 1243–1248.

Hayward, C.R., 2004. Doxa and deliberation. Crit. Rev. Int. Soc. Polit. Philos. 7 (1), 1–24.

Holmes, T., Scoones, I., 2000. Participatory Environmental Policy Processes: Experiences from North and South. Institute of Development Studies, University of Sussex, Brighton.

Huntjens, P., Lebel, L., Pahl-Wostl, C., Shulze, R., Kranz, N., 2012. Institutional design propositions for the governance of adaptation to climate change in the water sector. Glob. Environ. Change 22, 67–81.

Ives, J.D., 2004. Himalayan perceptions: environmental change and the well-being of mountain peoples. Routledge, London.

Johnson, C., Forsyth, T., 2002. In the eyes of the state: negotiating a "rights-based approach" to forest conservation in Thailand. World Dev. 30 (9), 1591–1605.

Kimbrough, E., Vostroknutov, A., 2015. The social and ecological determinants of common pool resource sustainability. J. Environ. Econ. Manag. 72, 38–53.

Kothari, U., 2001. Power, knowledge and social control in participatory development. In: Cooke, B., Kothari, U. (Eds.), Participation: The New Tyranny? Zed Books, London, pp. 139–152.

Larson, A., Ribot, J., 2004. Democratic decentralization through a natural resource lens: an introduction. Eur. J. Dev. Res. 16 (1), 1–25.

Lawoti, M., 2008. Exclusionary democratization in Nepal, 1990–2002. Democratization 15 (2), 363–385.

Leach, M., Mearns, R., Scoones, I., 1999. Environmental entitlements: dynamics and institutions in community-based natural resource management. World Dev. 27 (2), 225–247.

Li, T., 1996. Images of community: discourse and strategy in property relations. Dev. Change 27 (3), 501–527.

Li, T.M., 2002. Engaging simplifications: community-based resource management, market processes and state agendas in upland Southeast Asia. World Dev. 30 (2), 265–283.

Lind, E.A., Kanfer, R., Early, P.C., 1990. Voice, control and procedural justice - instrumental and non-instrumental concerns in fairness judgements. J. Pers. Soc. Psychol. 59 (5), 952–959.

Lind, E., Earley, P., 1992. Procedural justice and culture. Int. J. Psychol. 27 (2), 227–242.

Mahanty, S., Fox, J., Nurse, M., Stephen, P., McLees, L., 2006. Hanging in the Balance: Equity in Community-Based Natural Resource Management in Asia. RECOFTC and East-West Centre, Honolulu.

Malinowski, B., 1926. Crime and Custom in Savage Society. Routledge and Kegan Paul, London.

Mansbridge, J., 1999. Everyday talk in the deliberative system. In: Macedo, S. (Ed.), Deliberative Politics: Essays on Democracy and Disagreement. Oxford University Press, New York, pp. 211–239.

Mansbridge, J., Bohman, J., Chambers, S., Christiano, T., Fung, A., Parkinson, J., Thompson, D.F., Warren, M.E., 2012. A systemic approach to deliberative democracy. Deliberative Syst., 1–26.

Mansuri, G., Rao, V., 2004. Community-based and-driven development: a critical review. World Bank Res. Obs. 19 (1), 1.

Mansuri, G., Rao, V., 2013. Localizing Development: Does Participation Work? World Bank, Washington, DC.

McCay, B.J., Jentoft, S., 1998. Market or common failure? Critical perspectives on common property research. Hum. Organ. 57 (1), 21–29.

McCay, B.J., 2000. Post-modernism and the management of natural and common pesources. Common Prop. Resour. Dig. 54, 1–8.

McGinnis, M., Ostrom, E., 2014. Social-ecological system framework: initial changes and continuing challenges. Ecol. Soc. 19 (2), 30–32.

Messerschmidt, D., 1981. Nogar and other traditional forms of cooperation in Nepal: significance for development. Hum. Organ. 40 (1), 40–47.

Mitchell, W.C., 1988. Virginia, Rochester, and Bloomington: twenty-five years of public choice and political science. Public Choice 56 (2), 101–119.

Mohan, G., Hickey, S., 2004. Relocating participation within a radical politics of development: critical modernism and citizenship. In: Mohan, G., Hickey, S. (Eds.), Participation – From Tyranny to Transformation? Exploring New Approaches to Participation in Development. Zed Books, London, New York, pp. 59–74.

I. INTRODUCTION AND CONCEPTUAL BACKGROUND

Mohan, G., Stokke, K., 2000. Participatory development and empowerment: the dangers of localism. Third World Q. 21 (2), 247–268.

Mosse, D., 1997. The symbolic making of a common property resource: history, ecology and locality in a tank-irrigated landscape in south India. Dev. Change 28 (3), 467–504.

Mosse, D., 2005. Cultivating Development: An Ethnography of Aid Policy and Practice. Vistaar Publications, New Delhi.

National Research Council, 1986. In: Proceeding of the Conference of Common Property resource Management. April 21–26. National Academy Press, Washington, DC.

Nelson, N., Wright, S., 1995. Participation and power. In: Nelson, N., Wright, S. (Eds.), Power and Participatory Development: Theory and Practice. Intermediate Technology Publications, London, pp. 1–18.

Newmann, R.P., 1997. Primitive ideas: protected area buffer zones and the politics of land in Africa. Dev. Change 28, 559–582.

Norman, R., 1999. Equality, priority and social justice. Ratio XII, 178–194.

Nozick, R., 1974. Anarchy, State and Utopia. Blackwell, Oxford.

Ojha, H., 2008. Reframing Governance: Understanding Deliberative Politics in Nepal's Terai Forestry. Adroit, New Delhi.

Ojha, H., Timsina, N., Kumar, C., Belcher, B., Banjade, M., Chhetri, R., Nightingale, A., 2008. Community based forest management programmes in Nepal - an overview of issues and lessons communities. In: Ojha, H., Timsina, N., Kumar, C., Belcher, B., Banjade, M. (Eds.), Forests and Governance: Policy and Institutional Innovations from Nepal. Adroit, New Delhi, pp. 25–54.

Ojha, H., 2013. Counteracting hegemonic powers in the policy process: critical action research on Nepal's forest governance. Crit. Policy Stud. 7 (3), 242–262.

Ojha, H., 2014. Beyond the 'local community': the evolution of multi-scalepolitics in Nepal's community forestry regimes. Int. For. Rev. 16 (3), 339–353.

Olson, M., 1965. The Logic of Collective Action: Public Goods and the Theory of Groups. Cambridge University Press, Cambridge.

Ophuls, W., 1973. Leviathan or oblivion. In: Daly, H.E. (Ed.), Towards a Steady State Economy. Freeman, San Francisco, pp. 215–230.

Ostrom, E., 1990. Governing the Commons: The Evolution of Institutions for Collective Action. Cambridge University Press, Cambridge.

Ostrom, E., 1992. The rudiments of a theory of the origins, survivals, and performance of common property institutions. In: Bromley, D.W. (Ed.), Making the Commons Work: Theory, Practice and Policy. ICS Press, San Francisco, CA, pp. 293–318.

Ostrom, E., 2000. The danger of self-evident truths. Pol. Sci. Polit. 33 (1), 33–44.

Ostrom, E., 2002. Reformulating the commons. In: Burger, J., Norgaard, R., Ostrom, E., Policansky, D., Goldstein, B. (Eds.), The Commons Revisited: An Americas Perspectives. Island Press, Washington, DC.

Ostrom, E., 2003. How types of goods and property rights jointly affect collective action? J. Theor. Polit. 15 (3), 239–270.

Ostrom, E., 2005. Understanding Institutional Diversity. Princeton University Press, Princeton.

Ostrom, E., 2009. A general framework for analyzing sustainability of social–ecological systems. Science 325 (5939), 419–422.

Ostrom, E., 2010. Beyond markets and states: polycentric governance of complex economic systems. Am. Econ. Rev. 100 (3), 641–672.

Ostrom, E., 2011. Background on the institutional analysis and development framework. Policy Stud. J. 39 (1), 7–27.

Ostrom, E., Cox, M., 2010. Moving beyond panaceas: a multi-tiered diagnostic approach for social-ecological analysis. Environ. Conserv. 37 (4), 451–463.

Paavola, J., Adger, W.N., 2002. Justice and Adaptation to Climate Change. Tyndall Center for Climate Change Research, University of East Anglia, Norwick, UK.

Pandey, S., Yadama, G.N., 1990. Conditions for local level community forestry action: a theoretical explanation. Mt. Res. Dev. 10 (1), 88–95.

Parfit, D., 1991. Equality or Priority? The Lindley Lecture, University of Kansas, Kansas.

Peet, R., Watts, M., 1993. Introduction: development theory and environment in an age of market triumphalism. Econ. Geogr. 69 (3), 227–253.

Peters, P., 1987. The grazing lands of Botswana and the commons debate. In: McCay, B.J., Acheson, J. (Eds.), The Questions of the Commons: The Culture and Ecology of Communal Resources. The University of Arizona Press, Tuscan.

Petrzelka, P., Bell, M., 2000. Rationality and solidarities: the social organization of common property resources in the Imdrhas Valley of Morocco. Hum. Organ. 59 (3), 343–352.

Polanyi, K., 1957. The Great Transformation. Beacon Press, Boston.

Poteete, A.R., Janssen, M.A., Ostrom, E., 2010. Working Together: Collective Action, the Commons, and Multiple Methods in Practice. Princeton University Press, Princeton, NJ.

Przeworski, A., 1998. Deliberation and ideological domination. In: Elster, J. (Ed.), Deliberative Democracy. Cambridge University Press, Cambridge, pp. 19–43.

Pulhin, J., Dressler, W., 2009. People, power and timber: the politics of community-based forest management. J. Environ. Manage. 91, 206–214.

Rangan, H., Lane, M.B., 2001. Indigenous peoples and forest management: comparative analysis of institutional approaches in Australia and India. Soc. Nat. Resour. 14, 145–160.

Rawls, J., 1971. A Theory of Justice. Harvard University Press, Cambridge.

Raz, J., 1986. The Morality of Freedom. Oxford University Press, Oxford.

Reich, R.B., 1990. Public Management in Democratic Society. Prentice-Hall, New Jersey.

Ribot, J.C., 2004. Waiting for Democracy: The Politics of Choice in Natural Resource Decentralisation. World Resource Institute, Washington, DC.

Ringquist, E.J., 1998. A question of justice: equity in environmental litigation, 1974–1991. J. Polit. 60 (4), 1148–1165.

Rorty, R., 2009. Philosophy and the Mirror of Nature. Princeton University Press, Princeton, NJ.

Runge, C.F., 1986. Common property and collective action in economic development. World Dev. 14 (5), 623–635.

Sabatier, P.A., 1991. Toward better theories of the policy process. Political Sci. Politics 24 (2), 147–156.

Sanders, L., 1997. Against deliberation. Polit. Theor. 25 (3), 347–376.

Sassen, S., 1996. Losing Control? Sovereignty in an Age of Globalization. Columbia University Press, New York.

Schatzki, T., 1993. Theory at bay: Foucault, Lyotard, and politics of the local. In: Jones, J.I., Natter, W., Schatzki, T. (Eds.), Postmodern Contentions: Epochs, Politics, Space. The Guilford Press, New York, London, pp. 39–64.

Sen, A., 1999. Development as Freedom. Oxford University Press, Oxford.

Seymour-Smith, C., 1986. Dictionary of Anthropology. MacMillan Reference Books, London, Basingstoke.

Shrestha, K.K., McManus, P., 2007. The embeddedness of collective action in Nepalese Community Forestry. Small-Scale For. 6, 273–290.

Shrestha, K., McManus, P., 2008. The politics of community participation in natural resource management: Lessons from community forestry in Nepal. Aust. Forest. 71 (2), 135–146.

Shrestha, K.K., 2009. Rethinking participatory forestry for equity: views from Nepal. In: XIII World Forestry Congress: Forests in Development - A Vital Balance, Buenos Aires, Argentina, 18–25 October, 2009.

Sillitoe, P., 1998. The development of indigenous knowledge - a new applied anthropology. Curr. Anthropol. 39 (2), 223–252.

Smith, G., 2003. Deliberative Democracy and the Environment. Routledge, London.

Steins, N.A., Edwards, V.M., 1999. Collective action in common pool resource management: the contribution of a social constructivist perspective to existing theory. Soc. Nat. Res. 12, 539–557.

Stott, P., Sullivan, S., 2000. Political Ecology: Science, Myth and Power. Arnold, London.

Taylor, C., 1988. The nature and scope of distributive justice. In: Lucash, F.S. (Ed.), Justice and Equality Here and Now. Cornell University Press, Ithaca, New York.

Thiel, A., Adamseged, M., Baake, C., 2015. Evaluating an instrument for institutional crafting: how Ostrom's social–ecological systems framework is applied. Environ. Sci. Policy 53, 152–164.

Tucker, C.M., 1999. Private versus common property forests: forest conditions and tenure in a Honduran community (statistical data included). Hum. Ecol. 27 (2), 201–218.

Tyler, T., 1994. Psychological models of the justice motive: antecedents of distributive and procedural justice. J. Pers. Soc. Psychol. 67 (5), 850–863.

Uphoff, N., 1992. Learning from Gal Oya: Possibilities for Participatory Development and Post-Newtonian Social Science. Cornell University Press, Ithaca, London.

Uphoff, N., 1993. Grassroots organisations and NGOs in rural development: opportunities with diminishing states and expanding markets. World Dev. 21 (4), 607–622.

Vernooy, R., McDougall, C., 2003. Principles for good practice: reflecting on lessons from the field. In: Pound, B., Snapp, S., McDougall, C., Braun, A. (Eds.), Managing Natural Resources for Sustainable Livelihoods: Uniting Science and Participation. EarthScan and IDRC, London.

I. INTRODUCTION AND CONCEPTUAL BACKGROUND

Wade, R., 1988. Village Republics: Economic Conditions for Collective Action in South India. Cambridge University Press, Cambridge.

Walzer, M., 1983. Spheres of Justice: A Defence of Pluralism and Equality. Basil Blackwell, Oxford.

WCED, 1987. Our Common Future, Report of the World Commission on Environment and Development (WCED). Zed Books, London.

Willis, P., 2012. Engaging communities: Ostrom's economic commons, social capital and public relations. Public Relat. Rev. 38 (1), 116–122.

Wollenberg, E., Merino, L., Agrawal, A., Ostrom, E., 2007. Fourteen years of monitoring community-managed forests: learning from IFRI's experience. Int. For. Rev. 9 (2), 670–684.

Woolcock, M., Narayan, D., 2000. Social capital: implications for development theory, research, and policy. World Bank Res. Obs. 15 (2), 225–249.

Yanow, D., 2007. Interpretation in policy analysis: on methods and practice. Crit. Policy Anal. 1 (1), 110–122.

Young, I.M., 1997. Difference as a resource for democratic communication deliberative democracy. In: Bohman, J., Rehg, W. (Eds.), Essays on Reason and Politics. MIT Press, Massachusetts and London, pp. 383–406.

Young, D.A., 2003. The democratic chorus: culture, dialogue and polyphonic paideia. Democr. Nat. 9 (2), 221.

3

Governing the Commons Through Understanding of Institutional Diversity: An Agenda for Application of Ostrom's Framework in Managing Natural Resources in Asia

R.C. Bastakoti*, G. Shivakoti[†,‡]

*International Water Management Institute, Lalitpur, Nepal †The University of Tokyo, Tokyo, Japan ‡Asian Institute of Technology, Bangkok, Thailand

3.1 INTRODUCTION

Some of the key issues of inquiry in managing common pool resources (CPRs) include how institutions are formed; how they operate and change; and how they influence behavior in society. Elinor Ostrom's detailed Institutional Analysis and Development (IAD) framework and design principles for robust performance of CPRs have become among the most widely used research frameworks for scholars and practitioners studying and managing CPRs (Ostrom, 1990, 2005).

In all her works, Professor Ostrom put efforts into exploring the mechanisms for tackling common problems that communities face across the world. Her works often looked beyond both state- and/or market-oriented approaches. She argued that societies and groups regularly devise rules and enforcement mechanisms that stop the degradation of CPRs.

In this chapter we review her two seminal works, *Governing the Commons* (1990) and *Understanding Institutional Diversity* (2005). First we bring in the major theoretical contribution, and then discuss empirical applications of her theoretical framework describing how such frameworks were instrumental in shaping the policy-relevant researches. At the end of the chapter, we suggest some further agenda and implications to guide the natural resources management-related activities in Southeast Asia.

3.2 THEORETICAL CONTRIBUTION

3.2.1 Understanding When Collective Action Can Be Effective

In her influential book *Governing the Commons: The Evolution of Institutions for Collective Action* (1990) Elinor Ostrom built on empirical studies of commons including fisheries, forests, and irrigation systems to provide a broader theory of institutional arrangements related to the effective governance and management of CPRs. In this book, Ostrom begins by noting the problem of natural resource depletion and provision of collective goods. She discusses three models that had been used to explain the many problems that individuals face in managing CPRs.

The three models discussed are Hardin's "tragedy of the commons"; the "prisoners dilemma model"; and Olson's "logic of collective action." Those three models were often used to provide a foundation for recommending state or market solutions.

Hardin's model has been very influential (Hardin, 1968). As Ostrom pointed out, it has been used as a metaphor for the general problem of overpopulation. She noted that much of the world is dependent on resources that are subject to the possibility of a tragedy of the commons (Ostrom, 1990). Hardin's model has often been formalized as a prisoner's dilemma game. She develops alternative games to open up the discussion of institutional options for solving commons dilemmas (Ostrom 1990, 2000). The third model is Olson's "The Logic of Collective Action" (Olson, 1965), which Ostrom later criticized as a theory of collective inaction. The main point was that Olson's model specifically excludes altruistic groups from his theory.

Professor Ostrom proposed theoretical and empirical alternatives to these models to begin to illustrate the diversity of solutions that go beyond states and markets. She pointed out that these models are not wrong; rather they are special models that utilize extreme assumptions rather than general theories. These models can successfully predict strategies and outcomes in fixed situations approximating the initial conditions of the models, but they cannot predict outcomes outside that range. She tried to find solutions to the "tragedy of the commons," exploring the prisoner's dilemma and using game theoretic reasoning in an innovative way. She emphasized the need to incorporate actors' self-understanding of their roles in particular contexts.

Her efforts provided the alternatives to standard and narrowly constructed models of rational choice. Ostrom shows that rational choice thinking can be deployed and developed to explore cooperative solutions. She argued that long-term sustainability needs rules that match the attributes of resource systems and users. Large-scale governance systems may facilitate such arrangements, but they could also be destructive; for example, colonial powers did not recognize local resource institutions that had developed over centuries.

3.2.2 Design Principles to Govern the CPRs

Elinor Ostrom describes that when the individuals involved in CPR management gain a major part of their economic return from the CPRs they are strongly motivated to try to solve common problems to enhance their own productivity. Therefore, the cooperative governance of CPRs, all other things being equal, could be more effective in formulating and enforcing rules than governance by either a government agency or a corporation.

In *Governing the Commons* she sought to identify the underlying design principles of the institutions used by those who have successfully managed their own CPRs over extended periods of time. This book included surveys in diverse field settings that covered (1) how appropriators have devised, applied, and monitored their own rules to control the use of their CPRs and (2) that the resource systems, as well as the institutions, have survived for long periods of time. Almost all the sets of institutions analyzed were more than 100 years old, among which some were even more than 1000 years old.

The rules for governing CPRs, in the instances Ostrom examined, worked in situations where game theory would have predicted incentives to defect were strong and negative consequences of defection were weak (for example, in common governance systems for irrigation water where monitoring was relatively weak and fines were low compared to the benefits of defection). Governance systems for CPRs have typically reflected close empirical reasoning from historical experience. Based on her survey, Ostrom proposed design principles that characterize robust self-governing CPR institutions. The design principles are defined as "… an essential element or condition that helps to account for the success of these institutions in sustaining the CPRs and gaining the compliance of generation after generation of the appropriators to the rules-in-use."

According to Ostrom herself, these design principles are not a blueprint for analyzing CPR management. Their presence has been found consistently in many long-enduring CPRs. The eight design principles that characterize the configuration of rules devised by long-enduring CPR institutions are presented in Table 3.1.

3.2.3 Understanding the Institutional Diversity

Elinor Ostrom, in another important book, *Understanding Institutional Diversity* (2005), provides a coherent method for undertaking the analysis of diverse economic, political, and social institutions. This book further explains the IAD framework enabling researchers to choose the most relevant level of interaction for a particular question.

The IAD framework (Fig. 3.1) examines the arena within which interactions occur, the rules employed by participants to order relationships, the attributes of a biophysical world that structures and is structured by interactions, and the attributes of a community in which a particular arena is placed. The book explains and illustrates how to use the IAD framework in the context of both field and experimental studies. Concentrating primarily on the rules aspect of the IAD framework, it provides empirical evidence about diversity of rules, the calculation process used by participants in changing rules, and the design principles that characterize robust, self-organized resource governance institutions.

In the book, Professor Ostrom describes analytical concepts useful in examining institutions; focuses on the hierarchy of rules that govern human behavior; and summarizes some key case studies of how institutions have been developed to mitigate the losses of the commons. One of the key features of this book is the inclusion of detailed discussion on a grammar of the institutions, which was proposed earlier by Crawford and Ostrom (1995). She describes the "grammar of institutions" [ADICO syntax] to define and classify three essential components of institutional analysis: strategies, norms, and rules. The ADICO syntax helps analyze the rule formation process and includes Attributes, Deontic, Aim, Conditions, Or else (Table 3.2).

TABLE 3.1   Ostrom's Institutional Design Principles for Long-Enduring CPR Institutions

Design Principles	Explanation
1. Clearly defined boundaries	Individuals or households who have rights to withdraw resource units from the CPR must be clearly defined, as must be the boundaries of the CPR itself
2. Congruence between appropriation and provision rules and local conditions	Appropriation rules restricting time, place, technology, and/or quantity of resource units are related to local conditions and to provision rules requiring labor, material, and/or money
3. Collective choice arrangements	Most individuals affected by the operational rules can participate in modifying the operational rules
4. Monitoring	Monitors, who actively audit CPR conditions and appropriator behavior, are accountable to the appropriators or are the appropriators themselves
5. Graduated sanctions	Appropriators who violate operational rules are likely to be assessed graduated sanctions by other appropriators, by officials accountable to these appropriators, or by both
6. Conflict resolution mechanisms	Appropriators and their officials have rapid access to low-cost, local arenas to resolve conflicts among appropriators or between appropriators and officials
7. Minimal recognition of rights to organize	The rights of appropriators to devise their own institutions are not challenged by external government authorities
8. Nested enterprises (for CPRs that are parts of larger systems)	Appropriation, provision, monitoring, enforcement, conflict resolution, and governance activities are organized in multiple layers of nested enterprises

Adapted from Ostrom, E., 1990. Governing the Commons: The Evolution of Institutions for Collective Action. Cambridge University Press, New York.

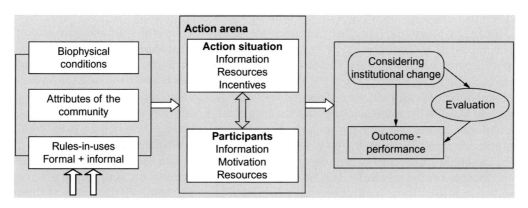

FIG. 3.1   Institutional Analysis and Development (IAD) framework.

Ostrom (2005) suggested classifying the rules into seven broad types based on the "Aim" element of rules as discussed in the ADICO syntax (Table 3.3). She classifies rules according to whether they define actors' positions, the resource's boundary, choice variables, aggregation, information, payoffs, scope, succession, and exit. All of these conventions are critical aspects of any institution, and how they are implemented or how they constrain decision

TABLE 3.2 ADICO Syntax to Analyze Rule Formation Process

ADICO Component	Explanation
"Attributes" (A)	Defines values of participant-level variables distinguishing to whom the institutional statement applies
"Deontic" (D)	Refers to the three modal verbs: "may" (permitted); "must" (obliged); and "must not" (forbidden)
"Aim" (I)	Describes particular actions or outcomes in the action situation
"Conditions" (C)	Defines when and where an action or outcome is permissible, obligatory, or forbidden
"Or else" (O)	Sets out the consequence for not following rule

Source: *Crawford, S., Ostrom, E., 1995. A grammar of institutions. Am. Polit. Sci. Rev. 89(3), 582–600; Ostrom, 2005.*

TABLE 3.3 The "Aim" Component of Seven Types of Rule

Type of Rule	Basic "Aim" Verb	Regulated Component of Action Situation
Position	Be	Positions
Boundary	Enter or leave	Participants
Choice (allocation)	Do	Actions
Aggregation	Jointly affect	Control
Information	Send or receive	Information
Payoff	Pay or receive	Costs/benefits
Scope	Occur	Outcomes

Source: *Ostrom (2005).*

makers importantly affects the institution's success in improving welfare. She notes that implementing and enforcing rules are costly processes, and, in the case of boundary rules, the costs of demarcating property rights can determine their extent and usefulness in addressing the commons.

3.2.4 Understanding the Complexity of Socioecological Systems

In more recent paper, Professor Ostrom provided a general framework for analyzing the sustainability of socioecological systems (Ostrom, 2009a,b). The paper identified 10 subsystem variables that affect the likelihood of successful self-organization of efforts to achieve a sustainable socioecological system by the communities involved. A simple form of the 10 variables identified is presented below:

1. The size of resource system—a moderate territorial size is most conducive to self-organization.
2. The productivity of system—self-organization is less likely to work if a resource is either overabundant or already exhausted.
3. The predictability of system dynamics—for example, some fishery systems approach mathematical chaos, making self-organization infeasible.

4. Resource unit mobility—self-organization becomes more difficult with mobile rather than stationary units; for example, in a river versus a lake.
5. The number of users—transaction costs can be higher with larger groups, but such groups can also mobilize more resources. The net effect depends on other variables and on the tasks undertaken.
6. Leadership—high skills and an established track record among leaders aids self-organization.
7. Norms and social capital—in terms of shared moral and ethical standards.
8. Knowledge of the socioecological system—more is better.
9. The importance of resource to users—where the resources are vital, self-organization becomes easier.
10. Collective choice rules—can lower transaction costs.

3.3 EMPIRICAL APPLICATIONS OF THE FRAMEWORKS

3.3.1 Overview of Long-Term Research Collaboration With Professor Ostrom

Professor Ostrom's theoretical foundations have made significant contributions during the last 25 years toward policy interventions in governing and managing natural resources, mainly irrigation and forestry resources, across the world and particularly in developing countries. In selected countries, her theoretical frameworks have been widely used in policy-relevant works through the effective use of results from the analysis of meta-database both cross-sectional and overtime using the IAD framework (Ostrom, 1986).

She is well known to have collaborated with scholars from different parts of the world over very long periods, including colleagues from Africa, Asia, and Latin America. The Workshop in Political Theory and Policy Analysis at Indiana University (Ostrom Workshop), which she cofounded with her husband, Vincent Ostrom, has served as an inviting place for visiting scholars from around the world. We, both coauthors, have been fortunate to have had the opportunity to work with her. Our first affiliation dates back to 1988 while one coauthor of this chapter (Ganesh Shivakoti) was preparing for dissertation research. He received a call from Professor Ostrom, who surprisingly mentioned that she had read his dissertation proposal "Organizational Effectiveness of User and Non-user Controlled Irrigation Systems in Nepal" which was recommended by Bob Yoder and Prachanda Pradhan during a meeting at the then International Irrigation Management Institute (IIMI) country office in Nepal (Shivakoti, 1991).

The coauthor had the opportunity to work with Ostrom to visit irrigation systems as a consultant for the USAID/Nepal project on "Decentralization, Finance and Management" (Ostrom, 1988). During the summer of 1988 the team visited several irrigation systems in the Western mid-hill region of Nepal and Ostrom was so impressed with the craftsmanship and self-governance arrangement of farmers in managing centuries old irrigation systems that the die was cast for long-term detailed research with the creation of the Nepal Irrigation Institutions and Systems (NIIS) database at the Ostrom Workshop with a multidisciplinary team.

During 1991–2002, the Ford Foundation office in New York and Delhi provided a parallel grant to the Ostrom Workshop and the Irrigation Management Systems Study Group

(IMSSG) of the Institute of Agriculture, Rampur, in Nepal to develop the NIIS database and add on several more irrigation systems. During the same time, the Asian Development Bank (ADB) had planned a massive irrigation system construction ignoring 88 farmer-managed irrigation systems (FMISs) in the Chitwan District of Nepal. Professor Ostrom, with support from IMSSG members and IIMI staff, used the NIIS data, which provided enough evidence of excellent governance and management of irrigation systems by the farmers themselves to convince the ADB to redesign the intervention by improving and rehabilitating the existing FMISs (IMSSG, 1993). When she was later invited by the World Bank to talk about the puzzles of underperformance in the irrigation sector in a global context, she based her argument on the findings from Nepal irrigation data and case studies (Ostrom, 1996). This later became a pillar foundation for World Bank and ADB-assisted participatory irrigation management and irrigation management transfer program implementation in developing countries.

During the early 1990s, the Ostrom Workshop and IMSSG organized two international workshops at Rampur, Nepal, on the themes "From Farmers' Fields to Data Fields and Back" (Sowerwine et al., 1994) and "People and Participation in Sustainable Development of Natural Resources" (Shivakoti et al., 1997) whereby international scholars, development practitioners, irrigation officials, academic scholars, students, and farmers participated in the discussion. The successive policy documents on governance and management of irrigation systems drafted by the government of Nepal used Ostrom's theory, which was backed by field evidence. Subsequently, during 1996–2002 two important volumes outlining policies for irrigation governance and management improvement in Nepal were published by ICS Press based on NIIS data analysis (Lam, 1998; Shivakoti and Ostrom, 2002).

Our collaboration with Professor Ostrom and the Ostrom Workshop continued and expanded after the coauthor of this chapter joined the Asian Institute of Technology (AIT) as faculty in Thailand. A research center was established in joint collaboration with the Ostrom Workshop and other institutes in Asia focusing the research agenda on the study of self-governing institutions in the management of irrigation and forestry resources in Asia. AIT later became an International Forestry Resources and Institutions (IFRI) collaborative research center (CRC). Several doctoral dissertations, referred journal publications, and two major volumes (*Asian Irrigation in Transition and Decentralization*, and *Forest and Rural Communities in Asia*) were published by Sage Publications (Shivakoti et al., 2005; Webb and Shivakoti, 2008).

The University of Hong Kong and AIT collaborated in a research project "Asian Irrigation Institutions and Systems (AIIS) Dynamics Study and Database Management." Later, the project aimed to expand the NIIS database to the Asian Irrigation Institutions and Systems (AIIS) database using the IAD framework, Ostrom's design principles, and ADICO syntax as the main theoretical framework. The research comprises the first author's dissertation, covering case studies from Nepal and Thailand (Bastakoti, 2010).

We continued our collaboration with Professor Ostrom. During 2011, Ostrom, Lam, Pradhan, and Shivakoti authored the book *Improving Irrigation in Asia: Sustainable Performance of an Innovative Intervention in Nepal* based on a longitudinal study over two decades on innovative intervention for sustained performance of irrigation systems, which has identified key factors that can help explain the performance of interventions (Ostrom et al., 2011). In addition, some of our recent works are also based on the application of Ostrom's design principles and ADICO syntax (Bastakoti et al., 2010; Bastakoti and Shivakoti, 2009, 2012; Kamran and Shivakoti, 2013a,b).

3.3.2 Application of Design Principles

Ostrom's (1990) design principles have been used in many research works. The design principles were tested in decision-making arrangements in a community-based watershed management program in northern Thailand (Wittayapak and Dearden, 1999) and community forestry in Nepal, with suggestions for modifications or expansions (Gautam and Shivakoti, 2005). Other cases such as irrigation in Japan (Sarker and Itoh, 2001) and Bulgaria (Theesfeld, 2004) and agropastoral, irrigated, and rain-fed farming villages in semiarid Tanzania (Quinn et al., 2007) also proposed modifications in design principles for their applicability in specific conditions. Yet other studies based on community fisheries argue that the presence of design principles for collective action does not necessarily lead to "successful" collective action (Steins and Edwards, 1999) and design principles have ignored the effects of contextual (biophysical, socioeconomic, and cultural) factors on CPR governance and management (Agrawal, 2002). Ostrom (2009a,b)) herself examined the validity of the design principles after nearly two decades. One major concern was on the study of the robustness of CPRs, particularly on self-governing irrigation systems.

In this context, Bastakoti and Shivakoti (2009) assessed the applicability of design principles in irrigation water management comparing the similarities and differences from Nepal and Thailand (considering the difference in economic situation and nature of irrigation systems). The analysis was done using the primary information collected from 50 irrigation systems from each country, in total 100 irrigation systems. According to Ostrom herself, these design principles are not a blueprint for analyzing CPR management. Their presence has been found consistently in many long-enduring CPRs. In our analysis we looked at the comparative applicability of these design principles in cases of irrigation management in Nepal and Thailand. Efforts were given to analyze whether robustness is influenced by the differences in physical attributes of the system, farming, and other contextual settings. The applicability has been presented in five categories: not applicable, rarely applicable, partially applicable, mostly applicable, and fully applicable. Table 3.4 presents the summary of how the irrigation systems in Nepal and Thailand meet the conditions of Ostrom's design principles.

TABLE 3.4   Comparative Applicability of Design Principles in the Management of Irrigation Systems of Nepal and Thailand

Design Principles	Nepal	Thailand
Clearly defined boundaries	Fully applicable	Fully applicable
Congruence between appropriation and provision rules and local conditions	Mostly applicable	Partially applicable
Collective choice arrangements	Mostly applicable	Partially applicable
Monitoring	Mostly applicable	Partially applicable
Graduated sanctions	Partially applicable	Partially applicable
Conflict resolution mechanisms	Partially applicable	Partially applicable
Minimal recognition of rights to organize	Mostly applicable	Partially applicable
Nested enterprises	Rarely applicable	Partially applicable

The findings showed that most of the design principles were applicable in irrigation management in Nepal and Thailand. But the level of applicability of each design principle varied depending on the local context and other factors across these two countries. The applicability varied from rarely applicable to fully applicable among different design principles. The findings showed that basically the irrigation systems with highly autonomous water users' associations (WUAs) fulfilled most of the conditions as described in the design principles. As described by Ostrom (1990), satisfying the design principle conditions largely ensures the success of the local institutions. In our case, some of the design principles—design principle two "congruence" and design principle five "graduate sanctions"—could not fully capture the existing context and institutional settings.

Regarding design principle 2 (congruence between appropriation and provision rules and local conditions), we agree with Gautam and Shivakoti (2005) that Ostrom's definition of local condition does not fully explain the problem of congruence in the studied irrigation systems. This design principle does not explicitly discuss the economic and social context. In our cases, the appropriation rules in most case are affected by economic and social conditions. Thus, it needs some expansion incorporating "topography, economic, and social conditions" to represent the local conditions explicitly. The need to give attention to the local variations was also noted by a study of spate irrigation systems in Pakistan (Kamran and Shivakoti, 2013a).

Regarding design principle 5 (graduated sanctions), it was noted that it cannot fully capture the existing situations in irrigation management in Nepal. In some cases the users have devised special arrangements to include a strong enforcement mechanism in case of repeated violations of the rules. For example, some cases showed that if fines are not paid in cases of repeated water stealing they are referred to the local government or even to the police station. Thus design principle 5 needs expansion incorporating the clause for "enforcement mechanisms" in case of repeated violations of the rules.

A recent detailed review of applicability of design principles in managing various CPRs also pointed to generally strong support for the design rules, as well as some specific shortcomings and the need for modifications (Cox et al., 2010). The review included 91 case studies covering forestry, irrigation, fisheries, pastureland, and other CPRs. They suggested the reformulations of design principles 1, 2, and 4.

3.3.3 Application of ADICO Syntax

Studies show that CPR users' decision-making authority related to devising rules is crucial (Bardhan, 2000; Meinzen-Dick, 2007; Regmi, 2008). Literature on collective choice discusses rules and their effect on collective action (Ostrom, 1990; Ostrom et al., 1999). Analytical approaches such as ADICO syntax have been offered to study the evolution of rules and norms (Crawford and Ostrom, 1995; Ostrom, 2005; Ostrom and Basurto, 2011). An understanding of the grammar of the rules facilitates a better understanding of the distinction between rules, norms, and strategies (Vatn, 2005) and can be useful in discerning the formation and evolution of institutional statements (Siddiki et al., 2011).

However, Bastakoti and Shivakoti (2012) noted that the empirical evidence examining the relevance of those theories was still limited. Taking the case of irrigation systems in Nepal, we assessed rule formation, enforcement mechanisms, and collection action across different governance modes. Our main interest was on two empirical puzzles related to irrigation management in Nepal: (1) the relationship of governance structure with rule formation, enforcement, and rule following; and (2) the effect of rule enforcement on irrigation system performance.

We have used a revised version of the IAD framework as a guiding framework for analysis (Ostrom et al., 1994, 1999). We start our analysis by focusing on the rules-in-use component part of the IAD framework. We then consider the action arena as the main unit of institutional analysis, where the users interact with other components of the whole system, and at the end affect the outcome (Ostrom, 2005, 2007).

The ADICO syntax was found useful in analyzing the rule formation process (Bastakoti and Shivakoti, 2012). At the beginning of any irrigation system, the users start with no rule situation. When people come together and start some work related to an irrigation system, most often the rule formation starts immediately. Farmers start with an "informal shared strategy" that includes the "AIC" part of ADICO syntax. In the process, the farmers determine the actions that participants "may," "must," and "must not" do. Then they add some component that permits/obliges/forbids certain actions making it "Norms," which includes the "ADIC" part of the syntax. Finally, farmers also add consequences for not following agreed upon norms making it formal "Rules," having all part of ADICO syntax.

The ADICO syntax in general was applicable to understand how the rule-formation process starts with a simple statement with only a few components, and then the farmers add other components making it a formal rule. Therefore, the ADICO syntax was useful in analyzing the evolution of the rules (see also Kamran and Shivakoti, 2013b). But we found that it does not fully capture the conditions related to the problem of rule compliance and the role of autonomy. Kamran and Shivakoti (2013b) recently noted that externally crafted rules failed to fully incorporate context-specific needs and enforceability due to complex and unclear institutional statements as compared to locally evolved customary rules. Some other scholars have also noted the need for this additional component in rule analysis to capture the enforcement mechanisms (Mittenzwei and Bullock, 2001; Smajgl, 2007). This is very important in ensuring collective action.

3.4 FURTHER APPLICATIONS/AGENDAS

Based on the review of applicability situations of Ostrom's key analytical frameworks, some modifications are proposed for their wider applicability in specific conditions. Past studies show that Ostrom's design principles were basic and well configured. But, all design principles were not equally applicable in diverse settings. It is important to note here that the design principles have wider applicability in analyzing governance of natural resources in Southeast Asia. The design principles can help analyze the robustness of irrigation, forestry, fisheries, and coastal resources. However, it is advisable to consider

- Rephrasing the definition of local condition, including social and economic context.
- Establishing an enforcement mechanism to ensure rule compliance.

Another key analytical framework, the ADICO syntax, was also found useful in studying the rule-formation process in managing the CPRs. However, it should be noted that it does not fully incorporate context-specific situations and enforceability. It could be due to complex and unclear institutional statements as compared to locally evolved customary rules. The ADICO syntax can be used to study the evolution of institutional rules in management of irrigation systems, fisheries, community forestry, and coastal resources in Southeast Asia. But, we suggest the need of refinement and/or inclusion of additional components in ADICO syntax to capture local condition and CPR specific context:

- Rule compliance—enforcement mechanism
- Autonomy of local institution

References

Agrawal, A., 2002. Common resources and institutional sustainability. In: Ostrom, E., Dietz, T., Dolak, N., Stern, P.C., Stovich, S., Weber, E.U., National Research Council, Committee on the Human Dimensions of Global Change (Eds.), The Drama of the Commons. National Academy Press, Washington, DC, pp. 41–85.

Bardhan, P., 2000. Irrigation and cooperation: an empirical analysis of 48 irrigation communities in South India. Econ. Dev. Cult. Change 48 (4), 847–865.

Bastakoti, R.C., 2010. Institutional Dynamism and Performance of Irrigation Systems: A Comparative Study Between Nepal and Thailand (Unpublished PhD Dissertation). Asian Institute of Technology, Bangkok, Thailand.

Bastakoti, R.C., Shivakoti, G.P., 2009. Context and institutions in irrigation management: robustness and sustainability implications. In: Paper presented at Workshop on Workshop panel "Water resource governance and Design Principles" held at Indiana University in June 3–6, 2009.

Bastakoti, R.C., Shivakoti, G.P., 2012. Rules and collective actions: an institutional analysis of irrigation systems in Nepal. J. Inst. Econ. 8 (2), 225–246.

Bastakoti, R.C., Shivakoti, G.P., Lebel, L., 2010. Local irrigation management institutions mediate changes driven by external policy and market pressures in Nepal and Thailand. Environ. Manage. 46, 411–423.

Cox, M., Arnold, G., Tomás, S.V., 2010. A review of design principles for community-based natural resource management. Ecol. Soc. 15 (4), 38. http://www.ecologyandsociety.org/vol15/iss4/art38/.

Crawford, S., Ostrom, E., 1995. A grammar of institutions. Am. Polit. Sci. Rev. 89 (3), 582–600.

Gautam, A.P., Shivakoti, G.P., 2005. Conditions for successful local collective action in forestry: some evidence from the hills of Nepal. Soc. Natur. Resour. 18 (2), 153–171.

Hardin, G., 1968. The tragedy of the commons. Science 162 (3859), 1243–1248.

IMSSG, 1993. Irrigation Resource Inventory of East Chitwan. Irrigation Management Systems Study Group (IMSSG)/ Institute of Agriculture and Animal Science, Rampur, Chitwan, Nepal.

Kamran, M.A., Shivakoti, G.P., 2013a. Design principles in tribal and settled areas spate irrigation management institutions in Punjab, Pakistan. Asia Pacific Viewpoint 54 (2), 206–217.

Kamran, M.A., Shivakoti, G.P., 2013b. Comparative institutional analysis of customary rights and colonial law in spate irrigation systems of Pakistani Punjab. Water Int. 38 (5), 601–619.

Lam, W.F., 1998. Governing Irrigation Systems in Nepal: Institutions, Infrastructure, and Collective Action. ICS Press, Oakland, CA.

Meinzen-Dick, R., 2007. Beyond panaceas in water institutions. Proc. Natl. Acad. Sci. 104 (39), 15200–15205.

Mittenzwei, K., Bullock, D.S., 2001. Towards a framework for institutional analysis. In: 16th Annual Congress of the European Economic Association, Lausanne, Switzerland.

Olson, M., 1965. The Logic of Collective Action: Public Goods and the Theory of Groups. Harvard University Press, Cambridge.

Ostrom, E., 1986. A method of institutional analysis. In: Kaufmann, F.-X., Majone, G., Ostrom, V. (Eds.), Guidance, Control, and Evaluation in the Public Sector. Walter de Gruyter, Berlin/New York, pp. 459–475.

Ostrom, E., 1988. Decentralization, Finance and Management Project Document. Associates in Rural Development, Burlington, VT.

Ostrom, E., 1990. Governing the Commons: The Evolution of Institutions for Collective Action. Cambridge University Press, New York.

Ostrom, E., 1996. Incentives, rules of the game, and development. In: Proceedings of the Annual World Bank Conference on Development Economics 1995. World Bank, Washington, DC, pp. 207–234.

Ostrom, E., 2000. Collective action and the evolution of social norms. J. Econ. Perspect. 14 (3), 137–158.

Ostrom, E., 2005. Understanding Institutional Diversity. Princeton University Press, Princeton, New Jersey.

Ostrom, E., 2007. Challenges and growth: the development of the interdisciplinary field of institutional analysis. J. Inst. Econ. 3 (3), 239–264.

Ostrom, E., 2009a. A general framework for analyzing sustainability of social-ecological systems. Science 325 (5939), 419–422.

Ostrom, E., 2009b. Design principles of robust property rights institutions: what have we learned? In: Ingram, G.K., Hong, Y.-H. (Eds.), Property Rights and Land Policies. Lincoln Institute of Land Policy, Cambridge, MA, pp. 25–51.

Ostrom, E., Basurto, X., 2011. Crafting analytical tools to study institutional change. J. Inst. Econ. 7 (3), 317–343.

Ostrom, E., Gardner, R., Walker, J., 1994. Rules, Games and Common-Pool Resources. University of Michigan Press, Ann Arbor, MI.

Ostrom, E., Burger, J., Field, C., Norgaard, R.B., Policansky, D., 1999. Revisiting the commons: local lessons, global challenges. Science 284 (5412), 278–282.

Ostrom, E., Lam, W.F., Pradhan, P., Shivakoti, G.P., 2011. Improving Irrigation Performance in Asia: Sustainable Performance of an Innovative Intervention in Nepal. Edward Elgar Publishers, Cheltenham, UK.

Quinn, C.H., Huby, M., Kiwasila, H., Lovett, J.C., 2007. Design principles and common pool resource management: an institutional approach to evaluating community management in semi-arid Tanzania. J. Environ. Manage. 84, 100–113.

Regmi, A.R., 2008. Self-governance in farmer-managed irrigation systems in Nepal. J. Dev. Sustain. Agr. 3 (1), 20–27.

Sarker, A., Itoh, T., 2001. Design principles in long-enduring institutions of Japanese irrigation of common-pool resources. Agr. Water Manage. 48, 98–102.

Shivakoti, G., 1991. Organizational Effectiveness of User and Non-user Controlled Irrigation Systems in Nepal (Unpublished PhD Dissertation). Department of Resource Development, Michigan State University, East Lansing, MI.

Shivakoti, G., Ostrom, E. (Eds.), 2002. Improving Irrigation Governance and Management in Nepal. ICS Press, Oakland, CA.

Shivakoti, G., Varughese, G., Ostrom, E., Shukla, A., Thapa, G., 1997. People and participation in sustainable development: understanding the dynamics of natural resource system. In: Proceedings of an International Conference held at the Institute of Agriculture and Animal Science, Rampur, Chitwan, Nepal. 17–21 March, 1996.

Shivakoti, G., Vermillion, D., Lam, W.F., Ostrom, E., Pradhan, U., Yoder, R., 2005. Asian Irrigation in Transition-Responding to Challenges. Sage, New Delhi.

Siddiki, S., Weible, C.M., Basurto, X., Calanni, J., 2011. Dissecting policy designs: an application of the institutional grammar tool. Pol. Stud. J. 39 (1), 79–103.

Smajgl, A., 2007. Modelling evolving rules for the use of common pool resources in an agent-based model. Interdiscipl. Description Complex Syst. 5 (2), 56–80.

Sowerwine, J., Shivakoti, G., Pradhan, U., Shukla, A., Ostrom, E., 1994. From farmers' fields to data fields and back—a synthesis of participatory information system for irrigation and other resources. In: Proceeding of the International Workshop held at the Institute of Agriculture and Animal Science, Rampur, Nepal, March 21–26. Colombo, Sri Lanka, IIMI and Rampur, Nepal, IAAS.

Steins, N.A., Edwards, V.M., 1999. Collective action in common pool resource management: the contribution of a social constructivist perspective to existing theory. Soc. Nat. Res. 12, 539–557.

Theesfeld, I., 2004. Constraints on collective action in a transitional economy: the case of Bulgaria's irrigation sector. World Dev. 32 (2), 251–271.

Vatn, A., 2005. Institutions and the Environment. Edward Elgar, Northampton, MA.

Webb, E., Shivakoti, G.P. (Eds.), 2008. Decentralization, Forests and Rural Communities: Policy Outcomes in South and Southeast Asia. Sage, New Delhi.

Wittayapak, C., Dearden, P., 1999. Decision-making arrangements in community-based watershed management in northern Thailand. Soc. Nat. Res. 12, 673–691.

THEORETICAL ISSUES

Challenges of Polycentric Water Governance in Southeast Asia: Awkward Facts, Missing Mechanisms, and Working with Institutional Diversity

B. Bruns[a],*

*Independent Consulting Sociologist, PO Box 176, Warm Springs, VA, United States

4.1 INTRODUCTION

In *Governing the Commons*, Ostrom (1990) analyzed how communities had successfully managed irrigation systems, fisheries, forests, and other shared resources. While local rules were extremely diverse, she identified institutional design principles favoring local self-governance, one of which was nested enterprises, multiscale organizations in which smaller units federated to govern at wider scales. In *Understanding Institutional Diversity*, Ostrom (2005) applied the Institutional Analysis and Development framework to analyze autonomous rule-making by resource users, polycentric governance, and ways of coping with threats to commons governance.

Zanjera irrigation systems in the northern Philippines were among the examples of long-enduring locally governed commons studied by Ostrom. Subaks in Bali and communal muang fai schemes in northern Thailand are other influential examples of self-organization along hydrological lines in farmer-managed irrigation systems. Ostrom also examined groundwater governance in Southern California to explore how polycentric governance could develop at

a The author works as an independent consultant on participatory irrigation management and water resources management. In 2009–2010 he was a Visiting Scholar at the Ostrom Workshop in Political Theory and Policy Analysis at Indiana University, Bloomington. Views expressed in this paper are the responsibility of the author and do not represent any organization with which he is or has been affiliated.

a larger scale. However, conditions in much of Southeast Asia differ substantially from those that Ostrom used to develop her concepts of self-organization and polycentric governance. This chapter examines some of the main differences and implications for the development of polycentric water governance in Southeast Asia. The following sections review some awkward facts that need to be considered in the development of water governance in Southeast Asia, point to missing mechanisms whose absence impedes the development of polycentric governance, and discuss opportunities for working with institutional diversity.

4.2 AWKWARD FACTS

Ostrom emphasized the importance of autonomous organization in accordance with resource boundaries, such as canal networks and river basins rather than territorial jurisdictions of villages, districts, and provinces. However, conditions in much of Southeast Asia contradict assumptions that autonomous organizations can be easily organized or reorganized along hydraulic lines. This affects the feasibility of prominent policies for reforming water governance such as development of water users' associations (WUAs), participatory irrigation management (PIM), irrigation management transfer (IMT), formation of river basin organizations (RBOs), and integrated water resources management (IWRM).

4.2.1 Village Irrigation

Ostrom's work, including *Governing the Commons* and her shorter, more specific book *Crafting Institutions for Self-Governing Irrigation Systems* (1992) helped to popularize a narrative of spontaneous, autonomous, bottom-up management of irrigation, and other natural resources distinct from village governments.[1] Balinese subaks, communal muang-fai schemes in northern Thailand, and zangjeras in the Philippines have been key examples of such autonomous, hydraulically organized irrigation systems. Research by Ostrom and colleagues in Nepal (Ostrom et al., 1994; Lam and Ostrom, 2009; Ostrom, 2011) showed the relative effectiveness of community management compared to agency-managed schemes. Pursuit of the hoped-for benefits of such devolution drove projects for PIM and IMT, including those in the Philippines, Indonesia, Thailand, Cambodia, Laos, and Vietnam. Empirically, the situation usually has been and remains more complicated. In many places village governments play a strong or even predominant role in local irrigation management, in stark contrast and contradiction to a stereotyped narrative of autonomous, hydraulically based organization.

Village water masters, ulu-ulu, had been a standard part of village government in Java since the time of colonial administration (Hutapea et al., 1978; Duewel, 1984).[2] Reforms to village administration in 1979 abolished the formal position of ulu-ulu. However, villages usually continued the practice, often putting the *ulu-ulu* in the more general position of Head of Economic and Development Activities, *kaurekbang,* or informally continuing to designate

---

[1] For a more general discussion of community and natural resources management, see Agrawal and Gibson (2001).

[2] In addition to the cited sources, the discussion in this paper draws on the author's research and consulting experience in Thailand, Laos, Cambodia, Vietnam, the Philippines, and Indonesia.

and oversee the work of ulu-ulu. In Aceh, the heads of blocks of irrigated fields, *keujreun blang*, had been and continued to be part of village government (Coward, 1988). Development projects in Indonesia usually tried to set up WUAs, *perkumpulan petani pemakai air* (P3A), along hydraulic lines. Communities might go along with the formalities of P3A, while perpetuating ulu-ulu and other customary practices (Duewel, 1984; Bruns, 2013). Furthermore, in much of Java, irrigation systems had often been developed so that they largely followed village boundaries, reducing the need for any crosscutting pattern of governance.

Beginning in the middle 1970s, Thai governments channeled funds for rural infrastructure through subdistricts, which evolved from a council of village heads to today's Tambon Administrative Organizations with substantial offices, staff, and budgets (Bruns, 1991). These funds became increasingly attractive for repairing or improving weirs, tanks, and other small-scale water resources infrastructure, reinforced by the increasing availability of earth-moving equipment and by politicians offering benefits in exchange for votes. Thus, village and subdistrict authorities came to control more funds, and have more influence over the decisions about construction and use of water resources, disrupting and discouraging more autonomous and self-reliant efforts by farmers. In this context, villages and subdistricts were becoming more, not less, important in water governance.

During collectivized agriculture in Vietnam, irrigation was managed by agricultural co-operatives. Irrigation teams controlled not only canals but also water deliveries and water levels in individual fields. After decollectivization, cooperatives continued to play a strong role in many areas where gravity irrigation and drainage were important. In northern and central Vietnam corporate villages had historically played a major autonomous role in local governance, the phrase "the emperor's rule stops at the village gate" reflected the ability of villages, with village elites in a leading role, to manage their internal affairs as long as they satisfied external demands such as taxes, conscripts, and keeping local order. After decollectivization, many cooperatives continued to provide irrigation services, often still through irrigation teams managing water down to the field level. Cooperatives functioned under the guidance of commune government, and satisfactory performance of leadership roles in cooperatives was a path for becoming a commune leader. In the 1990s, cooperatives were formally restructured to be more democratic and accountable, but continued to be active in irrigation, and to work closely with commune governments. Efforts to form WUAs along the hydraulic lines of canals often made little headway against the strength and interests of commune and cooperative leaders (Mcgrath, 2005). Any changes that may have seemed to occur often quickly faded after external projects were over. The relative strength of communes, lack of established institutional arrangements for intercommune coordination, and interest of district governments in retaining their roles created infertile conditions for efforts to develop higher WUA federations or transfer irrigation management at wider scales.

Irrigation self-organized according to hydraulic connections has certainly been part of the picture in Southeast Asia, and continues to be important in many areas. However, in many cases, historic trends and current trends maintain and even strengthen the role of village authorities. Access to project benefits is often sufficient to create the appearance of hydraulically based WUAs. However, these tend to quickly fade away after projects end, and may represent a facade under which communities continue to manage irrigation in old ways, or result in institutional bricolage where communities pragmatically mix components of old and new institutions.

II. THEORETICAL ISSUES

4.2.2 Big Basins

Ostrom and Blomquist's research on polycentric water governance in Southern California looked at a number of relatively small, distinct basins. Some were more successful and others less successful in efforts to solve problems resulting from groundwater depletion. These were important for those involved, but not crucial for larger regional or national economies. However most efforts to develop basin institutions in Southeast Asia have focused on large basins. Bangkok and Hanoi lie in the deltas of major river basins, and other cities such as Phnom Penh, Ho Chi Minh City, and Surabaya are part of major basins. Water governance easily becomes a subject of national concern, including funding and media attention. At the same time, national institutions have often done little to support mechanisms for horizontal coordination and problem solving between provinces, preferring instead to keep a central role in dealing with any problems of broader scope. In such an environment, it becomes easier to understand why efforts to promote governance institutions at the river basin level have often made little progress, and been subordinated to national bureaucracies. Trying to set up a Chao Phaya basin organization as an alternative to officials attached to the prime minister's office is an uphill battle. A series of projects, with support from the Asian Development Bank (ADB), have sought to develop river basin management institutions in Vietnam, with limited results. As discussed further in the next sections, the availability of international funding, and the ability of national bureaucracies to control such funds, also puts more local efforts at a disadvantage.

4.2.3 Hydraulic Bureaucracies

Post–World War II international development put national governments in the leading role. Officials from countries such as the United States and Japan forgot or ignored their own history of locally managed irrigation and channeled funds through national bureaucracies that not only built irrigation systems but then continued to manage them. This set the stage for a vicious cycle where the interests of donors, politicians, and irrigation agencies align on the large expenditures in building and rehabilitating infrastructure, with weak incentive to deliver good services or maintain facilities (Araral, 2005). Power and money stay with the center, in construction, with little accountability to farmers for operation and maintenance. This historic pathway strengthened the power and influence, budget, staffing, and technical capabilities of centralized irrigation agencies.

In most of Southeast Asia, the language of PIM has been adopted, including the formalities of WUAs. However, much less has occurred in terms of stronger empowerment. Efforts to promote IMT, including in Indonesia and Vietnam, have shown not only the disinclination of bureaucracies to surrender power or resources, but also the lack of any vigorous local demand to take over irrigation management. Despite the apparent success of some pilot efforts, there has been little spontaneous spread, lobbying from farmers, or other indication of such demand. Ideas of self-organization and polycentric governance have helped inspire efforts to change irrigation governance through PIM and IMT, showing the potential for acceptance of more participatory rhetoric and processes, and the apparent limits to stronger devolution.

Irrigation agencies have grown out of the historic pathway followed by centralized development, and are now a major part of the institutional landscape, able to protect and pursue their interests, and unlikely to quickly wither away. Even in Thailand, where a vigorous civil society has been able to stymie some plans to build reservoirs or move water between basins, the hydraulic bureaucracy continues its efforts to build more reservoirs and link rivers.

One can ask when agencies might see reforms as in their interests. In the case of IMT in Mexico, senior officials seem to have seen benefits from getting rid of the hassles of lower-level water distribution, so the agency could focus on larger works, and of carrying out these changes as part of broader reforms that would restore agency autonomy that had been lost when it was put under the Ministry of Agriculture (Rap, 2006). This occurred within the context of a regime that was undertaking major policy reforms in many sectors, staking its prestige on making reforms rather than on protecting existing bureaucratic interests. Similarly, in Australia, IMT was part of broader policy reforms. On a more political level, one can also ask when farmers might want to take over increasing authority, and when they would have enough political power to do so. Bates's (1981) arguments about the effectiveness of narrower, more specialized interests are relevant. Obviously such interests may pursue subsidies, and only be interested in management reforms if there are political checks and balances that would limit subsidies and make other options seem more feasible.

In Vietnam, Laos, and Thailand, governments have reassigned formal responsibility for water regulation to different ministries, taking them away from irrigation agencies. This follows the logic of establishing an independent regulator, a neutral referee to ensure that all players follow the rules. As a bureaucracy, the regulatory agency does have an interest in using its mandate to gain recognition, authority, budget, staff, and other bureaucratic desiderata. However, the reality is that agencies such as the Royal Irrigation Department in Thailand and the Ministry of Agriculture and Rural Development in Vietnam still have vastly bigger budgets, stronger technical capabilities, and better political connections than those who are supposed to be overseeing them. The putative regulators are faced with the daunting task of trying to control an "eight-hundred pound gorilla" inclined to continue doing what it wants. This may not change unless there are countervailing interests, such as cities and businesses demanding more water or better flood protection. This may change, in time, and perhaps incrementally on one issue at a time, but shows the limits on any attempts to quickly shift away from the current structure and toward one that would be more open to a polycentric plurality of players.

4.2.4 Subsidy Politics

Ostrom's work highlighted the potential for self-organization, often by communities with no choice but to rely on their own resources. As Ostrom discusses in *Understanding Institutional Diversity*, external funding can be a threat to commons governance. The availability of outside assistance, usually fully subsidized, makes self-reliance less attractive, and encourages dependence on external, politically driven funding. Up to a point, communities may use such aid opportunistically, without necessarily succumbing to dependence or institutional distortion. Vote buying in Thailand has sometimes taken the form of explicit exchanges, quid pro quo; for example, a block of village votes in exchange for a bridge. More commonly, exchanges are not so specific, but provide a politician with a way to claim credit for dispensing benefits.

High levels of subsidies may include mechanisms that still give resources users a voice in governance; thus, for example, Japan's highly subsidized irrigation development still includes requirements for local requests and a small percentage of cost-sharing that can function to give communities a voice and a veto in the process. Rather than fantasizing about abolishing subsidies, it may be more feasible to structure them so that they are at least compatible with, or even reinforce, water governance that involves water users. As middle-income countries in Southeast Asia provide more resources for rural areas, it becomes more important to think about smart subsidies that are well targeted and provide incentives for good governance and increased productivity. Polycentric governance is not incompatible with subsidies, but subsidies may disrupt and disable institutions that were premised on local self-reliance and lack of alternatives. The political reality is that subsidies are unlikely to go away, so the challenge is how to improve water governance within a context where subsidies play an important role.

In summary, WUAs are relatively easy to establish if they offer access to project benefits, but hard to sustain after projects end, although communities may well continue to cooperate on activities such as canal cleaning and distributing water during periods of scarcity. Policies for irrigation management turnover have not led to any vigorous process of community-driven takeover at the level of secondary canals or schemes serving multiple communities, although they may have opened space for communication and cooperation, and helped to block or somewhat reverse tendencies for further takeover by government. RBOs can be established, and may help with issues such as water allocation and planning, but generally have not demonstrated major achievements in solving basin-scale problems. The strength of village institutions, hydraulic bureaucracies, subsidy politics, and national government intervention in big basins are part of a context that challenges the hypothesis that it is easy to establish new, autonomous organizations along hydraulic boundaries using simple one-size-fits-all models. Instead, they suggest the need for diversity and political feasibility, including hybrid institutional arrangements that local governments will support and sustain, which offer meaningful but not monopolistic roles for hydraulic bureaucracies. The next section looks further at some of the conditions that have been less than favorable for the development of polycentric water governance.

4.3 MISSING MECHANISMS

Polycentric governance in the case of groundwater in Southern California was enabled by a set of mechanisms that are largely absent or ineffective in Southeast Asia. This raises questions about the feasibility of transitions to more polycentric governance, unless similar mechanisms or functional equivalents become available, or alternative pathways are found.

4.3.1 Special Districts

Legislation in California and other Western states, supported by related national legislation, establishes a legal framework that enables irrigators to organize themselves and assume specific kinds of government authority. Special districts are a form of local government, responsible for a particular type of activity, such as irrigation districts. This means that, after a proper decision-making process, they are authorized to impose rules on their members about how water and irrigation infrastructure are used, require members to pay fees, and ultimately take their land if payments are not made.

There have been decades of policy research, analysis and debate, pilot projects, and institutional reforms to promote PIM and IMT in Southeast Asia. Nevertheless, WUAs usually still lack formal authority over their members, water, and infrastructure. WUAs are still usually treated as if they were little different from voluntary cooperatives or associations, with members free to join or leave, and no mandatory requirement to pay the costs required to keep benefits flowing. The rhetoric of empowerment turns into a reality of legal powerlessness. External authority is certainly not sufficient to make WUAs effective, but under changing conditions may become more necessary to cope with more diverse users and more formal patterns of interaction.

WUAs that can rely on local solidarity, the social capital of trust, and established patterns of cooperation may well be able to mobilize resources and work effectively. This usually depends on local social sanctions, not laws. Ostrom cites Levi's (1989) term, quasi-voluntary cooperation, to emphasize that most cooperation comes from people's willingness to comply with rules, even though violations will not always be detected or punished. In this, irrigation governance functions largely through the legitimacy of norms about how people should behave, through internalized values and through fear of social disapproval. However, the "quasi" part comes from the existence of mechanisms that can be used to compel obedience, especially in the face of persistent or conspicuous violation. The relative lack of progress in strengthening the legal authority of WUAs makes the prospects for future changes questionable, if user organizations lack many of the capacities needed to cooperate with other institutions in polycentric water governance, especially as communities become more diverse, economies become increasingly commercialized and governance relies more on formal legal mechanisms.

4.3.2 Courts

Many of the institutional arrangements for polycentric water governance in California grew out of agreements to settle disputes that had been taken to court (Blomquist, 2009). Courts encouraged disputants to negotiate rather than litigate. Experience showed that trying to solve water conflicts primarily based on legal reasoning, as a judge would be obligated to do, often yielded results that pleased no one, making negotiation a better option. Courts, particularly equity courts, provided authority that would bind disputants to their agreements, creating new institutional arrangements. Typically these may have included the arrangements for allocating water, importing additional water, monitoring flows to ensure that commitments were fulfilled, and resolving further conflicts. This provided a pathway through which water users could create their own institutions, and modify them over time, often learning from the experience and examples of others.

Top-down propagation of one-size-fits-all institutional arrangements provides less opportunity for such institutional innovation. The relative weakness of courts as forums for protecting rights or resolving disputes in most of Southeast Asia blocks this pathway for change. Similarly, the absence of well-defined water rights means that they are not used as a basis for framing or resolving disputes.

4.3.3 Water Rights

Although the Philippines has a formal framework of water rights, courtesy of Spanish and American colonial law, even there water rights tend to play little or no role in defining water allocations or resolving conflicts. This shows that the issue is not simply the formal presence of water rights, but instead the extent to which such resource rights become an effective part of water allocation and dispute resolution. Simply establishing a legal framework, or issuing formal rights, may have little impact, unless the rights become a meaningful part of how claims to water are disputed and defended.

4.3.4 Bonds

PIM and IMT policies have often been advocated and justified in terms of increasing the contribution of farmers toward the costs of irrigation, or at least the costs of operation and maintenance. However, typically little or nothing has been done to increase the financial capacity of WUAs, in particular to borrow money to finance investments. The implicit assumption seems to be that investment will still come through subsidies. Farmers' cash constraints and the prospect of subsidized rehabilitation further discourage local maintenance. The small scale of WUAs, and reluctance to transfer responsibility and authority above the level of tertiary canals further restrict the capacity of WUAs and WUA federations.

Without the ability to make major capital investments and hire the services of engineers and other professionals, the scope for greater involvement of farmers in polycentric governance is limited. There may be other pathways, for example where agriculture becomes highly commercialized and farmers are well-organized for other activities, such as marketing, and then use that capacity to also act in water governance. However, such dynamics do not yet seem to be strongly apparent in Southeast Asia.

4.3.5 Functional Equivalents and Feasible Pathways

Comparison with the case of Southern California provides an illustration of mechanisms that facilitated the development of polycentric water governance, but should not be taken to mean that those are the only mechanisms or the only pathway to more polycentric water governance. As discussed, where communities have strong social capital, and where water is scarce and valuable, communities may use the institutional form of WUAs effectively, despite the lack of legal authority. Agreements with agency officials or dispute settlements mediated by political leaders can create new institutional arrangements. Farmers may use individual borrowing to finance collective goods. Water rights, in the more general sense of socially accepted and defensible claims to water, may take many forms. One priority for research should be precisely to identify and analyze the kind of institutional mechanisms that have facilitated more polycentric water governance.

The failure of IMT makes for a more dramatic story, and may overshadow more modest progress in terms of greater acceptance of local participation and consultation with farmers. Policy changes for participatory design and participatory construction are more aligned with agency interests in construction, and face fewer obstacles. At the secondary and scheme level, increased consultation on irrigation scheduling may be useful for an agency if it reduces complaints and conflict. Similarly, greater involvement of farmers and WUAs in identifying and prioritizing maintenance needs may be useful in obtaining funding, although this may still require overcoming the interests of those who benefit from manipulating their discretion over operations and maintenance budgets.

In river basins, specific problems, such as obtaining emergency reservoir releases during drought, are better suited to mobilizing coalitions of supporters than are complicated integrated plans. Saltwater intrusion into aquifers or up rivers that threatens the water supply of coastal cities provides a specific challenge. Responses to these may be ad hoc and episodic, but over time may also help develop more systematic arrangements, evolving institutions that are adapted to address other issues. Institutional diversity means institutional creativity; that there are lots of potential solutions that may be crafted. The challenge that faces those governing water in Southeast Asia is to find workable solutions, which will be more achievable if it includes a broader range of alternatives that are politically feasible, and can use a variety of mechanisms.

4.4 WORKING WITH INSTITUTIONAL DIVERSITY

4.4.1 Adaptive Comanagement

To the extent that standardized ideas such as WUAs, IMT, and RBOs have the potential to be effective, they have been tried in most of Southeast Asia. There seems little reason to expect big new breakthroughs from simply trying more of the same reforms and policies. More fundamentally, the potential for standardized models to lead to dramatic reforms seems limited, especially if change is primarily driven by international funding agencies, and not supported by irrigation agencies. The historical pathways through which water resources development has occurred, and in particular the strong role of centralized hydraulic bureaucracies, have entrenched particular patterns of interests that are unlikely to change easily or quickly. Nevertheless, it is clear that some forms of participation and cooperation have become established as part of rhetoric and policy and, at least in some places and to some extent, as part of practice. In larger irrigation systems there seems to be room for a strong role of WUAs at the lowest level, and a degree of cooperation or comanagement at secondary canal and higher levels, where users have more information and more voice in decisions about matters such as maintenance priorities and cropping schedules, even if the final formal authority may remain with a water agency (Bruns, 2013). The organizations that operate major hydraulic structures, such as reservoirs and barrages, may not be ready to cede much control, but may be willing to at least listen and negotiate; for example, to notify those downstream about releases or to make special emergency releases during periods of drought.

4.4.2 Weaving Network Governance

IWRM has reigned as the hegemonic set of ideas framing how water resources should be managed. The intellectual arguments for IWRM are persuasive: the need to escape silos of usage sectors, academic disciplines, and bureaucratic agencies to deal with the multiple ways in which water is used. Water resources are linked upstream and downstream, and between surface water and groundwater. Water quantities are unusable without adequate water quality, and the impact of pollutants depends on water quantities. Water management needs not only knowledge from engineering, but public health, economics, political science, and other academic disciplines. The need for integration has justified efforts to reform policies and develop suitably comprehensive plans and institutional arrangements.

However, it is now more than 10 years since Biswas's (2004) article asserting, essentially, that the emperor of IWRM has no clothes, that IWRM has little to show in the way of practical results, and there is little reason to expect much more to be coming. The World Bank's 2003 water sector policy called, somewhat more diplomatically, for "principled pragmatism," being aware of desirable directions to move, and selective about where opportunities seem ripe; a sharp contrast from the comprehensive ambitions of much of IWRM. Others have called for more expedient, adaptive management (Lankford et al., 2007). Schlager and Blomquist (2008), two of Ostrom's students and colleagues, have pointed out the impracticality of many of the ambitions of IWRM and the need to "embrace watershed politics." These kinds of recommendations point toward more modest, incremental changes. While the world of water policy and expertise will certainly be susceptible to further intellectual fads and fashions, they may well be less hegemonic, and more specific, and so open more space for particular changes, including ones that may in some cases move toward more polycentric forms of water governance, as a form of network governance (Carlsson and Sandström, 2008). There is a need for institutions that can weave linkages across hydrological and administrative boundaries, facilitating the sharing of information and collective action, without mandating a single form. Exchange of ideas within epistemic communities may be more influential than formal structures, and create shared ideas and contacts that then facilitate working together to address shared concerns.

4.4.3 Coalitions for Problem Solving

Elinor Ostrom's design principles were based on empirical studies of the details of individual cases of governing commons, and local diversity of institutions. However, her ideas have often been applied within the context of larger attempts at social engineering, such as that associated with international development projects. However, application is not necessarily straightforward, easy, or effective. Hunt (1989) long ago cautioned that WAUs set up by projects in large irrigation systems would differ in important ways from small irrigation systems embedded in village societies, and that it was unrealistic to expect them to share the strengths of such self-organized institutions. Broader experience with induced community organization sponsored by development projects has confirmed that it differs substantially from more spontaneous organization, and needs detailed project-specific learning (Mansuri and Rao, 2012). The process of change is likely to be much messier, contingent, and improvisational than that suggested by talk of institutional design and design principles. The results are more likely to be institutional bricolage (Cleaver, 2012), a messy patchwork of old and new. Assembling coalitions and harvesting learning may be more important than diagnosis and prescription by policy analysts (Andrews, 2013). Institutional diversity implies a diversity of pathways for change, some of which may converge, but many of which may diverge.

4.5 CONCLUSIONS

Polycentric water governance in Southeast Asia faces many challenges. In important ways governance is already at least partially polycentric, with a multitude of stakeholders and organizations involved, linked at different scales and in different sectors. Flows of information, accelerated and expanded by the Internet, flow through larger epistemic communities whose ideas and debate frame the discourse and define problems and acceptable solutions within patterns of network governance. Concepts of polycentricity point out the potential for alternative ways of organization, in contrast to assumptions of top-down hierarchical command and control, or markets. Nevertheless, the historic pathways have entrenched a very large role for national hydraulic bureaucracies, and experience indicates that although there is some space for change, dramatic shifts are not easily achieved.

The Ostrom's own model of polycentric governance drew heavily from experience in California, in particular institutional circumstances that enabled and facilitated self-organization through mechanisms that are largely absent or weak in Southeast Asia, including legal authority for specialized governmental units such as irrigation districts, courts as forums for resolving disputes and giving legal force to agreements that bred new institutions, and financial mechanisms that enabled substantial investments by irrigators organizations and other water management authorities. This is not to say that polycentric governance is absent or impossible in Southeast Asia, but rather that it takes different forms, and may well evolve through different pathways than those that shaped either the long-enduring community institutions of *Governing the Commons* or the fashionable ideas of WUAs and RBOs defined along basin boundaries that built on such concepts. Instead, changes may be more diverse, and less predictable. They seem likely to be heavily shaped by subsidy politics and by the goals and instruments of broader national policies toward rural areas and environmental protection. There are no guarantees that those affected in rural areas will have a strong voice in defining or redirecting the course of change. Nevertheless, there is scope for farmers, village residents, and other stakeholders to influence the discourse and direction of change, including the development of polycentric water governance through which they may have a measure of autonomy and opportunities to work together with others in pursuing their aims.

References

Agrawal, A., Gibson, C.C., 2001. The role of community in natural resource conservation. In: Agrawal, A., Gibson, C.C. (Eds.), Communities and the Environment: Ethnicity, Gender, and the State in Community-Based Conservation. Rutgers University Press, New Brunswick, NJ.

Andrews, M., 2013. The Limits of Institutional Reform in Development: Changing Rules for Realistic Solutions. Cambridge University Press, Cambridge.

Araral, E., 2005. Bureaucratic incentives, path dependence, and foreign aid: an empirical institutional analysis of irrigation in the Philippines. Policy Sci. 38, 131–157.

Bates, R.H., 1981. Markets and states in tropical Africa. The Political Basis of Agricultural Policies. University of California Press, Berkeley.

Biswas, A.K., 2004. Integrated water resources management: a reassessment. Water Int. 29 (2), 248–256.

Blomquist, W., 2009. Crafting water constitutions in California. The Practice of Constitutional Development: Vincent Ostrom's Quest to Understand Human Affairs, 105.

Bruns, B., 1991. The Stream the Tiger Leaped: A Study of Intervention and Innovation in Small Scale Irrigation Development in Northeast Thailand. Cornell University, Ithaca, New York.

Bruns, B., 2013. Bureaucratic bricolage and adaptive co-management in Indonesian irrigation. The Social Life of Water, 255.

Carlsson, L., Sandström, A., 2008. Network governance of the commons. Int. J. Commons 2 (1), 33–54.

Cleaver, F., 2012. Development Through Bricolage: Rethinking Institutions for Natural Resource Management, first ed. Routledge, London.

Coward, E.W., 1988. Baskets of stones: government assistance and development of local irrigation in a district of Northern Sumatera. Water Management Synthesis Project.

Duewel, J., 1984. Central Java's Dharma Tirta WUA 'model': peasant irrigation organization under conditions of population pressure. Agr. Adm. 17 (4), 261–285.

Hunt, R., 1989. Appropriate social organization? Water user associations in bureaucratic canal irrigation systems. Hum. Organ. 48 (1), 79–90.

Hutapea, R., Dirjasanyata P., Nordholt, N.G.S., 1978. The organization of farm-level irrigation in Indonesia. In Irrigation Policy and Management in Southeast Asia, pp. 167–174. IRRI.

Lam, W., Ostrom, E., 2009. Analyzing the dynamic complexity of development interventions: lessons from an irrigation experiment in Nepal. Policy Sci. 43, 1–25.

Lam, W.F., Pradhan, P., Ostrom, E., 2011. Improving irrigation in Asia: Sustainable Performance of an Innovative Intervention in Nepal. Edward Elgar, Cheltenham, UK/Northampton, MA.

Lankford, B.A., Douglas, J.M., Julien, C., Nick, H., 2007. From Integrated to Expedient: An Adaptive Framework for River Basin Management in Developing Countries. International Water Management Institute; Colombo, Sri Lanka.

Levi, M., 1989. Of Rule and Revenue. University of California Press.

Mansuri, G., Rao, V., 2012. Localizing Development: Does Participation Work? World Bank Publications, Washington, DC.

Mcgrath, T., 2005. Irrigated Agriculture in Vietnam: Responses to Policy Change 1975–1995. http://espace.library.uq.edu.au/view/UQ:107350.

Ostrom, E., 1990. Governing the Commons: The Evolution of Institutions for Collective Action. Cambridge University Press, Cambridge.

Ostrom, E., 1992. Crafting Institutions for Self-Governing Irrigation Systems. Institute for Contemporary Studies Press, San Francisco.

Ostrom, E., 2005. Understanding Institutional Diversity. Princeton University Press, Princeton, NJ.

Ostrom, E., Lam, W.F., Lee, M., 1994. The Performance of Self-Governing Irrigation Systems in Nepal. Hum. Syst. Manage. 13 (3), 197–207.

Ostrom, E., Lam, W.-F., Pradhan, P., 2011. Improving Irrigation in Asia: Sustainable Performance of an Innovative Intervention in Nepal [Internet]. Edward Elgar Publishing. [cited 2016 Apr 17].

Rap, E., 2006. The success of a policy model: irrigation management transfer in Mexico. J. Dev. Stud. 42 (8), 1301–1324.

Schlager, E., Blomquist, W.A., 2008. Embracing Watershed Politics. University Press of Colorado, Boulder, Colorado.

Modeling Effect of Conservation and Livelihood Policies on Community Land Use and Management in Yogyakarta

Partoyo*, R.P. Shrestha[†]

*Universitas Pembangunan Nasional Veteran Yogyakarta, Yogyakarta, Indonesia [†]Asian Institute of Technology, Pathumthani, Thailand

5.1 INTRODUCTION

Human population growth is one of the greatest causes of complex problems facing the world, along with climate change, poverty, and resource scarcity (Collodi and M'Cormack, 2009). The urban population has grown more rapidly than rural populations around the world in the last two decades, especially in developing countries. For the first time in 2008, the total urban population (3.4 million) was the same as the number of rural area inhabitants (UN, 2008a). According to current projections, the world's urban population is expected to increase by 3.1 billion people by 2050, whereas the rural population will peak at approximately 3.5 billion in 2019 and subsequently decline (UN, 2008b). Almost all the increase of the urban population is expected to occur in developing countries. Natural growth continues to contribute significantly to the urban population size, which often represented 60% or more of the growth (UN, 2008a).

As the population increases, food production should continuously increase to support global food security. In 2008, approximately 40 million more people suffered from hunger, and the total number reached 963 million (FAO, 2008). According to the United Nations, the world population will increase to approximately 9.1 billion people in the year 2050 (UN, 2008b). Feeding this enormous world population requires the increase of overall food production by approximately 70% from 2005 to 2050, which translates into significant increases in the production of several key commodities. For example, the annual production of cereal would need to grow by nearly 1 billion tons and the production of meat by more than

200 million tons to a total of 470 million tons by 2050, with 72% from developing countries, which is an increase of 58% (FAO, 2009).

Additionally, loss of prime farmland area due to land conversion is also becoming a major concern. Rapid farmland loss is generally due to the combined effect of rapid economic development, population growth, urbanization, agricultural restructuring, government-stimulated conversion of marginal croplands (to forest and pastures), natural hazards, and land degradation (Braimoh and Onishi, 2007; Cheng and Masser, 2003; Ding, 2003; Firman, 2004; Heilig, 1999; Seto et al., 2002; Verburg et al., 1999; Yang and Li, 2000). Brown (1995) estimated that a decrease in grain production would occur in China during the country's industrialization era. Furthermore, Brown (1995) showed that the loss of agricultural land is greatest in densely populated countries before industrialization, and such countries are rapidly becoming net grain importers. As industry grows, countries lose significant amounts of valuable land for factories, warehouses, roads, parking lots, and houses. As incomes increase, rural farmers leave for cities, which augments food demands (Brown, 1995). Some estimates indicate that there is a production loss of 1–3 million hectares of arable land in developing countries each year to meet the demand for housing, infrastructure, industrial, and recreational land (Doos and Shaw, 1999).

In Indonesia, the National Development Program led to significant economic growth from 1990 to 1998, a period marked by economic deregulation to ease foreign investments and domestic investment to boost the development of nonoil industries and property. As a consequence of the economic structural transformation (from agriculture to nonagricultural) in addition to the change in demographic aspects (rural to urban areas), this process has affected the sustainability of agricultural land (mainly irrigated farmland) (Widjanarko et al., 2004). Meanwhile, loss of farmland has been a national issue since 1994. The Ministry of Agriculture reported that the conversion of agricultural land to residential areas and other urban facilities is greater than 0.15 million hectares per year (DGWLM, 2008).

Due to climate change, the challenge for food security is becoming more complicated. Climate change threatens agricultural production through higher and more fluctuating temperatures, changing patterns of precipitation, and increased frequency of extreme events such as droughts and floods. The effect of population growth, climate change, land degradation, loss of crops and agricultural land on nonfood production, water scarcity, desertification, and urban expansion would result in a 25% decrease in food production in 2050 (Nellemann et al., 2009). Hence, stabilizing global population and protecting the agricultural base, including farmland protection, are prime concerns (Brown, 1995).

Agricultural land protection issues during the industrialization era are raising major dilemmas. Indonesia has become increasingly industrialized, and such increased population and economic growth competes with scarce land for the food supply. Land-use conversion in Java is unavoidable due to high demand for services and jobs; therefore, formulating proper land-use strategies to address such conflicting interests is critical. This study analyzed agricultural land conversion and suggested land-use options for farmland preservation in various contexts.

5.1.1 Land-Use Change in Indonesia

Human population growth and urbanization has significantly impacted spatial planning in Indonesia. Urban development occupied prime agricultural land, notably in the urban fringe area. With a population of over 206 million people in 2003, Indonesia is one of the most highly populated countries in the world after China, India, and the United States. Furthermore, the urban population proportion of Indonesia is still growing rapidly and already reached 30% in 1990 and 42% in 2000 (Firman, 2004). On the basis of the 2010 census, the population of Indonesia was 237.5 million people, and approximately half of them (118 million) live in urban areas (BPS-RI, 2010). This means that urban development is becoming a more pressing problem for the country (Firman, 2011). For this reason, a policy of urban spatial development capable of responding to rapid urbanization is crucial in Indonesia.

In Indonesia, most of the encroachment on agricultural land by urban growth has occurred on Java Island, which is the most populated island, with 58% of the total population of the country in 2010 (BPS-RI, 2010). In Java and Bali, 1.7 million hectares have been converted during the last decade, particularly in the provinces of West, East, and Central Java, which is the most fertile land in Java Island. Because Java Island has the most fertile soil and the highest level of agricultural infrastructure among other islands, farmland loss in the urban fringe area may jeopardize the local, regional, or even national agricultural base of the economy (Yunus, 1990). In 2000, agricultural land in Indonesia was approximately 50 million hectares (ha) or 26% of the total land area. Of the total land area, 54% was in Java and Bali, and approximately 10% of land is still under forest cover. However, forest cover is still the predominant land use in the outer Java and Bali islands, but deforestation is high (Undang, 2003).

Yogyakarta is a province located in the southern part of central Java Island, Indonesia (Fig. 5.1). Land-use change analysis is of extreme interest in this region due to the high dynamics, rapid urbanization, and substantial loss of prime agriculture area (Ritohardoyo, 2001). Land-use change monitored from 2002 to 2006 by the Provincial Board of Planning and Development of Yogyakarta revealed a significant enlargement in the Yogyakarta urban area. This urban development has increasingly encroached on farmland especially rice fields (BAPPEDA-DIY, 2007). Interpretation of a satellite image from 2009 revealed that wet agriculture land and dry agriculture land covered an area of 438.61 and 220.78 km^2, respectively (Partoyo and Shrestha, 2013). Land-use conversion from rice fields to another land-use type is threatening local rice production in Yogyakarta (Syamsiar, 2013). In addition, Yogyakarta has anticipated a transition period toward full implementation of the newly promulgated

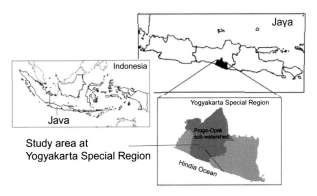

FIG. 5.1   Study area at Yogyakarta.

II. THEORETICAL ISSUES

Indonesian Law No. 41/2009 on farmland protection. This law requires data inventory and supporting regulation to be prepared during the transition period. Geographically, the province is situated between south latitude 7°3′–8°12′ and east longitude 110°00′–110°50′. The province consists of one municipality and four regencies, which are the Yogyakarta municipality, Sleman regency, Bantul regency, KulonProgo regency, and Gunungkidul regency. Dealing with the most land-use change, this study focused on the Progo-Opak subwatershed, which primarily belongs to Yogyakarta municipality, Sleman regency, and Bantul regency. The study area covers approximately 1502 km^2 of land between the altitude of zero in the Bantul alluvial plain bordered by the Hindia Ocean in the south to 2968 m in Mount Merapi in the north. The watershed represents the most urbanized region in the southern part of central Java.

5.1.2 Land Regulation in Indonesia

Indonesian land regulation has developed significantly since the issuance of the Basic Agrarian Law (Undang-Undang Pokok Agraria, UUPA) in 1960. This law is remarkable as a completion of preexisting ordinances, some of which are products of colonial government. The first implication of the law was a land reform program launched in the 1960s with the goal of distributing land in a fair, equitable manner as a source of livelihood for farmers.

UUPA mandated that a certificate should be issued to ensure land ownership. However, this larger program was not very successful in implementation. There are at least five possible reasons for this failure, namely, lack of political will and support, financial shortage, lack of reliable data and information, and weakness in the operating rules and lack of trained human resources (Nasoetion, 2003).

Over the past 42 years (1960–2002), the National Land Agency (BPN) authority only redistributed 885,000 ha of land, less than 2% of the total agricultural land area in the country. This land was distributed to 1.3 million farmer households or only 7% of the total farmer households in the country (Ali, 2006). The agricultural census in 1993 reported that 70% of rural households occupied less than 0.5 ha. The latest agricultural census in 2003 reported that farmer households with land holdings less than 0.5 ha increased from 10.8 million households in 1993 to 13.7 million households in 2003 (BPS-RI, 2004).

The UUPA was developed with the intention to addressing issues related to rural land and, therefore, it neglects urban land problems. However, urban land development is now an emerging issue that must be addressed in light of socioeconomic conditions that have greatly changed. In 1960, the UUPA clearly stated that land has both social and functional values and, therefore, it is not a commodity that can be traded for profit. In this case the UUPA is not able to address the recent trend of land commoditization (Ali, 2006).

In addition, there is much evidence that land ownership has not been protected by land certificate. The ownership is still only registered in a book called Letter C, which is held by the village government. This type of ownership is weaker in law than a land certificate, and it is difficult to transfer. However, several land ownerships have been granted to children because of the inheritance process, but the certificate is still registered under the parents. This is called family ownership. For example, ownership is difficult to transfer to the person who bought the land because the family collectively owns the land. For that purpose, the family should apply to the BPN for the issuance of a new certificate for each family member.

Intention to reimplement land reform was raised later in 2001 and was induced by the House of Representative Decree (TAP MPR) IX/MPR/2001 on land reform and natural resource management. The decree mandated that the President of the Republic of Indonesia serve as the executive government to develop a land reform policy. Land reform policies were then issued as Presidential Decree No. 34/2003 on national land policy. With this decree, the BPN is charged with two tasks to accelerate agrarian reform. First, the draft amendment of the Basic Agrarian Law (UUPA 1960) and the draft of a law about land ownership were prepared. Second, a land information and management system was developed (Ali, 2006).

Currently, agrarian reform is still underway. The essence of agrarian reform is in terms of ownership and control of land redistribution (Winoto, 2005). In 2009, the government redistributed 310,000 ha of land to 17 provinces (Santosa and Idris, 2009). Redistribution of land is part of the two agrarian reform measures by BPN. Another step is to set the political system and land-related law (Winoto, 2010).

After UUPA 1960, several policies and regulations have been developed regarding the utilization of natural resources in Indonesia, but these are fragmented in terms of objectives, orientation, and institution. Not surprisingly, these regulations and policies have been inefficient, inconsistent, and conflicting at times. Several laws and regulations are listed in Table 5.1.

5.1.3 Land-Use Policy Anticipating Farmland Conversion to Nonagricultural Use

The national economic development policy of Indonesia focusing on industrialization has placed pressure on land resources. Land has been exploited for short-term needs and benefits enjoyed by only a small portion of society to support economic development. Economic policies focused on growth have facilitated the allocation of land to large investors and given less access to the people for the acquisition and utilization of land (Firman, 2004; Widjanarko et al., 2004). According to Nasoetion (2003), the causes of this situation were various sectoral laws that overlap or even contradict one another and inconsistencies in implementation. In any circumstance, land development authorities provide the easiest way for investors to obtain land for operation; even by relaxing the requirements for land development permits, if necessary. Consequently, land conversion is not regulated because of the uncontrolled issuance of development permits (Firman, 2000).

Discussion and countermeasures of farmland conversion have been initiated since the 1990s. Several regulations have been established to control farmland conversion and to prevent anticipated conversion of prime farmland to nonagricultural use. However, this implementation has not been effective. There have been many violations of Presidential Decree No. 53/1989, which clearly states that developmental activity should not occur in the preservation and conservation area or on prime and irrigated agricultural land (Firman, 2004). Some regulations related to anticipation of agricultural land conversion are shown in Table 5.2.

TABLE 5.1   Several Laws and Regulations Related to Land Resources in Indonesia

Title	Content
Law 26/2007	Spatial Planning, updated from Law No 24/1992 for the same subject
Law 7/2004	Water Resources, updated from Law No. 11/1974 about Irrigation
Law 3/2002	State Land
Law 41/1999	Forestry, updated from Law No. 5/1967 about Basic Terms of Forestry
Law 23/1997	Environmental Management
Law 4/1992	Housing and Settlement
Law 5/1990	Conservation of natural resources and its ecosystem
Law 5/1960	Basic Agrarian Law
Government Regulation 26/2008	National Spatial Planning, updated from Government Regulation 47/1997 for the same subject
Government Regulation 16/2004	Land Use
Government Regulation 44/2004	Forestry Planning
Government Regulation 68/1998	Area of natural resources preservation and conservation
Presidential Decree 4/2009	Coordinating Board for National Spatial Planning
Presidential Decree 34/2003	National Policy on Land
Presidential Decree 33/1990	Land Use for Industrial Area
Presidential Decree 32/1990	Management of Protected Area
Presidential Instruction 1/1976	Synchronization Task on Agrarian, Forestry, Mining, Transmigration, and Public Work
Regional Regulation	Detailed District/Municipality Land Use Planning
Traditional regulation	Sultan Ground in Province of Yogyakarta Special Region

Source: *Ministry of Public Work, Republic of Indonesia.*

According to Nasoetion (2003), three fundamental constraints were the reason that land conversion control included difficult to implement regulations: (i) a contradictory policy, (ii) the limited scope of the policy, and (iii) the inconsistency of planning. First, the government attempted to ban agricultural land conversion due to contradictory policies, but policies regarding industrial/manufacturing and other nonagricultural sector development encouraged the conversion of agricultural land. Second, the limited scope of the policy implies that regulations were imposed mainly on the companies/legal entities that will occupy land and/or will convert agricultural into nonagricultural land. Meanwhile, conversion of agricultural to nonagricultural land by individuals has not been addressed by these regulations, even though land-use conversion by individuals is estimated to be significant. Third, planning inconsistency was found in the Regional Spatial Plan (RTRW), which was followed by a permit issuance mechanism for the permissible land conversion area. In fact, RTRW involves the conversion of irrigated paddy fields to nonagricultural use. Data from

TABLE 5.2 Several Land Regulations Anticipating Farmland Conversion

No.	Regulations Descriptions	Description
1	Law No.24/1992	Indonesian law ordered development of regional spatial planning should consider irrigated paddy field
2	Presidential Decree No.53/1989	Development of industrial area is prohibited to convert prime agricultural land
3	Presidential Decree No.33/1990	Prohibition of permit issuance for conversion of wetland and irrigated land for development of industrial area
4	SE MNA/KBPN/410-1851/1994	Prevention of conversion of irrigated paddy field to nonagricultural use by development of spatial planning
5	SE MNA/KBPN/410-2261/1994	Location permit should not convert irrigated paddy field
6	SE/KBAPPENAS/5334/MK/9/1994	Prohibition of conversion of irrigated paddy field to nonagriculture use
7	SE MNA/KBPN/5335/MK/9/1994	Spatial plan for district level should not convert irrigated land to non-agricultural use
8	SE MNA/KBPN/5417/MK/10/1994	Land use efficiency in residential development
9	SE MENDAGRI/474/4263/SJ/1994	Preservation of irrigated land to support food self sufficiency
10	SE MNA/KBPN/460-1594/1996	Prevention of wet agricultural land conversion to dry land

Source: *Murniningtyas, E., 2006. Strategy for controlling agricultural land conversion, Jakarta: Directorate of Food and Agriculture, Ministry of National Development Planning – BAPPENAS (in Indonesian); Widjanarko, B.S., Pakpahan, M., Rahardjono, B., Suweken, P., 2004. The agrarian aspect in controlling the conversion of agricultural (paddy field) land. Paper Presented at the National Seminar on Multifunction of Rice Field, Bogor, Indonesia (in Indonesian).*

the Directorate of Land Use of the BPN indicated that not reviewing such spatial planning would result in only about 4.2 million hectares (57.6%) that can remain functional. The remaining area of approximately 3.01 million hectares (42.4%) is threatened by conversion to other uses (Winoto, 2005).

Other weaknesses in the existing legislation are the following: (i) agricultural land protected from the conversion process is based on the physical condition of the land, but the physical condition can be manipulated relatively easily, and land conversion can occur without violating the regulations; (ii) the existing regulations are most often an appeal and are not equipped with clear sanctions for both penalties and determination of the sanctioned party; and (iii) in the event of agricultural land conversion that violates existing regulations, tracing the responsible institutions is difficult because conversion permits are a collective decision by various institutions (Simatupang and Irawan, 2002).

In addition, there are two other strategic factors. First, farmers as landowners and agents in local institutions have not been actively involved in various efforts to control agricultural land conversion. Second, there has been a lack of commitment, coordination system improvements, and competence development of formal institutions in dealing with agricultural land conversion. Some of the abovementioned weaknesses and limitations have caused the existing policy for agricultural land conversion to not directly address critical field components (Iqbal and Sumaryanto, 2007; Simatupang and Irawan, 2002).

In some cases, agricultural land conversion to other uses is a dilemma. Increased population and rapid economic activity growth in some areas require sufficient land for nonagricultural use. However, population growth also requires a bigger food supply, which means that more area is needed for agricultural land, while the total land area is fixed. As a result, there has been intense competition in land use resulting in increased land value (land rent) and that the use of land for agriculture will always be defeated by other uses, such as industry and housing (Nasoetion and Winoto, 1996), although the intrinsic value of farmland, mainly paddy fields, was much higher than its market value (Pakpahan et al., 2005; Sumaryanto and Sudaryanto, 2005).

On the internal side of the agricultural sector, various farm characteristics have not fully supported the implementation of existing agricultural land preservation. The narrowness of the average cultivated land area for a farmer due to the fragmentation caused by the inheritance system has increasingly marginalized farming activities. The narrowness of the land holding resulted in inadequate revenue of the agricultural business activities to cover the needs of daily life and does not encourage the application of new technologies to increase productivity. Instead of the technological applications, the farmland is sold to other users for profit (Murniningtyas, 2006).

5.1.4 Land-Use Policy in Yogyakarta

In the case of the Yogyakarta Special Region (Daerah Istimewa Yogyakarta, DIY), the primary land-use issue is the conflict of interest between urban development and agricultural land preservation. At least three factors are considered to drive urban development. First, high population growth in the urban area significantly increases population density. With denser populations, the environmental quality becomes inadequate to provide good living conditions, and groups of people are motivated to move to other places. Land encroachment has happened because most selected places are suburban or rural areas, which are predominantly agricultural land. In fact, people with a higher level of wealth and educational background prefer to build houses separate from their parents, while economically weak groups typically live with their parent families.

Second, Yogyakarta municipality sprawl is a combination of concentric, ribbon, and frog leap patterns. This has directed the new settlements to grow rapidly and not only along the road or surrounding the city but also in many other locations (Yunus, 2009). The policy of national government called "Perumnas" (Perumahan Nasional, national public housing) aimed at creating suburban residential neighborhoods for lower-middle income was launched in 1974 (Perumnas, 2010), which was designed with well-organized space and good quality construction. However, this policy initiated an urban sprawl because location selection is primarily based on the low cost of land, which is mostly in rural areas. This emerging settlement has accelerated the conversion of surrounding agricultural land into a built-environment to provide supporting services for the new inhabitants.

Lastly, the high settlement demand in Yogyakarta is caused by the preference of people to live in this city. In addition to high demand for student boarding, many people living outside of Yogyakarta bought houses as an investment to live in the city after their retirement (Azhar and Roozanty, 2010). Yogyakarta is the Indonesian city with the longest life expectancy of 73 years compared to the Indonesian average of 67 years (BPS-RI, 2005).

Statistics show that the increased urban development has lasted for nearly four decades since 1970 (Ritohardoyo, 2001). Furthermore, the loss of agricultural land has increasingly threatened food availability. Land-use policies have been applied such that this phenomenon continues to occur unexpectedly.

Anticipating this condition in 1992, the provincial government of DIY formulated a Planning for Provincial Spatial Structure (Rencana Struktur Tata Ruang Propinsi, RSTRP) 1992–2006. This is a policy to develop proper land use and involve sustainable urban development and agricultural land protection. In relation to urban development, the RSTRP provides strategies and rules to develop an improved urban area. In terms of agriculture protection, RSTRP has a strategic role in determining the success of agricultural land protection for two primary reasons. First, RSTRP is a structured spatial plan that arranges and directs the use of whole land, including agricultural land. Secondly, the role of RSTRP is to influence the pattern of agricultural land, particularly the connection with urban planning. Unfortunately, this policy was not effective (Irham, 1993), as evidenced by scattered urban development, unprecedented urban expansion, and high loss of prime agricultural land.

Furthermore, the policy did not effectively protect the land, and previous analysis identified several weaknesses (Irham, 1993). First, RSTRP is a too broad policy. The guideline was aimed at the provincial level only. The detailed planning for the regency/municipal level has not yet been formulated. In this case, the policies seem to be vague, which in turn confuses local officials. Second, RSTRP had no flexibility to accommodate changes in socioeconomic conditions. Parts of the guideline were no longer applicable for a certain area, and no evaluation could be performed immediately. Third, monitoring and control was difficult to conduct. The document can only be comprehended by provincial-level officials, while the land-use conversion occurs at the regency level. Fourth, the BPN is in charge of managing land-use issues and experienced difficulties applying the RSTRP to these problems. The RSTRP should coordinate many related department/institutions, particularly the office of agriculture, office of public work, and Board of Local Development and Planning (Badan Perencanaan Pembangunan Daerah, BAPPEDA). Successful coordination is a common constraint in developing bureaucracy, including the government of the Province of Yogyakarta Special Region. Fifth, there was inconsistency between RSTRP and the spatial planning of regencies/municipalities because they were formulated separately and did not refer each other. Sixth, there was no agricultural zoning ordinance in the RSTRP and no clear legal basis for BPN to prohibit any nonagricultural development proposal on agricultural lands.

Regarding agricultural regulation, Indonesia initiated national laws related to agriculture in 1992 after the issuance of Law No. 12/1992 on the crop cultivation system. This law assures the rice self-sufficiency level, which was reached in 1984, and supersedes all predecessors, including more than 10 ordinances issued by the colonial government. In the following year, many supporting regulations were created to implement the law for various pesticides and fertilizers but not for related land-use issues.

The escalating conflict of interest on land utilization between sectors/subsectors forced the government to renew the outdated regulation by issuing Law No. 26/2007 on spatial planning. This new law reaffirmed the mandate of Law No. 12/1992 on the crop cultivation system to publish legislation for the protection of agricultural land. Beginning in 2007 with the draft entitled Permanent Agricultural Land Regulation. the document has undergone much revision based on objections and criticisms. Finally, the approved draft has been issued as Law No. 41/2009 for the protection of land for sustainable food crop farming on Oct. 14, 2009. This latest law is the primary reference related to agricultural land conversion issues, which refers to the law, the Government of Yogyakarta Special Region has issued Provincial Regulation No. 10/2011 and several supported Regency Regulations to sustain agriculture land in the area.

Based on this analysis, several gaps were identified that resulted in land-use policy failure, particularly inhibiting the encroachment of agricultural land by settlement development. These weaknesses include the lack of applicable regulations, weakness of law enforcement, and lack of a supporting policy to encourage farmers to keep their agricultural land. This example demonstrates that agricultural land conversion involves three stakeholders, which are the government as regulator, the farmer as actor in agricultural land use, and the housing developer/individual as demander/user.

5.1.5 Driving Factors of Land-Use Change in Yogyakarta

Land-use changes are usually modeled based on the biophysical and socioeconomic variables chosen to serve as driving forces (Turner-II et al., 1993). Driving forces are usually divided into three groups: socioeconomic drivers, biophysical drivers, and associated land management variables (Turner-II et al., 1995). Although most biophysical factors do not drive land-use changes directly, these factors can cause changes in land cover (eg, through climate change) and affect decisions of land-use allocation (eg, soil quality). On another scale of analysis, the dominant driving factor in land-use systems could be different. This can be a local policy or the presence of small areas of ecological value at the local level, while the distance to ports, markets, or airports may be the primary determinant of land-use patterns at the regional level (Turner-II et al., 1995).

Driving forces are generally considered to be exogenous to the land-use system to facilitate modeling (Verburg et al., 2004). However, in some cases this assumption hampers the proper description of the land-use system. For example, population pressure is often regarded as an important driver of deforestation. Pfaff (1999) noted that the population is not always endogenous to forest conversion depending on the local issue context. If increased population is facilitated by forest clearing, involving population as an exogenous driver of land-use change would produce a biased estimate and lead to a misleading policy conclusion. If the population is collinear to the forest conversion process, then the estimates would be unbiased but inefficient, leading to a potential false interpretation of the significance of variables in explaining deforestation (Verburg et al., 2004). Another example of endogeneity of driving forces in land-use study is given by Irwin and Geoghegan (2001).

The choice of driving forces is highly dependent on simplifications and theoretical and behavioral assumptions in land-use modeling. In the economic approach, most economic models of land-use change are related to the land rent theories of Von Thunen and Ricardo (Nelson, 2002). In the simplest form (ie, the monocentric model), the distance to the urban center is the most important driving variable. Other models, such as the hedonic model, that try to explain land values, combine variables that measure the distance to the urban center and specific location features of the land parcel (Bockstael, 1996).

The temporal analysis scale is important for determining which driver is endogenous to the model. In economic models of land-use change, a function of supply and demand is the driving force of land-use change. Price can be considered exogenous to changes in land use in the short term, but it is endogenous at longer timescales (Doygun, 2009; Verburg et al., 2004).

Summarizing several reported case studies (Lambin et al., 2003; Pfaff, 1999; Serneels and Lambin, 2001), land-use change is affected by a combination of factors, including biophysical, economic, technological, demographic, institutional, and cultural/social factors.

In the case of Yogyakarta, land-use change was related to biophysical, socioeconomic, and policy factors. With respect to agricultural land conversion, evidence of land-use change to built-up land mostly occurred at locations far from the district's capital city, close to the road, low in altitude, and low in population (Partoyo, 2010). Agricultural land conversion is expected to continue. Recently, agricultural land conversion to a built-up area occurred in areas with an irrigation facility and highly suitable for rice cultivation (Tarigan, 2013).

Regarding socioeconomic factors, agricultural land conversion was related to decisions and perceptions about farmland conversion. The study of Partoyo (2011) revealed that the household decision to convert farmland was significantly related to revenue earned from farming activity, socioeconomic background of the household, farming sustainability, access to agricultural information, perception about farmland protection, and land tenure. The possibility of the household deciding on wet agricultural land conversion will be higher for a household with higher revenue from farming activity; that is, large land-holding farmers, higher socioeconomic status of the household, less access to information on land-related regulation, less possibility to sustain farming activity, negative perceptions of farmland protection, and secure land tenure (Partoyo, 2010; Partoyo and Shrestha, 2013).

Related to policy, several land-related regulations have been launched, but without effective enforcement. Regulations related to land have not been widely recognized by people, including the prohibition of prime agricultural land conversion.

A household survey in the study area (Partoyo, 2011) indicated that farmers' livelihoods were vulnerable to decisions on agricultural land conversion. The average surveyed family size is typically small with three members. Although farming is a major occupation, income from farming activity contributes only 54% of household expenditures. Nevertheless, these households are not categorized as poor, and the average per capita income in the study area is slightly higher than the national poverty limit. People in the area seek minor occupations to obtain additional income for financial betterment.

Typically, farmer households in the study area have very small agricultural land (on average 0.24 ha), which is not feasible as a primary source of household income. The four most common farming problems were lack of financial capital, low price of farm products, high price of fertilizer, and marginally small profit. When surveyed about farming revitalization, the four most significant requests were subsidized inputs, financial capital access, farming insurance, and upgraded farming technology. This result implied that technology was not a major problem, but financial factors were unmanageable (Partoyo, 2011).

Regarding land conservation, the respondents agree that sustaining viable crop cultivation and preserve land resources is important. Several conservation measures are practiced with farming activity, although almost two-thirds do not recognize regulations governing this issue by the public authority. For agricultural land conversion, approximately one-third were aware of one or two regulations that addressed this issue. Regulation to prevent prime agricultural land conversion to nonagricultural use already existed but was poorly enforced. Households have their own reasons other than regulation enforcement to sustain agricultural land, such as recognizing the agricultural land as a family inheritance or as valued household assets. However, 37% of household respondents that experienced agricultural land conversion indicated the reason was monetary needs for consumptive use and financial capital (Partoyo, 2011).

Regarding access to information, households are commonly informed mainly during farmer group meetings by colleagues, agricultural extension officers, or other government officials. Farmer group meetings are effective forums that can enhance discussions of new agricultural technology and government regulations.

Although current rice cultivation in the existing paddy field is sufficient to fulfill the food demand, farmland zoning is desirable for identifying the prime agricultural land to preserve a reliable food production system. Due to land-use change, wet agricultural land situated on the high-potential land suffered a $1.14\,km^2$ loss per annum. In 2009, 20% of high-potential land in the study area was occupied by built-up land use. Zoning of the study area based on land potential into four land classes: high, moderate, low, and no potential, which is required to develop spatial policy on farmland protection. Conversion of high-potential land should be controlled to prevent potential loss of rice production. Farmland with high potential should be preserved.

5.2 LAND-USE CHANGE MODELING

Modeling of land-use change allows for analysis of the causes and consequences of land-use change to better understand the function of the land-use system and to support land-use planning and policy. The model aims to simulate the function of land-use systems and to conduct spatially explicit simulation of land-use patterns in the future (Verburg et al., 2004).

The models are designed for running several scenarios for policy choices. Scenarios should not be confused with forecasts because these do not allow for prediction but rather the exploration of technical options based on explicit assumptions for a set of goals. The strength and weakness of each will be demonstrated by running scenarios. Land-use models may help to make potential choices more visible. Then, policy makers and land users can decide more easily on explicit choices.

Regarding selection of variables for the model, there has been advanced development of the inductive approach involving empirical selection of many suspected variables until deductive selection based on a firm theoretical background (Overmars et al., 2007). The simulation was more visually depicted under the contribution of geographical information systems (GIS) and the remote sensing technique (Irwin and Geoghegan, 2001; Schweik and Thomas, 2002). Land-use change simulation has also been developed to include dynamic factors in the scenario to facilitate top-down and bottom-up approaches (Xiang and Clarke, 2003). This allows for the involvement of socioeconomic variables, which are usually not spatial in nature, instead of biophysical variables. Many land-use modeling scholars have been more interested in the process of land-use change and not only the pattern of change (Bakker and van Doorn, 2009; Nagendra et al., 2004). The combination of remote sensing and household data became more popular after "socializing the pixel" and "pixelizing the social" were described by many scholars (Geogeghan et al., 1998). Therefore, land-use change modeling resulted in a more acceptable output for the decision maker and stakeholder to develop land-related policy.

A model should be developed that integrates disciplinary approaches and models studying urban and rural land-use change to better support the analysis of land-use dynamics and policy formulation (Verburg et al., 2004).

Regarding the involvement of broad drivers of change, a new concept of Land Use and Cover Change (LUCC) has been developed to account for the interaction between biophysical characteristics of land and socioeconomic conditions. Landowners or users as well as institutions are considered agents for decision-making processes that influence the land-use/land-cover change (Rajan and Shibasaki, 1998).

The decade since the initiation of the LUCC project in 1995 has witnessed considerable advances in the field of LUCC modeling. During this period, the combined use of simulation models, expert systems, GIS, various types of databases and multiple goal planning techniques has allowed for technical land-use options to be formulated in a more precise and varied way.

Reviews have characterized and classified land-use models (Verburg et al., 2004), a model based on economic theory (Bockstael, 1996), a model for deforestation (Kaimowitz and Angelsen, 1998; Lambin, 1997), integrated urban models (Hilferink and Rietveld, 1999), and agricultural intensification models (Lambin et al., 2000).

A wide range of techniques are available, and each has its own strengths and limitations. Several techniques, including microeconomic (Caruso et al., 2005) and multiagent-based simulation (Brown et al., 2005), deal with land-use change as a consequence of socioeconomic conditions. The increasing concern among LUCC scholars about land-use models that are spatially explicit has resulted in several approaches, such as logistic regression (Overmars and Verburg, 2005), neural networks (Pijanowski et al., 2005), cellular automata (Jantz and Goetz, 2005), and Markov chains (Pontius and Malanson, 2005).

Cellular automata are frequently used in land-use modeling with spatially explicit approaches. Land-use change is simulated as a function of neighborhood land use and a set of driving-factor relationships (Balzter, 2000). Neighborhood functions and transition rules are specified either based on user expert knowledge or empirical relationships between land use and driving factors (Pontius et al., 2001).

The Dyna-CLUE model (Dynamic Conversion of Land Use and Its Effects) (Verburg and Overmars, 2009) is based on spatial allocation of demands for different land-use types to individual grid cells. This model has the advantage of allocating land use simultaneously based on the highest possibility among competitive land-use types based on prescribed scenarios. Simulation will be shown in a spatially explicit result to visualize the projected effect of land-use policy, which provides a wide-scale of simulation based on the analysis level, even until the watershed level with the highest accuracy, depending on the available data.

There is an adapted version of the CLUE-s model (Castella et al., 2007; Verburg et al., 2002). The predecessor CLUE has been used and validated in multiple case studies of land-use change in many regions including Europe (Britz et al., 2011), Costa Rica (Veldkamp and Fresco, 1996), Ecuador, Central America, Honduras, China (Luo et al., 2010), Java-Indonesia (Verburg et al., 1999), Sibuyan Island (Philippines) and Malaysia (Verburg et al., 2002), and Vietnam (Castella and Verburg, 2007).

Dyna-CLUE has been successfully implemented in Thailand, as Trisurat et al. (2010) used the model to project land use change from 2002 to 2050 due to deforestation. Improved from the former version of CLUE, Dyna-CLUE combines more dynamic modeling and empirical quantification of the relations between land use and its driving factors. Probability for each location to be changed is estimated on the basis of actual land use and the competitiveness of different types of land use. Scenarios can be prescribed to evaluate different land-use change situations cause by differences in land-use requirements and spatial policies (Verburg and Overmars, 2009).

This study applied the Dyna-CLUE platform for modeling the effect of land-use policy on the future land use of the study area. The procedures applied are described as follows.

5.2.1 Preparation of Land-Use Base Map

We derived a base map of land use of the study area from a Landsat ETM+ image acquired in 2004 and a reference map of land use from ASTER Terralook images acquired in 2009. The standard procedure for image preparation was done prior to image interpretation. Using maximum likelihood classification, we classified seven classes of land use: wet agriculture land (WetA), dry agriculture land (DryA), mixed garden (MixG), high-density built-up land (HiBu), low-density built-up land (LoBu), forest, and miscellaneous.

5.2.2 Developing Inputs for Land-Use Modeling

Four inputs were prepared for land allocation procedure in this land-use modeling: logistic regression model, land demand, conversion elasticity index, and conversion sequence matrix.

5.2.2.1 Logistic Regression Model

The regression models link the functional relationship between land-use type and location factors. Location factors were variables contributing to the probability of certain land-use types existing at certain locations. In this study, we used 14 location factors as follows: Elevation (=E), Slope (=S), Distance to main road (=R), Distance to capital city of district (=D), Distance to capital city of regency (=Rg), Distance to capital city of province (=P), Land suitability for rice (=Rs), Irrigation support (=I), Population density (=Pd), Land property right (=Pr), Land utilization right (=U), Land owned by village (=V), Land owned by state (=St), and Land of SG/PAG (=SG). Based on land-use maps and other GIS data prepared, we have developed binary logistic regression models for each land-use type as follows (Partoyo, 2012):

$$HiBu = 0.84 + 0.004E - 0.13S - 0.0005R - 0.0001D + 0.0001Rg - 0.0003P + 0.00001Pd - 0.17Pr - 0.55U + 0.50SG + 0.58Rs + 1.03I$$

$$LoBu = -3.78 - 0.002E + 0.01S - 0.00008D - 0.00006Rg + 0.0001P + 1.18U + 0.54St - 0.32Rs - 0.26I$$

$$WetA = -1.73 - 0.003E - 0.038S + 0.00008R + 0.00008D - 0.00006R + 0.00009P - 0.000004Pd + 0.971U + 0.366St - 0.852Rs - 1.236I$$

$$DryA = -5.57 - 0.002E + 0.098S + 0.0002R + 0.0002D - 0.00005R + 0.00007P - 0.00001Pd + 1.807U + 1.311St - 0.205I$$

$$MixG = -6.82 + 0.034S + 0.0002R - 0.00007D - 0.00006R + 0.0001P + 2.238U + 0.721St + 0.165I$$

$$Forest = -6.71 + 0.006E - 0.221S - 0.0003R + 0.0001D - 0.0001R + 0.0002P - 0.0001Pd + 1.596St + 3.111I$$

Logistic regression models were calibrated using relative operating characteristics (ROC) value and validated by running the model to project land use from 2004 to 2009 using prescribed input and parameters under an ongoing trend scenario. A simulated land-use map then was compared to a reference land-use map from 2009.

5.2.2.2 Land Demand

Land demand declared areas that will be occupied by each land-use type. The area was determined regarding prescribed scenarios of land-use change and calculated for every year until the end of the simulation year.

For this study, land demand was prepared under three scenarios. The scenarios were prescribed considering the implication of farmland loss to rice production, food security and food self-sufficiency, as well as protection of the landscape for better environment. The first scenario was intended to project future land use if change followed the ongoing trend. The calculated land demand was based on empirical data derived from the satellite interpretation and statistical data. The first scenario is called "business as usual." The second scenario was developed to explore the impact of land-use policy on future land use. It used a similar trend of land-use change as the first scenario, but applied farmland protection as spatial policy. This scenario is called the "farmland protection" scenario. The third scenario was prepared to provide enough paddy field area for 400 km² by 2030. This area was predicted based on the rate of rice demand and paddy field productivity. The rate of change for each land-use type was determined for the minimum area of farmland required. This scenario is called the "minimum required farmland" scenario. Table 5.3 describes details of the three scenarios.

5.2.2.3 Spatial Policy

For some scenarios, areas of land-use change restriction can be defined because of spatial policies, such as the preservation area or protected area. In this study, Merapi National Park and/or the high suitable land for rice were considered a restricted area (Table 5.3).

TABLE 5.3 Characteristics of Land Demand Scenarios for 2030

	Scenario 1: Business as Usual	Scenario 2: Farmland Protection	Scenario 3: Minimum Required Farmland
Projected land use (2030)	Follows the trend of existing conversion rate (high density built-up, +2.7%; low density built-up, +1.8%; wet agriculture, −0.4%; dry agriculture, −1.2%; mixed garden, −1.0%; forest, −0.7%)	Follows the trend of existing conversion rate (same as Scenario 1)	At minimum 400 km² (high density built up, +2.6%; low density built up, +1.8%; wet agriculture, −0.3%; dry agriculture, −1.2%; mixed garden, −0.9%; forest, −1.0%)
Spatial policy	No land conversion allowed in Merapi National Park	No land conversion allowed in Merapi National Park, and in irrigated area with a high land suitability class for rice	No land conversion allowed in Merapi National Park and limited farmland conversion as allowed

5.2.2.4 Conversion Elasticity Index

The conversion elasticity is related to reversibility of land-use changes. This factor is based on expert judgment, ranging from 0 (easy conversion) to 1 (irreversible change). We applied an index of 1 for built-up land, forest, and miscellaneous, instead of wet agriculture land, dry agriculture land, and mixed garden with an index of 0.2, 0.2, and 0.8, respectively.

5.2.2.5 Conversion Sequence Matrix

The conversions that are possible and impossible are specified in a land-use conversion matrix. For each land-use type it is indicated what the land-use type can be converted to during the next time step. Value "1" means land-use conversion possible to take place, while "0" means not possible.

5.2.3 Model Calibration and Validation

Calibration of the predicted land use by the model was done based on the ROC value. The spatial distributions of the allocated land-use types are explained moderately to well by the selected location factors, as indicated by the ROC values that ranged from 0.76 to 0.99. The lowest ROC of low-density built-up land (0.76) is caused by a wide spreading of location of this land-use type in the study area. The high ROC value of high-density built-up land implies an accurate prediction of the location of high-density built-up area. Based on these good ROC values, it means that a model of all land-use types can be implemented in the Dyna-CLUE framework.

Validation of the model was conducted either by visual or statistical methods to compare projected map and reference map year 2009. Under visual examination, both of these maps look similar. High-density built-up land clearly shows similar extent and pattern (Fig. 5.2A and B). Some high-density built-up areas that can be tracked have been developed along the main roads. Wet agriculture land, mixed garden, and forest are also spread in a similar pattern. This good conformity implies the validity of the Dyna-CLUE model for simulating future land-use change in the study area.

Furthermore, a statistical approach was performed to validate the model. We calculated ROC for each land-use type between the projected map and the reference map. The ROC answers on how well the projected map matches with the location map showed on the reference map (Pontius and Schneider, 2001). ROC of 0.5 indicates conformity equivalent to random chance, when the grid cells are hard to classify. ROC of 1 indicates perfect conformity (Pontius and Chen, 2006).

The analysis of ROC resulted in a value of all land-use type ranging between 0.54 and 0.83. The lowest ROC value (0.54) evidences a slightly better conformity of low-density built-up with random locations. The higher ROC of all other land-use types implies that, as a whole, the model projects a valid future land-use map. Based on ROC value, the predictions have better accuracy, respectively, for high-density built-up, forest, mixed garden, wet agriculture land, dry agriculture land, and low-density built-up.

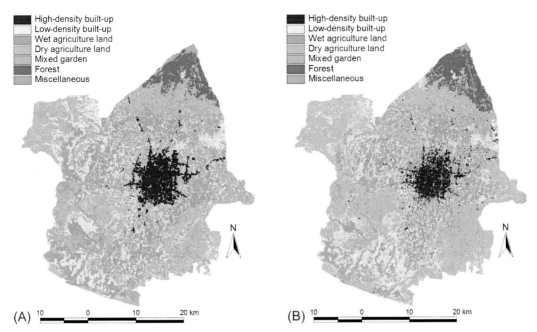

FIG. 5.2   Land-use map: (A) Reference map 2009; (B) Projected map 2009. *Adapted from Partoyo, 2012. Application of spatial modeling for simulating land use conversion to support sustainable agricultural land planning. Proceedings of National Seminar on Agriculture and Fisheries Research, Year 2012. Yogyakarta, 15 September 2012. Faculty of Agriculture Gadjah Mada University. pp. 575–582, ISBN: 978-979-8678-25-7 (in Indonesian).*

5.3 LAND-USE PROJECTION BY 2030

Regarding the satisfied calibration and validation result of the model, we proceeded to use the model to project land use by 2030.

Under scenario 1, simulation resulted in the expansion of the urban area. Both high-density and low-density built-up land expanded to the west instead of northeast and southeast as observed between the reference maps in 2009 (Fig. 5.2A) and 2030 (Fig. 5.3A). It is also shown clearly that some mixed garden and dry agriculture land have been changed into low-density built-up land in the southwest part.

Under scenario 2, urban sprawl is not as much expanded as scenario 1. Increased high-density built-up land mostly occurred within the municipality boundary. This development resulted in a massive area of high-density built-up. It implies that preventing conversion of high-potential agriculture land has hindered urban sprawl from occupying the surrounding area.

Actually both scenarios 1 and 2 allocated a similar area of each land-use type as both assume the same land demand. However, implementation of farmland protection policy in scenario 2 clearly resulted in a spatially different pattern of land use. The projected map proves that land policy governs the pattern of land-use change.

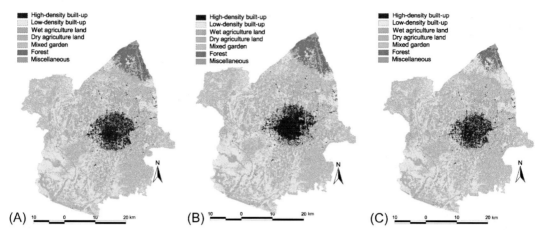

FIG. 5.3   Projected land-use map year 2030 based on the three scenarios. (A) Projected map: scenario 1; (B) Projected map: scenario 2; (C) Projected map: scenario 3.

In term of farmland preservation objective, scenario 2, is successful to preventing high-potential agriculture land conversion against urban sprawl. However, under the ongoing trend of land-use change, both scenarios allocated area for wet agriculture land as only 38,919 km², which is less than the area required as 400 km² by 2030. This result implies that additional land policy is needed to ensure a sufficient area of land is available for agricultural development. For scenario 3, an area of 400 km² was projected for the wet agricultural land. The urban area of scenario 3 (Fig. 5.3C) will be similar in shape to scenario 1 (Fig. 5.3A), but with the former having less massive distribution of high-density built-up land (Fig. 5.3B). The National Park of Merapi Mountain will be preserved from land-use conversion but not the surrounding forest and shrubs, which are converted into low-density built-up land or dry agricultural land.

When comparing the results of all scenarios, scenario 2 yields the most acceptable result. In terms of farmland preservation, this scenario resulted in wet agricultural land that was prevented from becoming urban sprawl. Both scenarios 1 and 2 allocated a similar area for each land-use type as both assume the same land demand. However, implementation of farmland protection in scenario 2 clearly resulted in a spatially different pattern of land use.

Therefore, allocation of adequate farmland area remains a concern because scenario 1 and 2 failed to preserve at least 400 km² of wet agricultural land. This result suggests that the current trend of farmland conversion is high and should not continue. Furthermore, future built-up land should not be allowed to occupy wet agricultural land.

This model facilitates incorporating of any additional policy inputs of the study area provided that all other location factors and assumptions remained unchanged and valid.

5.3.1 Proposed Policy

In relation to food availability and land conservation issues, increased land demand for nonagricultural use and high rates of farmland conversion are very threatening. Accordingly, combined land-use options should be implemented to control land-use change and to manage the remaining farmland with appropriate land use.

First, control of land-use change should aim to maintain enough farmland area by reducing farmland conversion. Approaches to decrease farmland conversion include promoting farmer households to not convert their farmland and reducing nonagricultural land demands to avoid competition with farmland.

In addition to enforcing legal sanctions, strategies to empower farmers might be more effective to persuade households to sustain their farming activity. As summarized in Fig. 5.4, the topmost incentives proposed by farmer household to revive their farming activity were subsidized farming inputs, better financial capital access, guaranteed income from farming activity, high-yielding farming technology, and reduced land tax. Therefore, strategies providing those incentives will effectively sustain farming activity, which will result in the maintenance of farmland.

To reduce land demand for nonagricultural use and particularly in built-up areas, strategies for developing building that require less land will be beneficial, such as residential houses. Zoning for residential or industrial areas will also be a useful strategy to suppress competition with agricultural land use. Robust spatial policy will be required to regulate zoning for urban development, including high-density built-up area.

There is projected to be a lack of available wet agricultural land in 2030, and strategies to better manage existing farmland will be useful to secure food production and to conserve environmental functioning. Increasing land productivity is an obligatory strategy to produce sufficient food within the reduced farmland area. High-yielding varieties of rice and related farming technology are required components. The pressure of high rice demand might also be managed through more diversified staple food sources. Food crops other than rice, such as cassava, sweet potatoes, and taro, can be grown on nonirrigated farmland. Therefore, food availability can be secured as rice field area decreases.

Based on the projection result, there will be a lack of wet agricultural land for rice field available in the study area by 2030. To secure food production, land demand for built-up area must be suppressed and not compete with agricultural land. If agricultural land is reduced, food production will be insufficient without increasing land productivity or reducing dependence on rice as the staple food.

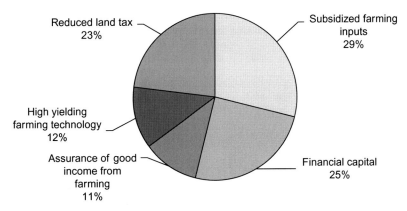

FIG. 5.4 Incentives proposed by households for sustaining viable farming activity. *From Partoyo, 2011. Land Use Options and Strategies for Food Availability in Yogyakarta, Indonesia. Unpublished PhD Dissertation, Asian Institute of Technology, Thailand.*

With the assumption of constant productivity and based on the scenarios tested, protection of wet agricultural land against conversion is necessary. High-potential land must be prioritized for protection. However, any policy should be accompanied to control land demand for agricultural or nonagricultural use.

As concluded above, farmers' decision making is affected by several factors that encourage farmers to convert agricultural land to other uses, and it is important to empower farmers in providing higher revenue from farming activities that will help discourage further conversion of wet agricultural land.

5.4 SUMMARY

Yogyakarta has experienced significant land-use change in the past. The major land-use change includes a decrease of wet agricultural land and dry agricultural land, and an increase of high-density built-up land and low-density built-up land. Urban development has expanded for low- and high-density built-up areas combined by nearly fourfold between 1992 and 2009. Urban sprawl occurred in the former low-density built-up areas and agricultural land around the municipality area. At least 14% of the high-density built-up land in 2009 was converted from former high-potential land for rice cultivation.

Projections of future land use by 2030 indicated that under the ongoing trend of land demand, land conversion will reduce wet agricultural land to be less than the required area ($400 \, km^2$). Meanwhile, under the scenario of farmland protection, all high-potential land will be secured from conversion, but still less area will be available for wet agricultural land. With the scenario of providing agricultural area of at least $400 \, km^2$, it needs to reduce land occupation for high-density built-up upon wet agricultural land by 0.1% per year.

Farmland loss due to land conversion has implications for rice production, food security, as well as food self-sufficiency. Meanwhile, land conversion is unavoidable due to higher demand for services and jobs. Spatial modeling figured out what will be the land-use map in the future. As resulted from the simulation, future land-use projections for 2030 indicate that policies on farmland protection and urban development are necessary. It should be coupled with the objective to preserve productive farmland and develop urban expansion in less important farmland so that the land-use change will be controlled to maintain the productive function of farmland and nonagricultural function of other land uses.

The spatially explicit projection methods like Dyna-CLUE give a spatial view of the simulation under different scenarios. Such a projected pattern makes possible an assessment of spatial impact due to different applied scenarios being done in relation to land-use development. As the scenario can be dynamically adjusted considering the recent situation and updated available data, this simulation method can be tailored following any newer prescribed land-use policy.

Factors that drive land-use change, particularly agricultural land conversion to nonagricultural land use, are distance to city, distance to road, population density, elevation, terrain slope, irrigation availability, land tenure, and land suitability for rice cultivation. In relation to factors affecting household decision to convert farmland, six factors, namely, revenue from farming activity, socioeconomic status of household, access to land-related regulation, sustainability of household farming, perception about farmland protection, and land tenure were found to affect decision making.

Irrigation plays an important role to increase rice cultivation intensity, thus implying higher potential to increase production. Considering combined irrigation availability and land suitability as criteria, only 22% of the study area falls into the high-potential class, which calls for immediate protection of these areas. Given the land area of 400 km² needed to reserve for rice cultivation by 2030, as per a projection of the Provincial Agricultural Agency, this study showed that the area of wet agricultural land will be less than the required area of 400 km². Under the farmland protection scenario in this study, even after preserving all high-potential land from conversion, the availability of wet agricultural land remains below the requirement.

References

Ali, T.H., 2006. Development of Land Institution and Its Relation to History of Spatial Planning in Indonesia. Directorate General of Spatial Planning, Department of Settlement and Infrastructure, Republic of Indonesia, Jakarta (in Indonesian).

Azhar, Y.K., Roozanty, A.M., 2010. Real estate in Yogyakarta targeting students and elderly people. Housing estate, January edition.

Bakker, M.M., van Doorn, A.M., 2009. Farmer-specific relationships between land use change and landscape factors: introducing agents in empirical land use modelling. Land Use Policy 26 (3), 809–817.

Balzter, H., 2000. Markov chain models for vegetation dynamics. Ecol. Model. 126 (2), 139–154.

BAPPEDA-DIY, 2007. Profile of Yogyakarta Special Region. Board of Regional Development Planning, Yogyakarta (in Indonesian).

Bockstael, N.E., 1996. Modeling economics and ecology: the importance of a spatial perspective. Am. J. Agric. Econ. 78 (5), 1168–1180.

BPS-RI, 2004. Distribution of agricultural household in provincial level of Indonesia. Berita Resmi Statistik, VII (14), 1-9 (in Indonesian).

BPS-RI, 2005. Estimation of life expectancy based on province in Indonesia, 2000–2025. Central Bureau of Statistics – Republic of Indonesia. Retrieved 11 September 2010, from http://www.datastatistik-indonesia.com/content/view/922/938/ (in Indonesian).

BPS-RI, 2010. Result of Population Census 2010: Aggregate Data Per Province. Board of Statistics, Republic of Indonesia, Jakarta (in Indonesian).

Braimoh, A.K., Onishi, T., 2007. Spatial determinants of urban land use change in Lagos, Nigeria. Land Use Policy 24 (2), 502–515.

Britz, W., Verburg, P.H., Leip, A., 2011. Modelling of land cover and agricultural change in Europe: combining the CLUE and CAPRI-Spat approaches. Agric. Ecosyst. Environ. 142 (1), 40–50.

Brown, L.R., 1995. Who Will Feed China? Wake-up Call for a Small Planet. W.W. Norton, New York.

Brown, D.G., Page, S., Riolo, R., Zellner, M., Rand, W., 2005. Path dependence and the validation of agent-based spatial models of land use. Int. J. Geogr. Inf. Sci. 19 (2), 114–123.

Caruso, G., Rounsevell, M., Cojocaru, G., 2005. Exploring a spatio-dynamic neighbourhood-based model of residential behaviour in the Brussels periurban area. Int. J. Geogr. Inf. Sci. 19 (2), 103–123.

Castella, J.C., Verburg, P.H., 2007. Combination of process-oriented and pattern-oriented models of land-use change in a mountain area of Vietnam. Ecol. Model. 202 (3-4), 410–420.

Castella, J.C., Pheng-Kam, S., Dinh-Quang, D., Verburg, P.H., Thai-Hoanh, C., 2007. Combining top-down and bottom-up modelling approaches of land use/cover change to support public policies: application to sustainable management of natural resources in northern Vietnam. Land Use Policy 24 (3), 531–545.

Cheng, J., Masser, I., 2003. Urban growth pattern modeling: a case study of Wuhan City, PR China. Landsc. Urban Plan. 62 (4), 199–217.

Collodi, J., M'Cormack, F., 2009. Population Growth, Environment and Food Security: What Does the Future Hold? IDS Knowledge Series, Horizon, pp. 1-4.

DGWLM, 2008. Draft of Governmental Regulation on Permanent Foodstuff Land Protection. Retrieved May 2009, from http://setjen.deptan.go.id/pla/berita_detail.php (in Indonesian).

Ding, C., 2003. Land policy reform in China: assessment and prospects. Land Use Policy 20, 109–120.

Doos, B.R., Shaw, R., 1999. Can we predict the future food production? A sensitivity analysis. Glob. Environ. Change 9, 261–283.

Doygun, H., 2009. Effects of urban sprawl on agricultural land: a case study of Kahramanmaraş, Turkey. Environ. Monit. Assess. 158 (1–4), 471–478.

FAO, 2008. The State of Food Insecurity in the World 2008. Food and Agriculture Organization of the United Nations, Rome.

FAO, 2009. Global agriculture towards 2050. High Level Expert Forum – How to Feed the World in 2050. Agricultural Development Economics Division Economic and Social Development Department, Rome.

Firman, T., 2000. Rural to urban land conversion in Indonesia during boom and bust periods. Land Use Policy 17 (1), 13–20.

Firman, T., 2004. Major issues in Indonesia's urban land development. Land Use Policy 21 (4), 347–355.

Firman, T., 2011. Indonesia's Urban Development Challenges. Retrieved 11 February 2011, from http://www.thejakartapost.com/news/2011/02/05/indonesia%E2%80%99s-urban-development-challenges.html.

Geogeghan, J., Pritchard, J.L., Ogneva-Himmelberger, Y., Chowdhury, R.R., Sanderson, S., Turner II, B.L., 1998. Socializing the pixel and pixelizing the social in land-use and land-cover change. In: Liverman, D.M., Moran, E.F., Rindfuss, R.R., Stern, P.C. (Eds.), People and Pixels Linking Remote Sensing and Social Science. National Academy Press, Washington, DC, pp. 51–69.

Heilig, G.K., 1999. Can China feed itself? – a system for evaluation of policy options. International Institute for Applied System Analysis, 18 September 2010. Retrieved from http://www.iiasa.ac.at/Research/LUC/ChinaFood/index.htm, accessed September 2010.

Hilferink, M., Rietveld, P., 1999. Land use scanner: an integrated GIS based model for longterm projections of land use in urban and rural areas. J. Geogr. Syst. 1, 155–177.

Iqbal, M., Sumaryanto, 2007. Strategy for Controlling Agricultural Land Conversion Relied on Societies' Participation. Analisis Kebijakan Pertanian 5 (2), 167–182 (in Indonesian).

Irham, 1993. Efficacy of Government Policy on the Preservation of Agricultural Land Use: A Case Study in Yogyakarta, Indonesia. Master of Science Thesis No. HS-93-1 ed. Asian Institute of Technology, Bangkok.

Irwin, E.G., Geoghegan, J., 2001. Theory, data, methods: developing spatially explicit economic models of land use change. Agric. Ecosyst. Environ. 85 (1–3), 7–23.

Jantz, C.A., Goetz, S.J., 2005. Analysis of scale dependencies in an urban land use change model. Int. J. Geogr. Inf. Sci. 19 (2), 124–133.

Kaimowitz, D., Angelsen, A., 1998. Economic Models of Tropical Deforestation – A Review. Center for International Forestry Research (CIFOR), Bogor.

Lambin, E.F., 1997. Modelling and monitoring land-cover change processes in tropical regions. Prog. Phys. Geogr. 21 (3), 375–393.

Lambin, E.F., Rounsevell, M.D.A., Geist, H.J., 2000. Are agricultural land-use models able to predict changes in land-use intensity? Agric. Ecosyst. Environ. 82 (1–3), 321–331.

Lambin, E.F., Geist, H.J., Lepers, E., 2003. Dynamics of land-use and land-cover change in tropical regions. Annual Review of Environment and Resources 28, 205–241.

Luo, G., Yin, C., Chen, X., Xu, W., Lu, L., 2010. Combining system dynamic model and CLUE-S model to improve land use scenario analyses at regional scale: a case study of Sangong watershed in Xinjiang, China. Ecol. Complex. 7 (2), 198–207.

Murningtyas, E., 2006. Strategy for Controlling Agricultural Land Conversion. Directorate of Food and Agriculture, Ministry of National Development Planning – BAPPENAS, Jakarta (in Indonesian).

Nagendra, H., Munroe, D.K., Southworth, J., 2004. From pattern to process: landscape fragmentation and the analysis of land use/land cover change. Agric. Ecosyst. Environ. 101 (2–3), 111–115.

Nasoetion, L.I., 2003. Land problematics and its relation to soil science education in Indonesia Berita HITI. Indonesian Soc. Soil Sci. News 10 (26), 9–19 (in Indonesian).

Nasoetion, L.I., Winoto, J., 1996. Problem of agricultural land conversion and its impact on sustainability of food self-suficiency. In: Paper Presented at the Workshop of Competition in Utilization of Land and Water. Bogor (in Indonesian).

Nellemann, C., MacDevette, M., Manders, T., Eickhout, B., Svihus, B., Prins, A.G., et al., 2009. The Environmental Crisis-The Environment's Role in Averting Future Food Crises-A UNEP Rapid Respond Assessment. UNEP/GRID-Arendal, Arendal, Norway.

Nelson, G.C., 2002. Introduction to the special issue on spatial analysis for agricultural economists. Agric. Econ. 27 (3), 197–200.

Overmars, K.P., Verburg, P.H., 2005. Analysis of land use drivers at the watershed and household level: linking two paradigms at the Philippine forest fringe. Int. J. Geogr. Inf. Sci. 19 (2), 133–144.

Overmars, K.P., de Groot, W.T., Huigen, M.G.A., 2007. Comparing inductive and deductive modeling of land use decisions: principles, a model and an illustration from the Philippines. Hum. Ecol. 35 (4), 439–452.

Pakpahan, A., Kartodihardjo, H., Wibowo, R., Nataatmadja, H., Sadjad, S., Haris, E., et al., 2005. Developing Indonesian Agriculture: Work, Prestigious, and Prosperity. IPB Alumni Society, Bogor (in Indonesian).

Partoyo, 2010. Spatial and statistical analysis of the determinants of farmland conversion in sub-urban area of Yogyakarta, Indonesia. In: Paper Presented at the International Seminar "The Role of Indonesian Students in the Scientific Development in Indonesia", Bangkok, 24–25 September 2010. ISBN: 978-616-90749-0-8, pp. 126–130.

Partoyo, 2011. Land Use Options and Strategies for Food Availability in Yogyakarta, Indonesia. Unpublished PhD Dissertation, Asian Institute of Technology, Thailand.

Partoyo, 2012. Application of spatial modeling for simulating land use conversion to support sustainable agricultural land planning. In: Proceedings of National Seminar on Agriculture and Fisheries Research, Year 2012. Yogyakarta, 15 September 2012. Faculty of Agriculture Gadjah Mada University. ISBN: 978-979-8678-25-7, pp. 575–582 (in Indonesian).

Partoyo, Shrestha, R.P., 2013. Monitoring farmland loss and projecting the future land use of an urbanized watershed in Yogyakarta, Indonesia. J. Land Use Sci. 8 (1), 59–84.

Perumnas, 2010. History of Perum Perumnas (National Public Housing). Retrieved 18 September 2010, from http://www.perumnas.co.id/site/profile.

Pfaff, A.S.P., 1999. What drives deforestation in the Brazilian Amazon? Evidence from satellite and socioeconomic data. J. Environ. Econ. Manag. 37, 25–43.

Pijanowski, B.C., Pithadia, S., Shellito, B.A., Alexandridis, K., 2005. Calibrating a neural network-based urban change model for two metropolitan areas of the upper midwest of the United States. Int. J. Geogr. Inf. Sci. 19 (2), 145–153.

Pontius, R.G., Chen, H., 2006. GEOMOD Modeling: Land-Use & Cover Change Modeling. Retrieved from http://www.clarku.edu/~rpontius/pontius_chen_2006_idrisi.pdf, accessed August 2010.

Pontius, R.G., Malanson, J., 2005. Comparison of the accuracy of land change models: Cellular Automata Markov versus GEOMOD. Int. J. Geogr. Inf. Sci. 19 (2), 103–113.

Pontius, R.G., Schneider, L.C., 2001. Land use change model validation by an ROC method for the Ipswich watershed, Massachusetts, USA. Agric. Ecosyst. Environ. 85, 239–248.

Pontius Jr., R.G., Cornell, J., Hall, C., 2001. Modeling the spatial pattern of land-use change with Geomod2: application and validation for Costa Rica. Agric. Ecosyst. Environ. 85 (1-3), 191–203.

Rajan, K.S., Shibasaki, R., 1998. A New Concept in Modelling Land Use Land Cover Change. GIS Development, ts8, 1–7.

Ritohardoyo, S., 2001. Housing development and land conversion at the surrounding area of Yogyakarta city. J. Geografi UMS 15 (1), 74–89.

Santosa, U.A., Idris, U., 2009. Land Reform: Government to Distribute Land. Retrieved 19 August 09, from http://regional.kompas.com/read/xml/2008/05/29/11104460/2009.pemerintah.bagikan.tan ah (in Indonesian).

Schweik, C.M., Thomas, C.W., 2002. Using remote sensing to evaluate environmental institutional designs: a habitat conservation planning example. Soc. Sci. Q. 83 (1), 244–262.

Serneels, S., Lambin, E.F., 2001. Proximate causes of land-use change in Narok district, Kenya: a spatial statistical model. Agric. Ecosyst. Environ. 85 (1–3), 65–81.

Seto, K.C., Woodcock, C.E., Song, C., Huang, X., Lu, J., Kaufmann, R.K., 2002. Monitoring land-use change in the Pearl River Delta using Landsat TM. Int. J. Remote Sens. 23 (10), 1985–2004.

Simatupang, P., Irawan, B., 2002. Controlling agricultural land: review on permanent agricultural land policy. Paper Presented at National Seminar on Multifunction and Conversion of Agricultural Land, 25 October (in Indonesian).

Sumaryanto, Sudaryanto, T., 2005. Comprehension of negative impacts of rice field conversion as the basis of formulation of control strategy. In: Paper Presented at Seminar on Managing Land Conversion and Accomplishment of Permanent Agricultural Land. Jakarta, 13 December (in Indonesian).

II. THEORETICAL ISSUES

Syamsiar, S., 2013. Rice production and land resources availability for strengthening food sufficiency in Yogyakarta Special Region. J. Soc. Econ. Agric. Agribussiness 9 (2), 183–189 (in Indonesian).

Tarigan, W.N., 2013. Mapping of Agricultural Land Use Change in Banguntapan Sub-District, Bantul Regency, Yogyakarta, Year 1999–2012. Unpublished Thesis, Universitas Pembangunan Nasional "Veteran" Yogyakarta, Yogyakarta (in Indonesian).

Trisurat, Y., Alkemade, R., Verburg, P.H., 2010. Projecting land-use change and its consequences for biodiversity in Northern Thailand. Environ. Manag. 45 (3), 626–639.

Turner-II, B.L., Moss, R.H., Skole, D.L., 1993. Relating land use and global land-cover change: a proposal for an IGBP-HDP core project. Stockholm: IGBP-HDP Working Group on Land-Use/Land-Cover Change. HDP Report No. 5/IGBP Report No. 24.

Turner-II, B.L., Skole, D., Sanderson, S., Fischer, G., Fresco, L., Leemans, R., 1995. Land-use and land-cover change: Science/Research Plan IGBP Report 35/HDP Report 7 (Vol. Report No. 35). HDP Secretariat, Geneva.

UN, 2008a. Shifting populations. DESA News 12 (04), 3–6.

UN, 2008b. World population prospects: The 2008 revision. Retrieved 01 October, 2009, from http://esa.un.org/unpp.

Undang, K., 2003. Indonesia. In: Willardson, L.S. (Ed.), Impact of Land Utilization Systems on Agricultural Productivity. Asian Productivity Organization, Tokyo, pp. 174–185.

Veldkamp, A., Fresco, L.O., 1996. CLUE-CR: an integrated multi-scale model to simulate land use change scenarios in Costa Rica. Ecol. Model. 91 (1–3), 231–248.

Verburg, P.H., Overmars, K.P., 2009. Combining top-down and bottom-up dynamics in land use modeling: exploring the future of abandoned farmlands in Europe with the Dyna-CLUE model. Landsc. Ecol. 24 (9), 1167–1181.

Verburg, P.H., Veldkamp, T.A., Bouma, J., 1999. Land use change under conditions of high population pressure: the case of Java. Glob. Environ. Change 9 (4), 303–312.

Verburg, P.H., Soepboer, W., Veldkamp, A., Limpiada, R., Espaldon, V., Mastura, S.S.A., 2002. Modeling the spatial dynamics of regional land use: the CLUE-S Model. Environ. Manag. 30 (3), 391–405.

Verburg, P.H., Schot, P.P., Dijst, M.J., Veldkamp, A., 2004. Land use change modelling: current practice and research priorities. GeoJournal 61 (4), 309–324.

Widjanarko, B.S., Pakpahan, M., Rahardjono, B., Suweken, P., 2004. The agrarian aspect in controlling the conversion of agricultural (paddy field) land. In: Paper Presented at the National Seminar on Multifunction of Rice Field, Bogor, Indonesia (in Indonesian).

Winoto, J., 2005. Land Holding Must Be Restructured. Retrieved 8 February 2011, from http://us.detiknews.com/read/2005/12/26/074033/508731/158/penguasaan-tanah-harus-ditata-ulang (in Indonesian).

Winoto, J., 2010. Over 50 Years No New Agrarian Law Established. Retrieved 9 February 2011, from http://regional.kompas.com/read/2010/09/24/03401714/Selama.50.Tahun.Tak.Ada.Produk.Hukum.Agraria.Baru (in Indonesian).

Xiang, W.N., Clarke, K.C., 2003. The use of scenarios in land-use planning. Environ. Plan. B 30, 885–909.

Yang, H., Li, X.B., 2000. Cultivated land and food supply in China. Land Use Policy 17, 73–88.

Yunus, H.S., 1990. Searching new strategies for managing and controlling urban land growth: a preliminary outlook on Indonesia. Indones. J. Geogr. 20 (60), 1–10.

Yunus, H.S., 2009. Urban Land Development. Gadjah Mada University Press, Yogyakarta.

Social Insecurity, Natural Resources, and Legal Complexity

F. von Benda-Beckmann[*,†,a], *K. von Benda-Beckmann*[*,†]

*Max Planck Institute for Social Anthropology, Halle, Germany †Martin Luther University Halle-Wittenberg, Halle, Germany

6.1 INTRODUCTION

With Étienne Le Roy we share an interest in the social and political consequences of law in plural legal settings. His work on land issues in West Africa shows many comparable parallel interests with our own work in Indonesia. We therefore consider it appropriate to discuss some issues related to land and legal complexity in a volume dedicated to him. This contribution builds upon the work we started in Wageningen and Rotterdam and on which we have worked in collaboration with a number of universities in Indonesia.[1] Each of the topics of the title—legal complexity, social insecurity, and sustainable use of natural resources—stands by itself for a very complex set of issues. We think that there are compelling reasons to bring these issues together and not treat them in isolation. Indonesia faces important policy decisions on how and by whom natural resources are to be controlled and exploited. This was one of the most sensitive political issues instrumental in bringing down the Suharto regime and now is a crucial issue in renegotiating the relations between the center and the regions, as well as Indonesia's international relations. In these political fights concerns for social security hardly play a role at all. Yet the issues are intimately related. Despite the increasing body of literature on these issues, there are important empirical gaps as well as many shortcomings in the theoretical work on these issues. There is a need to question and clarify the theoretical plausibility of assumptions built into policy objectives and legislation that connect specific types of land and water rights to specific patterns of economic performance or to more or less sustainable

---

[a] Deceased.

[1] The research project "Legal Complexity, Ecological Sustainability and Social (In)Security in the Management and Exploitation of Land and Water Resources in Indonesia" was carried out by the Department of Agrarian Law of Wageningen Agricultural University and the Faculty of Law of Erasmus University Rotterdam in cooperation with Universitas Andalas in Padang, Universitas Pattimura in Ambon, Universiats

uses of natural resources. Theories on the relationship between economic growth and redistribution of wealth for more social security for the population also should be scrutinized.

We shall start by sketching the socioeconomic background and a brief outline of the main problems in each of the three themes. After that we present some of the main debates around these issues and formulate our critique. Then follows an outline of an appropriate approach[2] followed by conclusions.

6.2 THE PROBLEMS: SOCIAL AND LEGAL INSECURITY, UNSUSTAINABLE RESOURCE USE

6.2.1 A Functional Approach to Social (In)security

To overcome these shortcomings we have advocated an approach that provides a more adequate basis for discussing the actual social security needs of people, the totality of arrangements upon which they can draw, and analyzes possible changes. To that end we define social insecurity/security and the ways of dealing with it as an analytical field of problems that empirically is filled by those arrangements developed in a society that provide or aim to provide goods and services for overcoming these insecurities for those who cannot do this on their own: the arrangements become "functional" for social security. Social security manifests itself at different, analytically distinguishable "layers" of social organization, in the realm of cultural ideas, legal regulation, and social relationships that become interrelated in social practices. Such an analysis includes both relationships and institutions specifically set up for social security purposes, and relationships developed for and serving different purposes while also serving social security purposes. As with property relations, relationships upon which social security is based are often multifunctional. Relationships of help and care are structured among other things on the basis of kinship and shared rights to natural resources (eg, to *pusako* land or village land), but are possibly also connected to membership in religious organizations or political parties. Ordinary economic cooperation cannot always sharply be distinguished from social security. Sharing risks may go hand in hand with economic gain, whether at the level of village social organization or at the level of national policy making. State social security policies and programs are usually also multifunctional and depend on multifunctional relationships. Whether or not a village receives government help or becomes an Inpres Desa Tertinggal (IDT) village often depends more on the relationships of influential village leaders with state agencies than on the actual economic conditions in the village. And governments and other political organizations often use social security projects as a means to win political support or votes in elections. The idea of multifunctionality of relationships and institutions allows us to critically assess whether and to what extent certain institutions perform a social security function, rather than taking such functions for granted. It also allows us to look into the specific mixes of social security people have or try to make on the basis of the various relationships with potential providers of social security, whether these are kinsmen, neighbors, religious institutions, nongovernmental organizations (NGOs), state agencies, or commercial insurance companies.

---

[2] These ideas build upon a number of earlier publications, in particular von Benda-Beckmann and von Benda-Beckmann (1994a,b, 1995, 1998, 1999).

This approach captures the many variations we encounter in this field. Notions of neediness vary according to socioeconomic conditions. Middle-class Jakarta people will differ greatly from peasants in Sulawesi, Ambon, Minangkabau or in Kalimantan, Irian Jaya, or Sumba, in their conceptions of who needs to be taken care of. Different societies also have different standards according to which a person is considered to be able to work or to earn an income from work and which person cannot earn an income out of work. Among other things, this has to do with notions of health, work pressure, wage labor and other types of work, and competition on the labor market. It also has to do with the legal regulations regarding work and social security. Moreover, in an economy in which a large part of the population derive an income and make a living from several jobs or several different sources, there is often no clear distinction between working and unemployment or disability. More often it is a question of more or less. An old peasant gradually reduces the amount of work; a young person losing one job may still have several other odd jobs. This does not mean this person can fully make a living on the remaining jobs, but he or she is not entirely unemployed either. As a result, the distinction between healthy and sick is sharper in Europe or for the highly paid full-time employed in the formal sector in Indonesia than in rural Indonesia.

Moreover, ideas of what constitutes "need," under which conditions persons have a right to support, which persons and institutions are obliged to provide help and support in such situations, and what kind of support this should be, are defined differently in government regulations, religious law, and the various adat systems. The national legal system, through the Department of Social Affairs, set up social programs for needy categories: widows, disabled persons, orphans, and the elderly. In Ambon, widows participated in the projects despite the fact that they were not at all considered needy according to adat, because they inherited their own family land (von Benda-Beckmann, 1988). Islamic law has yet another concept of neediness for widows. The condition of widowhood in itself is not considered a reason for support. Because women are allowed to have property or a business of their own during marriage, a widow needs only support for a brief period of 3 months, needed to adjust to the new situation. Beyond that, she is supposed to continue her business as before. There is no need for further support, unless for some reason she cannot work or has no business.

For a proper assessment of social security, it is useful to distinguish the mechanisms of social security from the conditions for social security; that is, the availability of resources in sufficient quantity and quality that can be used for providing support. The social security mechanism that requires close kin to take care of each other is meaningless in the absence of close kin. Likewise, the mechanism that *pusako* or village land can be used for social security of a kin group or village is of no use for those who have no *pusako* or village land at all, or if the soil is too infertile. And the provision of a lump-sum payment by governmental or commercial agencies may be meaningless if inflation rates are too high. In addition, for a well-functioning system it is critical that available resources are indeed channeled into social security arrangements, instead of being withdrawn from them. The conditions for social security pose a problem at all levels of social organization because of resource competition. There is competition among ministries between the social security sector and other sectors. Within the ministries these budgets must be divided further and allocated to different sections, and further down to province and district offices. Similar competition for resource allocation occurs within villages and families. For example, a family may be faced with the choice between using land for agricultural production or selling it to finance a good education for their children, who may later be able to take better care of their parents. Converting *pusako* rights into private ownership channels resources away from social security of the larger kin group. Migration of young adults may leave old and needy parents' home alone without the necessary care. Resource competition may even take place between different organizational mechanisms with a social security function: for fulfilling kinship obligations, for investment in rotating credit organizations, for the investment in one's childrens' educational career, between social assistance programs or contributions to insurance schemes.

Security is also a state of mind. Someone's feelings of security or insecurity are based on a combination of past experiences and on promises for the future embodied in existing mechanisms that reach into the future, in particular rights and the continuing availability of resources. The evaluation of material and psychological conditions hinge on what people experience in the present and on potentialities; on the perceived likelihood of future threatening occurrences and on the social relationships and resources that might be tapped into if needs arise, now and in the future. These potentialities are a component of a person's social security although they may never be mobilized in social interaction and never lead to actual transfers of goods and services. Social security thus combines past, present, and future.[3] To make a relevant assessment of the social security situation in a country or place, we have to look at it ex post facto, consider the conditions, including the functioning of social security mechanisms, as they have performed up till the present, and use this for future projections. But we also have to look ex ante, try to get a better grasp of the potentialities, and spell out the scenarios for possible futures.

To have such an insurance, whether it is formed by reserve assets, by land rights, by kinship relationships and their attached obligations, or by a commercial insurance contract, is one thing, but whether these can be maintained or successfully mobilized tomorrow or in 10 years is uncertain. None of these relationships functions automatically. The help and care that is provided through kinship ties is generally not voluntary and for free, it is usually not an act of charity or of pure affection, though charity and affection of course do play a considerable role, if only in the quality and extent of care. The idea of sharing risks is based on the idea that everyone contributes and only those for whom the risk becomes an actual problem will receive benefits, whether in modern industrial societies or in traditional, family-based systems of mutual help. All such arrangements require investments, though the investments are of different kinds. In social security within the family or village setting, the investment is not primarily in cash, though it often does involve cash payments, but mainly in creating and maintaining good relationships. This requires paying "social" or "political premiums," and those who do not pay them are likely to be excluded from the networks of support (see Vel, 1994).

Migration poses particular problems for social security, both for migrants and those who stay behind. There are clear connections with rights to natural resources as well. When people look for other sources of income and migrate to other places, common origin from the same village or region is often a basis on which migrants structure migrant organizations. In some respects migration may become an important source of social security for those staying behind: Many new and more comfortable houses are built with money from migrants. Many

---

[3] Sen (1987) talks about "opportunity sets" and "outcome sets" to indicate the difference between actualized and potential resources for social security.

migrants support relatives or members of their home village in various ways: by paying doctors' bills, by securing places in schools, by redeeming pawned property, and by mediating all sorts of services. Influential migrants provide information and often take a leading role in disputes about family land, whether with other families in the community or with government. It is not infrequent that relationships with NGOs dealing with problems of sustainability or land claims are mediated through migrants from the village. Since the fall of the Suharto regime, many migrants have set up foundations related to social security in their home village. Migration may be induced by very different motives. In some areas it is a response to demographic pressures and poverty: people are forced to leave because there is not enough land to make a living. Minangkabau, however, has a long tradition of migration for education and learning. Many families sell, pawn, or rent out land to finance the education and migration of their sons and daughters. Those who are successful may later redeem that land, but not all migrants are successful and many sever all connections over a long period of time. The hope and promise that investments will be returned in the form of social security for the remaining do not always become true. The family that stays behind has less land and fewer strong hands to work the land. For them migration means more stress and sometimes hardship. Migration also may decrease social security for those staying behind, especially because the age categories that are the most important providers of social security care move away.

Migrants also often continue to rely on their home village for social security. Rights to natural resources are an important basis for social security for the migrants. Wealthier migrants often build houses in their home village in which relatives are allowed to live, often with the tacit understanding that they themselves will have a place to live back home in their old age (von Benda-Beckmann and von Benda-Beckmann, 1995, 1998). Motives and needs of social security glue migrants and their relatives back home together as much as love and belonging.

6.2.2 A Brief Sketch of the Socioeconomic Background

Over the past 50 years, socioeconomic developments in Indonesia have been characterized by a mixture of market-oriented and self-supporting agriculture, industrialization and urbanization, large-scale forest exploitation, migration, increasing demographic pressures, and concomitant changes in family relationships and kinship ties. Under the Suharto regime, the Indonesian state obtained a leading and characteristic role in economic development, which is being reevaluated since the reforms started. Despite the dominance of neoliberal ideologies with their decentralization and privatization policies, there has been an increase rather than a decrease of government intervention in natural resource management, redirecting large amounts of resources away from local communities to the government. In the past 25 years, this increase has only accelerated. The increasing outreach of the state administrative system, the discovery and greater feasibility in exploitation opportunities of natural resources caused by better infrastructure and technology have put resource environments under pressure, which until the 1960s and 1970s had been relatively untouched by the world market economy, and which had been subject to more moderate exploitative demands of the population groups living close to these resources and earning their subsistence from them. At the same time, more and more elements of the natural environment have obtained an explicit economic value outside the realm of agriculture. They are now regarded as "resources" and have been brought under new property regimes. Even when these new valuables are made marketable, the government retains a strong influence in defining the extent and range of such rights and the way such markets operate. These developments call for an analysis of the implications for social security and sustainable resource use.

As in many other developing countries, processes of economic growth in Indonesia are characterized by continuous processes of differentiation and redistribution. But the direction of this development has been from the poorer to the richer, not the other way around as proponents of the trickle-down theory predicted. Social policies have in general added to this differentiation (see Midgley, 1984). These developments have implications for each of the larger sets of problems we are dealing with in this project. The benefits and burdens of these developments are not equally distributed, not among the present generation, nor between the present and future generation of Indonesians. For example, the government's decision to grant a concession to a logging or mining company, or even the aggressive pressure of international NGOs to create a natural reserve may all of a sudden deprive a local community of part of its resource base, upsetting the ways in which they had locally organized resource use and imposing new constraints on their attempts to preserve the sustainability of these resources. Those having lived off these natural resources, and claiming these resources as their own, often gain little while big profits are made by small, more remote political and economic elites and by transnational corporations. Only a fraction of these profits is rechanneled into the state budget. Moreover, state expenditures for social services in Indonesia have remained very limited compared with other states in the third world (Mesa-Lago, 1992). It is unlikely that this will change, given the structural adjustment policies by the World Bank and International Monetary Fund (IMF), which advocate decrease state expenditures in the social sector. All these developments together reduce the available resources—in terms of quantity as well as quality—that can be used in local transfers of goods and services for social security. Obviously, this affects populations in different ways, at different scales and localities. The political and economic crisis following the end of the Suharto regime has exacerbated the situation for many, although the effects differ widely across Indonesia.

6.2.2.1 *Social Security*

Generally speaking, these developments have led to changing social and material conditions for social security: population pressure on land in combination with the large extracting companies and the establishment of natural reserves decreases the possibility to keep land in reserve for times of need for the local population. Education has become increasingly important, as a way to advance economically, and at the same time as a way to ensure social security for the dependents of those who advance economically. Education thus has become an important social security strategy for parents, but it has proven to be an insecure strategy.

While large sums are paid to provide education, it no longer automatically leads to good jobs (von Benda-Beckmann and von Benda-Beckmann, 1998). Likewise, migration has important implications for the social security situation both of migrants themselves and of the relatives who stay behind. Among the most important consequences is the fact that the circles of solidarity have changed in size and character. As relationships among the wider kin have become weaker, the bonds of mutual rights and obligations of social security have also weakened. For some, this has led to a decrease in the level and quality of social security. Others have managed to build new networks that fill in these functions. For example, patron-client networks between civil servants and relatives in home villages, or some rotating savings groups, prayer groups of women, or religious or secular foundations have taken over some of the social security functions that were earlier organized in kin groups. The problem is that not everyone has equal access to these new types of solidarity networks that all have their specific rules of inclusion and exclusion and lead to new forms of social stratification. The extent to which the Indonesian state has assumed responsibility for the social security of the population is limited and the provisions are very unevenly distributed. For the rural population and those working in the informal economic sector in urban and semiurban areas, the state provides very little security. In fact, for most ordinary people, the state has in many respects proven to be a major source of insecurity.

6.2.3 Sustainability

Looking at the developments in Indonesia, there are immense changes in the quantity, quality, and distribution of Indonesia's "natural capital." Increasing overexploitation of natural resources is not always compensated by regenerating them. Deforestation is not always followed by reforestation. Many agricultural areas have been subject to overapplication of chemical fertilizers. Land and water resources in many areas are increasingly polluted by industrial plants and mining operations. Improvements of infrastructure and extraction techniques and expanding markets have created additional opportunities for extraction by others than concession holders and local populations, all of which has exacerbated the stress upon natural resources. In various ways, the quality of land, water, and forests, soil fertility is threatened, the renewability of resources is no longer guaranteed, and the depletion of nonrenewable resources is approaching at great speed. As a result, the resource base, both for economic development and for social security purposes, is contracting and severely weakening and the negative economic and ecological externalities are passed on to future generations.[4]

6.2.4 Conflicts

These developments have led to new kinds of conflicts, at different levels of social organization and in rapidly changing constellations of interest groups. At small-scale local levels, pressure on the resources has increased and fueled conflicts over natural resources as a means of production for the market as well as a means for subsistence. The political and economic value of natural resource property rights has also increased interethnic tensions and struggles. As a consequence of spontaneous and planned transmigration, members of different ethnic groups are directly involved in struggles over natural resources. Although the interethnic conflicts in Indonesia cannot be reduced to economic conflicts, and economic conflicts cannot be reduced to conflicts over natural resources only, it is nevertheless clear that such conflicts do play an important role, and may continue to do so in the future. Conflicts between local populations or communities and state agencies, and with private enterprises ("investors") who derive their exploitation rights from government concessions and licenses have also increased. Moreover, the question to whom and in which regions profits from natural resources exploitation are to be reallocated puts the relationship between central and

---

[4]See Berkes and Folke (1998: 4). WCED (1987) defined sustainability as "development that meets the needs of the present without compromising the ability of future generations to meet their own needs."

regional government under great stress. In some regions, such struggles for control over natural resources are increasingly connected with struggles for self-determination or political autonomy. The Indonesian nation-state is challenged by the emergence of assertive ethnic and local politics striving for the recognition and consolidation of regional and/or ethnic economic and political interests. Since the end of the Suharto regime, conflicts over natural resources are openly carried out in a wide range of public arenas such as the media, political fora, and the courts.

6.2.5 Legal Complexity and Legal Insecurity

These problems are aggravated by the complex constellations of law and the high extent of legal insecurity. Legal pluralism, the concept indicating the condition that more than one legal system or institution coexists with respect to the same set of activities and relationships, has a long history in Indonesia. Legal pluralism shows up clearest in the coexistence of government law, adat, adat law, and religious law, mainly Islamic law, but there are also new forms of local legal regulations ("unnamed law") that cannot be subsumed under these larger systems. In addition, international law and conventions that have introduced human rights issues, regulate environmental issues, access and exploitation rights, and with respect to the legal status and political and economic rights of "indigenous peoples" have come to play an increasingly important role.[5] The legal demarcation of the spheres of respective validity of these subsystems has always been dynamic and contested, and the actual significance of the legal systems for socioeconomic organizations has changed with the economic and political power of their proponents. The reform policies and the ensuing decentralization have triggered intensive debate and contestation of the relative place of adat and state law (von Benda-Beckmann and von Benda-Beckmann, 2001, 2004).

We speak about legal insecurity in three meanings. First, given such plural legal order it is often not certain what legal concepts and regulations become really relevant for defining the legal status of natural resources. It is often unclear what rights and obligations, based in which type of law, people have with respect to natural resources or with respect to social security. Adat notions about rights over village land, for instance, not only conflict with state legal ideas about *tanah negara* (national land) but also with *hutan negara* (state forest). Two or more different state departments, each with its own legal concepts and regulations, may be involved in defining rights and obligations to land. The pluralism of legal procedures and substantive law provides opportunities for "forum shopping" in which parties seek the optimal solution for their legal problem (von Benda-Beckmann, 1981).

Second, even if it is known which legal order will prevail, it is not always clear what exactly the substantive rules of these subsystems are. This is true as much for state legislation as for rules of adat. The legal subsystems themselves are by no means well-integrated and consistent wholes. At the state level, property law, agrarian law, agricultural development, mining, forestry, nature protection, and conflict management are entrusted to different policy making and implementing institutions whose laws and regulations often express distinct and conflicting interests. Also, local forms of customary and folk legal regulation

---

[5] On UN and ILO conventions and the Rio Declaration, see Van de Fliert (1994); von Benda-Beckmann et al. (1994); von Benda-Beckmann (2001b).

are full of ambiguities. There are often different and contradictory versions of such laws, depending on the context in which they are used. In many areas, hybrid legal institutions have developed that amalgamate elements of adat and Islamic law, or adat and state law. Finally, in individual legal struggles, parties as well as decision makers usually make a very selective and strategic use of all kinds of legal arguments to justify their claims, not caring much whether such argumentation would fit the logic of lawyers, *ulama*, or adat experts.

Third, even if the relevant rules should be rather clear, there is a high level of insecurity with respect to the sincere application of such rules to decide on problematic situations and conflicts. Undue political and economic pressure may influence decisions and subvert the law (corruption). And generally, even when court decisions have been taken in the proper way, it often is still uncertain whether they can be executed.

Despite the complexities and insecurities surrounding law and its manifold uses and abuses, law and legal procedures—whether based in state legislation, in adat, or in religion—are factors of importance in the constellation of problems we are discussing. Legal concepts, institutions, and procedures legitimate social power. They constitute crucial social resources in people's strategies, negotiations, and struggles over the rights to natural resources and over the function natural resources should have, both at the village level and at the level of national and regional policy making and legislation. Whether law is looked at with idealism or cynicism and even if it is abused and corrupted, it remains a factor taken seriously by all actors and interest groups concerned, as the heated discussions over legislation on decentralization and autonomy, about hak ulayat, and about the elections clearly show.

6.3 DEBATES

Of course these issues are not new and they have been frequently discussed; but they are discussed in three quite separate debates:

1. The relationship between property rights and economic development.
2. The relationship between property rights and sustainable natural resource management.
3. The relationship between economic development, social security, and natural resources.

Despite the important insights that have come out of these debates, we feel that the way they have been discussed is problematic. Often the themes are discussed in relative isolation and fail to grasp the ways in which they are connected. Besides, they are frequently guided by unwarranted assumptions. Briefly touching upon the historical roots of the present debates, we discuss these main theoretical assumptions and make some constructive suggestions for the development of research questions and comparative analysis.

6.3.1 Economic Development and Property Rights

Academic and policy discussions are strongly dominated by the assumption that economic modernization and growth needs to be based upon individual ownership rights to the means of production. This has its historical background in the economic developments in the late

18th and 19th century in Europe and formed the core of colonial land policies.[6] After independence and decolonization, and with increasing appreciation of social and racial equality, there were attempts to create a more equal distribution of property rights. It was one of the original objectives of the Basic Agrarian Law of 1960 to introduce minimal and maximal sizes for land ownership. These processes, however, were mostly stopped or at least slowed down, both from the outside by international trade and development policies, as well as from within by the emergence of local economic and political elites who were inimical to the economic and political instability that such large-scale redistribution might have (Bachriadi et al., 1997). Large-scale redistribution of rights to natural resources was associated with socialism or communism and with political instability. Since the beginning of the New Order government by Suharto in 1965, neoliberal economic modernization and development became the dominant paradigm. Rights to natural resources as defined in various adat systems were regarded as an obstacle to economic development. Modernization was pursued along two roads. First, an ever-growing number of laws and regulations expanded and consolidated state ownership and control over natural resources.[7] Second, like most other colonial and postcolonial governments in the third world, the Indonesian government pursued an agrarian legal policy aiming at converting traditional or customary adat rights to land into new categories of rights largely modeled after European legal systems, in this case Indonesian Dutch law. The assumption was that legal reform creating marketable individual private ownership rights would significantly contribute to economic growth. It would create greater legal security, free individuals from communal constraints, and provide collateral to obtain productive credit.

The objectives pursued with these policies were rarely reached and Indonesia largely shares experiences with the land rights reform of many third world countries.[8] Registration has generally met with strong opposition from farmers and the land rights conversion and registration program has remained an "empty dream" for the majority of the rural population outside Java (von Benda-Beckmann, 1986; see also Slaats, 1999). The efforts to accelerate registration through the Proyek Operasi Nasional Agraria (PRONA) introduced in 1981 have not yielded the intended results. To be sure, there are categories of persons who do prefer registration, notably civil servants and immigrants, outsiders to local property networks. These categories are relatively free from the constraints of local networks but at the same time they are insecure about the rights they might have under adat law (see von Benda-Beckmann and Taale, 1992). They cherish the greater freedom and security that an individual ownership title under state law promises to give them. But these new ownership rights do not mean that land held in hak milik (private ownership) or hak guna usaha (long lease) is used for more efficient agricultural production, nor have they necessarily encouraged large-scale credit transactions. In addition, the introduction of new legal rights often merely adds to the already existing legal insecurity, by creating more ambivalence and more room for manipulation. Local village

[6] Developments in Indonesia were very similar to those in other countries. So-called wastelands were declared to be "state domain" and given the status of quasi-state ownership (see Van Vollenhoven, 1919. For other colonies, see Lynch and Talbott, 1995; Okoth-Ogendo, 1984; Peters, 1987; Wiber, 1993).

[7] See Bachriadi et al. (1997). For forest policies, see Peluso (1993).

[8] von Benda-Beckmann (1979); see for Ambon, von Benda-Beckmann and Taale, (1992); for the Batak area, Simbolon (1997. Similar developments have been reported from many African countries after land reform (see Okoth-Ogendo, 1984; Fisiy, 1992, 1997; Hitchcock, 1980; Bruce and Migot-Adholla, 1994).

versions of customary property law thus still coexist with state regulation and continue to influence people's dealings with property.[9]

On the other hand, and contrary to these economic and policy assumptions, there is evidence that economic production can very well be quite effective when based upon adat rights to natural resources. Indonesian producers, before and after colonization, have produced crops for the world market quite successfully. Small holder production has often led to more yields than estate production. When production based on adat rights declined, it was usually more a result of governments' constraints on their production, of bureaucratic marketing organizations, and of price fluctuations than of any constraints inherent in their land rights. There is no clear-cut correlation between a specific type of property right and economic performance.

6.3.2 Property Rights and Sustainable Management of Resources

The assumptions about the relationship between types of property rights and ecological and environmental developments are equally problematic. Since Hardin's influential essay on the "Tragedy of the Commons" appeared in 1968, discussions of property rights have shifted toward the merits of different types of property rights for sustainable resource management. Common property, Hardin had argued, induced rational actors to ever increase resource use because they would gain the full benefits of each additional labor or capital input while sharing the costs of resource use (degradation) with all other users. Without internalization of such environmental costs or coercive action by government, the tragedy of overexploitation was programmed in. But while common property in earlier economic theories had been held responsible for underexploitation, it was now held to lead to overexploitation and degradation of natural resources. Hardin's ideas have been used to advocate an increasing government role in dealing with property and the environment; at the same time they have been used to propagate breaking up common property regimes and converting them into private individual ownership. It was assumed that private individual property rights would internalize costs and benefits, thereby increasing individual responsibility for the environment and rational use of its resources.[10] These assumptions are still widely held, especially by economic development theorists, and continue to guide policies.

At the same time, and largely in critical reaction to Hardin's ideas,[11] a new mainstream literature and policy discourse has emerged, which specifically focuses on the relationships between common property and resource management by local communities on the classical commons such as grazing lands, forests, water, and fishing grounds. This body of literature

[9] The Indonesian government, in collaboration with the World Bank, has since 1995 been engaged in a new massive land administration program to improve the existing agrarian legislation and the registration system (see World Bank, 1994; Slaats, 1999). Thus far the problems of multiple regulation as sketched above have not been effectively redressed.

[10] (Demsetz, 1967; Furobotn and Pejovich, 1972).

[11] See McCay and Acheson (1987) for an early critique of these assumptions. It is now generally accepted that Hardin failed to distinguish open access as a theoretical condition in which there are no relevant institutions from common property as a social institution. Most resource complexes in third world countries such as grazing lands, forests, or water, or the Indonesian *tanah ulayat* (communal land) are more or less institutionalized forms of common or communal property or have been declared state property.

attempts to establish that, and how, communal forms of resource management can be used as a more sustainable, equitable, and effective alternative to both top-down government intervention or radical individual, private property rights.[12] Partly influenced by this literature, the past 20 years have also seen important changes in development cooperation. In particular in circles of NGOs, but also among national donor agencies, a lot of work has been done in developing resource management arrangements that are more adequate than the straightforward and often top-down social engineering that was common before. "Participation," "new partnerships," "comanagement," and "community-based rights" are the various terms of development policies and international conventions that, each with a different focus, reflect the search for new and more adequate policies (Chambers, 1983). Even large donor agencies such as the World Bank have started to be open to these new approaches. The increasing international concern for sustainable use of natural resources has been particularly conducive for reappraising indigenous environmental knowledge and resource management skills.[13] The change has also been a reaction to the many failures of top-down programs that have led to the insight that cooperation with user and interest groups in a more respectful manner than was usual might be a basis for more successful programs. Common or communal ownership and resource management, at least in large parts of the NGO community, thus have lost their scapegoat role, to make room for an opposite image. Local communities seem to be the "natural" protectors of the environment. There have been many attempts to make use of communal management, under the assumption that local communities are usually better equipped than other organizations to manage water or forests.[14]

Though communal property has lost its negative image, the discussions about the relationship between property rights and economic development suffer from oversimplifications. Empirical evidence suggests that there are no clear correlations between types of property rights and sustainable resource use. Overexploitation and degradation can and does occur under all property rights regimes. There are examples of overexploitation of land and forest areas by local people, as a consequence of their traditional economic activities and on the basis of their traditional land rights. But probably the most degrading activities take place on land held in private ownership or in licenses or concessions extended by the government to private individuals or firms.[15] The expectation that private ownership or similarly strong economic use rights would indeed internalize negative environmental effects has often proven to be wishful thinking. In many states, including Indonesia, there is sufficient evidence to show that state property, land, forests, or mineral deposits, whether exploited directly by the state or by private enterprises, is often used rather unsustainably (Balland and Platteau, 1999: 774).

[12] (See McCay and Acheson, 1987; Ostrom, 1990; Schlager and Ostrom, 1992; Spiertz and Wiber, 1996; and many others).

[13] See, for instance, Huijsman and Savenije (1991) and Ghai (1994).

[14] IUCN (1984) and Lynch and Talbott (1995). This development may have reached its peak and may be on the decline (see Agrawal and Gibson, 1999; Leach et al., 1999; see also von Benda-Beckmann and von Benda-Beckmann, 1999; von Benda-Beckmann and von Benda-Beckmann, 1995; von Benda-Beckmann, 1997).

[15] von Benda-Beckmann (2001a). The polluted soils of individually owned Dutch agriculture, the exploitation of oil by Shell in Nigeria, Freeport in Iran, Russian oil pipelines, or the increased overexploitation through overgrazing due to privatization of formerly communal resources in Botswana (see Hitchcock, 1980; Peters, 1987) are all cases in point.

II. THEORETICAL ISSUES

6.3.2.1 Rethinking Property Rights

That there is no clear correlation between types of property rights and sustainability is not surprising. In the first place, sustainability always depends on a combination of factors of which the nature of property rights is but one. The categories of property rights used in these discussions—"open access resources" (or no one's or everyone's property), "private, often individual, ownership," "state or public ownership," and "common/communal property" (see Berkes, 1996: 89)—are too wide and undifferentiated to be helpful for description and analysis.

First, they are selective. The legal property status of a natural resource, such as individual ownership, communal property, state ownership, or open access only, captures a limited number of those rights and obligations that pertain to the resource. It does not give a full account of the total bundle of rights related to the resource in question. While the individual aspects are overstressed in private property, the individual aspects are virtually left out altogether in common or communal property while the communal aspects are overdramatized. What remains are caricatures of both individual and communal property rights.[16]

Second, these categories are umbrella terms that encompass a variety of quite different bundles of rights, which can change greatly over time. For example, the category state or public ownership comprises at least four very distinct types of rights states may have: (1) states on the basis of their sovereignty claim overriding political and regulatory rights with respect to land, water, forests, and subsoil resources; (2) states have assumed not only political but also proprietary rights over large tracts of natural resources termed crown land or state domain and mineral resources[17]; (3) states can also be owners in private law and then are legally treated like any other citizen/owner; and (4) states are an ever-increasing source of "governments largess," out of which quotas, licenses, or concessions are allocated by government on its own terms and held by recipients under conditions that express the "public interest." This has created a new mix of public and private rights with strong political functions (Reich, 1964: 745).

Common or communal property is probably the most varied category. Contemporary discussions are heavily dominated by the proverbial commons, grazing lands, forests, and fishing grounds. But communal properties such as inherited property of larger kin groups such as *harta pusaka*—an important form of property holdings of agricultural land in many parts of Indonesia—usually consist of quite different mixes of individual rights and rights of other social entities. Moreover, individual ownership can mean quite different bundles of rights over time. In European legal systems, for instance, individually owned farmland in the 19th century was quite a different form of "individual ownership" than in the late 20th century, where the range of individual rights has been severely restricted by imposing constraints on the transferability and economic use of the land in the name of the social function of ownership and the protection of the market and the environment.[18]

[16] (See von Benda-Beckmann and von Benda-Beckmann, 1999).

[17] See for the Dutch East Indies, von Benda-Beckmann and von Benda-Beckmann (1999) and Bachriadi et al. (1997).

[18] See for traditional environment protection on Ambon F. von Benda-Beckmann et al., 1995.

Third, discussions about rights are often confusing because they fail to distinguish between "categorical" and "concretized rights." Categorical rights are legal concepts that construct a general relationship of rights, options for behavior, between categories of persons or groups with respect to categories of resources. Examples are the categories of "ownership," *hak milik* (private ownership), *harato pusako* (inherited family property), state land, or state forest. Categorical rules include the general rules and principles that state in general terms under which conditions property rights can be acquired and transferred. In concretized rights, on the other hand, a rights relationship is established between actual persons or groups and an actual resource, where the legal criteria of the rights category are inscribed into a concrete social relationship. We no longer speak of *hak milik* as a type of right, but of Mr. A's *hak milik* in a sawah field, the multinational's concession to a demarcated area of forest or mineral deposit, a descent group's rice land and upland fields. This distinction is important because it enables one to look into the possible relations between categorical rights, say individual private ownership, and the distribution of concretized property rights (ie, wealth). While categorical rights in legal systems tell us little about their social, economic, or ecological significance, concretized rights give us a perspective on the constellation and distribution of property rights (and the wealth that they embody) over the categories, over people, and over resources in actual environmental space. Moreover, it is the constellation of concretized property rights rather than the types of property rights categories that plays an important role for the ways in which people use their property. Only on that basis can one begin to understand how property rights may be related to sustainable resource use and economic development.

Fourth, many of the erroneous economic and policy assumptions turn out to be a result of overlooking the multifunctional character that property rights have for most people or organizations, and certainly for farmers. In many Indonesian villages, land rights in the first place serve as the material basis for the social continuity of groups and not so much immediate maximum production. This is expressed in the Minangkabau saying about "the fruits may be eaten, the water may be drunk, but the stem remains." Natural resources often also have a function for long-term economic and social security; that is, they are kept as a reserve for future needs of the living or future generations.[19] Thus, many civil servants living in their home village in West Sumatra have returned to part-time farming, as their salaries did not rise with the price increases resulting from the economic crisis. Sago palm gardens in Ambon are another example (see von Benda-Beckmann, 1990; Brouwer, 1996). People revert to eating sago and go out fishing in times of economic decline, while eating rice and fish bought from professional fishermen in times of economic growth. The existence of a food reserve not used for immediate commercial production appears to be of great value for social security and has mitigated the otherwise severe consequences of Indonesia's recent economic crisis. Communal village lands in many parts of Indonesia also have the function of providing an area that can be used when the population increases and new land has to be taken into cultivation.

6.3.2.2 *Time Perspective, Distribution of Property, and Sustainable Resource Use*

The significance of law and culture changes and varies with social and economic changes. Looking at concrete property relationships reveals that pusako is not always preserved and

---

[19] (See Agarwal, 1991; Van de Ven, 1994; Chambers and Leach, 1989).

much is given out in *pagang gadai* or rent, or is sold. These relationships also influence what conservation measures may be taken and how land is being used. The rise of smaller family units also tends to reduce people's conceptions of the circle of persons for whom the property should be maintained. In some areas, land fragmentation due to demographic pressures has reached disturbing levels. Many of the present conflicts over land use stem from conflicts between those who want to put land to private use, and other members of the family who stress the importance of land for the family as a whole, in the past, the present, and the future. Which factors then influence the actual functions that property rights have in socioeconomic life? An important factor for the sustainable use of natural resources is a long-time perspective that takes into account not only immediate needs but also those of future contingencies and those of future generations. Generally speaking, the longer the time perspective inherent in the relation between rights holders and property objects, the more likely it is that they will aim at preserving the resource. Variations in time horizon are related to the distribution and composition of wealth and to the availability of opportunities for alternative income. The time perspective a person adopts in relation to specific natural resources depends to a great extent on the range of property objects he has. For instance, a person's use of communal land will be affected by other property rights he may have (such as private ownership or concessions), and vice versa. The extent to which people can use communal land depends on whether they have the technological means to do so, whether they live sufficiently close, and whether they are physically able to do so. Those who have a relatively large amount of production factors (capital equipment, control over labor power, etc.), in general are those who are wealthier, better educated, have better contacts, and have privileged access to these resources and are more capable of exploiting them (Balland and Platteau, 1999: 785; McCay, 1987: 207). In addition, those who have various kinds of property have more freedom in how to use their property and will be able to put some to short-term but highly profitable extraction. People who have fewer resources must employ a longer-term perspective. Short-term overexploitation will lead to a loss of natural resources that cannot be afforded by those fully dependent on them, unless at the expense of falling into poverty. If there is a tragedy of resource depletion or overexploitation, it will usually be a result of a combination of individual rights to appropriate these resources, social relationships that enable people to use their rights in practice, and the individual ownership of other means of production (animals, chain saws, and technical equipment) as well as the availability of alternative sources of income. The more sophisticated and efficient these resources are, the more capital intensive, the greater the danger of overexploitation and degradation of the natural resources and the tragedy.

We typically find people looking for a quick but exploitative gain among the wealthier groups in society, because the wealthy always can revert to other resources. Those who intend to remain dependent on the property they are cultivating, either by lack of capital to shift to other sources or for other social and economic constraints, develop an entirely different attitude toward that property and have a greater interest and invest more in sustainable exploitation. The poorest, however, may be forced to overexploit their land due to survival constraints that force them to reduce their time horizon, even though they intend to continue to live of that land. Wealthier landowners may not need the land for agricultural production and have alternative incomes. But they may use land as a form of old age insurance, as an object of speculation and/or accumulation of wealth.

II. THEORETICAL ISSUES

These considerations do not pertain to communal forms of property rights only but are equally and perhaps even more pertinent with respect to natural resources that are used and controlled under individual ownership or state control (see von Benda-Beckmann, 2001a; von Benda-Beckmann and von Benda-Beckmann, 2004). According to the dominant theories, private ownership derives its eminent position from two characteristics. Because it is in principle unlimited in time, it provides a secure legal basis for the preservation of the resource owned, and because it is marketable, it allows for efficient and therefore sustainable use. Both assertions are problematic in their generality. A long-time perspective is indeed conducive for sustainable use, though, as we have mentioned, there are circumstances of poverty that may force owners into unsustainable use necessary for immediate survival. Farmers who intend to live off their land only tend to follow sustainable strategies. But they have no intention to sell their land and its marketability bears more negative than positive connotations. Selling their land would mean a marked decrease in their quality of life rather than economic profit, a nightmare for many farmers. For large owners who do not live primarily off their land, marketability makes frequent transfers possible with the aim of quick economic profit without any consideration of the ecological quality of the resource. For large companies, land may be no more than a minor asset. For example, once the forest is depleted, it may be more profitable for a logging company to sell the land and move on to new tracts of highly profitable forest, than it is to engage in laborious reforestation. The legal obligation to reforest may be paid off by bribing all levels of the controlling authority. As long as profitable alternatives are available, there is little incentive to exploit the land sustainably, irrespective of the type of property right (Balland and Platteau, 1999: 774).

While the continuation of the company may not depend on a particular resource, the social security of its workers may be a different matter. Local workers are as easily laid off as the land is abandoned after depletion. Shifting to a new location may involve hiring new local workers. The old workers remain with the remnants of the exploitative extraction and when they revert to agricultural use, the old workers often pay much of the cost of recovery that the companies refused to pay. Another category of land users aggravates the situation. Migrant farmers often follow logging companies and start cultivating depleted forestland. But other than the local population, who tend to remain where they have lived, these new migrant shifting cultivators often keep following the logging companies to newly deforested areas. They use the land as long as it is still fertile and move on to newly depleted forest as soon as the loggers have lifted their heels. They leave the further degraded fields to the local population. For them, a new cycle of sustainable use may begin, but starting from a more problematic baseline. All in all, it seems that the availability of resources through market transactions as often as not facilitates irresponsible resource use and transfers to irresponsible users.

The picture is not much different when we look at state property. Although the existence of a state structure is based on a long-time horizon, government officials and politicians have a notoriously short time perspective. They have to capitalize on their election period or the term they serve in one particular place, and rarely seriously look beyond that time. Democratic decision-making, job rotation schemes, and the constant pressures of the immediate present foster a short-term time horizon and override a long-time perspective. For politicians and government officials, variations in time horizon and commitment to the common good are also likely to vary with the distribution of wealth, dependence on resources, and alternative economic opportunities. Many high- and low-ranking civil servants cannot earn their livelihood without additional and conscious direct or indirect appropriation of state controlled resources. For lower-ranking civil servants, this may indeed be a survival imperative (see von Benda-Beckmann and von Benda-Beckmann, 1998). But this attitude is certainly not conducive for responsible long-term use of state owned resources, neither in terms of sustainable resource use nor in terms of social security.

6.3.3 Social Security

As in the field of property rights, economic and policy discussions about social security have been dominated by evolutionist and modernization biased assumptions. The models taken over, or forced upon third world governments in what Midgley (1996) calls "welfare imperialism," were developed in Europe in the rise of industrialism and the later welfare state. They were until recently completely dominated by concerns for loss of wages, as becomes clear, for example, from ILO convention No. 102 of 1952 regarding social security. Moreover, they focus on institutions set up by state governments. The view of most social policy and development economists on the respective value of state regulated or nonstate village social security in the third world was what we call "the no longer—not yet perspective."[20] According to this perspective, the traditional systems of social security no longer are capable of preventing poverty among the rural population, especially in those situations and periods in which they are especially vulnerable due to physical, mental, or socioeconomic conditions. Moreover, these village arrangements of social security with their emphasis on communal obligations to assist kin and village members were seen as an obstacle to growth-oriented economic development, to be substituted by modern forms of social security provided by the state or the market. However, these new forms are not yet in place and do not yet cater sufficiently for the social security needs of rural people. Therefore, social policies aim at creating and "extending" formal social security mechanisms.

In line with other third world countries, Indonesia also started out with social security for its civil servants and armed forces (Esmara and Tjiptoherijanto, 1986). Furthermore, arrangements were developed for the workers in the large, formal, private, and semiprivate sector, but only reaching a very small percentage of the employed population. Only very recently Indonesia has issued a law on social security that aims to cover also the self-employed of the rural population (see McLeod, 1993). Besides, various departments have initiated targeted programs, such as the IDT program and various small grant projects based on a Presidential Instruction.[21] In general, much emphasis is put upon projects that concentrate on income-generating activities. This serves those who are physically and mentally capable of working but for some reason lack the possibilities to earn an income. But there is a serious lack of programs catering to those who are incapable of working because of sickness or disability.[22] The era of deregulation and privatization has brought some changes in this respect. Confronted with the apparent inability to prevent further erosion of such "traditional" social security mechanisms, and the inability of state or other donor agencies to take over this burden, the value of traditional arrangements of social security has been reevaluated in a more positive

[20] (von Benda-Beckmann and von Benda-Beckmann, 1994b; Gilbert, 1976: 365; Platteau, 1991: 163).

[21] See for IDT, Harjono (1983); see Ravaillon (1988) for In(struksi) Pres(iden) projects.

[22] See, eg, von Benda-Beckmann (1988) for a discussion on income-generating programs on the Moluccas.

light and there are increasingly calls for "strengthening" local institutions of social security and for finding innovative linkages between them and externally (state, NGO) supported forms of social security.[23] The two characteristics of the official social policies—exclusive emphasis on public or state institutions as providers of social security and on loss of income as the primary situation of need, or contingency—have for a long time made it difficult to get to an adequate understanding of the conditions of social insecurity and security of rural populations. First, the dominant approach obscures the fact that social security, even in the most advanced welfare states, has always been a mixture of arrangements by the state, family relationships, voluntary organizations, and the market sector (Rose, 1989). The focus should therefore not lie entirely on the question of what official social security institutions do or should do. The basic question to address is what sets of social security relations people have, how they form their specific "mixes" of social security, and what changes in these mixes take place as a result of economic developments. Second, instead of an exclusive focus on loss of income in the form of wages, the focus should be on the more basic contingencies and risks people are confronted with, whether they are involved in wage labor, self-employment, or farming. Clearly the contingencies with which different categories of people are confronted vary, as do cultural and legal notions of what appropriate living entails and notions of what "poverty" is. And what "need" and "poverty" are cannot be reduced to statistical criteria.

6.4 CONCLUSIONS

Given the multiple and often conflictive functions property and social security relations have for different categories of people and agencies, it is of crucial importance to consider property, economic growth, social security, and sustainability in their interdependence. This helps to redress some of the shortcomings of the theoretical debates that discuss these issues in isolation and without considering the implications of legal pluralism sufficiently. And we have shown that it has important policy implications as well. Some of the negative consequences of land policies result from neglecting the social security aspects of land and land rights. The use farmers, business enterprises, and state agencies make of natural resources is not so much determined by the type of property rights, but by a set of interconnected economic, political, and social factors, as well as by the individual interests and aspirations of the actors involved. The distribution of concretized property rights, the dependence on the resources for survival/livelihood, profit making, alternative economic possibilities they have or do not have, and the time horizon in which they view the functions of property play an important role.

We have seen that notions of neediness are intimately connected to access and exploitation rights to natural resources. In general, the availability and rights to natural resources are important indicators of wealth, and thereby of who is needy and who is not.

They are also important resources that can be put to use in providing social security and thus provide a necessary condition for social security for the rural population. However, the relative importance of these rights for the social security situation of a person has to be considered against the totality of a person's mixes of rights and social relations. For those

---

[23] See Midgley (1994) and von Benda-Beckmann and Kirsch (1999).

economically engaged in nonfarming activities, labor market conditions, employment opportunities, or prices for cash crops may be more important for their social security than rights to land. Moreover, land rights do not always directly translate into more social security. One cannot eat land rights or money. Wealth as well as monetary income are always mediated by intervening processes, a chain of conversions of land rights or money into work, food processing, and care before actual needs are overcome (see Sen, 1981). Migration shapes the relationships of social security in complex ways. Education and migration are two important strategies of social security. Families sacrifice considerable amounts of resources to make migration possible, often causing considerable hardship at first, softened only by the expectation of a greater security for the future.

Socioeconomic conditions as much as the plural legal structure of a society, including inheritance and family law, determine who is needy and who can take care of himself or herself. This means that in rural areas notions of neediness and social security are intimately related to rights to natural resources. But while the general availability and distribution of natural resources and other forms of wealth in a community are an important basis for social security, the actual mix of sources of wealth each person has is as important for the overall situation of social security.

Most social relationships through which goods and services for social security are transferred are multifunctional. In Indonesia (as in other countries) the same social relationships, such as kinship, neighborhood, or patron-client relationships, may serve several functions at the same time. Only when looking at concretized rights as actual relationships, can we see the actual embeddedness of property relationships; that is, the ways in which property rights are distinguished from, or interwoven with, other social, economic, and political relationships—whether this is among farmers or top bureaucrats and politicians.

We have seen that there is a stronger relationship between the time horizon and interest in sustainable use of property objects than between specific types of rights and sustainable use. The time perspective and the sustainable use are highly dependent on the extent to and ways in which users depend on the resources and their possibilities for alternative sources of income. For many, social security is one important factor in the considerations about sustainable use.

The sense of insecurity or security depends on material and psychological factors. Social security is about potentialities and expectations for the future, as well as concrete provisions of care in the present. Experiences in the past and present are crucial for the sense of security that in the future care will be provided if needed. Thus, social security links past, present, and future in a peculiar way. We have seen that security of tenure is an important factor in the overall sense of security. However, the constellations of legal pluralism, but more so the erratic land policies of the government, have proven to be important sources of insecurity. A comparative perspective on social security as outlined here might contribute to a better insight into the characteristic nature that issues of property, social security, and legal pluralism have in the diverse regions. A comparative analysis will also be helpful for getting a better understanding of the conditions under which specific factors shape the extent of social insecurity and security as well as a more or less sustainable use of natural resources.

Acknowledgments

This chapter is a slightly adapted version of the published paper "Benda-Beckmann, F. v. and K. v. Benda-Beckmann (2006). Social insecurity, natural resources and legal complexity. La quete anthropologique du droit. Autour de la démarche d'Étienne Le Roy. C. Eberhard and G. Vernicos. Paris, Karthala: 221–248," which is duly acknowledged.

References

Agarwal, B., 1991. Social security and the family: coping with seasonality and calamity in rural India. In: Sen, A., Ahmad, E., Drèze, J., Hills, J. (Eds.), Social Security in Developing Countries. Clarendon Press, Oxford, pp. 171–244.

Agrawal, A., Gibson, C.C., 1999. Enchantment and disenchantment: the role of community in natural resource conservation. World Dev. 27, 629–649.

Bachriadi, D., Faryadi, E., Setiawan, B., 1997. Reformasi Agrarian. Universitas Indonesia, Jakarta.

Balland, J.M., Platteau, J.P., 1999. The ambiguous impact of inequality on local resource management. World Dev. 27, 773–788.

Berkes, F., 1996. Social systems, ecological systems, and property rights. In: Hanna, S., et al. (Eds.), Rights to Nature. Island Press, Washington, DC, pp. 87–107.

Berkes, F., Folke, C., 1998. Linking social and ecological systems resilience and sustainability. In: Berkes, F., Folke, C. (Eds.), Linking Social and Ecological Systems: Management Practices and Social Mechanisms for Building Resilience. Cambridge University Press, Cambridge, pp. 1–25.

Brouwer, A., 1996. Natural resources, sustainability and social security: simplifying discourses and the complexity of actual resource management in a Central Moluccan village. In: Mearns, D., Healey, D. (Eds.), Remaking Maluku: Social Transformation in Eastern Indonesia. Centre for Southeast Asian Studies, Northern Territory University, Darwin, pp. 64–79. Special Monograph No. 1.

Bruce, J.W., Migot-Adholla, S. (Eds.), 1994. Searching for Land Tenure Security in Africa. Kendall and Hunt, Dubuque.

Chambers, R., 1983. Rural Development: Putting the Last First. Longman Scientific & Technical, Harlow Essex.

Chambers, R., Leach, M., 1989. Trees as savings and security for the rural poor. World Dev. 17, 329–342.

Demsetz, H., 1967. Toward a theory of property rights. Am. Econ. Rev. 57, 347–359.

Esmara, H., Tjiptoherijanto, P., 1986. The social security system in Indonesia. ASEAN Econ. Bull. 3, 53–69.

Fisiy, C., 1992. Power and Privilege in the Administration of Law: Land Reform and Social Differentiation in Cameroon. Africa Studies Centre, Leiden.

Fisiy, C.F., 1997. Conflicting approaches to the management of a natural resource base: the case of land colonization on the slopes of Mount Oku, Camaroon. In: von Benda-Beckmann, F., von Benda-Beckmann, K., Hoekema, A. (Eds.), Natural Resources, Environment and Legal Pluralism. Martinus Nijhoff, The Hague, pp. 124–145.

Furobotn, E.G., Pejovic, S., 1972. Property rights and economic theory: a survey of recent literature. J. Econ. Lit. 10, 1137–1162.

Ghai, D. (Ed.), 1994. Development and Environment: Sustaining People and Nature. United Nations Research Institute for Social Development (UNRISD)/Blackwell, London.

Gilbert, N., 1976. Alternative forms of social protection for developing countries. Soc. Serv. Rev. 50, 363–387.

Hardin, G., 1968. The tragedy of the commons. Science 162, 1234–1248.

Harjono, J., 1983. Rural development in Indonesia: the top-down approach. In: Lea, D.A.M., Chauduri, D.P. (Eds.), Rural Development and State: Contradictions and Dilemmas in Developing Countries. Methuen, London, pp. 38–65.

Hitchcock, R.K., 1980. Tradition, social justice and land reform in Central Botswana. J. Afr. Law 24, 1–34.

Huijsman, B., Savenije, H., 1991. Making haste slowly. In: Huijsman, B., Savenije, H. (Eds.), Making Haste Slowly: Strengthening Local Environmental Management in Agricultural Development. Royal Tropical Institute, Amsterdam, pp. 13–34.

IUCN, 1984. Traditional life-styles, conservation, and rural development. Commission on Ecology Papers no. 7, Gland.

Leach, M., Mearns, R., Scoones, I., 1999. Environmental entitlements: dynamics and institutions in community-based natural resource management. World Dev. 27, 225–247.

Lynch, O.J., Talbott, K., 1995. Balancing Acts: Community-Based Forest Management and National Law in Asia and the Pacific. World Resources Institute, Washington, DC.

McCay, B.J., 1987. The culture of the commoners: historical observations on old and new fishermen. In: McCay, B.J., Acheson, J.M. (Eds.), The Question of the Commons. University of Arizona Press, Tucson, AZ, pp. 195–216.

McCay, B.J., Acheson, J.M., 1987. Human ecology of the commons. In: McCay, B.J., Acheson, J.M. (Eds.), The Question of the Commons. University of Arizona Press, Tucson, AZ, pp. 1–35.

McLeod, R.H., 1993. Workers' social security in Indonesia. In: Manning, C., Hardjono, C. (Eds.), Indonesia Assessment 1993. Labour: Sharing in the Benefits of Growth? Department of Political and Social Change, Research School of Pacific Studies, Australian National University, pp. 88–107.

Mesa-Lago, C., 1992. Comparative analysis of Asian and Latin American social security systems. In: Getubig, I.P., Schmidt, S. (Eds.), Rethinking Social Security: Reaching Out to the Poor. APDC and GTZ, Kuala Lumpur/Eschborn, pp. 64–105.

Midgley, J., 1984. Social Security, Inequality and the Third World. John Wiley, Chichester.

Midgley, J., 1994. Social security policy in developing countries: integrating state and traditional systems. In: von Benda-Beckmann, F., von Benda-Beckmann, K., Marks, H. (Eds.), Coping With Insecurity: An 'Underall' Perspective on Social Security in the Third World, pp. 219–229. Special issue Focaal 22/23.

Midgley, J., 1996. Social Welfare in Global Context. Sage, London.

Okoth-Ogendo, H.W.O., 1984. Development and legal process in Kenya: an analysis of the role of law in rural development administration. Int. J. Sociol. Law 12, 59–83.

Ostrom, E., 1990. Governing the Commons. Cambridge University Press, Cambridge.

Peluso, N.L., 1993. Coercing conservation: the politics of state resource control. In: Lipschutz, R.D., Conca, K. (Eds.), The State and Social Power in Global Environmental Politics. Columbia University Press, New York, pp. 46–70.

Peters, P., 1987. The grazing lands of Botswana and the commons debate. In: McCay, B.J., Acheson, J.M. (Eds.), The Question of the Commons. University of Arizona Press, Tucson, AZ, pp. 171–194.

Platteau, J.P., 1991. Traditional systems of social security and hunger insurance: past achievements and modern challenges. In: Ahmad, E., Drèze, J., Hills, J., Sen, A. (Eds.), Social Security in Developing Countries. Clarendon Press, Oxford, pp. 112–170.

Ravaillon, M., 1988. Inpres and inequality: a distributional perspective on the centre's regional disimbursements. Bull. Indones. Econ. Stud. 24 (3), 53.

Reich, C., 1964. The new property. Yale Law Rev. 72, 734–787.

Rose, R., 1989. Ordinary People in Public Office: A Behavioural Analysis. Sage, London.

Schlager, E., Ostrom, E., 1992. Property-rights regimes and natural resources: a conceptual analysis. Land Econ. 68 (3), 249–262.

Sen, A., 1981. Poverty and Famines: an Essay on Entitlements and Deprivation. Clarendon Press, Oxford/New York.

Sen, A., 1987. Reply. In: Hawthorn, G. (Ed.), The Standard of Living. Cambridge University Press, Cambridge, pp. 103–112.

Simbolon, I.J., 1997. Understanding women and land rights in the context of legal pluralism: the case of Toba Batak, Indonesia. In: De Bruijn, M., Van Halsema, I., Van den Hombergh, H. (Eds.), Gender and Land Use: Diversity in Environmental Practice. Thela, Amsterdam, pp. 69–86.

Slaats, H., 1999. Land titling and customary rights: comparing land registration projects in Thailand and Indonesia. In: Van Meijl, T., von Benda-Beckmann, F. (Eds.), Property Rights and Economic Development: Land and Natural Resources in Southeast Asia and Oceania. Kegan Paul International, London, pp. 88–109.

Spiertz, J., Wiber, M.G. (Eds.), 1996. The Role of Law in Natural Resource Management. Vuga, The Hague.

van de Fliert, L., 1994. Indigenous Peoples and International Organisations. Spokesman, Nottingham.

van de Ven, J., 1994. Members only: time-sharing rice fields and food security in a Sumatran valley. In: von Benda-Beckmann, F., von Benda-Beckmann, K., Marks, H. (Eds.), Coping With Insecurity: An 'Underall' Perspective on Social Security in the Third World. pp. 85–96. Special issue Focaal 22/23.

van Vollenhoven, C., 1919. De Indonesier en zijn grond. Brill, Leiden.

Vel, J., 1994. Manu Wolu and the birds' nests: the consequences of a deviant way to cope with insecurity. In: von Benda-Beckmann, F., von Benda-Beckmann, K., Marks, H. (Eds.), Coping With Insecurity: An 'Underall' Perspective on Social Security in the Third World, pp. 35–46. Special issue Focaal 22/23.

von Benda-Beckmann, F., 1979. Property in Social Continuity: Continuity and Change in the Maintenance of Property Relationships Through Time in Minangkabau, West Sumatra. M. Nijhoff, The Hague.

von Benda-Beckmann, K., 1981. Forum shopping and shopping forums. J. Leg. Pluralism 19, 117–159.

von Benda-Beckmann, F., 1986. Leegstaande luchtkastelen: over de pathologie van grondenrechtshervormingen in ontwikkelingslanden. In: Brussaard, W., et al. (Eds.), Recht in ontwikkeling—Tien agrarisch-rechtelijke opstellen. Kluwer, Deventer, pp. 91–109.

von Benda-Beckmann, K., 1988. Social security and small-scale enterprises in Islamic Ambon. In: von Benda-Beckmann, F., von Benda-Beckmann, K., Casiño, E., Hirtz, F., Woodman, G.R., Zacher, H. (Eds.), Between Kinship and the State: Law and Social Security in Developing Countries. Foris Publications, Dordrecht, Holland/Cinnaminson, USA, pp. 451–471.

von Benda-Beckmann, F., 1990. Sago, law and food security on Ambon. In: Bakker, J.I.H. (Ed.), The World Food Crisis: Food Security in Comparative Perspective. Canadian Scholars' Press, Toronto, ON, pp. 157–199.

II. THEORETICAL ISSUES

von Benda-Beckmann, K., 1997. Environmental protection and human rights of indigenous peoples: a tricky alliance. In: von Benda-Beckmann, F., von Benda-Beckmann, K., Hoekema, A. (Eds.), Law and Anthropology, vol. 9. Special Issue on Natural Resources, Environment and Legal Pluralism. Martinus Nijhoff, The Hague, pp. 302–323.

von Benda-Beckmann, F., 2001a. Between free riders and free raiders: property rights and soil degradation in context. In: Heerink, N., Van Keulen, H., Kuiper, M. (Eds.), Economic Policy Analysis and Sustainable Land Use: Recent Advances in Quantitative Analysis for Developing Countries. Physica Verlag, Heidelberg, NY, pp. 293–316.

von Benda-Beckmann, K., 2001b. Transnational dimensions of legal pluralism. In: Fikentscher, W. (Ed.), Begegnung und Konflikt—eine kulturanthropologische Bestandsaufnahme. Verlag der Bayerischen Akademie der Wissenschaften, C.H. Beck Verlag, München, pp. 33–48.

von Benda-Beckmann, F., Kirsch, R., 1999. Informal social security systems in Southern Africa and approaches to strengthen them through policy measures. J. Soc. Dev. Afr. 14, 21–38.

von Benda-Beckmann, F., Taale, T., 1992. The changing laws of hospitality: guest labourers in the political economy of rural legal pluralism. In: von Benda-Beckmann, F., Van der Velde, M. (Eds.), Law as a Resource in Agrarian Struggles. Pudoc, Wageningen, pp. 61–87.

von Benda-Beckmann, F., von Benda-Beckmann, K., 1994a. Property, politics and conflict: Ambon and Minangkabau compared. Law Soc. Rev. 28, 589–607.

von Benda-Beckmann, F., von Benda-Beckmann, K., 1994b. Coping with insecurity. In: von Benda-Beckmann, F., von Benda-Beckmann, K., Marks, H. (Eds.), Coping With Insecurity: An 'Underall' Perspective on Social Security in the Third World. pp. 7–31. Special issue Focaal 22/23.

von Benda-Beckmann, F., von Benda-Beckmann, K., 1995. Rural populations, social security, and legal pluralism in the Central Moluccas of Eastern Indonesia. In: Dixon, J., Scheurell, B. (Eds.), Social Security Programs: A Cross-Cultural Perspective. Greenwood, Westport, pp. 75–107.

von Benda-Beckmann, F., von Benda-Beckmann, K., 1998. Where structures merge: state and off-state involvement in rural social security on Ambon, Eastern Indonesia. In: Pannell, S., von Benda-Beckmann, F. (Eds.), Old World Places, New World Problems: Exploring Issues of Resource Management in Eastern Indonesia. Centre for Resource and Environmental Studies, The Australian National University, Canberra, pp. 143–180.

von Benda-Beckmann, F., von Benda-Beckmann, K., 1999. A functional analysis of property rights, with special emphasis to Indonesia. In: Van Meijl, T., von Benda-Beckmann, F. (Eds.), Property Rights and Economic Development: Land and Natural Resources in Southeast Asia and Oceania. Kegan Paul International, London, pp. 15–56.

von Benda-Beckmann, F., von Benda-Beckmann, K., 2001. Recreating the nagari: decentralisation in West Sumatra, Working paper series no. 31, Max Planck Institute for Social Anthropology, Halle/Saale.

von Benda-Beckmann, F., von Benda-Beckmann, K., 2004. Struggles over communal property rights and law in Minangkabau, West Sumatra. Working paper series no. 64, Max Planck Institute for Social Anthropology, Halle/Saale.

von Benda-Beckmann, K., Flinterman, C., Oostenbrink, T., 1994. Human rights and indigenous peoples: how the Netherlands could contribute. In: Morales, P. (Ed.), Indigenous Peoples, Human Rights and Global Interdependence. International Centre for Human and Public Affairs, Tilburg, pp. 155–166.

von Benda-Beckmann, F., von Benda-Beckmann, K., Brouwer, A., 1995. Changing 'indigenous environmental law' in the Central Moluccas: communal regulation and privatization of Sasi. Ekonesia 2, 1–38.

WCED, 1987. Our Common Future: The Report of the World Commission on Environment and Development. Oxford University Press, Oxford.

Wiber, M.G., 1993. Politics, Property and Law in the Philippine Uplands. Wilfried Laurier University Press, Waterloo, Canada.

World Bank, 1994. Staff Appraisal Report: Indonesia—Land Administration Project. World Bank, Washington, DC.

LEARNING FROM THE FIELD CASES/ISSUES

High Resolution of Three-Dimensional Dataset for Aboveground Biomass Estimation in Tropical Rainforests

W.V.C. Wong*,†, *S. Tsuyuki**

*University of Tokyo, Tokyo, Japan †University of Malaysia Sabah, Kota Kinabalu, Sabah, Malaysia

7.1 INTRODUCTION

Tropical forests contain high biomass compared to other forest ecosystems, with approximately half of the total living biomass of the world's major ecosystem (Houghton et al., 2009). In the Global Forest Resources Assessment 2010 (FAO, 2010), the total carbon stock in the living forest biomass for Southeast Asia was estimated at 22 Gt C, or approximately 8% of the global total. Indonesia accounted for more than half of the carbon stock with the value of 13.0 Gt C, followed by Malaysia (3.2 Gt C) and Myanmar (1.7 Gt C). Additionally, Slik et al. (2010) reported that aboveground biomass (AGB) per unit area in Borneo island is relatively 60% higher than in the Amazon. However, the forest area in Southeast Asia declined by 31 Mha from 267 Mha in 1990 to 236 Mha in 2010 with a two-thirds majority occurring in insular Southeast Asia where the main drivers were attributed to forest conversion to cash crops plantations, logging, and conversion to forest plantations (Stibig et al., 2014). Concurrently, the Global Forest Resources Assessment 2010 also reported the carbon stock declined by 3.3 Gt C in the same period of 1990–2010 (FAO, 2010). In the Fifth Assessment Report (AR5), activities from forestry and other land use (FOLU) contributed the total greenhouse gases emission by 11% or 5.4 Gt CO_2-eq/year in 2010 (IPCC, 2014).

Recognizing the importance for developing countries along with industrialized countries for the total emissions reductions from all major sources, reducing emission from deforestation and forest degradation and the role of conservation, sustainable management of forests on enhancement of forest carbon stocks in developing countries (REDD+) was introduced and proposed during the 11th session of the Conference of Parties (COP) to the United Nations Framework Convention on Climate Change (UNFCCC) in Montreal, 2005 (UNFCCC, 2005) and adopted in COP 13, Bali, 2007 (UNFCCC, 2007). To implement the REDD+ scheme, an estimation and monitoring system of forest biomass with reliable accuracy along with a robust and transparent system is one of the major technical issues under discussion. This activity is discussed mainly under the measurement, reporting, and verification (MRV) system of REDD+ (eg, UNFCCC, 2014). Field-based inventory alone will be resource intensive and yield higher uncertainties in the biomass estimation. The development of remote sensing technology with a combination of ground-based inventory approaches for estimating forest carbon stocks and forest area changes was accepted in the methodological guidance for activities relating to REDD+, which contribute to the robust and transparent forest monitoring system or MRV system (Decision 4/CP. 15).

There have been successes in estimating forest biomass on a regional scale (eg, Brown et al., 1993; Saatchi et al., 2011; Baccini et al., 2012; Avitabile et al., 2016); however, the resolutions were coarse of 1 km (Saatchi et al., 2011) or 500 m (Baccini et al., 2012) derived using a low-resolution optical data set such as the moderate resolution imaging spectroradiometer (MODIS). Biomass estimation using only an optical sensor data set (ie, multispectral or hyperspectral data) will yield an estimation accuracy problem, especially for high biomass stands, and it is recommended it be combined with other types of remote sensing data sets (Koch, 2010). Recently, the use of a high-resolution three-dimensional data set (ie, airborne laser scanning (ALS) and structure from motion (SfM) photogrammetry) have been demonstrated to yield good estimation, especially with height-related forest variables such as stem volumes, stand height, and biomass (eg, Gobakken et al., 2015; Ioki et al., 2014; Ota et al., 2015). This type of high-resolution three-dimensional data set offers great improvement on estimation accuracy and reliability and reduces uncertainties for forest biomass estimation in accordance with the Intergovernmental Panel on Climate Change (IPCC)'s Tier 3 for the land use, land-use change and forestry (LULUCF) sector (IPCC, 2006). In addition to the accuracy issue, a cost-effective system is also a major consideration when developing a biomass monitoring system for the national or subnational level.

Thus, in this chapter, we discuss the technical issues in estimating forest biomass for tropical rainforest using a combination of a remote sensing data set and ground samples. We also present an example of a case study in estimating forest biomass using a high-resolution three-dimensional data set of ALS and an aerial photogrammetry data set in tropical montane forest in northern Borneo. We then discuss technical challenges, large-scale applications, and how integrating this method can contribute to forest biomass estimation in an effective way for the Southeast Asia region.

7.2 ESTIMATING ABOVEGROUND BIOMASS USING A COMBINATION OF REMOTE SENSING DATA SETS AND GROUND SAMPLES

The interest in forest biomass studies in Southeast Asia can be tracked back to the late 1980s (eg. Brown et al., 1989; Yamakura et al., 1986; Yoneda et al., 1990). Since then, studies in many aspects of AGB such as allometric equation (eg, Yamakura et al., 1986; Brown et al., 1989; Ketterings et al., 2001), biomass dynamic (eg, Nakagawa et al., 2012; Toma et al., 2005), estimation approach (eg, Okuda et al., 2004; Ioki et al., 2014), and regional estimation (eg, Brown et al., 1993; Langner et al., 2015) studies have been developed.

AGB is one of the major components of carbon pools in forestland together with below-ground biomass (BGB), dead organic matter, and soil organic matter. Estimating AGB is rather straightforward compared to other components of carbon pools, although a default value of 0.37 for the ratio of BGB to AGB can be employed to estimate BGB for tropical rainforest as recommended by IPCC (2006).

The remote sensing technology with a combination of ground samples has enabled wall-to-wall estimation of AGB. There are several guidelines that have been published in estimating forest biomass using a remote sensing data set such as can be found in *REDD+ Cookbook* (Hirata et al., 2012) or "Integrating remote-sensing and ground-based observations for estimation of emissions and removals of greenhouse gases in forests" (GFOI, 2013). In many of the guidelines and research studies, the technical aspects, which are still undergoing research and development and discussion, are the allometric equation, ground sample, remote sensing data set, and prediction method, which in one or more combinations may influence the estimation accuracy (eg, Fassnacht et al., 2014; Koch, 2010).

7.2.1 Ground Sample

To develop a biomass estimation model, a ground sample data set is used, of which the predictor variables derived from the remote sensing data set are related to the corresponding ground biomass values and by using a prediction method (ie, regression or machine learning algorithm), a biomass estimation and map can be produced for the target area. Many of the biomass models are developed using the plot-based technique (eg, Ioki et al., 2014). Collecting a ground sample is resource intensive and challenging especially in remote areas. Thus, the number of ground samples is often limited to 100 plots or less in many AGB studies conducted in tropical rainforest (eg, Fassnacht et al., 2014; Basuki et al., 2013; Ioki et al., 2014). Fassnacht et al. (2014) demonstrated that the ground sample is important after the type of remote sensing data set and prediction method. In addition to the number of ground samples, the size of sample plots is also an important criteria (eg, Mauya et al., 2015) as is sampling strategy.

7.2.2 Allometric Equation

The estimation of AGB using remote sensing technology relies heavily on the selection of an allometric equation of AGB. Many allometric equations have been proposed that have been developed in different types of forest and growth environments and can be categorized into generic models (eg, Brown, 1997; Pearson et al., 2005; Chave et al., 2014) or regional models (eg, Basuki et al., 2009; Yamakura et al., 1986). The AGB can be influenced by forest changes and species composition (eg, Culmsee et al., 2010). Thus, it is recommended to evaluate the estimation error of each allometric equation if there are two or more applicable equations that are available (eg, Hirata et al., 2012), which may reduce uncertainties in biomass assessment (eg, Rutishauser et al., 2013).

The allometric equation for Southeast Asia was developed as early as the 1980s (Yamakura et al., 1986) for lowland tropical forest. Because of the highly diverse forest type along with the degradation and plantation type in the Southeast Asia region, there has been continuous development of allometric equations (eg, Kenzo et al., 2014; McNicol et al., 2015; Basuki et al., 2009). Yuen et al. (2016) reviewed biomass allometric equations for 12 land-cover types in Southeast Asia, including Papua New Guinea and Southern China, and uncovered a total of 402 AGBs along with 138 BGBs for 12 major classes. Despite the high number of allometric equations available, the review concluded that there is a pressing need to address the insufficient number of allometric equations. However, optimizing the specific allometric equation will require a good data set of vegetation class at first.

7.2.3 Type of Remote Sensing Data Set

Remote sensing data sets varied in terms of the type of information and accuracy depending on the sensor type and altitude or distance from the land surface. Basically, categorization can be grouped by sensor type (ie, optical, light detection and ranging (LiDAR) or synthetic aperture radar (SAR)) or platform (ie, spaceborne or airborne). Applications in forest biomass estimation have been attempted using a variety of remote sensing data sets of optical systems (eg, Hirata et al., 2014), LiDAR (eg, Ioki et al., 2014; Fassnacht et al., 2014), SAR (eg, Dobson et al., 1992), or combination of several remote sensing data sets. The limitation to the use of a spaceborne data set is largely due to cloud cover and loose correlation between spectral information and biomass, especially in a high biomass of forest (Koch, 2010). The cloud cover for the tropic area was estimated at 58–70% according to the International Satellite Cloud Climatology Project (ISCCP). Spaceborne LiDAR such as the Ice, Cloud, and land Elevation Satellite (ICESat)/Geoscience Laser Altimeter System (GLAS) with a large footprint of 70 m was still limited due to the laser spots separated by nearly 170 m along the satellite's ground track while the spaceborne radar faced a saturation problem (Koch, 2010). Height information that is derived from the airborne platform has been demonstrated to be superior in estimating forest variables related to height such as volume, tree height, and biomass (eg, Nurminen et al., 2013; Rahlf et al., 2014; Vastaranta et al., 2013). Among the types of remote sensing data set, the ALS data set was found to rank first, followed by the SfM data set, interferometry SAR, and radargrammetry in estimating stem volume (eg, Rahlf et al., 2014).

7.2.4 Prediction Method

In biomass estimation, a linear model is among the most popular choices (ie, more than 50%) when compared to other prediction methods such as support vector machine, nearest neighbor-based methods, random forest, and Gaussian processes as evaluated by Fassnacht et al. (2014) over 113 studies. The results show that random forest has the highest R^2 value followed by linear models when using an ALS data set. The root mean square error (RMSE) values were also the lowest for random forest and the second lowest value when using linear models. Fassnacht et al. (2014) concluded that the prediction method had a considerable impact on the accuracy of the AGB estimate, nearly equally important as data type, and more important than sample size. However each prediction method may have its own advantage or disadvantage. For example, random forest models are likely to overestimate small value and underestimate high value (eg, Baccini et al., 2004; Chen, 2015).

7.3 CASE STUDY OF ESTIMATING FOREST BIOMASS IN TROPICAL MONTANE FOREST

We present here an example of estimating aboveground forest biomass in a tropical montane forest, northern Borneo. The full result will be published in a forthcoming journal. In this example of biomass estimation, we selected the Yamakura allometric equation (Yamakura et al., 1986) predicted by a linear regression model using both ALS and aerial photograph data sets.

7.3.1 Study Site

The study area is located in the Ulu Padas forest area (approximately 4°26'N, 115°45'E; Fig. 7.1) of northern Borneo, Malaysia, inside the Heart of Borneo Initiative area that forms part of an important mountain ecoregion representation of Borneo together with Pulong Tau National park in Sarawak and Kayan Mentarang National Park in Kalimantan, Indonesia. The area is covered by rugged terrain ranges approximately between 1000 and 1908 m in altitude at Bukit Rimau, while the vegetation of this region consists of several types (ie, dominant montane oak/chestnut forest with Agathis, hill dipterocarp forest, stunted montane mossy forest, and high-level swamp forest) (SBCP, 1998). The land use consists of both small- and large-scale logging activities as well as small-scale farming activities by the local people with some portion remaining as old growth forest. There is no weather station located in the study area; however, the rainfall is suggested to be ranging from 1500 to 2300 mm (Sinun and Suhaimi, 1997).

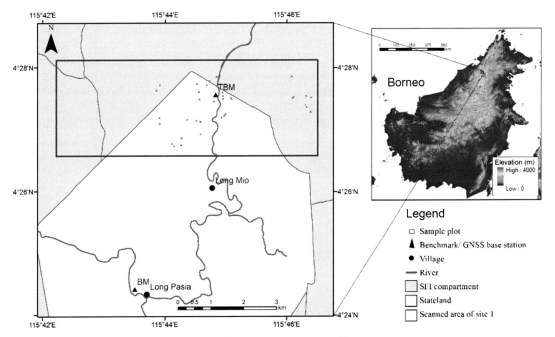

FIG. 7.1   The location of study area inside the mountain ecoregion of Borneo.

7.3.2 Field Data and Remote Sensing Data Set

Field data consisted of individual tree information, which was measured between 2011 and 2012. We used 35 plots located in site 1, with each plot size of $900\,m^2$ ($30\,m \times 30\,m$; $n = 33$) or $400\,m^2$ ($20\,m \times 20\,m$; $n = 2$). The position of each individual plot was determined using differential global navigation satellite system (DGNSS) receivers of Ashtech ProMark 100 (Spectra Precision, Westminster, CO, USA) and JAVAD Triumph-1 (JAVAD GNSS Inc., CA, USA). Diameter at breast height (DBH) and tree height within the plot area were recorded for all trees with DBH $\geq 10\,cm$. Tree height was measured using an electronic hypsometer of TruPulse® 360 Laser RangeFinder with foliage filter (Laser Technology Inc., Colorado, USA) or Haglöf Vertex IV (Haglöf Sweden AB, Västernorrland, Sweden). More than 300 species were recorded such as *Bischofia* sp. (Euphorbiaceae), *Trema* sp. (Ulmaceae), *Lithocarpus* sp. (Fagaceae), *Litsea* sp. (Lauraceae), and *Tristaniopsis* sp. (Myrtaceae). The aboveground biomass for all individual trees with DBH ≥ 10 cm were calculated using Yamakura allometric equation and converted to unit Mg/ha within each plot (Table 7.1).

The ALS data was acquired from a flight mission in October 2012 using airborne laser scanner Riegl LMS-Q560 (Riegl LMS GmbH, Horn, Austria). The ALS system was attached to a helicopter platform, Bell 206B3 Jet Ranger Helicopter, and the flight mission was conducted in 4 days with a flying altitude of approximately 400m above ground level. However, due to the terrain ruggedness and also the strong wind on the data acquisition days, the pilot had difficulty in maintaining the preplanned flying altitude of 400m above ground level during the entire flight mission. The flying altitude above ground level was calculated ranging from as low as 73 to 791m with an average of 390m. The system was operated with 45° of field of view and 240kHz of pulse repetition rate with beam divergence of less than 0.5mrad. The side overlap was within 30–50%. The processed data were delivered in the coordinate system of WGS84 UTM Zone 50N/WGS84 ellipsoid, with classification into ground and nonground returns comprising 832 million point clouds in LAS 1.2 format (for processing information, see Ioki et al., 2014) with classification into ground and nonground points. The average density was 14.9 pulses/m^2 and vertical accuracy (RMSE) of ALS point cloud was estimated within 25cm. The average point density of ALS in plot level was computed at 26.6 points/m^2.

The aerial photographs were acquired simultaneously in the same flight mission with the ALS acquisition by using a small format digital single lens reflex camera Canon EOS-1D Mark III. The camera was fitted with a lens of 28mm focal length and the cross-track field-of-view and along-track field-of-view are 52.9° and 36.6°, respectively. Flying at an average speed of 100km/h and exposure interval of 3.5s resulted in 55–70% of forward overlap while the side lap was ~45%. Camera settings

TABLE 7.1   Statistic Summary of AGB for 35 Plots Using the Yamakura Model

	AGB (Mg/ha)
Lower montane ($n = 35$)	
Average	246.31
Min	47.55
Max	622.79
SD	133.40

were set with exposure time of 1/2500s, ISO-speed of 1250, and aperture range from f/1.8 to f/10. A total of 2400 aerial photographs for site 1 were delivered in 24-bit sRGB on large-size format of JPEG (3888×2592) with average ground sampling distance (GSD) of approximately 10cm together with Global Navigation Satellite System (GNSS)/Inertial Measurement Unit (IMU) data information. Aerial photographs were processed using the digital photogrammetry technique (also referred to as SfM) to generate dense photogrammetric point cloud (SfM point cloud). The SfM software used in this study was the Agisoft Photoscan Pro 1.0.3 software package (Agisoft LLC, Sankt-Petersburg, RU). Technical processing of the aerial photographs can be found in Wong et al. (2015).

7.3.3 Model Development

The model development comprised several steps from the extraction of predictor variables, linear regression analysis, to cross-validation. Then, the AGB map was developed using the selected model.

We extracted the points and derived the predictor variables using LAStools (rapidlasso GmBH, Gilching, Germany). For the SfM points, we merged with the ground points from the ALS data set. The ALS and SfM points were processed using the "lastile" of LAStools with the tile size of 900 m × 900 m. Then, the SfM points were normalized using the ground points from the ALS data set and canopy echoes (ie, ≥2m) were classified using "lasheight" of LAStools. Then, we used the canopy echoes to compute the predictor variables (eg, Nurminen et al., 2013) by deriving 16 height variables and 9 canopy cover percentile variables for each plot (Fig. 7.2). The height variables were maximum height (h_{max}), minimum height (h_{min}), mean height (h_{mean}), standard deviation of height (h_{std}), percentiles at 10% intervals ($h_{10}, h_{20}, ..., h_{90}$) and percentile at 25% (h_{25}), 75% (h_{75}), and 95% (h_{95}). Canopy cover percentile was computed as the proportion of returns below a certain percentage of total height with 10% interval ($d_{10}, d_{20}, ..., d_{90}$). We performed log transformation to all the predictor variables and thus doubled the number of predictor variables.

For the linear regression analysis, each of the predictor variables from both the ALS and SfM data sets were related to the corresponding surveyed reference values of each plot. We modeled the AGB using both original scale and log-transformed values of the response variables. We selected the model with the highest coefficient of determination (R^2) value.

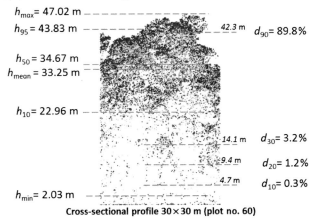

h_{max}= 47.02 m

h_{95} = 43.83 m

42.3 m d_{90}= 89.8%

h_{50} = 34.67 m
h_{mean} = 33.25 m

h_{10}= 22.96 m

14.1 m d_{30}= 3.2%

9.4 m d_{20}= 1.2%

4.7 m d_{10}= 0.3%

h_{min}= 2.03 m

Cross-sectional profile 30 × 30 m (plot no. 60)

FIG. 7.2   Example of several height variables and canopy cover percentile for plot no. 60.

Then, cross-validation was performed to assess the accuracy of the AGB estimation using leave-one-out-cross-validation (LOOCV). The LOOCV technique requires one of the training plots to be removed from the data set at a time, and the remaining plots ($n-1$) were fitted the using the selected model of AGB. The estimated AGBs were then predicted for the removed plot. This procedure was repeated until all estimated values were obtained for all plots. The accuracy of the estimations was assessed by the RMSE and relative RMSE (RMSE%) using the original scale values:

$$\text{RMSE} = \sqrt{\frac{1}{n}\sum_{i=1}^{n}(y_i - \hat{y}_i)^2}$$

(1)

$$\text{RMSE\%} = 100 \times \frac{\text{RMSE}}{\bar{y}}$$

(2)

where y_i is the surveyed reference value for plot i; $\hat{y}_i$ is the remote sensing-based prediction; $\bar{y}$ is the arithmetic mean of the surveyed AGB; and n is the number of the plot.

7.4 RESULTS

The single linear regression analysis found that a predictor variable percentile of 60% (h_{60}) yielded the highest R^2 value. In addition to that, both estimations showed that log-transformed variables resulted in the highest R^2 values. The RMSE, RMSE%, and R^2 for AGB estimation using the ALS data set are 73.92 Mg/ha, 30.0%, and 0.81, respectively; whereas, the RMSE, RMSE%, and R^2 for AGB estimation using the SfM data set are 71.88 Mg/ha, 29.2%, and 0.82, respectively (Table 7.2). The result demonstrated that estimation from both data sets yielded almost similar accuracy where the difference of RMSE and RMSE% are 2.0 Mg/ha and 0.83%, respectively.

Figs. 7.3 and 7.4 show the AGB map derived from using the linear regression model. The maximum estimation using the ALS and SfM data sets are 1018 and 1387 Mg/ha, respectively. One major limitation of the AGB map when derived using the SfM data set is no data area. In this study, only 86.1% of aerial photographs were aligned during the single image matching processing consisting of 2400 aerial photographs. The possible improvement of no data area is presented in the discussion section.

Fig. 7.5 shows the relative difference of estimation using SfM and ALS data sets. Although the RMSE and RMSE% are quite similar between ALS and SfM estimation, it seems that AGB estimation using the SfM data set tends to overestimate by more than 50 Mg/ha compared to AGB estimation using the ALS data set in some part of the study area (yellow and green raster).

TABLE 7.2   The Result of AGB Estimation Using ALS and SfM Data Sets

	ALS	SfM
Model	$\text{Ln(AGB)} = 1.956 \times \text{Ln}(h_{60}) - 0.59$	$\text{Ln(AGB)} = 2.145 \times \text{Ln}(h_{60}) - 1.388$
R^2	0.8075	0.8222
RMSE (Mg/ha)	73.92	71.88
RMSE%	30.01	29.18

FIG. 7.3   AGB estimation using the ALS data set.

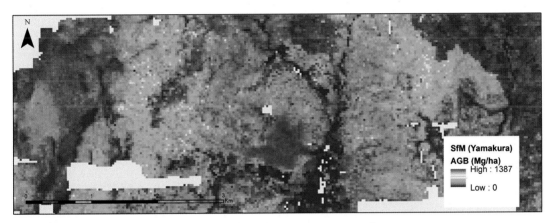

FIG. 7.4   AGB estimation using the SfM data set. Gray color represents no data area.

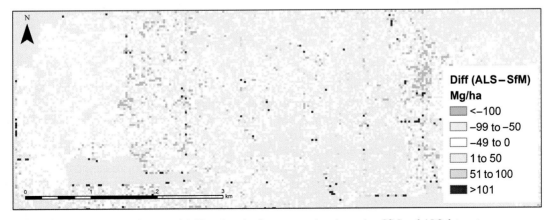

FIG. 7.5   The relative difference of AGB estimation between estimation using SfM and ALS data sets.

III. LEARNING FROM THE FIELD CASES/ISSUES

7.5 DISCUSSION ON BIOMASS ESTIMATION

7.5.1 Biomass Estimation Results

There have been successful studies using different types of remote sensing data sets of ALS (eg, Thapa et al., 2015; Ioki et al., 2014; Englhart et al., 2013; Kronseder et al., 2012), SfM (eg, Ota et al., 2015), SAR (eg, Solberg et al., 2014; Englhart et al., 2012), or a combination of the data sets (eg, Basuki et al., 2013; Morel et al., 2012). The biomass estimation accuracy varied between each other. Due to the influence of several factors, such as the prediction method, remote sensing data set, allometric equation, and ground sample, it is not feasible to make a direct comparison of the performance by comparing the RMSE or RMSE% values. At this time, there has been limited study that compares the performance using both ALS and SfM data sets in same study area (eg, Ota et al., 2015). Ota et al. (2015) demonstrated the estimation using ALS and SfM data sets also yielded almost similar accuracy in the Cambodian seasonal tropical forest when a digital terrain model from ALS is available. This suggests that there is a huge potential in using aerial photographs for a forest biomass monitoring program in the Southeast Asia region.

There have been several suggestions to improve estimation using remote sensing data sets mainly on the selection of allometric equation, forest type, and prediction method. Thapa et al. (2015) demonstrated that using a specific model could increase the estimation accuracy. In addition to that, using an allometric equation with height information (Feldpausch et al., 2012) and the use of wood-specific density (Manuri et al., 2014) can reduce the estimation error. Ota et al. (2015) demonstrated that adding the forest type information increases the accuracy of AGB estimation. The performance also can be attributed to different forest types. For example, Kronseder et al. (2012) demonstrated that the RMSE% was found to be lower in lowland dipterocarp forest compared to peat swamp forest.

7.5.2 Large Scale Application Using Aerial Photographs for AGB Monitoring

The main advantage of aerial photography is relatively low-cost data acquisition compared to an ALS data set (Leberl et al., 2010; White et al., 2013). Thus, there is the potential of using aerial photographs for a monitoring system considering there are several technical issues that should be addressed when applying the SfM data set for large-scale application of biomass estimation, namely: (1) availability of ALS data set for digital terrain model; (2) no data area; (3) camera system; (4) resolution; and (5) image matching processing.

The use of an ALS data set on a national scale is still limited to several countries; for example, Sweden, Denmark, Switzerland, and Finland (Ginzler and Hobi, 2015). In Southeast Asia, the Philippines is one of the leading countries for a LiDAR national program where up to November 2014, approximately 93,000 km² or one-third of the total land area have been acquired under the Disaster Risk and Exposure Assessment for Mitigation (DREAM) program started in 2011. In many cases, the product of a digital terrain model is delivered in raster format with typical resolution of 2 m or larger (eg, Ginzler and Hobi, 2015; Vesakoski et al., 2014). In the case where larger ALS-DTM was provided (eg, 2 or 5 m), resampling to lower resolution could reduce the error of the normalized point especially in a steeper area either by using the cubic convolution or bilinear interpolation technique (eg, Wong et al., 2014) or creating a triangulated irregular network (TIN) from the rasterized digital terrain model (DTM) before normalizing the points. Further development related to using different accuracy or resolution of ALS-DTM for biomass estimation in different environments (eg, forest type and slope gradient) would be an important area to be addressed. Jubanski et al. (2013) suggested the use of ALS density with four points per square meter resulted in the best cost-benefit ratio in a lowland forest ecosystem. However, the point density may be higher in an area of rugged or mountainous terrain such as in our study site.

The no data area can be improved significantly with a combination of a better data set of aerial photographs and image matching strategy. The use of an aerial photograph with higher radiometric resolution and overlap (both forward and sidelap) may contribute to image matching success (eg, Ginzler and Hobi, 2015). Nurminen et al. (2013) showed that using higher forward overlap of 80% can result in higher detection of forest gap. The forward overlap and sidelap in this study was estimated at between 55% to 70% and 45%, respectively. Ginzler and Hobi (2015) demonstrated success in creating a digital surface model (DSM) with a resolution of 1 m for the entire country of Switzerland, a first to be performed at a countrywide scale with 97.9% completeness using a push-broom camera, Leica ADS40/ADS80.

In this study, a higher resolution of 10 cm was used, which is higher than the conventional resolution used for national aerial photographs acquisition with a typical resolution of 25 or 50 cm (eg, Ginzler and Hobi, 2015). National flight programs take advantage of the large camera system where the swath width can reach up to 6 km using the ADS80 system for 50 cm resolution aerial photographs with a flying height about 4800 m above ground level. Flying in the mountain area would require the aircraft to fly higher and this would probably add to the cloud or haze problem. Meanwhile, Nurminen et al. (2013) demonstrated that there is no significant effect on estimation accuracy of forest variables when the off-nadir increases from 0° to 20°. Using different GSDs (ie, 12 and 48 cm) resulted in almost similar estimations when tested for forest biophysical characteristics estimation (Bohlin et al., 2012).

A SfM data set is derived indirectly using an image matching algorithm (eg, Photoscan Pro; Pix4Dmapper) unlike the ALS data set, which is derived directly from the active system using the emitted laser pulse from the sensor. The transition from analytical photogrammetry to fully digital photogrammetry has significantly increased the capability in deriving a detailed DSM as the process is fully automated. One of the major issues for large-scale operation is the processing capability. The first countrywide DSM was published by Ginzler and Hobi (2015) for the entire country of Switzerland (which is a relatively small country with a land area of about 4 million hectares), and was completed in 320 days using two workstations with 16 parallel processes. The Socet NGATE by BAE Systems was used for the processing of the entire country of Switzerland, and using other semiglobal matching software (eg, Trimble Match-T) resulted in up to four times higher processing times. A combination of ongoing image matching in the computer vision field and computer hardware development will certainly result in a continuous increase in computation capability for a large-scale forest monitoring system using an aerial photograph data set.

7.5.3 Importance of Effective Method in MRV Process of REDD+

The forest monitoring system must be consistent, transparent, comparable, accurate, and complete by following the IPCC's principles. To achieve this, remote sensing with a combination of ground-based inventory approaches is recommended for a robust and transparent forest monitoring system as adopted in COP 15 (UNFCCC, 2009). Ground samples remain important and when integrated with a remote sensing data set, better biomass estimation can be achieved with the capability of wall-to-wall estimation. Biomass estimation on the regional scale of Southeast Asia have been demonstrated in several studies since the 1990s. Brown et al. (1993) modeled carbon densities and pools in forest soils and vegetation in a tropical Asian forest using a geographic information system with several data sets (eg, pedon, climatic, and vegetation data sets) at a resolution of 3.75 km. Saatchi et al. (2011) and Baccini et al. (2012) provided a pan-tropical biomass map with 1-km and 500-m resolution, respectively. Further development was performed by Avitabile et al. (2016), of which a fusion method was proposed to improve biomass estimates by adjusting available data sets to local AGB patterns.

The current regional biomass estimates are at coarser resolution (ie, >500 m). In the case where higher resolution is needed to define the robustness of the monitoring system, higher resolution and/or with a combination of three-dimensional data set types may be needed to achieve it. In our case study, we demonstrated the capability of biomass estimation in high resolution of 30 m with greater resolution (ie, approximately 16 times higher compared to 500 m resolution). However, this accurate result may come in exchange for a higher cost of performance on a national or subnational scale. An MRV system must be feasible for the participating country and launching a regular flight program for aerial photographs or LiDAR may cost a substantial sum for some countries. In this case, using spaceborne imagery may be needed, and a high-resolution three-dimensional data set could be used either for calibration or verification purposes.

7.6 REGIONAL IMPLICATIONS (FOR AN EFFECTIVE TEMPORAL AND SPATIAL BIOMASS ESTIMATION METHOD FROM A REGIONAL PERSPECTIVE)

In this chapter, we have discussed several technical issues to be considered for estimating AGB in a forest environment, which include the allometric equation, ground sample, type of remote sensing data set, and prediction method. We presented a case study and compared the result by using two different data sets of ALS and SfM. Although the performance of RMSE and RMSE% are almost similar, further research must be performed to address the accuracy issues when applying the model for the entire area. Both data sets of ALS and SfM have their own advantages and disadvantages. However, high-resolution three-dimensional data sets have demonstrated great capability in estimating biomass in highly varied biomass values (ie, low biomass in degraded forest to high biomass in primary forest). Further research and development is needed to understand the error, reduce uncertainty, and improve the estimation; thus, a cost-effective method in yielding high accuracy can be further developed for forest monitoring of AGB in a tropical rainforest. If the flight program is not feasible for a country because of its high cost, at least the high-resolution three-dimensional data set can be acquired for a smaller area and used for calibration or verification purposes.

Acknowledgments

This study was supported by the Advanced Carbon Monitoring in Asian Tropical Forest by High Precision Remote Sensing Technology project of the Ministry of Agriculture, Fishery and Forestry (MAFF), Japan. The field data was collected by the members of Forestry and Forest Products Research Institute (FFPRI), Japan, the University of Tokyo, and Universiti Malaysia Sabah (UMS). We are thankful to Sabah Forestry Department (SFD) for the permission, to Sabah Forest Industries (SFI) and villagers of Kampung Long Mio for the field support.

References

Avitabile, V., Herold, M., Heuvelink, G.B.M., Lewis, S.L., Phillips, O.L., Asner, G.P., Armston, J., Asthon, P., Banin, L.F., Bayol, N., Berry, N., Boeckx, P., de Jong, B., DeVries, B., Girardin, C., Kearsley, E., Lindsell, J.A., Lopez-Gonzalez, G., Lucas, R., Malhi, Y., Morel, A., Mitchard, E., Nagy, L., Qie, L., Quinones, M., Ryan, C.M., Slik, F., Sunderland, T., Vaglio Laurin, G., Valentini, R., Verbeeck, H., Wijaya, A., Willcock, S., 2016. An integrated pan-tropical biomass map using multiple reference datasets. Glob. Change Biol. 22, 1406–1420. http://dx.doi.org/10.1111/gcb.13139.

Baccini, A., Friedl, M., Woodcock, C., Warbinghton, R., 2004. Forest biomass estimation over regional scales using multisource data. Geophys. Res. Lett. 31, L10501.

Baccini, A., Goetz, S.J., Walker, W.S., Laporte, N.T., Sun, M., Sulla-Menashe, D., Hackler, J., Beck, P.S.A., Dubayah, R., Friedl, M.A., Samanta, S., Houghton, R.A., 2012. Estimated carbon dioxide emissions from tropical deforestation improved by carbon-density maps. Nat. Clim. Change 2, 182–185. http://dx.doi.org/10.1038/nclimate1354.

Basuki, T.M., Skidmore, A.K., Hussin, Y.A., Van Duren, I., 2013. Estimating tropical forest biomass more accurately by integrating ALOS PALSAR and Landsat-7 ETM+ data. Int. J. Rem. Sens. 34, 4871–4888.

Basuki, T.M., van Laake, P.E., Skidmore, A.K., Hussin, Y.A., 2009. Allometric equations for estimating the above-ground biomass in tropical lowland Dipterocarp forests. For. Ecol. Manage. 257, 1684–1694. http://dx.doi.org/10.1016/j.foreco.2009.01.027.

Bohlin, J., Wallerman, J., Fransson, J.E.S., 2012. Forest variable estimation using photogrammetric matching of digital aerial images in combination with a high-resolution DEM. Scand. J. For. Res. 27, 692–699. http://dx.doi.org/10.1080/02827581.2012.686625.

Brown, S., 1997. Estimating Biomass and Biomass Change of Tropical Forests: A Primer. FAO Forestry Paper edition, UN FAO, Rome, Italy.

Brown, S., Gillespie, A.J.R., Lugo, A.E., 1989. Biomass estimation methods for tropical forests with applications to forest inventory data. For. Sci. 35, 881–902.

Brown, S., Iverson, L.R., Prasad, A., Liu, D., 1993. Geographical distributions of carbon in biomass and soils of tropical Asian forests. Geocarto Int. 4, 45–59.

Chave, J., Réjou-Méchain, M., Búrquez, A., Chidumayo, E., Colgan, M.S., Delitti, W.B.C., Duque, A., Eid, T., Fearnside, P.M., Goodman, R.C., Henry, M., MartínezYrízar, A., Mugasha, W.A., MullerLandau, H.C., Mencuccini, M., Nelson, B.W., Ngomanda, A., Nogueira, E.M., Ortiz-Malavassi, E., Pélissier, R., Ploton, P., Ryan, C.M., Saldarriaga, J.G., Vieilledent, G., 2014. Improved allometric models to estimate the aboveground biomass of tropical trees. Global Change Biol. 20, 3177–3190.

Chen, Q., 2015. Modeling aboveground tree woody biomass using national-scale allometric methods and airborne lidar. ISPRS J. Photogramm. Remote Sens. 106, 95–106.

Culmsee, H., Leuschner, C., Moser, G., Pitopang, R., 2010. Forest aboveground biomass along an elevational transect in Sulawesi, Indonesia, and the role of Fagaceae in tropical montane rain forests. J. Biogeogr. 37, 960–974.

Dobson, M.C., Ulaby, F.T., Le Toan, T., Beaudoin, A., Kasischke, E.S., Christensen, N.L., 1992. Dependence of radar backscatter on conifer forest biomass. IEEE Trans. Geosci. Rem. Sens. 30, 412–415.

Englhart, S., Jubanski, J., Siegert, F., 2013. Quantifying dynamics in tropical peat swamp forest biomass with multi-temporal LiDAR datasets. Rem. Sens. 5, 2368–2388.

Englhart, S., Keuck, V., Siegert, F., 2012. Modeling aboveground biomass in tropical forests using multi-frequency SAR data—a comparison of methods. IEEE J. Sel. Top. Appl. Earth Obs. Remote Sens. 5, 298–306.

FAO, 2010. Global Forest Resources Assessment 2010, Forestry Paper. ISBN: 978-92-5-106654-6.

Fassnacht, F.E., Hartig, F., Latifi, H., Berger, C., Hernández, J., Corvalán, P., Koch, B., 2014. Importance of sample size, data type and prediction method for remote sensing-based estimations of aboveground forest biomass. Remote Sens. Environ. 154, 102–114. http://dx.doi.org/10.1016/j.rse.2014.07.028.

Feldpausch, T.R., Lloyd, J., Lewis, S.L., Brienen, R.J.W., Gloor, M., Monteagudo Mendoza, A., Lopez-Gonzalez, G., Banin, L., Abu Salim, K., Affum-Baffoe, K., Alexiades, M., Almeida, S., Amaral, I., Andrade, A., Aragão, L.E.O.C., Araujo Murakami, A., Arets, E.J.M.M., Arroyo, L., Aymard, C.G.A., Baker, T.R., Bánki, O.S., Berry, N.J., Cardozo, N., Chave, J., Comiskey, J.A., Alvarez, E., de Oliveira, A., Di Fiore, A., Djagbletey, G., Domingues, T.F., Erwin, T.L., Fearnside, P.M., França, M.B., Freitas, M.A., Higuchi, N., Iida, Y., Jiménez, E., Kassim, A.R., Killeen, T.J., Laurance, W.F., Lovett, J.C., Malhi, Y., Marimon, B.S., Marimon-Junior, B.H., Lenza, E., Marshall, A.R., Mendoza, C., Metcalfe, D.J., Mitchard, E.T.A., Neill, D.A., Nelson, B.W., Nilus, R., Nogueira, E.M., Parada, A., Peh, K.S.-H., Pena Cruz, A., Peñuela, M.C., Pitman, N.C.A., Prieto, A., Quesada, C.A., Ramírez, F., Ramírez-Angulo, H., Reitsma, J.M., Rudas, A., Saiz, G., Salomão, R.P., Schwarz, M., Silva, N., Silva-Espejo, J.E., Silveira, M., Sonké, B., Stropp, J., Taedoumg, H.E., Tan, S., ter Steege, H., Terborgh, J., Torello-Raventos, M., van der Heijden,, G.M.F., Vásquez, R., Vilanova, E., Vos, V.A., White, L., Willcock, S., Woell, H., Phillips, O.L., 2012. Tree height integrated into pantropical forest biomass estimates. Biogeosciences 9, 3381–3403.

GFOI, 2013. Integrating remote-sensing and ground-based observations for estimation of emissions and removals of greenhouse gases in forests. In: Methods and Guidance from the Global Forest Observations Initiative. Group on Earth Observations, Geneva, Switzerland, 2014.

Ginzler, C., Hobi, M., 2015. Countrywide stereo-image matching for updating digital surface models in the framework of the Swiss national forest inventory. Remote Sens. 2015 (7), 4343–4370.

Gobakken, T., Bollandsås, O.M., Næsset, E., 2015. Comparing biophysical forest characteristics estimated from photogrammetric matching of aerial images and airborne laser scanning data. Scand. J. For. Res. 30, 73–86.

Hirata, Y., Tabuchi, R., Patanaponpaiboon, P., Poungparn, S., Yoneda, R., Fujioka, Y., 2014. Estimation of aboveground biomass in mangrove forests using high-resolution satellite data. J. For. Res. 19, 34–41.

Hirata, Y., Takao, G., Sato, T., Toriyama, J. (Eds.), 2012. REDD-plus Cookbook. REDD Research and Development Center, Forestry and Forest Products Research Institute, Japan, 156 pp, ISBN: 978-4-905304-15-9.

Houghton, R.A., Hall, F., Goetz, S.J., 2009. Importance of biomass in the global carbon cycle. J. Geophys. Res 114, G00E03. http://dx.doi.org/10.1029/2009JG000935.

Ioki, K., Tsuyuki, S., Hirata, Y., Phua, M.-H., Wong, W.V.C., Ling, Z.-Y., Saito, H., Takao, G., 2014. Estimating aboveground biomass of tropical rainforest of different degradation levels in Northern Borneo using airborne LiDAR. For. Ecol. Manage. 328, 335–341.

IPCC, 2006. 2006 IPCC Guidelines for National Greenhouse Gas Inventories, Prepared by the National Greenhouse Gas Inventories Programme, Agriculture Forestry and Other Land Use.

IPCC, 2014. Climate Change 2014: Synthesis Report. Contribution of Working Groups I, II and III to the Fifth Assessment Report of the Intergovernmental Panel on Climate Change, IPCC.

Jubanski, J., Ballhorn, U., Kronseder, K., Siegert, F., 2013. Detection of large above-ground biomass variability in lowland forest ecosystems by airborne LiDAR. Biogeosciences 10, 3917–3930.

Kenzo, T., Furutani, R., Hattori, D., Tanaka, S., Sakurai, K., Ninomiya, I., Kendawang, J.J., 2014. Aboveground and belowground biomass in logged-over tropical rain forests under different soil conditions in Borneo. J. For. Res. 20, 197–205. http://dx.doi.org/10.1007/s10310-014-0465-y.

Ketterings, Q.M., Coe, R., van Noordwijk, M., Ambagau', Y., Palm, C.A., 2001. Reducing uncertainty in the use of allometric biomass equations for predicting above-ground tree biomass in mixed secondary forests. For. Ecol. Manage. 146, 199–209. http://dx.doi.org/10.1016/S0378-1127(00)00460-6.

Koch, B., 2010. Status and future of laser scanning, synthetic aperture radar and hyperspectral remote sensing data for forest biomass assessment. ISPRS J. Photogramm. Remote Sens. 65, 581–590. http://dx.doi.org/10.1016/j.isprsjprs.2010.09.001.

Kronseder, K., Ballhorn, U., Böhm, V., Siegert, F., 2012. Above ground biomass estimation across forest types at different degradation levels in Central Kalimantan using LiDAR data. Int. J. Appl. Earth Obs. Geoinf. 18, 37–48.

Langner, A., Achard, F., Vancutsem, C., Pekel, J.-F., Simonetti, D., Grassi, G., Kitayama, K., Nakayama, M., 2015. Assessment of above-ground biomass of Borneo forests through a new data-fusion approach combining two pan-tropical biomass maps. Land 4, 656–669. http://dx.doi.org/10.3390/land4030656.

Leberl, F., Irschara, A., Pock, T., Meixner, P., Gruber, M., Scholz, S., Wiechert, A., 2010. Point clouds: Lidar versus 3D vision. Photogramm. Eng. Remote Sen. 76, 1123–1134.

Manuri, S., Brack, C., Nugroho, N.P., Hergoualc'h, K., Novita, N., Dotzauer, H., Verchot, L., Putra, C.A.S., Widyasari, E., 2014. Tree biomass equations for tropical peat swamp forest ecosystems in Indonesia. For. Ecol. Manage. 334, 241–253. http://dx.doi.org/10.1016/j.foreco.2014.08.031.

Mauya, E., Hansen, E.H., Gobakken, T., Bollandsås, O.M., Malimbwi, R.E., Næsset, E., 2015. Effects of field plot size on prediction accuracy of aboveground biomass in airborne laser scanning-assisted inventories in tropical rain forests of Tanzania. Carbon Balance Manage. 10, 1–14.

McNicol, I.M., Berry, N.J., Bruun, T.B., Hergoualc'h, K., Mertz, O., de Neergaard, A., Ryan, C.M., 2015. Development of allometric models for above and belowground biomass in swidden cultivation fallows of Northern Laos. For. Ecol. Manage. 357, 104–116.

Morel, A.C., Fisher, J.B., Malhi, Y., 2012. Evaluating the potential to monitor aboveground biomass in forest and oil palm in Sabah, Malaysia, for 2000–2008 with Landsat ETM+ and ALOS-PALSAR. Int. J. Rem. Sens. 33, 3614–3639.

Nakagawa, M., Matsushita, M., Kurokawa, H., Samejima, H., Takeuchi, Y., Aiba, M., Katayama, A., Tokumoto, Y., Kume, T., Yoshifuji, N., Kuraji, K., Nagamasu, H., Sakai, S., Nakashizuka, T., 2012. Possible negative effect of general flowering on tree growth and aboveground biomass increment in a bornean tropical rain forest. Biotropica 44, 715–719.

Nurminen, K., Karjalainen, M., Yu, X., Hyyppä, J., Honkavaara, E., 2013. Performance of dense digital surface models based on image matching in the estimation of plot-level forest variables. ISPRS J. Photogramm. Remote Sens. 83, 104–115. http://dx.doi.org/10.1016/j.isprsjprs.2013.06.005.

Okuda, T., Suzuki, M., Numata, S., Yoshida, K., Nishimura, S., Adachi, N., Niiyama, K., Manokaran, N., Hashim, M., 2004. Estimation of aboveground biomass in logged and primary lowland rainforests using 3-D photogrammetric analysis. For. Ecol. Manage. 203, 63–75. http://dx.doi.org/10.1016/j.foreco.2004.07.056.

Ota, T., Ogawa, M., Shimizu, K., Kajisa, T., Mizoue, N., Yoshida, S., Takao, G., Hirata, Y., Furuya, N., Sano, T., Sokh, H., Ma, V., Ito, E., Toriyama, J., Monda, Y., Saito, H., Kiyono, Y., Chann, S., Ket, N., 2015. Aboveground biomass estimation using structure from motion approach with aerial photographs in a seasonal tropical forest. Forests 6, 3882–3898. http://dx.doi.org/10.3390/f6113882.

Pearson, T., Walker, S., Brown, S., 2005. Sourcebook for Land Use, Land-Use Change and Forestry Projects. BioCarbon Fund & Winrock Internationa.

Rahlf, J., Breidenbach, J., Solberg, S., Næsset, E., Astrup, R., 2014. Comparison of four types of 3D data for timber volume estimation. Remote Sens. Environ. 155, 325–333. http://dx.doi.org/10.1016/j.rse.2014.08.036.

Rutishauser, E., Noor'an, F., Laumonier, Y., Halperin, J., Rufi'ie, Hergoualch, K., Verchot, L., 2013. Generic allometric models including height best estimate forest biomass and carbon stocks in Indonesia. For. Ecol. Manage. 307, 219–225.

Saatchi, S.S., Harris, N.L., Brown, S., Lefsky, M., Mitchard, E.T.A., Salas, W., Zutta, B.R., Buermann, W., Lewis, S.L., Hagen, S., Petrova, S., White, L., Silman, M., Morel, A., 2011. Benchmark map of forest carbon stocks in tropical regions across three continents. Proc. Natl. Acad. Sci. U. S. A. 108, 9899–9904. http://dx.doi.org/10.1073/pnas.1019576108.

Sabah Biodiversity Conservation Project (SBCP), 1998. Identification of Potential Protected Areas: Ulu Padas Final Report, 123 pp.

Sinun, W., Suhaimi, J., 1997. The Hydrological and Geomorphological Assessment of the Ulu Padas. Ministry of Tourism and Environmental Development Sabah, Kota Kinabalu.

Slik, J.W.F., Aiba, S.-I., Brearley, F.Q., Cannon, C.H., Forshed, O., Kitayama, K., Nagamasu, H., Nilus, R., Payne, J., Paoli, G., Poulsen, A.D., Raes, N., Sheil, D., Sidiyasa, K., Suzuki, E., van Valkenburg, J.L.C.H., 2010. Environmental correlates of tree biomass, basal area, wood specific gravity and stem density gradients in Borneo's tropical forests. Glob. Ecol. Biogeogr. 19, 50–60. http://dx.doi.org/10.1111/j.1466-8238.2009.00489.x.

Solberg, S., Næsset, E., Gobakken, T., Bollandsås, O.-M., 2014. Forest biomass change estimated from height change in interferometric SAR height models. Carbon Bal. Manag. 9, 5.

Stibig, H.-J., Achard, F., Carboni, S., Raši, R., Miettinen, J., 2014. Change in tropical forest cover of Southeast Asia from 1990 to 2010. Biogeosciences 11, 247–258. http://dx.doi.org/10.5194/bg-11-247-2014.

Thapa, R.B., Watanabe, M., Motohka, T., Shiraishi, T., Shimada, M., 2015. Calibration of aboveground forest carbon stock models for major tropical forests in central Sumatra using airborne LiDAR and field measurement data. IEEE J. Sel. Topics Appl. Earth Observ. Remote Sens. 8, 661–673.

Toma, T., Ishida, A., Matius, P., 2005. Long-term monitoring of post-fire aboveground biomass recovery in a lowland dipterocarp forest in East Kalimantan, Indonesia. Nutr. Cycl. Agroecosys. 71, 63–72.

UNFCCC, 2005. Eleventh Session of the Conference of the Parties (COP 11), November 2005, Montreal, Canada.

UNFCCC, 2007. Thirteenth Session of the Conference of the Parties (COP 13), December 2007. Bali, Indonesia.

UNFCCC, 2009. UNFCCC Resource Guide Module 3: National Greenhouse Gas Inventories for Preparing the National Communications of Non-annex I Parties Module 3 National Greenhouse Gas Inventories. Financial and Technical Support Programme of the UNFCCC, Bonn, Germany.

III. LEARNING FROM THE FIELD CASES/ISSUES

UNFCCC, 2014. Handbook on Measurement, Reporting and Verification for Developing Country Parties, United Nations Climate Change Secretariat, Bonn, Germany.

Vastaranta, M., Wulder, M.A., White, J.C., Pekkarinen, A., Tuominen, S., Ginzler, C., Kankare, V., Holopainen, M., Hyyppä, J., Hyyppä, H., 2013. Airborne laser scanning and digital stereo imagery measures of forest structure: comparative results and implications to forest mapping and inventory update. Can. J. Remote Sens. 39, 1–14. http://dx.doi.org/10.5589/m13-046.

Vesakoski, J.-M., Alho, P., Hyyppä, J., Holopainen, M., Flener, C., Hyyppä, H., 2014. Nationwide digital terrain models for topographic depression modelling in detection of flood detention areas. Water 6, 271–300.

White, J., Wulder, M., Vastaranta, M., Coops, N., Pitt, D., Woods, M., 2013. The utility of image-based point clouds for forest inventory: a comparison with airborne laser scanning. Forests 4, 518–536. http://dx.doi.org/10.3390/f4030518.

Wong, W., Tsuyuki, S., Ioki, K., Phua, M.H., 2014. Accuracy assessment of global topographic data (SRTM & ASTER GDEM) in comparison with lidar for tropical montane forest. In: 35th Asian Conference on Remote Sensing 2014, Myanmar.

Wong, W.V.C., Tsuyuki, S., Ioki, K., Phua, M.H., Takao, G., 2015. Forest biophysical characteristics estimation using digital aerial photogrammetry and airborne laser scanning for tropical montane forest. In: 36th Asian Conference on Remote Sensing 2015, Philippines. 10 p.

Yamakura, T., Hagihara, A., Sukardjo, S., Ogawa, H., 1986. Aboveground biomass of tropical rain forest stands in Indonesian Borneo. Vegetatio, 68, 71–82.

Yoneda, T., Tamin, R., Ogino, K., 1990. Dynamics of aboveground big woody organs in a foothill dipterocarp forest, West Sumatra, Indonesia. Ecol. Res. 5, 111–130.

Yuen, J.Q., Fung, T., Ziegler, A.D., 2016. Review of allometric equations for major land covers in SE Asia: uncertainty and implications for above- and below-ground carbon estimates. For. Ecol. Manage. 360, 323–340.

Integrating Social Entrepreneurship in the Design Principles of Long-Enduring Irrigation Management Institutions: A Lesson From the Karya Mandiri Irrigation System in West Sumatra, Indonesia

Helmi, B.A. Rusdi

Andalas University, Padang, Indonesia

8.1 INTRODUCTION

Self-governing and long-enduring common pool resource (CPR) institutions have been at the center of attention to sustain a benefits stream from any particular CPR. Ostrom (1990) has proposed eight "Design Principles" for self-governing CPR institutions and emphasized the importance of these principles in crafting institutions for self-governing irrigation systems (Ostrom, 1992). Since then, the efforts have been made to test the applicability of the design principles with mixed results, which then led to suggestions for modifications or expansions being proposed, especially in the context of self-governing irrigation systems.

This chapter attempts to provide an illustration from a long-enduring, self-governing, small-scale irrigation system in West Sumatra, Indonesia. The central lesson that this chapter presents is that there is a need to integrate the social entrepreneurship orientation with the design principles, which means moving from a social to a social entrepreneurship orientation in irrigation management. The concept of social entrepreneurship will be clarified first and then the evolution of the irrigation management institutions in the Karya Mandiri Irrigation System (KMIS) will be discussed with reference to the eight design principles proposed by Ostrom.

8.2 FROM A SOCIAL TO A SOCIAL ENTREPRENEURSHIP ORIENTATION: A PERSPECTIVE IN UNDERSTANDING FACTORS AFFECTING THE LONG-ENDURING IRRIGATION INSTITUTIONS

There were indications that institutions only concerned about their social mission without a strategy to support the achievement of that social mission would not be sustainable or would end up as "just enough organization" (Bruns, 1992; Helmi, 2002). Brinkerhoff and Goldsmith (1992) defined institutional sustainability as "the ability of an organization to produce output of sufficient value so that it can acquire enough input to continue production at a steady or growing rate." In this connection, Wilson (1992) and Cernea (1993) pointed out that it is both the weaknesses in institutional arrangements within which the organization responsible for production and provision of good and services operate and the insufficient attention to institutional sustainability that inhibits the continuation of a benefits stream from development activities. These arguments indicate that there is a need to move from merely an orientation toward achieving the social mission toward a social entrepreneurship orientation. The point is that the creation of social value requires a sustainability strategy in the form of income-generating activities embedded in the provision of social services. In short, this means achieving the social mission and generating sufficient income (and making money) to support the continuation of the provision of social (in this case irrigation water provision) services (benefits stream).

The concept of social entrepreneurship is rooted in the concept of entrepreneurship itself. The following definitions are presented to give a better understanding of the concept. First, Martin and Osberg (2007) argued that "entrepreneurship describes the combination of a context in which an opportunity is situated, a set of personal characteristics required to identify and pursue this opportunity, and the creation of a particular outcome." Based on the meaning of entrepreneurship, they define social entrepreneurship "as having the following three components:

(1) identifying a stable but inherently unjust equilibrium that causes the exclusion, marginalization, or suffering of a segment of humanity that lacks the financial means or political clout to achieve any transformative benefit on its own;

(2) identifying an opportunity in this unjust equilibrium, developing a social value proposition, and bringing to bear inspiration, creativity, direct action, courage, and fortitude, thereby challenging the stable state's hegemony; and

(3) forging a new, stable equilibrium that releases trapped potential or alleviates the suffering of the targeted group, and through imitation and the creation of a stable ecosystem around the new equilibrium ensuring a better future for the targeted group and even society at large."

Second, Alvord et al. (2002) pointed that "social entrepreneurship creates innovative solutions to immediate social problems and mobilizes the ideas, capacities resources, and social arrangements required for sustainable social transformation." Third, Dees (2001) describes social entrepreneurs as "one species in the genus entrepreneur, they are entrepreneurs with a social mission." Furthermore, he added that "for social entrepreneurs, the social mission is explicit and central and mission-related impact becomes the central criterion, not wealth creation, wealth is just a means to an end for social entrepreneurs." Fourth, Mort et al. (2002) defined social entrepreneurship as "a multidimensional construct involving the expression of entrepreneurially virtuous behavior to achieve the social mission, a coherent unity of purpose and action in the face of moral complexity, the ability to recognize social value-creating opportunities and key decision-making characteristics of innovativeness, proactiveness, and risk taking."

In this chapter, the case of KMIS in the context of the social entrepreneurship concept beside the eight design principles by Ostrom (1990, 1992) is analyzed and factors affecting the endurance of irrigation management institutions are identified.

8.3 THE KARYA MANDIRI IRRIGATION SYSTEM (KMIS): EVOLUTION AND ENDURANCE OF IRRIGATION MANAGEMENT INSTITUTIONS

The KMIS is an irrigation system located in Agam District, West Sumatra (precisely at the Sungai Janiah subvillage, Tabek Panjang). It has 87 ha service areas, which in 1995 extended to become 127 ha after a neighboring subvillage (named Salo) agreed to integrate their rice field into the service areas of KMIS and follow the rules of the irrigation service provision. In 1994, the population of the subvillage was more than 1200 and in 2008 it has increased to became more than 1500 people. The population is divided into four major clans, namely: (1) Suku Tanjung; (2) Suku Jambak; (3) Suku Koto; and (4) Suku Sikumbang. Resource mobilization for irrigation management was done based on the clan groupings. Rice paddy is the major crop with some farmers who might prefer to plant different kinds of vegetables. However, Rusdi (2008) reported that nowadays the farmers concentrate more on planting high-quality rice (with a relatively higher price) because they experienced diseases that infected their chili and tomato plants.

An irrigation management institution at the KMIS consists of three components: the council of clan leaders, the representatives of the clans or the group that is responsible for managing or performing irrigation management tasks, and the farmers and other water users. The council of clan leaders is the last resort for important decisions on irrigation management (like assigning the representative of the clans or approving the group that will be responsible for actually managing the system) and for conflict resolution. The representatives of the clans or the group are those who got the mandate from the council of clan leaders to perform irrigation management tasks and when they found they could no longer bear the tasks, they would return the mandate to the council of clan leaders. The farmers and other water users are the actual water users, be it for agriculture or for religious activities.

The KMIS management institutions have evolved through three different forms with different approaches in performing irrigation management tasks. The first form was using a collective action approach that was characterized by (1) labor mobilization by clan leaders for rehabilitation and maintenance of brush dam and canal cleaning and maintenance; and (2) contribution of materials required (to repair brush dam) proportional to land holding size. This approach was in place approximately until the 1950s and was replaced by new approach as it was no longer effective in facilitating labor mobilization to implement irrigation management tasks.

The second form was using a partial contractual approach,[1] which is characterized by (1) the implementation of major irrigation tasks (eg, operation and maintenance/rehabilitation of brush dam, and main canal maintenance), which were contracted to a group of farmers (known as the Group of Sixteen) representing farmers from all four clans in the village; (2) collective action in branch canals maintenance, which channeled water to the farmers' fields; and (3) payment of an irrigation service fee at the rate of 20% of rice yield every harvest season. This was practiced from the 1950s to 1988. This approach was replaced by the third one because the cost of replacing the brush dam every time it was washed away by a flood and rehabilitation of the main canal after a landslide occured, and other operational costs, could no longer be covered by the irrigation service fee collected.

The third form is (and still functioning until the current period) contracting the provision of irrigation services to a group of farmers (known as the Group of Eight). The third approach is characterized by (1) the Group of Eight being responsible for financing the construction of a concrete dam and main canal and performing all irrigation management tasks (the Group of Eight as service provider); (2) payment of an irrigation service fee at the rate of 20% of yield; and (3) opportunities for the service provider to access funds from other sources (parties) than the farmers (such as the government) and use it to repay the construction cost borrowed from the members of the group or other villagers.[2]

The agreement in the contract specifically covered the following points: (1) The Group of Eight will build the concrete dam by using the design provided by the district office of the Ministry of Public Works and all the construction costs will be mobilized by the Group; (2) The contract will last for 25 years (but was later revised and extended to be 30 years because the dam broke 2 years after construction and has to be rebuild) and the farmer water users will pay 20% of the yield as an irrigation service fee; (3) The Group of Eight is responsible for managing the system such that all the farmers could be provided with sufficient water every season; (4) If in any case the rice field is temporarily planted with vegetables, then the farmers still have to pay an irrigation service fee equal to the amount of 20% yield of the rice paddy; and (5) If in the future the irrigation system receives financial assistance from the government or other source, the Group of Eight is eligible to receive monies to repay the costs of construction for both the fund mobilized from the member of the group or borrowed from other village members (Helmi, 1994; Rusdi, 2008).

---

[1] There were four points of agreement in the contract: (1) handing over the responsibility for the management of the brush dam and main canal to the Group of Sixteen; (2) the main tasks of the group were to develop and/or rehabilitate the dam and the main canal, while the branch canals were still the responsibility of the respective farmers receiving water from them; (3) the period of the contract was 20 years; and (4) all the farmer water users were obliged to pay 20% of the yield every season as an irrigation service fee (Helmi, 1994).

[2] Rusdi (2008) found that the Group of Eight mobilized their own money and borrowed an additional amount from farmers or other village members, and they are responsible to return it by using part of the 20% irrigation service fee. He calculated at the time the contract was agreed upon, the level of interest the Group of Eight could give at that time was 13% per annum, lower than the bank interest, which was 15%.

The three forms of approaches clearly contained the eight principles developed by Ostrom (1990). If otherwise, it would not be possible for them to enforce the rules that were agreed upon. However, in addition to those eight principles the stakeholders at KMIS have moved further to develop social entrepreneurship principles/orientation (see Table 8.1 for a description of those principles), which consists of two aspects as follows:

- Aspect 1: make money from providing irrigation services and distribute the benefits through the cofinancing of the infrastructures (dam and main canal) development, and
- Aspect 2: build the mechanisms to ensure stable (or increased) revenue from the irrigation service provision: ensuring irrigation water availability at the on-farm level, developing a planting schedule, assisting the farmers with land preparation through the use of hand tractors (making sure that no rice field is uncultivated), providing additional services to the farmers in terms of availability of agriculture inputs (fertilizer), transferring agriculture technology (the use of high-yielding varieties (HYVs) of rice), and "selling" irrigation water service to the neighbor subvillage.

TABLE 8.1 Ostrom's Design Principles for Long-Enduring Irrigation Institutions and the Principles Adopted at the KMIS

No.	Ostrom's Design Principles (1990 and 1992)	Karya Mandiri Irrigation System (KMIS) 1994 (Helmi, 1994)	Karya Mandiri Irrigation System (KMIS) 2008 (Rusdi, 2008)
1.	Clearly defined boundaries	The service area of the system is 87 ha, within the Sungai Janiah subvillage with six clans (four bigger and two smaller)	The service area became 127 ha as a 40 ha rice field in neighboring Salo subvillage requested to get water and agreed to follow the rules
2.	Congruence between appropriation and provision rules and local conditions	The farmers would pay 20% of the harvest and their rice field guaranteed to get sufficient irrigation water to plant rice crop	The farmers would pay 20% of the harvest and their rice field guaranteed to get sufficient irrigation water to plant rice crop
3.	Collective choice arrangements	The management agreements were developed by clan leaders together with members of the clans/villagers and they can modify the agreement as necessary	The management agreements were developed by clan leaders together with members of the clans/villagers and they can modify the agreement as necessary
4.	Monitoring	The "Group of Eight" as service provider contactor is responsible for monitoring and ensuring proper service provision	The "Group of Eight" as service provider contactor is responsible for monitoring and ensuring proper service provision
5.	Graduated sanctions	The clan leaders are responsible for resolving any complaint both from the farmers and the service provider	The clan leaders are responsible for resolving any complaint both from the farmers and the service provider

Continued

III. LEARNING FROM THE FIELD CASES/ISSUES

TABLE 8.1   Ostrom's Design Principles for Long-Enduring Irrigation Institutions and the Principles Adopted at the KMIS—cont'd

No.	Ostrom's Design Principles (1990 and 1992)	Karya Mandiri Irrigation System (KMIS) 1994 (Helmi, 1994)	Karya Mandiri Irrigation System (KMIS) 2008 (Rusdi, 2008)
6.	Conflict resolution mechanisms	Conflicts are resolved through the meeting of clan leaders, service provider, and the farmers/villagers	Conflicts are resolved through the meeting of clan leaders, service provider, and the farmers/villagers
7.	Minimal recognition of rights to organize	The customary institutions are recognized by law and their roles in society are respected by the government	The customary institutions are recognized by law and their roles in society are respected by the government
8.	Nested enterprise		In 1995, the neighbor Salo subvillage requested to get irrigation services and agreed to follow the rules applied in Sungai Janiah subvillage
9.		Social entrepreneurship principle/orientation: provide irrigation services and making money out of that and distribute the benefits through the co-financing of the infrastructure development	Social entrepreneurship principle/orientation: provide irrigation services and making money out of that and distribute the benefits through the co-financing of the infrastructure development
10.		Social entrepreneurship principle/orientation: building the mechanisms to ensure stable (or increase) revenue from irrigation service provision by providing additional services to the farmers in term of availability of agriculture inputs (fertilizer)	Social entrepreneurship principle/orientation: building the mechanisms to ensure stable (or increase) revenue from service provision by providing additional services to the farmers in term of availability of agriculture inputs (fertilizer) and technology transfer (rice HYVs)

On top of those principles, everyone tried to maintain social trust among themselves through the role of the clan leader council and participatory processes in all decisions made. As mentioned earlier, the council of clan leaders is the last resort for important decisions regarding irrigation management, such as which group would be given responsibility or a contract for the management of the system; the group could also return the contract when the revenue they generated from the irrigation service fee could no longer cover the management and capital costs; the amount of the service fee that should be paid by the farmers; and resolution of conflicts. The application of those principles has enabled the institution managing the KMIS to develop major infrastructures (concrete dam and main canal lining) that cost them Rp.90 million (in 1992) and sustain the benefits stream from the system (see Fig. 8.1 about the condition of the concrete dam and Fig. 8.2 for piping irrigation water to the neighboring subvillage).

FIG. 8.1 The concrete dam of KMIS. *From Rusdi, B.A., 2008. Development With the Basis of Tradition and Cultural Capital at the Development and Management of Communal Irrigation System: A Case Study at Sungai Janiah Sub-village, Tabek Panjang, Agam District, West Sumatra, Indonesia. Andalas University, Padang (research report).*

FIG. 8.2 Piping irrigation water to the neighboring subvillage (Salo). *From Rusdi, B.A., 2008. Development With the Basis of Tradition and Cultural Capital at the Development and Management of Communal Irrigation System: A Case Study at Sungai Janiah Sub-village, Tabek Panjang, Agam District, West Sumatra, Indonesia. Andalas University, Padang (research report).*

III. LEARNING FROM THE FIELD CASES/ISSUES

8.4 CONCLUSION AND LESSONS LEARNED

The local community has crafted irrigation institutions at the KMIS that enabled them to adapt to the pressures and changes, make necessary investments, and perform various management functions. This has made the system continue to exist as a self-governing irrigation system and serve the farmers. The design principles of irrigation institutions at the KMIS are those principles proposed by Ostrom (1990, 1992). In addition to those eight principles the stakeholders at the KMIS have moved further to develop a social entrepreneurship principle and orientation, which consists of two aspects:

- Provide irrigation services and make money out of that and distribute the benefits through the cofinancing of the infrastructures (dam and main canal) development, and
- Build the mechanisms to ensure stable (or increased) revenue from the irrigation service provision: ensuring irrigation water availability at the on-farm level, developing a planting schedule, assisting the farmers with land preparation through the use of hand tractors (making sure that no rice field is uncultivated), providing additional services to the farmers in terms of availability of agriculture inputs (fertilizer), and agriculture technology transfer (the use of HYVs of rice), and "selling" irrigation water service to the neighboring subvillage.

The eight design principles and the social entrepreneurship orientation have enabled the irrigation institutions for the management of the KMIS to endure for a long period of time.

References

Alvord, S.H., Brown, L.D., Letss, C.W., 2002. Social Entrepreneurship and Social Transformation: An Exploratory Study. Hauser Center for Non-profit Organizations Working Paper No. 15.

Brinkerhoff, D., Goldsmith, A., 1992. Promoting the sustainability of development institutions: a framework for strategy. J. World Dev. 20 (3), 369–383.

Bruns, B., 1992. Just enough organization: water user association and episodic mobilization. J. Visi Irigasi Indonesia 6, 33–41.

Cernea, M., 1993. The sociologist's approach to sustainable development. Finance Dev. 30 (4), 11–13.

Dees, J.G., 2001. The Meaning of Social Entrepreneurship. Center for the Advancement of Social Entrepreneurship at Duke Universityís Fuqua School of Business. California, Stanford.

Helmi, 1994. From collective management to contractual system: farmers' response to the changing environment of small-scale irrigation management in West Sumatra, Indonesia. J. VISI Irigasi Indonesia 9 (4), 35–48.

Helmi, 2002. Moving Beyond water provision: strategic issues related to transition of irrigation management in Indonesia. In: A Paper Prepared for the Workshop on "Asian Irrigation in Transition – Responding to the Challenges Ahead" Asian Institute of Technology (AIT), Bangkok, Thailand, 22–23 April.

Martin, R.L., Osberg, S., 2007. Social Enterpreneurship: The Case for Definition. Stanford Social Innovative Review, Spring 2007.

Mort, G., Weerawardena, J., Carnegie, K., 2002. Social entrepreneurship: towards conceptualization. Int. J. Nonprofit Volunt. Sect. Mark. 8 (1), 76–88.

Ostrom, E., 1990. Governing the Commons: The Evolution of Institutions for Collective Action. Cambridge University Press, New York.

Ostrom, E., 1992. Crafting Institutions for Self-governance Irrigation Systems. Institute for Contemporary Studies Press, San Francisco.

Rusdi, B.A., 2008. Development With the Basis of Tradition and Cultural Capital at the Development and Management of Communal Irrigation System: A Case Study at Sungai Janiah Sub-village, Tabek Panjang, Agam District, West Sumatra, Indonesia. Andalas University, Padang (research report).

Wilson, F., 1992. Institutional sustainability—the beginning of a search for conceptual framework. In: Curry, et al. (Eds.), Sustainable Development and Environmental Issues: A Colloqium, New Series of Discussion Paper No. 22. University of Bradford, Bradford.

Land Rights and Land Reform Issues for Effective Natural Resources Management in Indonesia

M.T. Sirait*, B. White†, U. Pradhan*

*ICRAF-SEA, Bogor, Indonesia †Institute of Social Studies (ISS), Den Haag, The Netherlands

9.1 INTRODUCTION

In recent years (after Indonesia's 1998 reforms), through a long process of struggle between the Ministry of Forestry (MoF), National Land Agency (BPN), private sector, local government, and peasant movements, there have been some cases where upland peasant communities succeeded in being allocated individual land rights from the converted forest areas under the public land redistribution policy. For reasons of food security and bowing to pressure for land by the landless peasants, the MoF gave a "green light" to implement land reform through land redistribution to the tillers on a small scale in several densely populated areas of Indonesia in Java and Sumatra. The state (forest) land redistribution here is a process of redistribution of the so-called state (forest) land to the tillers that are already cultivating the land in traditional mixed farming or so-called agroforest. The "state" lands redistributed to the peasants were not an empty space, but land that had already been subject to an informal tenure system and provided them with individual land "ownership" (meaning that land may be bought and sold, and transferred from one generation to the next, even though it does not have formal private ownership status).

This chapter analyzes the changing agrarian structure in two hamlets in Indonesia (two upland hamlets practicing mixed farming or agroforest in Java and Sumatra) where so-called state forest areas were transferred to tillers after the 1998 political reforms, and looks back at the 8–10 years after a land reform process redistributed forest land under individual land title to peasants that had been cultivating the land. It aims to question whether individual private ownership tenure is the appropriate form of tenure in land redistribution programs, by analyzing the emergence of inequality in access to and control over land and resources from the early stage of the land redistribution process until approximately 8–10 years after land was classified as private lands in an Indonesia context.

In both cases, the land redistribution was able to provide access to land that potentially could bring benefits and security to all landless peasant households. But a few years after the land redistribution, significant numbers of landless peasant households had appeared, due to the effects of open competition and surplus appropriation by other classes. In both sites, there were also newcomer landless households that came to the hamlet and increased the number of landless households. They provide their labor as sharecroppers to the middle peasant and rich peasant class or as paid labor to nearby plantations. Rich peasant households have also appeared in both sites both during and after land redistribution, which reflects the fact that the land was not distributed equally, and those who controlled more land under informal tenure before the land redistribution received more formal individual land title. In both sites, the rich peasant households were predominantly those households that were rich before the land redistribution; they either owned other land in their own village/hamlet outside the areas of the land redistribution program where they were dominant actors in the struggle for land redistribution, and/or had a large number of household members. Some of them were able to accumulate more wealth and become landlord households through engaging in trading (opening small shops/kiosks in the village or hamlet).

The premise of the evolutionary theory of land rights that individual landownership would address the problems of tenure security, productive farm investment, access to credit, and so forth was only proved valid for some of the peasant household in both sites, especially the better-off peasant households, those who belong to the landlord class, and especially the absentee landowners from the cities. For those from the lower socioeconomic class, the near landless, and some of the middle-class peasant household, they were dispossessed and became landless only a few years after the land was redistributed. Those who could maintain control of their land are those few peasant households who are extra diligent, hardworking, and thrifty and those peasant household that have other income from nonfarming activities, such as those who received remittances from their household members in the cities and those who run small shops that extract surplus through usury and unequal terms of trade from the other peasant households in the hamlet.

9.2 PROBLEMS IN FOREST AND LAND GOVERNANCE IN INDONESIA

The Indonesian National Land Agency, *Badan Pertanahan Nasional* (BPN) is mandated to administer the entire land base of Indonesia according to Agrarian Law no. 5/1960. This law serves as the basis for defining and classifying land as public (state) land or private land, and for allocating land for large-scale plantations such as rubber, coffee, and tobacco, with a more recent emphasis on converting forest areas into oil palm plantations.[1] The MoF claims that the prevailing forestry legislation (Law no. 5/1967 and Law no. 41/1999) classifies two-thirds of Indonesia's total land area as state forest areas, and that this land is, therefore, under the

---

[1](Colchester et al., 2006).

MoF's jurisdiction. Land tenure arrangements implemented under each of these two sets of legislation conflict with each other,[2] resulting in the two concerned agencies being locked in competition over the control of vast areas of land in which 18,000 to 30,000 villages[3] and adat communities'[4] land are located. In practice, the BPN does not provide a land titling service to peasants located within areas classified as state forest areas, despite being mandated to do this by the Basic Agrarian Law 5/1960 (UPPA 1960). Upland peasants living outside state forest areas are often too remotely located to be reached by land agency officials. Thus, vast areas of the rural uplands, including privately held lands, are located in areas classified as state forest areas or "political forests" (Peluso & Lund, 2011), or have already been enclosed as corporate forests, as has happened in other parts of the world (Sikor and Tranh, 2007). This land includes settlement areas, productive agricultural lands, and some areas under shifting cultivation, none of which are actually forests.[5] Almost 50 million Indonesian peasants live in and around these political forests and, therefore, cultivate land under conditions of unclear and insecure land tenure. Of these peasants, 10 million (25% of all poor Indonesian households) are living below the poverty line, according to Indonesian Statistics 2005 and MoF data (Ministry of Forestry, 2006).[6] Access of local communities to land and natural resources is restricted through regulations that do not allow the cutting of trees, hunting, cultivating land, or house construction. In several cases, a local community's access to land and natural resources has been completely terminated by leasing rights over the land and resources to private-sector companies within production forests, or by classifying the area as protected forests (Rachman, 1997).

There are few options in Indonesia for peasants to gain secure tenure for their settlements, and for farming and mixed farming (agroforests) within areas classified as forest areas. Since 1995, the MoF has experimented with *Hutan Kemasyarakatan* (HKm), a collective community forestry stewardship program under state forest areas, to provide upland farmers with limited access to land and resources, and since 2007, *Hutan Desa* (HD), a village forest stewardship program (under state forest areas), has also been in effect. However, these initiatives

---

[2] (Moniaga, 1993; Soemardjono, 1998; Fay and Sirait, 2004).

[3] The Ministry of Forestry (2007, 2009) has released a publication based on statistical village data (Podes) that assesses the status of the villages located in and around forest areas, where complete villages or sections of them are located in areas designated as forest areas. A total of about 30,000 villages in Indonesia (40%) were classified as being in forest areas, from a total of 75,000 villages in Indonesia. According to MoF statistics in 2014, 12,000 villages have been formally excised from forest area for their housing and settlements, which means that their farm and forest land is still under the classification of forest areas.

[4] There are no data on the overlap of the land of adat communities (*masyarakat hukum adat*) with forest areas. AMAN (2013) stated that about 40 million hectares of adat communities' land overlapped with forest areas. A total of 7 million hectares was mapped in 2013 and a further 40 million hectares of adat land situated in forest areas will be mapped in 2020 (see http://www.hijauku.com/2013/08/26/aman-targetkan-40-juta-hektar-wilayah-adat-terpetakan-pada-2020/).

[5] It is evident from an analysis of satellite images taken in 2001 that 24 million hectares of the forest zone are not covered by forests (Santoso, 2002).

[6] This shows that even though the population living in and around forests is not considerable. As shown in MoF data, these inhabitants make up 25% of the Indonesian population living below the poverty line.

have resulted in very few pilot projects or forest stewardship contracts.[7] While indigenous cultural community forests (*Hutan Adat*) are an alternative tenure option for communal forest ownership (non–state forest areas), they entail complex administrative requirements that none of the intended beneficiaries are able to fulfill.

Land tenure security through stewardship contracts, which is provided for under both of the abovementioned programs (HKm and DII), does not permit residential settlements in these areas. These programs were exclusively designed for timber-based farming or forest protection, and do not, therefore, include food-based agriculture and indigenous mixed farming (agroforestry).[8] Moreover, the MoF has rejected a proposal by the local government and BPN for redesignating state forest areas—which have been actively managed by peasants as forestry-agriculture mosaics—as non–state forest areas.

9.3 POVERTY AND LAND CONFLICT

The 2013 agricultural census reveals that the Indonesia population has reached 259 million people, out of which 104.5 million people are peasants (40% of the total population). Of this number, 55.3% of peasants are almost or completely landless (controlling less than 0.25 ha of land), and most are living below the poverty line. Between 1998 and 2013, the percentage of the population living below the poverty line showed a significant decrease from 24.2% to 11.4%. However, the percentage of people living below the poverty line in rural areas remains quite high at 14.4% (17.92 million people) compared with 8.5% (10.63 million people) in urban areas (BPS, 2011, 2014). These figures reflect the location of the poor within rural areas in Indonesia, where most are engaged in the agricultural sector. The poverty gap index (GPI) and the poverty severity index (PSI) in Indonesia's rural areas are quite high (a GPI of 2.4 and a PSI of 0.6 in September 2013) compared with corresponding values for urban areas (a GPI of 1.4 and a PSI of 0.4 in September 2013). The statistical data shows increases in these indexes from time to time, which reflect growing inequality of expenditures in rural areas. Parallel to this trend, large-scale agriculture and forestry-based industries have rapidly penetrated the uplands. There is serious competition and conflict over land allocated for large-scale, commercially oriented concessions and upland peasant farming. There are also serious issues of inequality within upland peasant societies that are engaged in multiple opportunities provided by large-scale agricultural and forestry concessions.[9] The manifestations of this inequality

[7] Kemitraan (2014) has calculated that since the program was developed in 1995, only 0.5 million hectares were allocated under the community-based forest management (CBFM) scheme (0.51% of the forest area), and 99.5% of this area was allocated for large-scale concessions. Local communities registered their complaints with the ombudsmen in 2013, regarding the MoF's maladministration and discrimination in the allocation of forests for large-scale concessions compared with CBFM schemes (http://seg.mitra.or.id/2013/04/laporan-kemitraan-2012/).

[8] Safitri (2010). In practice, exceptions have occurred where mixed agroforestry was practiced in forest areas of Lampung under the HKm program.

[9] See White (1999) regarding inequality of access to land and resources as the consequences of the large-scale Nucleus Smallholder Estate Program implemented in upland Java. Only 0.25 million hectares were allocated under the CBFM scheme compared with 35.8 million of hectares allocated to 531 large-scale private and state forest concessions under terms of long lease (Sirait et al., forthcoming: 45)

are clear: land conflicts between peasants and the state over large-scale forestry concessions, plantations, and other allocations of state land for forest conservation and protection, mining, and oil palm plantations; a high rate of land degradation[10]; and a considerable proportion of households living below the poverty line, as indicated by the statistical data.

In recent years (subsequent to the 1998 reform), a long struggle between the forest agency, land agency, private sector, local governments, and peasant movements has culminated in some cases where upland peasant communities have succeeded in obtaining individual land rights within converted forest areas under the public land redistribution policy.[11] For reasons of food security and submission to pressure for land demands by landless peasants, the MoF gave the "green light" for the implementation of a land reform through small-scale state (forest) land redistribution to tillers in several densely populated areas in Java[12] and Sumatra.[13]

9.4 RATE OF DEFORESTATION 1995, 2000, 2005, AND 2015

The land cover trends provided in Fig. 9.1 shows the total area for each land-use type during three time periods (1990, 2000, and 2005).

Analysis shows that forest size is declining rapidly. In 1990 total forest area was about 68% (128 million ha) and was reduced to 53% (99.7 million ha) in 2005. At the same time, secondary forest increased from 12% (22.4 million ha) in 1990 to 21% (3.5 million ha) in 2005. Agroforests declined from 11% (19.7 million ha) in 1990 to 8.6% (16.3 million ha) in 2005. Agroforest lands were relatively unchanged during 1990–2000 and then rapidly declined at a rate of 2.13 million ha/year.

Three main changes in land use in Indonesia were observed during the period 1990–2005 (Ekadinata et al., 2011):

- Decline in the area of forest cover and increase in secondary forest and monoculture plantation areas
- Decline in agroforest land cover
- Increase in secondary forest cover and settlement.

Data that can identify the existence of agroforests and their dynamics, that separate them from secondary forest as well as monoculture plantations, are very useful for decision makers. The data can help them to see the trends of declining access of the communities to their resources, and the shift of control over resources to large-scale business, during 1995–2005 and subsequently. Such data was not available prior to 2005. Data was made available through the

---

[10] Indonesia was ranked poorly at 102 out of 149 countries in the 2008 Environmental Performance Index published by Yale and Columbia universities. In the Food and Agriculture Organization of the United Nations (FAO)'s State of the World's Forests (SOFO) Report for 2008, Indonesia was assessed as a global leader in land-based, or terrestrial carbon emissions, with approximately equal emission rates from aboveground carbon stock (mostly trees) and belowground carbon (mostly peat lands). See Ekadinata et al. (2010).

[11] Affif et al. (2005, p. 27) shows how the Indonesian state, as the biggest and most powerful landlord, was challenged during this era of reform.

[12] Rachman (2003).

[13] Fathullah et al. (2005) and Bachriadi and Sardjono (2006).

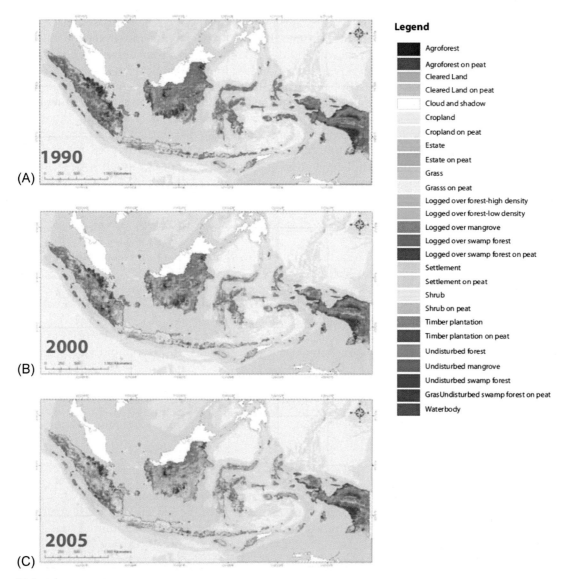

Legend

- Agroforest
- Agroforest on peat
- Cleared Land
- Cleared Land on peat
- Cloud and shadow
- Cropland
- Cropland on peat
- Estate
- Estate on peat
- Grass
- Grasss on peat
- Logged over forest-high density
- Logged over forest-low density
- Logged over mangrove
- Logged over swamp forest
- Logged over swamp forest on peat
- Settlement
- Settlement on peat
- Shrub
- Shrub on peat
- Timber plantation
- Timber plantation on peat
- Undisturbed forest
- Undisturbed mangrove
- Undisturbed swamp forest
- GrasUndisturbed swamp forest on peat
- Waterbody

FIG. 9.1   Map of Land Cover in Indonesia: (A) 1990, (B) 2000, (C) 2005. *From Ekadinata et al., 2011.*

contribution of technology development and research institutes; this was very useful, particularly when dealing with the increasing conflicts after 2005 (Fig. 9.2).

Data from Forest Watch Indonesia (FWI) in 2014 showed that the most deforestation happened in the two islands of Sumatera and Kalimantan. The deforestation rate per hour is 110% of the size of a football field ($90 \times 120 \, m^2$) dominated by logging concession, timber estate, plantation estate, open pit mining, road and building development, as well as agriculture expansion. The deforestation in Jawa, Sulawesi, and Papua islands were indicated in the lower rate.

Alternatives to protect the remaining forest and finding solutions to address the deforestation rate and carbon emission are crucial in forest and landscape management in Indonesia.

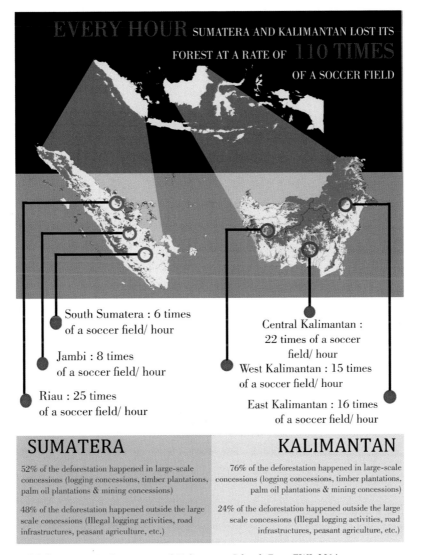

FIG. 9.2   Rate of deforestation in Sumatera and Kalimantan Island. *From FWI, 2014.*

Community-based natural resource management especially in managing landscape in agro-forest seems a good solution to address deforestation, conflict, and the poverty issue.

9.5 THE EVOLUTIONARY THEORY OF LAND RIGHTS (ETLR) TRAP

Without having a complete picture of past results of the kind of land redistribution process carried out in early 2007, the government subsequently announced a plan to redistribute 8 million hectares of former forest areas to peasants under a land reform program, the National

Program on Agrarian Reform (PPAN). This program prioritizes 33 districts in the southern regions of the island of Java[14] for land redistribution. However, in 2011 the pace of PPAN slowed down due to the shifting politics of land allocation, which resulted in priority being given to the granting of large-scale concessions, mainly outside Java, to support the National Master Plan for the Acceleration and Expansion of Indonesian Economic Development (*Masterplan Percepatan dan Perluasan Pembangunan Ekonomi Indonesia*) (MP3EI), established through Presidential Decree no. 32/2011. The MP3EI has been revoked by the new cabinet who took office in October 2014, and the land reform agenda has been put back on the table with a new target of 9 million hectares in 5 years, as promised by the elected president and vice president (Seknas Jokowi, 2014). Reformulation and reorientation of PPAN is, however, necessary.

Microlevel research is commonly neglected within mainstream discourses such as national policy and the international policy debate on forestry and agriculture.[15] Seven years before de Soto, Platteau (1996) already highlighted the problems in the policy trend of individualizing and formalizing property land rights for upland communities in Sub-Saharan Africa by criticizing the theory of evolutionary transition of communal land property rights to private property rights, the so-called evolutionary theory of land rights (ETLR). He unraveled four basic assumptions held by advocates of the ETLR: (1) security of individual land tenure; (2) expansion of land markets; (3) increasing credit and investments; and (4) increasing demands for land titling.

He argued that individual land tenure was not secure for those denied legal recognition of their customary rights to land during the registration process. Moreover, the insecurity of other contractual users, who could use the land only at the sufferance of the owners after the reform, was even greater. He further showed that the elites were able to adjudicate or manipulate the registration process to their advantage. Platteau (1996: 41–49) concluded that land

---

[14] The Jakarta Post Daily (2006) on PPAN (the National Programme on Agrarian Reform) as well as the BPN 2008–2009RAJASELA programme (Southern Java Agrarian Reform).

[15] With the exceptions of Scoones et al. (2012), Cliffe et al. (2011), and Moyo (2011), who carried out microlevel studies in Zimbabwe and compiled an edited book on the outcomes of post–2000 Fast Track Land Reform in Zimbabwe (Cliffe et al., 2013), most of the previous research and policy has neglected microlevel analysis. See the Revitalization Agenda of Indonesian Agriculture (2005) (*Agenda Revitalisasi Pertanian* 2005), as well as the Indonesian Agricultural, Fishery and Forestry Revitalization (2005) (*Revitalisasi Pertanian, Perikanan dan Kehutanan*, 2005). See also Menkoekuin (2005) and Jamal (2006) who emphasize the development of agroindustries and village-based industries. The Strategic Plan of the Indonesian Forestry Department (2005–2009) (*Rencana Startegis Kementrian Kehutanan*, 2005–2009) places emphasis on balancing the supply and demand of wood to support forest-based industries, and points to a lack of studies on peasant agroforesters. However, the World Bank (2008) did not place enough emphasis on describing the problems faced by peasant agroforesters. For details, see http://www.rimisp.org/consultwdr2008/. This report has been commented on by several experts, who still believe that wealth is redistributed to the poor by virtue of the market. The Asian Development Bank (ADB)'s (2006) report on Indonesia's Strategic Vision for Agriculture and Rural Development also emphasizes the problem of land tenure security and recommends land redistribution through individual and communal land titling. In contrast to these reports, the World Agroforestry Center (a Consultative Group for International Agricultural Research (CGIAR) research center, formerly known as the International Center for Research in Agroforestry (ICRAF)) has published reports since the 1980s that aim to foster a better understanding of peasants who practice agroforestry in third world countries (see www.worldagroforestrycentre.org).

registration created land disputes as people not in possession of land titles came under the threat of eviction from registered proprietors.

Platteau challenged the second assumption by providing evidence that the land market has not functioned effectively because the majority of land parcels continue to be transacted through customary channels (lending, gifts, inheritance, or nonregistered sales) and, therefore, markets for leaseholds appear to be relatively rare. Consequently, land sales tend to be the result of distress conditions due to the absence of insurance markets, imperfect credit markets, and declining self-insurance capacities on the part of rural dwellers. This situation tends to worsen the imbalance in factor proportions between larger and smaller holdings (Platteau, 1996: 49–60; see also Burchart et al., 2002, regarding the result of land reform in Nicaragua and Honduras).

Platteau further demonstrated the negative impact of the third assumption on credit and investment. He found no significant relationship between the percentage of households receiving credit and the proportion of land held with individual land rights. This situation mostly occurred as a result of administrative cost considerations that led many banks to set a minimum land area for obtaining credit, often exceeding the capital needs of small holders (Platteau, 1996: 60–66). On the investment side, there have been various findings, including those of Suyanto et al. (2007), that titling in Indonesia has increased capacities for investment but not willingness to invest in land improvements.

Lastly, Platteau challenged the fourth assumption, showing that demand and supply for land titling does not work because the main concern of poor peasants is clearly security of land tenure for their own use and for transmission to their children. Most vulnerable populations, namely, women, pastoralists, former slaves, and other groups, have traditionally had weak tenure arrangements as a result of being granted only subsidiary or derived rights to land. They may, therefore, prefer to have individual landownership granted through the state, if they can afford to pay for the land redistribution process to obtain the land through a fair and just process. However, if they believe that the local chief is trustworthy, they may alternatively prefer to have collective rights rather than individual titles to ensure continuance of customary overlapping rights (Platteau, 1996: 66–72). In summary, land registration or titling of individual property rights in land is not a neutral process. It clearly favors more influential classes of cultivators and excludes others (including women). Ultimately, it may end up creating less and not more security of tenure (Platteau, 1996: 39–49: in White et al., 2012).

A few years after the land was redistributed, landless peasant households had reemerged in both cases due to the impact of open competition and surplus appropriation by other classes and the arrival of landless peasants from other areas. Using the terminology of Cliffe et al. (2011) the "dropping off" peasant households are those who for some reason sold their plots and became landless. Some landless peasant households remain in the hamlet and work as sharecroppers and paid labor, but there are also others that left the hamlet, looking for other job opportunities as paid labor in the nearest plantation estates, and still others who found other job opportunities as agriculture extension workers or government officials in the village. Fifteen percent of the landless class in both cases were newcomers to the hamlets. They sell their labor as sharecroppers to the middle peasants and rich peasants, seek employment as paid labor in the nearest plantation, and at the same time save their money looking for affordable plots of farmland to buy or occupy forestland for their own use (see Table 9.1 in which this class is highlighted in red).

TABLE 9.1   Change in the Class Structure of Households in Ciniti and Simpang Duren, Based on Landholdings Before and After Land Redistribution

No.	Land-based class	Ciniti (Garut)		Simpang Duren (Lampung)	
		Before LR (%)	After LR	Before LR (%)	After LR
1.	Landless Peasants	0	9.3	0	26.7
2.	Near landless Peasants	5.2	13.1	1.3	1.3
3.	Middle Peasants (Category II)	50.4	30.2	34.7	28
4.	Middle Peasants (Category I)	30.4	27.9	41.3	32
5.	Rich Peasants	14	16.3	14.7	8
6.	Landlords	0	3.2	8	4

BPN Cadastral Map & Household Survey, 2010.

The cases also revealed that the (numerically) dominant class in this society, during the land redistribution process, and 10 years after its completion, were still middle peasant households (belonging to categories I and II).[16] However, their numbers in society were reduced by between 30% and 15% following land redistributions. Whereas previously they constituted 81% and 76% of the populations in Ciniti and Simpang Duren, respectively, these percentages were subsequently reduced to 58% and 60%. The total amount of land owned by these classes (middle peasant I and II categories) was also reduced by between 23% and 16% of their former amounts (see Table 9.1 in which these classes are highlighted in green).

Rich peasant households also emerged in both sites either during or after land redistribution, which reflects the fact that the land was not distributed equally, and that those who controlled more land before land redistribution under informal tenure formalized this through individual landownership. An examination of the actors from rich peasant households in both sites shows that these households were predominantly rich before the land redistribution. They maintained their wealth either through ownership of land in their own village/hamlet outside the areas of the land redistribution program, and their dominance in the process of struggle for land redistribution, or through their large size. Some of these households were former beneficiaries of the land redistribution programs that were able to accumulate more wealth and became rich peasants through their engagement in trading (opening small shops/kiosks in the village or hamlet) and those who were able to receive remittances from household members working abroad.

Eight to ten years after land redistribution in Ciniti and Simpang Duren, a landlord class had emerged, constituting 3.2–4% of the population. They owned a total of 30%–42% of the land in these areas (see Table 9.1 in which this class is highlighted in yellow). The "landlord class," however, consisted of only a few local households that were able to

[16] Rachman (2003).

upgrade their status to landlords through trade as their main means of transferring surplus from other peasant classes in the hamlets. Most of them were absentee landowners who accumulated land using their own savings acquired from work in the nearest cities.

9.6 NEW POLICIES TREND RELATED TO ACCESS TO FORESTRY AND LAND REFORM

In March 2013, Judicial Review no. 35 of Forestry Law no. 41/1999, regarding the status of customary forest, revoked the classification of customary forest as state land. This land is currently classified as non–state forest, or under private ownership. However, to be classified as private forest, there are still several administrative requirements for proving customary community status through local government regulations. Network for Participatory Mapping (*Jaringan Kerja Pemetaan Partisipatif*) (JKPP), a nongovernmental organization (NGO) consortium of community mappers, projected between 42 and 80 million hectares of customary forests in Indonesia that could potentially be classified as customary forests through this law, but that is currently delineated under state forest areas (JKPP, 2014).

A further law, Village Law no. 6/2014, was passed in early 2014. This gives autonomy rights to a village to be designated as a Customary Village or an Administrative Village with the required customary institutional structure or village administrative structure. This law gives a village the jurisdiction to manage its territories, including village-owned forests, land, and other natural resources (Article 76). It will be fully implemented within 1 year, and approved villages will also be able to manage their own annual budgets of 1.2 billion rupiah (approximately US$100,000).

In 2014, the MoF and the National Statistical Office, through its provincial spatial planning department, claimed to have reduced the number of villages having areas overlapping with forests from 28,689 villages in 2007 and 2009 to 18,239 villages in 2014 (MoF-BPS, 2014). This means that there are 10,450 upland villages that have had their land excised from state forest areas, and will be included in the land registration program under the BPN. As predicted by JKPP (2014) and Zakaria (2014), following the implementation of Judicial Review no. 35/2012 and the establishment of village autonomy under Village Law 6/2014, there will be more villages and customary forests excised from state forest areas, but still considered as forest area, that should maintain their forest functions. These communities will be trapped in the BPN's land registration program through a well-established system of individual landownership certification available at every BPN district office.

In the absence of a law, rules and regulations, as well as government offices that can accommodate the process of collective land registration, the pressure on peasants to acquire land tenure security within areas of land conflict is pragmatically following the mainstream course of formalizing the informal land tenure to individual landownership.

In 2013, Pokker SHK, a NGO in central Kalimantan, reported that during the implementation of Governor's Regulation 4/2012, allowing individual and communal ancestral land to be registered through the *demang* (adat chief), both the *demang* and the local government preferred to accommodate individual land claims rather than register communal ancestral land (Pokker, 2013). The Perkumpulan Pancur Kasih director, Mateus Pilin (2013, pers. comm., 24 Juni 2013), observed that the customary community that was being facilitated by his NGO to do community mapping succeeded in getting their communal land mapped, and some of these lands were recognized by the subdistrict government. However, the cohesion of the customary community fell apart when the private palm oil companies began to negotiate with the local elites, one by one. These local elites have tended to release their communal lands, treating them as if they were their individually owned lands, to the private sector, which has ruined the cohesion of the customary community.

This kind of trajectory has been observed in other parts of the Indonesian uplands, not only under programs of land redistribution in forest areas, as discussed in this chapter, but also in nonforested, postland conflict areas. In most of these situations, the NGO supporters and the local peasants or customary community movement were at the forefront during the conflict era, but were not able to sustain the momentum to prepare a long-term vision after the land conflict. The slogan, "land for the tillers" is widely used by most of the peasant movements to unify the movement against private large-scale concessions, or against the state (Setiawan, 2010). Changing an unjust agrarian structure was mostly considered from the external perspective of a community, viewed as a single entity, versus private/state actors, and not as an agrarian structure and agrarian relations that were internally unjust.

9.7 CHALLENGES AND OPPORTUNITIES

The evolutionary theory of land rights, discussed above, was seen by the NGO supporters in both sites, and in other parts of Indonesia, as a natural internal process, and in some cases NGO activists were reluctant to intervene. This study has shown that the market and state both play an important role in engineering the evolution of land rights toward privatization, rather than retaining their communal land tenure form or accommodating the household land tenure unit. The delaying tactics played by the previous government, to create a regulation that will establish a clearer process of land redistribution, create communal land rights titles, and secure the rights of the customary community, is a clear indication of the state's reluctance to regulate this issue. Thus, changes in the process are determined by the demands of the market for the benefit of the private sector.

It is apparent from this study that Indonesia's upland communities need security of tenure for their land and farming systems, which could protect them from land appropriation by state and local players. Approaches to land redistribution under individual ownership, as practiced so far, have not been appropriate for these upland communities that managed and possess the land in a household unit. The upland peasant households that were the focus of this study have been forced to follow the free market route, which includes farmland as one of the commodities that has been internally contested by peasants and other absentee land speculators. This process has been formative of a capitalist peasant society in the upland areas. As revealed by this study, and also by other studies (see Chrisantiny, 2007) the government land agency (BPN) was unable to provide support to every individual peasant, especially in the upland areas, to register their plots. Both this study and Chrisantiny (2007) have shown that land redistribution cannot lead to the registration of all lands. Only those who are sufficiently well off were able to acquire individual land certificates.

The civil society movement led by peasant organizations and unions that have been promoting land reform and redistribution should carefully examine the results of the land redistribution process several years after it occurred. Those in the civil society movement who promote recognition of adat rights through communal landownership should also observe internal processes of elite capture.

In this situation, there is a need for civil society organizations to work with peasant organizations and customary communities to create a long-term vision that extends beyond conflict resolution to enable local communities to identify problems relating to an unjust agrarian structure and to agrarian relations that are both internal and external to the communities. This requires us to be aware that agrarian differentiation is not a natural process, but a social construction process that needs to be addressed from the early steps in land redistribution and its land tenure choices. With good understanding of the trajectories of agrarian differentiation resulting from the design and implementation of land redistribution—as in the two cases we have studied—local communities should design their own land redistribution process. This would enable the local community through its own local leadership to periodically conduct a land redistribution, and decide when the land redistribution should be done. This is particularly important as an unjust agrarian structure will reappear as a result of market penetration and exploitative agrarian relations.

Knowing that there are always limitations to government rules and regulations, it is necessary to prepare a complete set of rules and regulations as well as build the capacities of state organizations to run genuine land redistribution programs that are suitable for the pluralistic Indonesian situation (in Java, the Outer Islands, forests, and nonforest areas and customary and peasant communities, upland and lowland communities). There is also a need to rethink generic rules and regulations according to clear principles for regulating the process.

It may be that individual or communal landownership tenures are not the critical need of upland communities, but rather local autonomy of communities and a clear vision of local leadership to regulate the land redistribution process. With a good understanding of the problems of agrarian differentiation arising from this study, there are several policy implications that need to be developed. These include

1. Communal land tenureship that can accommodate pluralistic local tenure forms and protect household farming units.
2. Guarantee of land allocation for all peasant households with maximum land under the control of peasant household units. This is important to limit accumulation by rich peasant households.
3. Sufficient allocation of land for women and youth as an incentive for them to farm and cultivate the land in their village or their ancestral land. A special effort is needed to emphasize women and youth as they are more innovative and will continue farming the land in the future.
4. Allocation of local budgets (that is, village funds under Law 6/2014) to buy back village or ancestral lands that were sold or released to absentee landowners, and the subsequent allocation of these lands for productive use.
5. Creation of consensus in allocating land for tillers, not under ownership, but under contractual bases with the village or adat communities. This should be periodically reviewed and allow for redistribution among the landless peasant households.

6. Creation of a comprehensive village or adat community land database with clear planning regarding the community's future. This should not only address the social aspects, but also involve economically viable and environmentally sound planning.
7. Government should have the right to interfere in a situation where village elite capture has happened at the village level, and also interfere to ensure the rate of deforestation and carbon emission is maintained low at the district level.

With this clear vision and planning conceptualized at the local level, there is a need to incorporate the lessons learned in the draft land law, draft law on the protection of customary communities, rules and regulation on land redistribution, and communal land tenure. Local autonomy at the village level, as defined in Village Law no 6/2014 means that village communities should be given the authority to lead the effective functioning of local land redistribution mechanisms, and these obligations should be clearly written into the implementation guidelines, including processes and mechanisms for resolving land conflicts.

References

ADB, 2006. Indonesia Strategic Vision for Agriculture and Rural Development. ADB, Manila.

Afiff, S., Noer, F., Gillian, H., Lungisile, N., Nancy, P., 2005. Refining Agrarian Power: Resurgent Agrarian Movements in West Java, Indonesia. Paper CSEASWP2-05. Berkeley: Center for Southeast Asia Studies, UC Berkeley.

AMAN, 2013. Indigenous World 2011. AMAN, Jakarta.

Bachriadi, D., Sardjono, A., 2006. Local Initiatives to Return Communities' Control Over Forest Land in Indonesia; Conversion or Occupation? Paper Presented during the 11th Biennial Conference of International Association for the Study of Common Property, Bali – Indonesia, June 19–23, 2006.

Biro Pusat Statistik, 2011. Statistik Indonesia. Biro Pusat Statistik, Jakarta.

Biro Pusat Statistik, 2014. Statistik Indonesia. Biro Pusat Statistik, Jakarta.

Burchart, T., le Grand, J., Piachaud, 2002. Introduction to social exclusion. In: Hills, J., Grand, J., Piachaud, D. (Eds.), Understanding Social Exlusion. Oxford University Press, Oxford.

Chrisantiny, P., 2007. Berawal dari Tanah; Melihat ke Dalam Pendudukan Tanah. Akatiga, Bandung.

Cliffe, L., Jocelyn, A., Ben, C., Rudo, G., 2011. An overview of fast track land reform in Zimbabwe. J. Peasant Stud. 38 (5), 907–938.

Cliffe, L., Alexander, J., Cousins, B., Gaidzanwa, R. (Eds.), 2013. Outcomes of post-2000 Fast Track Land Reform in Zimbabwe. Routledge, London.

Colchester, M., Martua, S., Boedhi, W., 2003. The Application of FSC Principles 2&3 in Indonesia; Obstacles & Possibilities. Walhi, Jakarta.

Colchester, M., Norman, J.A., Martua, S., Asep, F.S., Herbet, P., 2006. Promised Land: Palm Oil and Land Acquisition in Indonesia, Implication for Local Communities and Indigenous Peoples. FPP-ICRAF-HuMa-Sawit Watch, London-Bogor.

Ekadinata, A., Meine van, N., Sonya, D., Peter, M., 2010. Reducing emissions from deforestation, inside and outside the "forest". ASB Policy Brief 16. ASB Partnership for the Tropical Forest Margin, Nairobi, Kenya.

Ekadinata, A., Widayati, A., Dewi, S., Rahman, S., van Noordwijk, M., 2011. Indonesia's Land-Use and Land-Cover Changes and Their Trajectories (1990, 2000 and 2005). ALLREDDI Brief 01. World Agroforestry Centre - ICRAF, SEA Regional Office, Bogor, Indonesia, 6 p.

Fathullah, L.S., Cahyaningsih, N., Nuch, I., Sirait, M., 2005. Perubahan Status Kawasan Hutan, Guna Menjawab Permasalahan Kemiskinan dan Ketahanan Pangan; Studi Kasus dari Marga Bengkunat & Pekon Sukapura, Kabupaten Lampung Barat. In: Paper presented at the International Tenure Conference; Questioning the Answer. Jakarta 11–13 Oktober 2004.

Fay, C., Sirait, M., 2004. Kerangka Hukum Agraria dan Kehutanan Indonesia; Mempertanyakan Sistem Ganda Kewenangan atas Penguasaan Tanah. Paper presented at International Tenure Conference; Questioning the Answer. Jakarta 11–13 Oktober 2004, Jakarta.

Jakarta Post Daily, 2006. Government to offer land to the poor but experts unsure of program. Jakarta Post daily news, Jakarta. 23rd May 2006.

Jamal, E., 2006. Revitalisasi Pertanian dan Perbaikan Penguasaan Lahan di Tingkat Petani. Jurnal Analisis Sosial 11 (1), 105–121.

JKPP., 2014. Potensi Wilayah Adat Indonesia (PIWA). In: Presented during the JKPP Conference, 6 January 2014. JKPP, Bogor.

Kemitraan, 2014. Rencana dan Realisasi pengelolan Hutan Berbasis Masyarakat. Kemitraan, Jakarta.

Menkoekuin, 2005. Revitalisasi Pertanian, Perikanan dan Kehutanan Indonesia 2005, untuk Rakyat, Tanah Air dan Generasi Indonesia mendatang. Kantor Menko Bidang Perekonomian, Jakarta.

Ministry of Agriculture, 2005. Agenda Revitalisasi Pertanian. Departemen Pertanian, Jakarta.

Ministry of Forestry, 2005. Rencana Srategis Kementerian Negara/Lembaga (Renstra-KL) Departemen Kehutanan 2005–2009, (Stratgical Plan of the State Forestry Ministry 2005–2009). Kementrian Kehutanan, Jakarta.

Ministry of Forestry, 2006. Rencana Pembangunan Jangka Panjang Kehutanan 2006–2025. Kementrian Kehutanan, Jakarta.

Ministry of Forestry, 2007. Desa dalam Kawasan Hutan. Kementrian Kehutanan, Jakarta.

Ministry of Forestry, 2009. Desa dalam Kawasan Hutan. Kementrian Kehutanan, Jakarta.

Moniaga, S., 1993. Toward community-based forestry and recognition of adat property rights in outer Island of Indonesia. In: Fox, J. (Ed.), Legal Frame Works for Forest Management in Asia; Case Studies of Community/State Relations. East West Center, Hawaii.

Moyo, S., 2011. Changing agrarian relations after redistributive land reform in Zimbabwe. J. Peasant Stud. 38 (5), 939–966.

Peluso, N.L., Lund, C., 2011. New frontiers of land control: introduction. J. Peasant Stud. 38 (4), 683–701.

Platteau, J.P., 1996. The evolutionary theory of land rights as applied to Sub-Saharan Africa: a critical assessment. Dev. Change 27 (1), 29–86.

Pokker, S.H.K., 2013. Hasil Kajian Penerapan SKTA di Kalimantan Tengah. Pokker SHK, Palangkaraya.

Rachman, F.N., 1997. Tanah dan Pembangunan. Pustaka Sinar Harapan, Jakarta.

Rachman, F.N., 2003. Bersaksi untuk Pembaruan Agraria: Dari Tuntutan Lokal Hingga Kecendrungan Global. Insist Press, Jogyakarta.

Safitri, M., 2010. Forest tenure in Indonesia: the socio-legal challenges of securingcommunities' rights. Doctoral thesis, Faculty of Law. Leiden: Leiden University.

Santoso, H., 2002. Rasionalisasi kawasan Hutan di Indonesia; Kajian Keadaan Sumber Daya Hutan dan Reformasi Kebijakan. Badan Planology Dephut- World Bank, Jakarta.

Scoones, I., Nelson, M., Blasio, M., Jacob, M., Felix, M., Chrispen, S., 2012. Zimbabwe's Land Reform. Myths and Realities. Woodbridge, Johanesburg.

Seknas Jokowi, 2014. Jalan Kemandirian Bangsa. Gramedia Pustaka, Jakarta.

Setiawan, U., 2010. Kondisi Hutan dan Gerakan Petani in Setiawan, Usep 2010 Kembali Ke Agraria. STPN, Yogyakarta.

Sikor, T., Tranh, N.T., 2007. Exclusive versus inclusive devolution in forest management: insights from forest land allocation in Vietnam's Central Highlands. Land use Policy J. 24, 644–653.

Sirait Martua, T., Pradhan, U., Rachman, N.F., Safitri, M., forth coming. Strengthening Forest Management through Land Tenure Arrangement. ICRAF-SEA, Bogor, Bogor.

Soemardjono, M., 1998. Kewenangan Negara untuk mengatur Dalam Konsep Penguasaan Tanah oleh Negara, Presented during the inauguration of Professor-ship at University of Gajah Mada, 14 February 1998. UGM, Yogyakarta.

Suyanto, S., Thomas, T., Kenjiro, O., 2007. The role of land tenure in the development of Cinnamon Agroforestry in Kerinci Sumatera. In: Cairns, Malcom (Ed.), Voices from the Forest Integrating Indigenous Knowledge into Sustainable Upland Farming. RFF Press, Washington.

White, B., 1999. Inti dan Plasma: Pertanian Kontrak dan Pelaksanaan Kekuasaan di dataran Tinggi Jawa Barat. In: Li Tania, M. (Ed.), Proses Transformasi daerah pedalaman di. Yayasan Obor Indonesia, Jakarta, Indonesia.

White, B., Borras Jr., S.M., Hall, R., Scoones, I., Wolford, W., 2012. The new enclosures: critical perspectives on corporate land deals. J. Peasant Stud. 39 (3–4), 619–647.

World Bank, 2008. A yellow cover draft of the World Development Report 2008 on Agriculture for Development. World Bank, Washington DC.

Zakaria, Y., 2014. Sehari Membedah Undang-Undang Desa Bersama Yando Zakaria. Desa Membangun, Banyumas.

III. LEARNING FROM THE FIELD CASES/ISSUES

Dynamics and Effectiveness of the Multistakeholder Forum in Promoting Sustainable Forest Fire Management Practices in South Sumatra, Indonesia*

E. Achyar*, D. Schmidt-Vogt†, G. Shivakoti‡,§

*Institute of Participatory Approaches, Development & Studies, Padang, West Sumatra, Indonesia †Mountain Societies Research Institute, University of Central Asia, Bishkek, Kyrgyz Republic ‡The University of Tokyo, Tokyo, Japan §Asian Institute of Technology, Bangkok, Thailand

10.1 INTRODUCTION

The prevention of forest fires has become an emerging issue and global concern over the past three decades, primarily due to (1) the growing number of fires, and the global estimate of between 300 and 400 million hectares per year being affected by fires worldwide in 2002/2003 (FAO, 2005); (2) the widespread and negative impacts of forest fires on ecosystems, biodiversity, habitats, livelihoods, and economies (FAO, 2007; Qadri, 2001; Barber and Schweithelm, 2000); (3) an increased awareness of the emissions of carbon dioxide (CO_2) and other greenhouse gases by fires, and their contribution to climate change (FAO, 2007; Cochrane, 2003; Page et al., 2002); and (4) the transnational impacts of haze caused by aerosols emitted from forest fires (ASEAN Haze online, at http://haze.asean.org.). Forest fires are very complex phenomena with respect to their causes and effects, as is the issue of developing sustainable fire management solutions. Whether forest fires have a negative or positive impact depends, among other factors, on the ecosystem affected by them, and on their intensity and frequency.

*An earlier version of this chapter has been published in *Environmental Development*.

Indonesia contains one of the world's largest tropical forests. In terms of area under forest, Indonesia ranks third behind Brazil and the Republic of Congo (Zaire), and is considered one of the foremost carbon sinks in Asia (FWI/GFW, 2002). The tropical rain forest ecosystem of Indonesia is adapted to small-scale disturbances by localized and infrequent fires (Corlett, 2009). However, due to a combination of factors, such as droughts caused by the El Niño Southern Oscillation (ENSO) phenomenon, transmigration (resettlement schemes), and economic incentives to convert forest to commercial plantations (eg, of oil palms) using fire, parts of Indonesia have in recent decades become subject to some of the largest, most persistent, and destructive forest fires in recorded history. Fire is often the only available tool for land clearing for local people as it is cheaper and faster than mechanical methods (Varkkey, 2013; Simorangkir, 2006; Qadri, 2001; Barber and Schweithlem, 2000). For example, the large fires of 1982/1983, followed by the enormous burning that occurred during the summer of 1997/1998, destroyed 11,698,379 ha of forest, primarily in Kalimantan and Sumatra (Tacconi, 2003). According to data from the Ministry of Forestry (MoF), forest fires have decreased notably over the last 5 years (2007–2011), with a maximum area of 7619 hectares being burnt in 2009, and 2612 hectares in 2011 (Ministry of Forestry, 2012). However, in 2014 fires started early and, according to Greenpeace, by March had covered an area of at least 12,000 ha in Riau province of Sumatra alone (The Economist, 2014). The forest fires of Indonesia are not only a national problem, but have attained transnational significance due to haze impacting on neighboring countries such as Singapore and Malaysia (The Economist, 2013). The need to manage or prevent fires of this scale and intensity requires national and transnational collaboration, and has prompted the Indonesian government to introduce a number of measures to reduce fire risk and improve suppression efforts, including (1) increased collaboration with international institutions, (2) the development and improvement of new and existing institutions aimed at addressing fire management challenges, and (3) the implementation of a series of new laws and regulations related to fire management (Morgera and Cirelli, 2009; FAO, 2007; Goldammer, 2007; Tacconi et al., 2006; Simorangkir and Sumantri, 2002; Dennis, 1999).

However, past experience indicates that institutional arrangements among stakeholders involved in fire management are ineffective at preventing and controlling fires, both at the national and local levels. The reasons cited for this include conflicting roles, functions, and responsibilities; overly bureaucratic procedures; a lack of cooperation and coordination at all levels; and weak law enforcement (Herawati and Santoso, 2011; Morgera and Cirelli, 2009; Simorangkir and Sumantri, 2002). These are the main problems experienced at the ground level during emergency situations, especially when fire prevention and suppression is needed. Studies have, therefore, also highlighted the need to clarify fire management responsibilities and improve coordination among the agencies involved in fire management activities (Herawati and Santoso, 2011; Morgera and Cirelli, 2009; Qadri, 2001; Barber and Schweithlem, 2000). Appropriate and effective fire management needs to take multiple factors into account, including integration, intersectoral and multiparty cooperation, capacity-building initiatives, and human resources development, with a holistic approach being needed to address all these requirements (Morgera and Cirelli, 2009; Goldammer, 2007; FAO, 2006).

Currently, there is no specific act in place to handle forest fire management activities at the national level in Indonesia. Policies, strategies, and management activities related to forest fires are spread across four ministries: the Ministry of Forestry (RoI, 1999), Ministry of Agriculture (RoI, 2004), Ministry of Environment (RoI, 2009), and the National Agency for Disaster Management (RoI, 2007). All four of the above institutions have introduced some form of decentralization, as well as good governance and participatory initiatives, and the national government has also delegated some authority to local governments and other stakeholders involved in land and forest fire management. The various regulations that were introduced by different ministries as well as by the local governments to guide "who does what" in minimizing incidents and impacts of forest fires are not consistent and seldom enforced.

As of 2011, in increasing of forest fire management, the President of the Republic of Indonesia (RoI, 2011), in an effort to improve forest fire management, has instructed related ministries and other government institutions (ie, Forestry, Agriculture, Environment, Finance, Research and Technology, the Attorney General, the Chief of the National Army and Police) to aid forest fire management activities through cooperation and coordination, through increasing people's participation, as well as through law enforcement for actors causing forest fires.

Over the last two decades, the multistakeholder approach (MSA) has become a commonly used tool for dealing with complex development problems, as it brings together various stakeholders and interested parties to find sustainable development solutions through dialogue and consultation (Gilmour et al., 2007; Warner, 2005; Hemmati,2002). Multistakeholder processes (MSPs) are designed in such a way as to bring together different stakeholders in diverse ways, such as through the use of forums, partnerships, platforms, networks, consortiums, associations, and other organizational forms (Hemmati, 2002). MSPs can be operated over a number of years, depending on the problems encountered, objectives sought, resources available, and level of political will (Hemmati, 2002). Applying MSPs requires a shift to take place, from a centralized to a more decentralized decision-making structure, as well as more effective collaboration between stakeholders. Further requirements among stakeholders include better governance, continuous capacity building and innovation to be in place, the establishment of a suitable learning environment, and a more rapid decision-making process (Morgera and Cirelli, 2009; Gilmour et al., 2007; Goldammer, 2007; Wagner, 2005; Hemmati, 2002).

According to Faysse (2006), a number of challenges have to be overcome when trying to establish and promote effective MSPs, most notably (1) uneven power relationships, (2) determining the composition of platforms, (3) appropriate stakeholder representation and building the capacity needed to participate meaningfully in debate, (4) effective decision-making mechanisms, and (5) ascertaining the costs involved when establishing and operating an MSP approach.

In line with this trend for utilizing an MSP approach, and to develop an organizational setting that supports an integrated and coordinated fire management system, a multistakeholder concept was adopted and established for forest fire management activities in each of the three districts of South Sumatra, where the South Sumatra Forest Fire Management Project (SSFFMP) was active. Applying the MSP approach for the management not only of natural resources but of a natural hazard was at that time a novelty. The main goal of the SSFFMP, which received support from the European Union (EU), was to follow-up on the earlier project, Forest Fire Prevention and Control (1995–1999), the aim of which was to reduce the disastrous effects of land and forest fires[1] (Supriadi and Steinmann, 2007). A MSF

---

[1] Fire management in Indonesia commonly employs the terms "land fires" and "forest fires," which international organizations refer to as "wildland fires" (FAO, 2007). Forest fires that occur inside a forested area (state forests) are the responsibility of the MoF, while land fires, which occur outside forest areas, are the responsibility of the Ministry of Agriculture. The term "land" is used broadly and includes estate cropland, farmland, fisheries land, livestock land/grasslands, and bush land.

was established under the SSFFMP with a 5-year term (2003–2007). Unlike other forums, a MSF is a legally established entity. In the case of the district of Ogan Komering Ilir in South Sumatra, the MSF was established under a decree from the District Head/Bupati (Supriadi and Steinmann, 2007; Bupati, Ogan Komering Ilir, 2003). The parties involved in the MSF included representatives from government institutions and agencies, nongovernmental organizations (NGOs), academia, the local media and local community organizations, as well as from the private sector, such as forest and estate crop plantation companies.

Most of the abovementioned institutions played an important role in fire management activities under the MSF, such as investigating the underlying causes and impacts of fires, preventing and suppressing fires, and post-fire recovery and rehabilitation. The involvement of all major stakeholders can be seen as the most basic contribution of the MSF approach in terms of the effective, efficient, and sustainable management of fire risk. However, so far, no comprehensive empirical study has been carried out into how the MSF has contributed to forest fire management outcomes in Indonesia as a whole.

To fill this gap, we examine the MSF approach, as employed within the South Sumatra Forest Fire Management Project, and seek to answer the following questions:

1. What was the structure of the MSF? Who were the representatives, whom did they represent, and what were the decision-making and responsibility-sharing mechanisms used?
2. What were the dynamics, and what factors contributed to the effective coordination of the MSF in terms of managing land and forest fires?
3. How did the district government's policies and actions support the MSF in its objective to introduce sustainable fire management practices?

10.2 METHODS

10.2.1 Study Site

Our research was conducted in Ogan Komering Ilir District (OKI District; see Fig. 10.1), which is one of the most fire-prone districts in South Sumatra (Bompard and Guizol, 1999), and also one of the three districts that established a district-level MSF approach under the SSFFMP. This district is the largest in South Sumatra Province, has a total land area of 19,023.47 km², a population of 685,269, and an average population density of 36.02 inhabitants per km². Administratively, OKI District consists of 18 subdistricts, 299 villages, and 12 urban subdistricts (Bappeda, 2009).

The district is characterized by lowland topography, with elevations ranging from 0 to 14 m.a.s.l. Soils in the region are predominantly podzolic and alluvial. The average monthly rainfall ranges from 80.1 mm in August to 316.4 mm in December. May to October is the period during which the least rain falls, while November to April is the period of highest rainfall. Average annual temperatures during the period 1999 to 2008 ranged from 26.360°C to 27.790°C. Forest covers 974,430.05 ha, or about 51% of the district's area, while peat swamp forest covers nearly 40%. Since 1980, South Sumatra Province has suffered from severe droughts, and these have increased the frequency, extent, and intensity of fires (Bowen et al., 2001), with these fires occurring particularly in logged-over areas and peat swamp forests. In

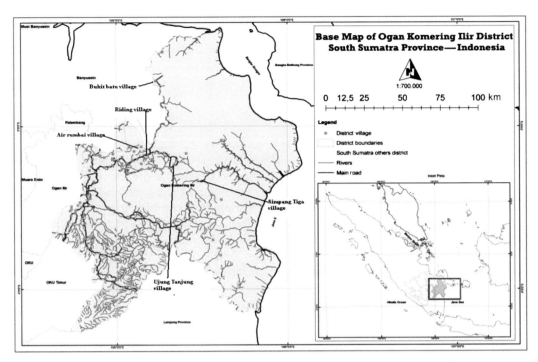

FIG 10.1   Study area.

addition to causing physical damage on the ground, fires produce haze and smoke that can cross international boundaries and affect neighboring countries. Commercial logging and the draining of wetlands has increased the area's susceptibility to fire, while the conversion of forests to commercial plantations using fire as a means of clearing land, and using fire within agriculture activities by local communities, have all resulted in an increase in the area at risk of fire (Chokkalingam et al., 2007; Bowen et al., 2001).

Since the late 1970s, about 500,000 ha of production forest has been handed over to seven logging companies (Bompard and Guizol, 1999), covering the areas most severely affected by the fire of 1997, and by recurrent dry season fires. Since 2004, three commercial forestry companies (under the Sebangun Bumi Andalas (SBA) wood companies group) have been operating on peat land to develop *Acacia* spp. over an area of 585,000 ha, and there are also 22 oil palm companies registered, with concessions estimated to cover 92,500 ha (Bappeda, 2009).

10.2.2 Data Collection and Analysis

This study uses primary and secondary data. Primary data were collected through the use of questionnaires and interviews. All members of the MSF and working groups registered by the district government decrees on the structure and composition of MSFs in 2003 and 2005 were selected as respondents. Members who rarely or never participated in the working group meetings were not selected. The number of questionnaire respondents from the MSF working groups was 39, or 98% of all members. The questionnaire consisted of four parts, focusing on (1) stakeholder representation and participation, (2) coordination mechanisms, (3) implementation of the MSF approach, and (4) communication and information flows.

The questionnaires were passed directly into the hands of the respondents, who were then requested to fill them in by themselves. All respondents returned the questionnaires within 1 to 4 weeks. Furthermore, all members of the MSF Board (10) were interviewed regarding MSF issues, and particularly fire-related policies and the institutional development of fire management activities.

Data were analyzed using quantitative and qualitative techniques. Quantitative data were analyzed using factor analysis (Hair et al., 1998), through Social Science Statistical Program (SPSS) 16.0 for Windows. Factor analysis was used to determine the factors and significant variables influencing the dynamics and effectiveness of MSF coordination and to reduce many of the original variables into a set of new variables or combination of variable dimensions (factors), and in particular using the principal component analysis (Hair et al., 1998). In general, there are four steps to the employment of factor analysis, namely, (1) determine quantitative variables; (2) identify inconsistencies or outliers, and then perform multiple correlation, (3) generate a grouping of significant variables, in which highly correlated variables are grouped within one component or factor, and (4) having established the interpretation of factors, give the name of the group factor that is to represent the variables.

Bivariate correlation techniques were used to reduce the number of variables to be used in subsequent analyses. Variables with a nonsignificant correlation (at the 0.05 level) were removed. We identified 18 out of 53 variables that met the principal component analysis criteria (Table 10.3). Qualitative analysis was used to supplement information not provided by quantitative data, such as the contribution of the MSF to forest fire management activities at the village and district levels, as well as policy and institutional developments on fire management.

10.3 RESULTS AND DISCUSSION

10.3.1 Policy and Institutional Development of the Multistakeholder Forum, Its Functions and Membership Dynamics

In 2003, the OKI District Government established a coordinating team to facilitate communication and coordination between the SSFFMP, the district government, and various relevant stakeholders. The coordinating team was also tasked with establishing an MSF to direct and support the development and implementation of programs for the improvement of natural resource management activities, and especially forest fire prevention and control, and fire impact mitigation. The district government had committed to allocate an annual budget for the operational costs of both the coordinating team and the MSF.

Forum membership was based on institutional or stakeholder representation. In principle, there were two selection mechanisms used to decide whether institutions became members of the MSF: election and appointment. In reality, however, all institutions/agencies that were members of the MSF organization were appointed by the SSFFMP team and the coordination team. In turn, it was the responsibility of each institution to decide on the individuals to represent them on the MSF.

There were four stages in the MSF establishment process.

i. Define the Concept and Strategy

Representatives of the SSFFMP and coordination team defined the concepts, objectives and strategy of the MSF, and designed criteria for evaluating and selecting the institutions to become MSF members.

ii. Select Institutions for MSF Membership

Six categories were selected for MSF membership: (1) District Government (Executive and Legislative), (2) NGOs, (3) Academic Institutions (University/Research), (4) the Private Sector, (5) Local Media, and (6) Local Community Organizations (Customary Institutions, the Association of Farmers and Fishermen).

iii. A Workshop on Understanding of MSF

A workshop was held on MSF guidelines concerning structure, organizational tasks, regulations, and mechanisms for actions, and to establish the working groups. The guideline workshop also established the following four thematic working groups: (1) Public Awareness of Fires (to help raise public awareness about the negative impacts of land and forest fires); (2) Institutional Fire Management (to help develop institutional capacity for fire management and to acquire equipment for fire suppression); (3) Land-Use Planning and Monitoring (to develop techniques for planning, for participatory development of sustainable natural resources management systems in fire-prone areas, and for a participatory approach to land tenure conflicts at the village level); and (4) Community Development (to help local communities improve income generation and reduce their dependence on the use of fire within their livelihood activities).

iv. Legal of MSF Institution

Finally, membership of the MSF was made legal by a decree given by the Ogan Komering Iir District Government (Bupati, Ogan Komering Ilir, 2003).

Administrative functions, which included establishing and directing working groups, together with planning, budgeting, and reporting, were conducted by the MSF Board. District government agency heads and other stakeholder-leaders related to fire management held a position on the MSF Board. Working group members prepared work programs, consulted with stakeholders, coordinated the programs and activities of the working groups, and attended various related meetings. Most significantly, the working groups assisted in fire management activities (technical and nontechnical assistance) at the subdistrict and village levels.

Prerequisites to increasing the effectiveness of coordination and responsibility sharing among MSF stakeholders included understanding the mission, providing a focus, and implementation.[2] In the case of OKI MSF, 26 different institutions were involved in total (Table 10.1). Of these, 53% represented district government agencies, while the remaining 47% were representatives from NGOs, local community organizations, the local media, and the private sector. Approximately 39% of all district government agencies were involved in the organization of the MSF, the largest number of government organizations at the district level involved in cross-sector management over a period of several years.

The functions of the respective institutions involved were discussed next.

---

[2] The term "sharing responsibility" refers to an increase in responsibility among owners and/or land users regarding fire management activities in their area, which includes prevention, response, mitigation, and recovery

TABLE 10.1 Functions of Each Institution Within the MSF With Respect to Fire Management Responsibilities

No.	Stakeholder Institutions	Fire Management Responsibilities			
		Prevention	Suppression	Mitigation	Recovery
I	*District Government Agencies*				
1	District Representative Council	x			
2	Agency for Regional Development Planning	x			
3	Environmental Agency	x		x	
4	Agency for Community Empowerment and Village Government	x			
5	Agency for National Unity, Political and People's Protection	x	x	x	x
6	Women's Empowerment Office	x			
7	Agency Food Security	x			
8	National Land Agency (district level)	x			
9	Forestry Service	x	x	x	x
10	Estate Crop Service	x		x	
11	Agriculture Service	x			
12	Livestock Service	x			
13	Marine and Fishery Service	x			
14	Education Service	x			
II	*Nongovernment Organizations (NGOs)*				
1	Kemasda	x			
2	KPB-SOS	x			
3	Spora	x			
4	Foslima	x			
5	Yayasan Masyarakat Mitra Mandiri,	x			
6	Lembaga Pemberdayaan Masyarakat	x			
7	P3L Oki	x			
8	OWA	x			
III	*Local Community Organizations*				
1	Farmers' and Fishermen's Association	x			
2	Customary Institution (district level)	x			

TABLE 10.1   Functions of Each Institution Within the MSF With Respect to Fire
Management Responsibilities—cont'd

		Fire Management Responsibilities			
No.	Stakeholder Institutions	Prevention	Suppression	Mitigation	Recovery
IV	*Local Media*				
1	Local Newspaper (SK lintas Timur)	x			
V	*Private Companies*				
1	Forestry Plantations	x	x	x	x

10.3.1.1 Institutions/Agencies of the OKI District Government

All district government agencies concerned with preventing and suppressing forest and land fires were involved in the MSF. There were three categories of institution present: 1.1 Institutions that assisted the district in formulating regulatory policies on the utilization and allocation of land, including the Regional Development Planning Agency, the National Land Agency, and the representative District Council. 1.2 Institutions that contributed directly to the prevention and suppression of fires, their role being to evaluate and assess the impact of fires, and provide people protection and disaster management activities (managing refugees and relocations). These institutions had direct access to land users or land managers such as plantation companies, industrial forestry plantations, cooperatives, as well as small holders. There were altogether four of these institutions: the Forestry Service, the Environment Agency, the Estate Crop Service, and the Agency for National Unity, Political and People Protection. 1.3 Institutions and agencies had duties and functions that were indirectly related to the management of fires. These institutions also had direct access to land users and conducted training for communities accustomed to using fire as a tool and as part of their farming systems. They included the Agency for Community Empowerment and Village Government, the Women's Empowerment Office, the Food Security Agency, the Agriculture Service, the Livestock Service, the Marine and Fishery Service, and the Education Service.

10.3.1.2 Nongovernment Organizations (NGOs)

Eight NGOs at the provincial and district level were involved in the MSF. None of these organizations focused on fire management; however, most of their target groups in rural areas were using fire as a land management tool.

10.3.1.3 Local Community Organizations

Local communities are key actors in the management of forest and land resources, as well as in terms of managing fire risk at the field level. They were represented on the MSF by two institutions at the district level, namely, (1) the Institute for Indigenous Peoples (a type of customary institution focused on traditional sociocultural life), and (2) the National Outstanding Farmers and Fishermen's Association (a group of select farmers and fishermen from across a large number of villages, who were engaged in social activities related to the agricultural sector). These institutions were expected to act as a bridge between communities and decision makers, as well as to contribute to the formulation of policies concerning fire management.

10.3.1.4 Local Media

The local newspaper, *Lintas Timur*, helped disseminate information on fires and fire hazards; a helpful service because government agencies had limited capacity to communicate and convey information quickly to the public, and especially to those people living in remote areas.

10.3.1.5 Private Sector

Fires on peat land are a serious threat, especially for companies that invest in crop businesses. In response to government policies, many private companies have adopted fire prevention and suppression as part of their crop protection strategies, and have increased their investment in fire management facilities, especially commercial forestry companies.

Changes in the membership of the MSF were a common issue, and part of the dynamics of stakeholder participation. Almost every year, the MSF conducted an evaluation of the activities, performance, and achievements of each individual working group member at workshops. Attendance and participation levels, and intellectual contributions during the various meetings held, were the main criteria used as part of the evaluation, but attendance was the most important criterion considered when revising MSF membership. An evaluation held in 2005 led to a revision of the composition of institutions and people on both the board and in the working groups. As a consequence of this, the composition of both the MSF Board and the working groups was no longer balanced, as the number of representatives from district government agencies had increased, mainly on the MSF Board. Academia, heads of subdistricts and village governments (Heads of Village and Village Representative Body) at selected subdistricts and villages were, as a result, no longer represented on the MSF Board or in any working group, and private sector company representation had decreased significantly, with only one forestry company involved. As a consequence of this imbalance, district government agency representatives dominate as MSF members, suggesting that policy formulation and decision making would be more influenced by such agencies. Edmunds and Wollenberg (2002) stated that the interests of state bureaucracies remained dominant, despite the increased involvement of diverse interest groups. Membership fluctuated for two main reasons, as discussed below.

a. Level of Attendance in the Coordination Meetings

About 10% of the questionnaire respondents revealed that they never attended a working group meeting, and 25% said that they attended only once or twice. The reasons cited for this were (a) the long distance and costly transportation to the district capital where the meetings frequently took place, (b) overlapping activities, (c) the lack of a clear and detailed schedule for each meeting, (d) a lack of information/no invitation, and (e) no clear instructions from the head of their respective institution telling them to attend the coordination meetings.

b. Termination of MSF Membership

The study revealed that 18% of MSF members functioned for the entire 5-year term (2004–2007), while the remaining 82%, or 32 different people, served on their MSF for just 1 or 2 years (Table 10.2)

TABLE 10.2 Component Factor Analysis Components and Results: Factors Influencing the Effectiveness of the Ogan Komering Ilir District MSF for Forest and Fire Management[a]

Principal component evaluation factors and variables defining each factor	Factors			
	1	2	3	4
Factor 1—Coordination between the MSF board and MSF working groups				
Eval_gap—perceived quality of performance evaluation coordination between the MSF board and working groups	0.755			
Comm_gap—perceived quality of communication between the MSF board and working groups	0.866			
Infor_gap—perceived quality of knowledge and information flows between the MSF board and working groups	0.679			
Deci_gap—perceived quality of decision-making processes between the MSF board and working groups	0.691			
Clear_ro—perceived clarity of roles and tasks in implementation of fire management activities	0.640			
Factor 2—Transparency in management decision making				
Bud_trans—perceived level transparency in budget release and control		0.723		
Imp_trans—perceived level of transparency applied in the implementation of activities		0.723		
Monev_trans—perceived level of transparency applied in monitoring and evaluation		0.656		
Mon_gap—perceived level of coordination in monitoring		0.715		
Budg gap—perceived level of coordination in budget management		0.739		
Factor 3—Collective learning quality				
Eff_impl—perceived ranking for overall effectiveness of the MSF coordinating the implementation mechanisms			0.706	
Eff_mon—perceived ranking for overall effectiveness of MSF monitoring			0.795	
Eff_eva—perceived ranking for overall effectiveness of MSF evaluation			0.631	
Eff_bud—perceived ranking for overall effectiveness of MSF budgeting coordination			0.698	
Budget_pr—perceived sufficiency of budget for implementation activities			0.689	
Factor 4—Decentralization in practical implementations				
Dece_bud—experienced level of decentralization applied in budget release and control				0.780
Dece_Imp—experienced level decentralization applied in implementation				0.917
Dece_mon—experienced level decentralization applied in monitoring and evaluation				0.918

Continued

TABLE 10.2   Component Factor Analysis Components and Results: Factors Influencing the Effectiveness of the Ogan Komering Ilir District MSF for Forest and Fire Management[a]—cont'd

Principal component evaluation factors and variables defining each factor	Factors			
	1	2	3	4
Eigenvalues	5.010	2.741	2.368	1.730
Total variance explained	30.560	15.226	13.158	9.611
Cumulative variance explained	30.560	45.786	58.944	68.555
Cronbach's Alpha	0.611	0.789	0.813	0.863
Kaiser Meyer Olkin Measure of sampling adequacy				0.594
Bartlett's Test of Sphericity	Approx Chi-square			472.003
	df			153
	Sig			0.000

[a] Data used in the factor analysis were collected using a questionnaire, with responses indicated using a four-point Likert scale.

The reasons why members left the MSF were (a) promotion or transfer to another position as part of career development both inside and outside the institution, (b) resignation due to workload, and (c) the head of the institution was no longer assigned as a representative on the MSF. The high turnover of members hampered MSF operations.

10.3.2 Factors Influencing the Dynamics and Effectiveness of Multistakeholder Activities

The ultimate goals of the MSF were to reduce the occurrence of forest and land fires, and to mitigate their negative impacts. However, it is not easy to assign success or failure to any particular agency within the fire management realm, as many sectors and institutions played a role. An exception to the overall low number of fires since 1997 was the peak in 2006, which, like 1997, was an El Niño year, and in which drought conditions increased the risk of large-scale fires occurring in the district. This is supported by government provincial data on the number of fires or fire hotspots, as shown in Fig. 10.2.

Nearly 26% of the questionnaire respondents stated that coordination and consultation among institutional stakeholders at the district and village levels increased as a result of MSF activities such as working group meetings, the preparation of annual plans for MSF/SSFFMP, field visits, training, and seminars.

Based on the quantitative analysis, four factors can be identified from the results of the factor analysis (Table 10.3), all with a satisfactory explanation in terms of cumulative total variance (68%). The accuracy of the model was confirmed by the value for sample adequacy (0.59) and the Bartlett test of spherity, with a Chi-square value produced of 472.003, at a significance level of 0.000.

Four highly significant factors explain the overall positive performance of the MSF: (1) intraforum coordination dynamics, (2) transparency in decision making, (3) collective learning quality, and (4) the use of decentralization.

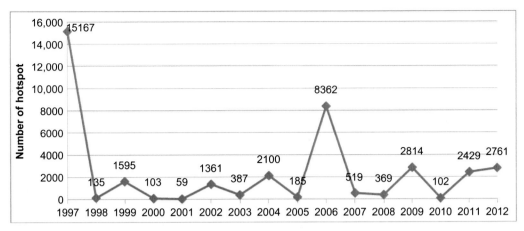

FIG 10.2   Number of hotspots in Ogan Komering Ilir District (1997–2012). *1997–2012—from NOAA-AVHRR and Sipongi, and 2003–2012—Terra & Aqua MODIS satellites, University of Maryland.*

TABLE 10.3   Length of Service on the Ogan Komering Ilir MSF and Reasons for Leaving

Reason for leaving the MSF	Length of Service		Total
	1–2 years	3–5 years	
Moved or transferred to another government agency-institution	4	0	4
Moved or transferred to another section within the same institution	9	0	9
Left the MSF for personal reasons	5	0	5
No comment	6	4	10
Others	8	3	11
Total number who resigned	32	7	39

The first factor represents coordination between the MSF Board and MSF Working Groups, and consists of five variables: communications dynamics, performance evaluation, the decision-making process, knowledge and information flows, and the clarity of roles or responsibility sharing. A lack of effective communication between the MSF Board and the working groups emerged as the most important variable (with the highest factor loading) shaping the factor, followed by the variable that signifies major shortcomings in the performance evaluation employed by the MSF Board and the working groups. The first factor, therefore, suggests that internal communication dynamics and the effective coordination of performance evaluations, along with a clarity of individual roles and collective responsibility sharing, would be critical if a new MSF wished to pursue the specified objectives of (sustainable) land and forest fire management. However, internal and external agency coordination will also need to be developed to increase the effectiveness of such communication. Communication will need to be improved in an effort to expedite information sharing about the vision and mission of stakeholder institutions, and also share data and up-to-date information about the progress of each organization's activities related to fire prevention and suppression. It will be important to synergize and coordinate programs and activities, in an effort to avoid overlapping of activities and help decision making for collective action. However, frequent turnover of personal representatives within the previous MSF, and internally within member organizations, led to a loss of data (missing link) and information. To avoid such losses in the future, intra-forum coordination will need to be improved in the following ways: (1) adjust the meeting schedule to suit participant availability, (2) select a location that is close by or preferred by most participants, and (3) provide funds to finance regular meetings or workshops.

The second factor shows the best linear combination of a set of variables related to transparency in decision making. Taking into account the configuration of the factor and the factor loading, we have called this factor "Transparency in Decision Making for Management Functions." It consists of five variables: coordination of budget management activities, transparency in terms of budget release and control, close the coordination gap between the MSF Board and the working groups, mobilize monitoring and evaluation, and apply transparency during monitoring and evaluation activities. The second factor, therefore, reveals that transparency, along with other principles of good governance, particularly in the implementation process and financial agreements, is one of the key factors that will make a new MSF work more effectively. Transparency in decision making will need to be supported by timely and relevant information, and should, in particular, help minimize the occurrence of "silent conflict" among MSF members. Transparent decision making when implementing the budget, as well as in terms of budget constraints, is something that needs to be considered.

The third factor was identified as collective learning quality. Five specific variables, those corresponding to the effectiveness of mechanisms used to establish coherent policies, and effective monitoring and evaluation, underlie this factor, together with a variable associated with financial resource availability (mainly budgetary constraints). From the highest to the lowest factor loading, these variables are effectiveness of monitoring, effectiveness of coordination when implementing policies and strategies, sufficient budget for implementation of policies and strategy, effective coordination of budgeting processes, and effective coordination of performance evaluation activities. We found that the quality of coordination, and the capacity to conduct regular performance measurements and evaluations, are critical elements of collective learning quality. Dynamics and number and frequency of changes within an organization, as well as new policies, all need to be learned through continuous collective learning quality, which needs to be integrated into each stage of the management cycle.

The fourth factor was identified as decentralization in practical implementation. Variables with the highest load factor are monitoring and evaluation of implementation, as well as budgetary release and control. In line with the organizational structure of the former MSF, the MSF Board delegated authority over the control of resources to MSF members, so that they could carry out the functions and organizational tasks.

10.3.3 The District Government's Policies to Introduce Sustainable Fire Management Practices

Although the MSF was conducive to institutional changes, its mandate was not extended beyond 2007. The District Government of OKI terminated its financial support at the end of the SSFFMP, claiming that, as the MSF was formed as part of the SSFFMP cooperation project, it should be terminated at the end of the project. This is in line with the comments of Hemmati (2002) that a MSP can occur over several years depending on the willingness of local governments to engage with stakeholders.

However, nearly 70% of the respondents suggested that the MSFs mandate should have been extended, with improvements made in the following areas: budget availability, transparency at all planning stages, better coordination and communication, and downsizing of membership. Data and information analysis point out that the main attributes and achievements of the MSF, those suggesting it should have continued, were as follows: (1) the MSF was a suitable instrument to overcome institutional boundaries, which are an inhibiting factor in the sectoral coordination and synergy between government and nongovernment agencies with respect to land and forest fire management; (2) a coordination system was developed within the MSF, which enhanced personal relationships and networking among working group and MSF Board member institutions; and (3) in the future, valuable assets should be made available for dealing with forest fires, such as specially trained human resources, more and better equipment, and last but not least, an infrastructure suitable for enhanced forest and land fire prevention and suppression activities

To institutionalize the concept of and strategy for a sustainable forest and land fire management approach, the government embarked on a number of initiatives. One of them was issuing the following important policies and regulations;

- Established new sections and agencies with duties and functions specifically related to forest and land fire management. Also, a new section focused on prevention and control of land and forest fires was established at the Forestry Service in 2006. In addition, in 2008 the district government established two new divisions, namely, a division on land and forest fire control within the Environment Agency, and a division on people protection and fire disaster management within the Agency for National Unity, Politic and People Protection (Sekretariat Daerah Kabupaten Ogan Komering Ilir, 2008). Later, the district government established a District Agency for Disaster Management that coordinates firefighting with government resources once the government has designated a forest or land fire as a disaster. These policies could be seen as positive indicators of the increased awareness and political commitment of the district government to institutionalize the prevention and reduction of land and forest fires in a sustainable manner at the lowest level, and to provide an increased annual budget.
- Strengthened the existing zero burning policy. Burning is routine during the dry seasons, so it was important to implement the edict, with criminal sanctions against those who started forest and/or grass, bush, and scrub fires. In addition, the impact of the edict was immediate, for in a short time the various components of the local community developed knowledge on the criminal penalties applied for those found to be responsible for forest and land fires. The law enforcement agencies (police and prosecutors) also increased their activities at the village level, in order to promote a zero burning policy, by monitoring and catching people or companies found to have violated the laws and regulations. An effect of the zero burning policy is that local communities or land users tend to burn secretly (hit-and-run burning) for land clearing to avoid punishment. This and the abandoning of the now effectively outlawed traditional rules creates a situation of uncontrolled burning. The zero burning policy was critically discussed among the public and MSF Working Group members. Our research shows that 30% of the questionnaire

respondents objected to the ban on the use of fire for agricultural activities among local communities. The reason for this is because they believe that (a) "better burning" is more appropriate than "zero burning," as many rural communities are not yet ready to apply alternative methods of farming; (b) rural communities have traditional and efficient rules in place to control burning; (c) there has been an increasing trend toward the use of herbicides as an alternative to the use of fire; and (d) the government should first provide cost-efficient technology and financial incentives. Some local communities said they do not agree with the "zero burning" policy and had lodged protests through the village government.

• Improved or introduced better coordination and consolidation of all the relevant stakeholders, to obtain joint commitment and agreement by holding a "special day" (*siaga api*), its aim being to create fire alertness and to anticipate the threat of fire in the dry season, as well as improve preparedness and mobilize resources (personnel, fire equipment, and funds).

These policies are the result of an increased commitment from decision makers at the district level, those who contribute to sustainable land and forest fire management activities.

10.4 CONCLUSIONS AND POLICY RECOMMENDATIONS

In this chapter, we noted that the local government played a catalytic role in developing a MSA, based on the establishment of an MSF. After being formed, the MSF acted as an umbrella organization for various cross-institution stakeholders (government, NGOs, local community organizations, private companies, and academia), the aim being to provide a foundation for learning and joint action. In Indonesia, districts and the higher authorities are the most important partners with regard to operational fire management activities, and especially those requiring preventive measures and suppression activities.

Our findings show that the MSF contributed actively to a reduction in fire risks and impacts, through increased coordination and consultation among MSF stakeholders, by proposing policies, and by helping to initiate the establishment of new functions for selected district agencies.

This discussed the MSF framework, those factors underlying its dynamics, and its effectiveness at achieving the stated goals and objectives, which can be summarized as "bringing together stakeholders for the purpose of reducing fire risks and fire impacts." Quantitative analysis has identified four underlying recommendations that OKI district should consider implementing to become more dynamic and effective at preventing and fighting fires using an MSF approach, namely, (1) improve intraforum coordination dynamics, (2) ensure transparency within the management function decision-making process, (3) introduce collective and quality learning, and (4) ensure the practical implementation of decentralization.

Although the MSF has now ceased as an institution, the district government, in pursuing sustainable forest fire management activities, has developed an integrated and comprehensive fire management strategy, by establishing specific fire management institutions and targeting financial resources at selected district government agencies, and by delegating responsibility to relevant stakeholders (landowners/users), for the prevention and management of fires in their areas. It has also enhanced law enforcement activities. Regarding the zero burning policy, we recommend the district government consider a controlled burning policy that permits controlled use of fire especially for local people so as to avoid the clandestine burning encouraged by the zero burning policy. In addition, district government should be prioritized to apply forest fire prevention policies in addition to firefighting including forest and peatland rehabilitation within the fire-prone region, and having fire management facilities at each commercial plantation.

We suggest that the district government should revitalize the MSF concept or develop mainstreaming MSF into the existing structure instead of project-based MSF, to further utilize the already-developed assets, especially human resources. The capacity of government agencies involved in the new role of fire management needs to be strengthened in terms of technical and nontechnical aspects, especially for coordination and collaboration to create cross-sectoral synergies for sustainable forest fire management. To revitalize the MSF as an institution, the district government should provide an opportunity for the private sector and NGOs to participate in any new MSF formed. In this regard, the financing mechanisms needed to manage a new MSF should also be discussed, such as fundraising opportunities like the corporate social responsibility mechanism, which would form the basis for greater decentralization to take place in the truest sense.

Acknowledgments

This study was funded by a grant from the Ford Foundation, Jakarta, to the Asian Institute of Technology. The authors gratefully acknowledge the South Sumatra Forest Fire Management Project for providing us an opportunity and facilities for research. The earlier version of this chapter was published in Achyar, Eris, D. Schmidt-Vogt, and Ganesh P. Shivakoti. 2015. "Dynamics of the multi-stakeholder forum and its effectiveness in promoting sustainable forest fire management practices in South Sumatra, Indonesia." *Environmental Development*. 13(2015):4–17, which is duly acknowledged.

References

Bappeda, 2009. Profil Kabupaten Ogan Komering Ilir 2009 (Profile of Ogan Komering Ilir District).

Barber, C.V., Schweithelm, J., 2000. Trial by fire: forest fires and policy in Indonesia's era of crisis and reform. World Resources Institute (WRI), Forest Frontiers Initiative. In collaboration with WWW-Indonesia and Telapak Indonesia Foundation, Washington, DC, USA.

Bompard, J.M., Guizol, P., 1999. Land management in the province of South Sumatra. Fanning the flames; the institutional causes of vegetation fires. Forest Fire Prevention and Control Project. European Union and Ministry of Forestry and Estate Crops, Palembang, Indonesia.

Bowen, M.R., Bompard, J.M., Anderson, I.P., Guizol, P., Gouyon, A., 2001. Anthropogenic fires in Indonesia: a view from Sumatera. In: Radojevic, M., Eaton, P. (Eds.), Forest Fires and Regional Haze in South East Asia. Nova Science, New York.

Bupati, Ogan Komering Ilir, 2003. Keputusan Bupati Ogan Komering Ilir, Nomor.303/KEP/D.KEHUT-MSF/2003 tanggal 22 Oktober 2003 tentang Pembentukan Multi Stakeholder Forum (MSF), Kabupaten Ogan Komering Ilir (A decree of Bupati on establish of the Multi Stakeholder Forum membership).

Chokkalingam, U., Suyanto, S., Permana, R.P., Kurniawan, I., Mannes, J., Darmawan, A., Khususyiah, N., Susanto, R.H., 2007. Community fire use, resource change and livelihood impacts: the downward spiral in the wetlands of southern Sumatra. Mitig. Adapt. Strat. Glob. Chang. 12, 75–100. http://dx.doi.org/10.1007/s11027-006-9038-5.

Cochrane, M.A., 2003. Fire science for rainforests. Nature 421, 913–919. http://dx.doi.org/10.1038/nature01437.

Corlett, R.T., 2009. The Ecology of Tropical East Asia. Oxford University Press, Oxford.

Dennis, R., 1999. A Review of Fire Projects in Indonesia (1982–1998). Center for International Forestry Research, Bogor, Indonesia.

Edmunds, D., Wollenberg, E., 2002. Disadvantaged Groups in Multi-Stakeholder Negotiations. Center for International Forestry Research, Bogor, Indonesia. http://www.cifor.cgiar.org/publications/pdf_files/.

FAO, 2005. State of the World's Forests 2005. FAO, Rome.

FAO, 2006. Fire management: review of international cooperation. Fire Management Working Paper 18, Rome (www.fao.org/forestry/site/firemanagement/en).

FAO, 2007. Fire management-global assessment 2006. FAO Forestry Paper 151. Food and Agriculture Organization of the United Nations, Rome.

Faysse, N., 2006. Troubles on the way: an analysis of the challenges faced by multi stakeholder platform. Nat. Res. Forum. 30, 219–229.

FWI/GFW, 2002. The State of the forest: Indonesia. Forest watch Indonesia, Bogor, Indonesia and Global Forest Watch, Washington, DC.

Gilmour, D.A., Durs, P.B., Shono, K., 2007. Reaching consensus. Multi Stakeholder Process in Forestry Experience From the Asia Pasific Region. FAO, Bangkok.

Goldammer, J.G., 2007. International Cooperation in Fire Management. South Sumatra Forest Fire Management Project, Indonesia.

Hair, J.F., Anderson, R.E., Tatham, R.L., Black, W.C., 1998. Multivariate Data Analysis. Prentice-Hall Inc, Upper Saddle River, New Jersey.

Hemmati, M., 2002. Multi-Stakeholder Processes for Governance and Sustainability: Beyond Deadlock and Conflict. Earthscan Publications Ltd., London.

Herawati, H., Santoso, H., 2011. Tropical forest susceptibility to and risk of fire under changing climate: a review of fire nature, policy and institutions in Indonesia. Forest Policy Econ. 13 (4), 227–233. http://dx.doi.org/10.1016/j.forpol.2011.02.006.

Ministry of Forestry, 2012. Forestry Statistics of Indonesia.

Morgera, E., Cirelli, M.T., 2009. Forest fires and the law; a guide for national drafters based on the Fire Management Voluntary Guidelines. FAO Legislative Study 99. FAO, Rome.

Page, S.E., Siegert, F., Rieley, J.O., Boehm, H.D.V., Jayak, A., Limin, S., 2002. The amount of carbon released from peat and forest fires in Indonesia during 1997. Nature 420, 61–65. http://dx.doi.org/10.1038/nature01131.

Qadri, S.T., 2001. Fire, Smoke and Haze: The Asean Response Strategy. Cooperation Association of South East Asian Nations with Asian Development Bank, Manila, Philippines.

Republic of Indonesia, 1999. Forestry Act no. 41/1999.

Republic of Indonesia, 2004. Estate Crop Act no. 18/2004.

Republic of Indonesia, 2007. Disaster Management Act no. 24/2007.

Republic of Indonesia, 2009. Environmental Protection and Management Act no. 32/2009.

Republic of Indonesia, 2011. Instruction of President no. 16/2011 on Increasing of Land and Forest Fire Control.

Sekretariat Daerah Kabupaten Ogan Komering Ilir, 2008. Lembaran DaerahKabupaten Ogan Komering Ilir Tahun 2008 (District regulation on establish the Organizations and functions of agencies at Ogan Komering Ilir District).

Simorangkir, D., 2006. Fire use: is it really the cheaper land preparation method for large-scale plantations? J. Mitig. Adapt. Strat. Glob. Chang. In Press.

Simorangkir, D., Sumantri, 2002. A Review of Legal, Regulatory and Institutional Aspects of Forest and Land Fires in Indonesia. Project Fire Fight South East Asia, Bogor, Indonesia.

Supriadi, D., Steinmann, K.H., 2007. SSFFMP; Summary report of results and achievements 2003–2007. E1, South Sumatra Forest Fire Management Project, Palembang, Indonesia.

Tacconi, L., 2003. Fires in Indonesia: causes, costs and policy implications. CIFOR, Occasional Paper No.38, Center for International Forestry Research, Bogor, Indonesia.

Tacconi, L., Moore, P.F., Kaimowitz, D., 2006. Fires in tropical forest-what is really the problem? Lessons from Indonesia. Mitig. Adapt. Strat. Glob. Chang. 12 (1), 55–56.

The Economist, 2013. June 29. South-East Asia's smog.

The Economist, 2014. March 22. Haze over Sumatra, p. 30.

Varkkey, H., 2013. Patronage politics, plantation fires and transboundary haze. Environ. Hazards 12 (3–4), 200–217.

Warner, J.F., 2005. Multi-stakeholder platforms: integrating society in water resource management? Ambiente Soc. 8 (2), 17–28. fi.uaemex.mx/luislalo/girh/31780201.pdf.

Collaborative Governance of Forest Resources in Indonesia: Giving Over Managerial Authority to Decision Makers on the Sites

M.A.Sardjono*, M. Inoue†

*Mulawarman University, Samarinda, Indonesia †The University of Tokyo, Tokyo, Japan

11.1 INTRODUCTION

Political change in Indonesia at the end of the 20th century, widely known as the 1998 reformation movement and followed by an agreement to implement regional autonomy in 2001, has brought high expectations for better forest governance. The main reasons for these expectations include the fact that forest covers at least 60% of the total land area of Indonesia (approximately 140.0 million hectares) and forest utilization through concessionaires, which started at the beginning of the 70s, contributed significantly to national revenues (ITTO, 2001), However, this change was unable to increase local communities' welfare and it even has limited their living space (see Mubyarto, 1991, 1992; Soetrisno, 1993).

The annual rate of deforestation and forest degradation in Indonesia increased from 700,000 hectares in the mid-1980s to approximately 2.4 million hectares at the beginning of 1999, although 10 years later it was reported that the rate could be reduced by about one-third (Ginting, 2001; FWI/GFW, 2001; FWI, 2011). Sardjono et al. (2012) reported that overexploitation, especially illegal logging and forest conversion for other utilizations (particularly extensive oil palm and coal mining industries) are leading causes of deforestation in East Kalimantan, which is one of the most forest-rich provinces in Indonesia.

The implementation of regional autonomy (*otda*) that was expected to bring decision making and policy formulation closer to the resources, and community needs for creating product sustainability and for forest functions, has generally created the opposite results. Research by Rahman et al. (2013) in nine districts in Indonesia drew the interesting conclusion that, in fact, forest and land governance by the local government in the *otda* era is far from good governance principles such as no transparency, less participation, low accountability, and weak coordination. In addition, Prayitno et al. (2013) stated that local governments were not able to optimize their authority for increasing community welfare and, to the contrary, their policies had degraded the remaining forests. Financial policies for stimulating regional incomes to support local economic development have also aggravated forest and land governance. In such a situation, a good approach is to implement collaborative forest governance (CFG) (Sranko, 2011; Weaver, 2012) where government (including local government) and nongovernment institutions (including the local community) are believed to be a key for successful resource management.

This chapter critically explores the potential and challenges of the Forest Management Unit (FMU) as a relatively new approach of CFG in Indonesia. The chapter will also present some recommendations for smarter strategies that can be adopted to accelerate FMU development.

11.2 INEFFECTIVE FOREST GOVERNANCE

Extensive forestry development in Indonesia, especially for bigger islands outside Java (eg, Sumatra, Kalimantan, and Sulawesi) was initiated in the 1970s. Although in the beginning the development had put Indonesia as a leading log, sawn timber, and plywood exporter among Southeast Asian countries (see eg, Sardjono and Inoue, 2011), it brought about ecological and economic disasters after three decades of development. According to Sardjono (2007), there are at least six characteristics of natural resource utilization policies, namely, centralized, capitalistic, monopolistic, exploitative, authoritarian, and homogenous, as shown in Table 11.1.

The above situation was considered possible because of several factors, especially (1) a dominant and powerful (central) government versus other inferior stakeholders (including the local community), which leads to practically uncontrollable policies and their implementation; and (2) the significant economic role of the forestry sector and no demand for sustainable forest management (SFM) by the global market, particularly in the first decades of the economic jump period of the country (see also Sardjono and Inoue, 2011). Those factors were considered to be significant reasons for the reformation movement at the end of the 90s, which was appealing to a forestry paradigm from state-based to community-based and from timber oriented to resource oriented.

Under regional autonomy, by which about 70% of the authority has already been handed over to local governments, forestry performance has not improved. As briefly mentioned earlier with the focus of Kalimantan, it even suffers from wider disturbances and concurrences against other land uses, particularly extensive development of the oil palm and coal mining industries. In East and North Kalimantan, mangrove forests in coastal regions have mostly converted into commercial fish ponds, while many suitable swampy areas have been planned for food-estate development (see Sardjono et al., 2012).

Based on a more detailed analysis it is assumed that factors affecting the ineffectiveness of forest governance under regional autonomy are (1) forestry issues are a low priority up to the district level, which has led to a wide span of control for policy and administrative implementation; (2) forestry is an optional (not main) duty of the local government and this has allowed the sector to become less prioritized and, in many districts, it has been merged with its competitors (crop-estate and mining sectors); and (3) forest administration and management is

TABLE 11.1 Six Characteristics of Indonesian Forest Policy During the New Order Regime 1967–1998 and their Implications

No	Policy Characters	Implications
1	Centralized	• Regulations issued by the central government. Positively, it is controllable but negatively most of them do not fit local needs • Because every, even small, case had to be consulted and decided at central levels, forest administration (including in financial affairs) was becoming bureaucratic and therefore time-consuming. In fact, some forestry cases needed immediate political responses (eg, forest fires)
2	Capitalistic	• Large investment or capital-intensive technology did not fit local traditional and cultural life, resulting in narrow opportunities for the local communities and local enterprises to participate • Capitalistic forest policies effectively innovated economic development, but at the same time introduced modern life and commercialization among traditional communities, which in many cases has destroyed the traditional culture and locally wise ethics
3	Monopolistic	• Monopolistic approach weakened the Indonesian macroeconomy and, following the 1997 Asian monetary crisis, led to business collapses including in the forestry sector. The collapse in forestry business caused unemployment and destroyed the local economic chains of many surrounding communities • Monopolistic conglomeration invaded not only timber businesses but also nontimber forest products (eg, rattan), which have primary roles in people's lives. It caused increasing middle traders, unstable prices, or even lowest prices at local farm levels
4	Exploitative	• Timber became a "major" product and others (including nontimber forest products and ecological services) were categorized as "minor." Therefore, the related forest policy was focused more on intensive timber exploitation/extraction rather than environmental resource management • Many companies tend to cut more carrying capacity of the forests (to avoid administrative punishment by decreasing the next annual allowable cut (AAC) or to get faster profits through data manipulation)
5	Authoritarian	• The justification of the decision makers at the central level, in the name of the national interest, ignored participatory processes (eg, public consultation and local people's aspirations) • To implement the top-down regulations, different technical guidance and operating procedures have been produced by the ministry and they have to be followed as a blueprint by forest operators
6	Homogenous	• To more easily control all businesses, there was a tendency among government agencies including the Forestry Department to produce homogenous policies for all of Indonesia (or at least just dividing Indonesia into Java and outer islands). In fact, the complexity of the tropical rain forest demands heterogeneous approaches • Homogenous policies do not even guarantee homogenous interpretation among the field apparatus and they lead to "illegal bargaining" at the field level (especially between operators and evaluators) following different difficulties of their implementation under specific field conditions

Sardjono, M. A., 2007. Trends of forest policies in Indonesia. A narrative material for an intensive lecture as a visiting professor at the University of Tokyo. Samarinda: Mulawarman University.

combined under one institution, namely local forestry-governed services, which technically do not have sufficient capacity, time, space, and financial support for doing their jobs. This last situation has been mainly influenced by the limited number of district forestry staff and professional forest site managers, as well as the fact that forestry nowadays is already becoming a sunset industry, particularly to support regional development.

III. LEARNING FROM THE FIELD CASES/ISSUES

In addition to the abovementioned reasons, there are also indirect causes of ineffective forest governance during the regional autonomy, including more attractive offers of nonforestry investments especially oil palm plantations and coal mining for local revenues or earning money, which have interested not only local governments, but also many villagers who already have other difficulties and have depended on forests and forest products for their livelihoods during the last two decades. This last trend has been investigated and reported by many researchers (see, eg, Sardjono et al., 2001; Sardjono and Inoue, 2007). Another indirect cause of ineffective forest governance has been increasing local community claims on forests and lands following the reformation euphoria, and unclear forest boundaries as an impact of an incomplete forest gazetting processes (see also Contreras-Hermosilla and Fay, 2005; Moeliono et al., 2009).

11.3 DEMAND OF FOREST COLLABORATIVE FOREST GOVERNANCE

Although autonomous regions, particularly at the district level, have been given most of the authority related to resource utilization, a complex problem of forest management under limited human resource capital as well as insufficient financial capital cannot be solved solely by the local government, without any collaborative support from the related actors, especially the existing forest concession holders and local communities. CFG is organized through collaboration among various stakeholders who have a range of interests in the use and management of local resources. The collaboration should be developed through a consensual "principle of involvement" (Inoue, 2009, 2011a).

In many districts (eg, East Kalimantan as one of the Indonesia's forest-rich provinces) where on the one hand forest area can cover more than 50% of the total land areas (see eg, Sardjono, 2014), but on the other hand most of the population in villages traditionally depend on the surrounding forest for their daily life, they have to compete with large-scale investors to find forest arable space for survival. To be able to do their jobs perfectly, even professional foresters have to get support from key governmental and nongovernmental stakeholders. Participation of governmental and nongovernmental parties with a common purpose of sustaining forest functions and their benefits should be followed by authority sharing, responsibilities, and indeed benefits based on the roles of each party. Inoue (2011a,b) proposed prototype design guidelines for CFG comprising nine design guidelines: (1) degree of local autonomy; (2) clearly defined resource boundary; (3) graduated membership; (4) commitment principle; (5) fair benefit distribution; (6) two-storied monitoring system; (7) two-storied sanctions; (8) nested conflict management mechanism; and (9) trust building. More detailed explanations of each design are presented in Table 11.2.

Based on some of the literature, Sardjono et al. (2013) underlined that following the design principles, appropriate arrangements should be assigned at the policy level to allow a sufficient degree of local autonomy in the use and management of local forest lands and the resources therein. While at the field implementation level it is imperative to identify relevant stakeholders who must get involved in the collaboration, particularly the local communities whose livelihood depends on the forests and in most cases also have traditional rights over the lands. For appropriate CFG arrangements, clarity and understanding of customary land tenure and the local communities' rights over the forestlands would be helpful and decisive (eg, Colfer, 1995; Tribowo and Haryanto, 2001; Carter and Gronow, 2005; Purnomo et al., 2005).

TABLE 11.2   Nine Design Guidelines of Collaborative Forest Governances

No	Design Guidelines	Remarks
1	Degree of local autonomy	• There is always room for designing CFG in accordance with the degree of local autonomy, unless the local community has no autonomous function
2	Clearly defined resource boundary	• Demarcation of resource boundary is often a difficult task because of obscure ownership
3	Graduated membership	• Based on "open-minded localism," some of the local people act as "core members" (first-class members), who have the strongest authority, cooperating with other graduated members who have relatively weaker authority (second-class and third-class members)
4.	Commitment principle	• This principle recognizes the authority to make decisions in a capacity that corresponds to their degree of commitment to forest use and management. Decision making is not equal, but should be fair and just
5.	Fair benefit distribution	• Benefit distribution is not necessarily equal, but is fair in accordance with cost bearing
6.	Two-storied monitoring system	• The core members of CFG monitor whether other members obey the rule. Then, local government monitors whether the rule itself is appropriate for sustainable forest management in a scientific way
7.	Two-storied sanctions	• The core members have responsibility, which is supported by the local government
8.	Nested conflict management mechanism	• Informal conflict resolution in the community with informal intercession by the local government supported by a formal mechanism at local and national levels
9.	Trust building	• For cooperation with outsiders, forming, maintaining, and strengthening social capital are essential

Inoue, M., 2009. Design guidelines for collaborative governance (kyouchi) on natural resources. In Murota, T (ed.) Local Commons in Globalized Era. Minerva-Shobou, pp. 3–25 (original article in Japanese); Inoue, M., 2011a. Prototype design guidelines for 'collaborative governance' of natural resource. In: Presented at 13th Biennial Conference of the International Association for the Study of the Commons, Hyderabad, India, January 12; Inoue, M., 2011b. Summary of the Design Guidelines: Instruction for the Authors of Multi-level Forest Governance in Asia—Recognizing Diversity. Tokyo: GSALS University of Tokyo.

The above nine principle designs of forest collaborative or cogovernance can certainly be further discussed for every scheme of forest management (see Sardjono et al., 2013), but for implementing, in the framework of regional autonomy in Indonesia, there should be an official organization (of central or local governments) that can get involved among other stakeholders (concession holders and/or the local community) and has the capacity (and opportunity) to professionally manage forest in the field. The FMU/KPH that has been nationally established since 2008 is considered to have the potential for solving the problems of ineffective forest governance.

11.4 POTENTIALS OF FOREST MANAGEMENT UNITS

Based on a generic definition, FMU/KPH is the smallest (lately it was defined as the optimum) unit of management based on its dominant main function and also its classification of the determined forest, which enables this unit to work efficiently and in a sustainable way (Kartodihardjo et al., 2011). The term "smallest" was considered an incorrect choice, because in fact the area of KPH ranges from thousands to hundreds of thousands of hectares, meaning that "optimum" can still be debatable. However, the KPH concept is that locating professional foresters on the sites where resources and local community exist just next to their office enables them to implement more effective management of the ecosystem under their responsibilities.

According to Indonesian policy the whole forested areas (or *Kawasan Hutan*) in Indonesia will be divided into KPHs under three categories based on the most dominant area, such as KPHP (production FMU), KPHL (protection FMU), and KPHK (conservation FMU). That understanding means, for instance under KPHP, it is possible to have different functions (protection and/or conservation). In addition to that it is possible to find different large- and small-scale utilization and use rights schemes (timber concessionaires, mining industries, etc.) in one KPH (including community-based forest management). If in the area of KPH there is still any area without use rights it can be directly managed by KPH independently or alternatively under partnership management with the local communities. These last areas can possibly be proposed for commercialization for financial sources of the KPH and even to contribute revenues for the local government (Fig. 11.1).

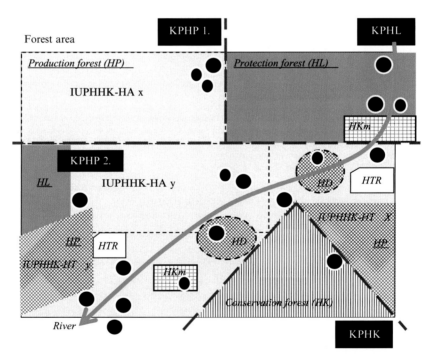

FIG. 11.1   Illustrated situation of KPHP, KPHL, and KPHK based on forest functions and different schemes of forest utilization [eg, IUPHHK-HA (*Izin Usaha Pemanfaatan Hasil Hutan Kayu—Hutan Alam* = Timber Utilization Permit in Natural Forests); IUPHHK-HT (*Izin Usaha Pemanfaatan Hasil Hutan Kayu—Hutan Tanaman* = Timber Utilization Permit in Planted Forests); HKm (*Hutan Kemasyarakatan* = Community Forests); HD (*Hutan Desa* = Village Forests); HTR (*Hutan Tanaman Rakyat* = Community-Based Planted Forests); = Villages]

Administratively, KPHP and KPHL are under the authority of local governments, either provincial or district government, while KPHK is still in the hands of the central administration. After the issue of Act No. 23/2014 on Regional Governance, stating that almost 99% of the authorities of the districts in a forestry sector have been taken over by the province, there will be political changes. Anyway, such changes cannot be discussed here, because government regulations to operationalize it have not been issued. Whatever the changes they will predictably not reduce the main functional tasks of FMU (especially for KPHP and KPHL) on the site level, namely, (1) executing all elements of forest management (production, protection, conservation, rehabilitation, etc.); (2) implementing national and regional forest policies; (3) conducting managerial aspects (planning; organizing; actuating; and controlling); and (4) monitoring and evaluation of forest management.

The Indonesian Ministry of Forestry (MoF) had a target at the end of 2014 to establish 120 (from a total of approximately 600 planned) KPHPs and KPHLs. The progress of the KPHs establishment until June 2014 can be seen in Table 11.3.

Table 11.3 shows how serious the central government is to encourage KPHs development, because it is widely realized that so far forest governance by relying on local forestry services for both forestry administrative tasks and technical services has not been able to achieve either resource sustainability or local community welfare. The central government National Development Planning Agency (BAPPENAS) has even stated "no KPH no budget for forestry institutions." There should be a management institution at the site level that can guarantee that on one hand critical forest lands could be rehabilitated and/or will not be invaded/occupied, and on the other hand potential forests could be managed sustainably and/or will not be illegally logged.

Potentially, there are at least three direct and indirect advantages of the existing KPHs, especially for promoting good forest cogovernance:

(a) *Resource conflicts resolution.* The existence of KPHs on the sites will make possible positive (associative) interaction that, according to a sociological theory, means enabling intensive contacts and communication for building mutual understanding (see eg, Soekanto, 1990) with other key stakeholders, especially the local community and forest concessionaires. Conflict resolution is an element of collaborative management.

(b) *De-bureaucratization for achieving low costs.* Limited capacity and an insufficient number of personnel in provincial as well as district forestry services have led to a long and weak span of the control and a bureaucratization chain that encourages corruption and high costs (see Kartodihardjo et al., 2011). KPHs that are located on-site and also equipped with professional foresters will guarantee better forest management and at the same time more optimum services for forest users. In a good working atmosphere, effective bilateral matching approaches in developing good forestry planning among stakeholders (vertically and horizontally) is very possible.

(c) *Socioeconomic facilitation of local institutions.* Conventional forestry has taught us that trickle-down effects expected to be created through obligating profit sharing of large-scale forest enterprises and other forest users to the surrounding local community did not happen. To the contrary, there was a subsidy from the local community to large-scale and stronger monetary actors in the form of donating their living space (see Sardjono, 2004). Two related issues are insufficient capacity of profit-oriented enterprises and/or weak control of unprofessional forestry apparatuses. Giving authority to KPHs as on-site management units assures direct and indirect solutions of those problems.

TABLE 11.3 Progress of KPHP/KPHL Establishment and Its Operationalization in Indonesia (Evaluation Results of the Ministry of Forestry, until June 2014).

Indicator/Sub Indicator		Output	Achievements	Remarks	Follow-Up Action
Criteria: Region/Managed Area					
Area designation		Decree (SK) of Minister of Forestry on *KPH* Model	120 *SKs*	40 KPHL: 3.550.855 Ha 80 KPHP: 12.888.863 Ha Total: 16.439.718 Ha	–
Criteria: Institution/Organization					
Organization		Regional/Governor/ Head of District/Major Regulations	12 Services 104 Technical Units	4 units under process (ie, KPHs of Memberamo, Lintas Sumut, Flores Timur and Wae Tina)	- Coordination and communication with the local government
Infrastructures	Office	Buildings	73 buildings	41 units established in 2014, 7 units by local government and 1 unit of additional building	- Communication with and facilitation for BPKH to finish the KPH office buildings
	Cars	1 unit for each KPH	89 units	30 units in 2014 and 1 unit KPHL Maria has not been budgeted	- To remind commitments of local governments when proposing KPH Models
	Motor-bikes	Some units per-KPH	88 units of KPH	32 KPHs in 2014 (incl. KPHL Maria and KPHL Sorong Selatan)	to build institution and to recruit/to place staffs
	Stationeries/Survey	Equipments for each KPH	90 units of KPH	30 units implemented in 2014	- Site visits

III. LEARNING FROM THE FIELD CASES/ISSUES

Human Resources/Personnel	a. Training for candidates of *KPH* heads	4 classes (120 participants)	- 97 KPHs have KKPHs (Head of KPH)
	b. Training for *KPH* planning process	1 class with 2 trainings: 165+145 participants	- Forestry High Schools for period July–Dec 2013=215 participants but 13 withdrawing, 37 accepted as civil servant candidates, so for period Jan–Dec 2014=165 participants, plus period July–Dec 2014=145 participants (financed by Directorate of Forest Plannology)
	c. Training for forest planning	118 participants	
	d. *SMKK*/Forestry High School		
	e. Job training for young foresters (*Basarhut*)		- Basarhut activities were located on 52 KPHs (financed by BP2SDM)

Criteria: Plans

| Forest Inventory | Documents and Maps | 86 Drafts | 34 are implemented in 2014 | - Technical advising
- Expert pool trainings and workshops |
| Management Plan | Documents and Maps | 83 Drafts | 37 are implemented in 2014; 24 are already authorized | |

MoE., 2014 Perkembangan Pembangunan KPH. Slides Presentation. Jakarta: Directorate General of Forest Plannology, Indonesian Ministry of Forestry

III. LEARNING FROM THE FIELD CASES/ISSUES

11.5 CHALLENGES OF FOREST MANAGEMENT UNITS IN IMPLEMENTING COLLABORATIVE FOREST GOVERNMENT

Although a long previous discussion has described significant roles of KPHs in promoting good forest cogovernance to achieve sustainable forestry, KPHs face many problems during their establishment and promotion in many districts and there are still big challenges. Those challenges can be classified into four groups (see also Kartodihardjo et al., 2011), as follows:

(a) *KPHs' Area (eg, size, boundary, accessibility).* There is still inconsistency in determining the optimum size of KPHs (KPHP, KPHL, KPHK), because the size of one KPH still ranges widely from below 7500 hectares to more than 750,000 hectares. Besides that, most KPHs use administrative boundaries (based on regions and existing permits), which are theoretically not ideal for optimum forest management. Finally, many KPHs are located in very remote or inaccessible areas and practically without a communication network, which makes it difficult to locate an office, assign staffs, and do productive collaboration.

(b) *Institution (eg, organization, administration, authority).* KPHs as newly promoted institutions are not yet well accepted, especially by many local governments for the following reasons: top-down design, taking over some authority from local forest services (recentralized), obligation for self-financing, and lack of human resources. In many cases, both in form of independent (as a services unit or Satuan Kerja Perangkat Daerah (SKPD)) or dependent (as a technical implementation unit or Unit Pelayanan Teknis Dinas (UPTD)) bodies that administratively do not fit with existing local structures. Specific authorities of KPHs are assumed to destroy decision-making processes, not to mention the financial structure of the local government.

(c) *Human resources (eg, number of personnel, qualifications, professionalism).* In general the districts in Indonesia, especially located in very remote areas, do not have sufficient and moreover qualified foresters. The best foresters have been placed in what are considered more strategic positions (eg, planning board) with an argument to accelerate regional development. In many cases, because of district authority as an autonomous government, local forestry services employ even nonforesters (or frequently called the wrong man at the wrong place) (eg, Mudhofir et al., 2014). Under such a situation expecting to get professional staffs for running KPHs is practically very difficult.

(d) *Financial/Business (eg, funding, core business, partnership).* As discussed, one of the issues faced during the establishment of KPHs was budget availability, especially in the districts. In the future, financing issues should be solved to ensure sustainability of the institution. For that purpose, the central government has given KPHs more space to do self-financing for their operational costs by optimizing their potential forestlands, products, and services. The goals can be achieved with the implementation of an official financial scheme, which is called Badan Layanan Umum (BLU) (public services). However, as discussed previously, many KPHs do not have qualified staff with excellent entrepreneurship. From a natural resource capital point of view, there are quite a few KPHs where their areas have been totally occupied by forest utilization permits (IUPHHK-HA and IUPHHK-HT) with similar core businesses, and therefore these KPHs do not have realistic chances for optimizing commercial programs. Possibly, partnership can only be developed if it is agreed to by existing concessionaires, which are considered having better capital, especially financial support.

Those challenges do not exist independently but are in complicated relationships with other problems faced in implementing good forest governance in the country (eg, political dynamics and rapid changes of local social issues). Therefore, there are still serious efforts to be made for KPHs as a potential institution for the implementation of coforest governance in forest management.

11.6 CLOSING REMARKS

The Constitution of Indonesia (Article 33) mandates the government take full control over natural resources, including forests, for the greatest community welfare. Huge and inaccessible forests in many regions of the country require an on-site forestry institution, and eventually the Ministry of Forestry (MoF) responded with the establishment of FMUs/KPHs. However, rapid sociopolitical dynamics in the last decade has resulted in more complex issues and therefore has demanded a better relationship among the main forestry stakeholders, particularly among the government levels (central, provincial, and district/municipality) to establish KPHs and make them functional. Consistent policies and highly committed decision makers are clearly suitable for accelerating KPHs to become a key factor for potentially implementing good forest cogovernance.

Decentralizing the authority for governing forests to the autonomous regions (province/district/city) has resulted in equitable distribution of responsibility to each governmental level. Recent political development in Indonesia, however, is considered risking more difficulties for promoting KPHs among district governments. The issuance of a new regulation, Act no. 23/2014 on regional governance (replacing Act no. 32/2004), where almost all authority of district governments in the forestry and mining sectors has been withdrawn and moved to provincial government, has become a disincentive factor for the districts to support any programs of the central government, including KPH. Anticipated actions, therefore, have to be taken immediately, otherwise hard efforts that have been made with much energy as well as expensive costs, as well as the significant progress that has been shown will be diminished and will have to be started over again from the beginning.

About the Authors

Mustofa Agung Sardjono is professor of social forestry at the University of Mulawarwan in Samarinda, Indonesia. His expertise covers a wide range of social issues on Indonesian forest management, such as community forestry, forest sociology, agroforestry, and forest politics. His book *Mosaik Sosiologis Kehutanan. Masyarakat Lokal, Politik dan Kelestarian Sumberdaya (Sociological Mosaics of Forestry: Local Community, Politics and Resource Sustainability)* (2004) has broken the long-standing conventional forestry science learned by many academicians in the country, who are concentrated more on biophysical aspects of the resources. Sardjono is currently getting opportunities to do more research and training with his position as the core team member of the Center for Social Forestry (CSF).

Makoto Inoue is professor of forest environmental studies at the University of Tokyo, Japan. His specialty is common-pool resource governance in accordance with field reality, based on the expertise of sociology and anthropology as well as forest policy science. He led an international research project and edited the book *People and Forest: Policy and Local Reality in Southeast Asia, the Russian Far East, and Japan* (Kluwer Academic Publishers, 2003), *Collaborative Governance of Forests: Towards Sustainable Forest Resource Utilization* (University of Tokyo Press, 2015), and *Multi-level Forest Governance in Asia: Concepts, Challenges and the Way Forward* (SAGE Publications, 2015).

References

Carter, J., Gronow, J., 2005. Recent Experience in Collaborative Forest Management. Center for International Forestry Research, Bogor. CIFOR Occasional Paper No. 43.

Colfer, C.J.P., 1995. Who Counts Most in Sustainable Forest Management. Center for International Forestry Research, Bogor.

Contreras-Hermosilla, A., Fay, C., 2005. Strengthening Forest Management in Indonesia Through Land Tenure Reform. ICRAF, Bogor.

FWI, 2011. Potret Keadaan Hutan Indonesia. Forest Watch Indonesia, Bogor. Periode Tahun 2000–2009.

FWI/GFW, 2001. Potret Keadaan Hutan Indonesia. Forest Watch Indonesia/Global Forest Watch, Bogor (Indonesia)/ Washington, DC.

Ginting, L., 2001. Otonomi Daerah dalam Konteks Pengelolaan Sumberdaya Alam. WALHI, Jakarta.

Inoue, M., 2009. Design guidelines for collaborative governance (kyouchi) on natural resources. In: Murota, T. (Ed.), Local Commons in Globalized Era. Minerva-Shobou, Kyoto, pp. 3–25 (original article in Japanese).

Inoue, M., 2011a. Prototype design guidelines for 'collaborative governance' of natural resource. In: Presented at 13th Biennial Conference of the International Association for the Study of the Commons, Hyderabad, India, January 12.

Inoue, M., 2011b. Summary of the Design Guidelines: Instruction for the Authors of Multi-level Forest Governance in Asia—Recognizing Diversity. GSALS University of Tokyo, Tokyo.

ITTO, 2001. Mewujudkan Pengelolaan Hutan Lestari di Indonesia. Laporan Misi Tekni ITTO untuk Indonesia, Jakarta.

Kartodihardjo, H., Nugroho, B., Putro, H.R., 2011. Pembangunan Kesatuan Pengelolaan Hutan (KPH). Konsep, Peraturan Perundangan dan Implementasi. Kementerian Kehutanan/GIZ Forclime, Jakarta.

Moeliono, M., Wollenberg, E., Limberg, G., 2009. Desentralisasi tata kelola hutan. Politik, konomi dan Perjuangan untuk Menguasai Hutan di Kalimantan, Indonesia. CIFOR, Bogor.

Mubyarto, L., Soetrisno, P., Sudira, S.A., Awang, S., Dewanta, A.S., Rejeki, N.S., Pratiwi, E., 1991. Kajian Sosial Ekonomi Desa-Desa Perbatasan di Kalimantan Timur. Penerbit Aditya Media, Yogyakarta.

Mubyarto, L., Soetrisno, P.S., Awang, S.A., Sulistya, A.S.D., Santiasih, E.P., Ismaryati, E.P., 1992. Desa dan perhutanan sosial. Kajian Sosial-Antro-pologis di Propinsi Jambi. Penerbit Aditya Media, Yogyakarta.

Mudhofir, I., Khairi, H., Sardjono, M.A., 2014. Penyusunan Standar Teknis Aparatur Kehutanan di Daerah. Naskah Akademik, GIZ Forclime and Ministry of Forestry, Jakarta.

Prayitno, H., Taufik, A., Fitriyani, R., Putra, R.A.S., 2013. Membongkar harta daerah. Analisis Kebijakan Anggaran Pengelolaan Hutan dan Lahan di 3 Provinsi dan 6 Kabupaten di Indonesia. Seknas FITRA/The Asia Foundation/ Ukaid, Jakarta.

Purnomo, H., Mendoza, G.A., Prabhu, R., Yasmi, Y., 2005. Developing multi-stakeholder forest management scenarios: a multi-agent system simulation approach applied in Indonesia. Forest Policy Econ. 7 (2005), 475–491.

Rahman, Y., Hartati, C., Maulana, M., Subagyo, H., Putra, R.A.S., 2013. Indeks kelola hutan dan lahan daerah. Kinerja Pemerintah Daerah dalam Pengelolaan Hutan dan Lahan di Indonesia. ICEL/Seknas FITRA/The Asia Foundation/UKaid, Jakarta (Studi Kasus Pada 9 Kabupaten).

Sardjono, M.A., Yasuhiro, Y., Wijaya, A., Kamaruddin, Ibrahim, 2001. Social Structure and Production Activities of the Community Surrounding Forest Concession Areas in Sangkulirang of East Kutai District (Indonesia). Unmul/ Tohoku Research Center/Yayasan Bioma, Samarinda/Tohoku.

Sardjono, M.A., 2004. Mosaik Sosiologis Kehutanan: Masyarakat Lokal, Politik dan Kelestarian Sumberdaya. Debut Press, Yogyakarta.

Sardjono, M.A., 2007. Trends of Forest Policies in Indonesia. A narrative material for an intensive lecture as a visiting professor at the University of Tokyo. Mulawarman University, Samarinda.

Sardjono, M. A., 2014. Dinamika Sektor Kehutanan di Kaltim. Tinjauan Periode 2009–2013 dan Perspektif ke Depan. Materials presented in Focused Group Discussion at Mesra Hotel Samarinda, November 19, 2014.

Sardjono, M.A., Inoue, M., 2007. Why do local community shift their orientation? Exploring important social values of tropical rain forests in East Kalimantan (Indonesia). In: A brief paper presented at the 16th Indonesian Scientific Conference 2007 in The University of Kyoto, Japan (August 25th, 2007).

Sardjono, M.A., Inoue, M., 2011. Resource conservation and people dynamics: a review of Southeast Asian tropical forest policies. Ecositrop Ecol. Conserv. J. Trop. Stud. 1 (2), 166–188 (July 2011).

Sardjono, M.A., Soedirman, S., Hardwinarto, S., Pambudhi, F., Diana, R., Heranata, W.W., Fallah, U., Makinudin, B.F., Wijaya, A., Momo, D., Rahmina, H., Fadli, M., Wahyuningtyas, D.C., 2012. Strategi dan Rencana Aksi Provinsi (SRAP) Implementasi REDD+ Kaltim. Badan Perencanaan Pembangunan Daerah Kalimantan Timur, Samarinda.

Sardjono, M.A., Devung, G.S., Imang, N., 2013. Local community dimension of Indonesian Forest Policy and Customary Land Tenure in East Kalimantan. In: Paper presented at The Commons: Commoners and the Changing Commons: Livelihoods, Environ-mental Security, and Shared Knowledge: Kitafuji Conference, June 3 to June 7, 2013 Mount Fuji, Japan.

Soekanto, S., 1990. Sosiologi. Suatu Pengantar. Rajawali Press, Jakarta.

Soetrisno, L., 1993. Problematika Sosial Masyarakat Sekitar Hutan di Indonesia dan Etika Pemanfaatan Sumberdaya Hutan. In: Sumardi, D., Setyarso, A., Suranto, Y., Iswantoro, H. (Eds.), Norma-norma Kelestarian Sosial, Ekonomi, dan Teknologi Pengolahan Sumberdaya Hutan. Fahutan UGM, Yogyakarta.

Sranko, G.R., 2011. Collaborative governance and strategic approach to facilitating chance: lesson learned from forest agreement in Southeast Queensland and the Great Bear Rainforest. Interface 3 (1), 210–239.

Tribowo, D., Haryanto, 2001. Disappearing diversity: an overview on Indonesia's degrading forest and its biodiversity. In: Paper for the Workshop on Integration of Biodiversity in National Forestry Planning Programme, CIFOR Bogor 13–16 August 200.

Weaver, L., 2012. Collaborative Governance. Innovating Tamarack—An Institute for Community Engagement Oct 1–5, 2012. Kithener, Ontario: Communities Collaborating Institute http://tamarackcci.ca/files/collaborative_governance_cci_2012.pdf. retrieved on 07.01.2014.

Coastal Water Pollution and Its Potential Mitigation by Vegetated Wetlands: An Overview of Issues in Southeast Asia

R. Cochard*,†

*Institute of Integrative Biology, Swiss Federal Institute of Technology, Zurich, Switzerland
†Asian Institute of Technology, Klong Luang, Pathumthani, Thailand

Abbreviations

BOD	biological oxygen demand
C	carbon
COD	chemical oxygen demand
DO	dissolved oxygen
DOC	dissolved organic carbon
EHTM	exceptional high-tide mark
GBRMP	Great Barrier Reef Marine Park
HAB	harmful algal bloom
HTM	high-tide mark
LTM	low-tide mark
MS	municipal sewage
MTM	mid-tidal mark
N	nitrogen
OD	oxygen demand
P	phosphorus
PAH	polycyclic aromatic hydrocarbons
TBT	tributyltin
TKN	total Kjeldahl nitrogen
TP	total phosphorus
TSS	total suspended solids

12.1 INTRODUCTION: VALUES OF COASTAL ECOSYSTEMS UNDER POLLUTION EXPOSURE

Coastal ecosystems are of high economic and societal value for providing a multitude of services to humans. For example, mangroves, coral reefs and seagrass beds provide stability to coastlines and thus act against erosion and inland saltwater intrusion (cf. Thampanya et al., 2006). The ecosystems provide shelter and nutrients for marine life and are therefore important feeding, spawning, and nursery grounds for fishes and crustaceans (Ikejima et al., 2003; Chong et al., 1990; Islam and Wahab, 2005). Products derived from mangrove areas include timber and charcoal, medicines and pharmaceuticals, honey and other plant and animal products, and coral reefs are particularly valuable for diving tourism.

Many of these ecosystem services that sustain livelihoods of coastal communities on an "everyday" basis can become particularly important in the wake of natural disasters (UNEP-WCMC, 2006; Cochard, 2011a). Over the last few decades coastal disasters due to storm waves or tsunamis have been on the rise throughout Southeast Asia. This is partly because of growing numbers of people living in danger zones along the coast, but it appears that some coastal hazards have also become increasingly frequent and intense, probably as a result of global warming and associated sea level rise (Cochard, 2013). In this context, the potentially important wave buffering functions of coastal ecosystems have attracted increased interest from scientists. After the tragic Asian tsunami disaster of December 26, 2004, it was argued, for example, that tsunami wave impacts were generally higher behind mined and degraded coral reefs, and behind mangrove vegetation that had already been affected by "cryptic anthropogenic degradation" (Dahdouh-Guebas et al., 2005; Fernando et al., 2005; Marris, 2005). Even though many claims were not substantiated on a firm scientific basis (cf. Cochard et al., 2008; Cochard, 2011b), it is nonetheless clear that in postdisaster situations communities are often particularly dependent on local natural resources. This in turn tends to increase the pressures and threats to resources and compounds the risks of significant ecosystem degradation in postdisaster situations (Cochard, 2011a). Therefore, it represents a fundamental task to maintain these ecosystems in as healthy a state as possible, and to restore fundamental ecosystem functions after any significant degradation, such as may be caused by a tsunami.

Coastal pollution by sewage and other wastewater runoff is a major "slow hazard" (but often intensified during disaster situations) posing a threat to human health and contributing to cryptic degradation of coastal ecosystems, in particular coral reefs and seagrass beds (Peters et al., 1997; cf. Cochard et al., 2008). Coastal wetlands can thereby contribute in mitigating pollution via the filtering and trapping of sediments and the absorption and reduction of nutrients and other pollutants in domestic waste effluents. Depending on their nature and composition, wastewater effluents can, however, also lead to significant compounding degradation of the mangrove greenbelt, with ensuing losses of mangrove ecosystem biodiversity, functions, and services. Destructive impacts are then often highest during extreme events, as discharges of sediments and wastewater often reach peaks during storms and floods (Mallin et al., 1999). While wastewater "tsunamis" hardly ever receive any media attention (except perhaps if posing an eminent risk to human health) they may be lethal in their summed-up environmental impact over time by increasing coastal ecosystem and community vulnerability and decreasing resilience to natural fast hazard impacts (Cochard, 2011a). On the other hand, it is not always clear what constitutes "pollution" and/or whether "pollution" is actually harmful. Some types and limited levels of sewage may actually be used as a mangrove fertilizer, whereby sewage may potentially increase the strength and robustness of mangrove stands. Wastewater may also be used to enhance and support seedling establishment and growth, for example, in mangrove replanting schemes on abandoned aquaculture ponds. It is increasingly being recognized that alternative integrated approaches to coastal management are needed (Cochard, 2011a).

Here in this chapter an overview is first provided of sewage pollution hazards: types of pollutants, volumes of pollution runoff, and available assessments of their impacts on coastal ecosystems and fisheries. A special focus is set on Thailand where several fairly detailed studies have been conducted; this includes some field observations by the author at Bang Pu mangrove forest south of Bangkok. However, general findings of the review also apply to many other regions throughout Southeast Asia, where industrial and agricultural developments have occurred at a similarly fast pace during the past decades. A short summary is then provided of wastewater treatment practices, specifically focusing on the role and function of natural and constructed wetlands in general. Following this is a discussion of the role of mangrove forests as a last physical (waste-) water catchment system and a potential "pollution mitigation facility" in terms of biological nutrient absorption and sequestration of other potentially hazardous contaminants. A specific focus is set on studies that actually have considered mangroves as a wastewater "pollution filter." This includes consideration of the impacts (both negative and positive) of pollutants on the mangrove ecosystem.

12.2 WASTEWATER POLLUTION HAZARDS AFFECTING THE COASTAL ENVIRONMENT

12.2.1 Wastewater Pollution: Some Definitions and Risk Assessment

"Waste" has been defined as a misplaced resource or a resource that has lost its value to its owner. Any type of liquid waste mixed with clean water at any concentration may produce wastewater (sewage mainly refers to domestic wastewater); that is, water that has been adversely affected in quality by anthropogenic influence (Wikipedia, 2014). Liquid waste may be emanating from various sources, for example, domestic residences, commercial and industrial establishments and/or agriculture or aquaculture, highway drainage, and so on, and be carried by groundwater or surface water. It may be composed of various contaminants, for example, organic and inorganic soluble and particulate material (including emulsions and toxins), bacteria, and other microorganisms (including pathogens) (Metcalf and Eddy, 2003).

Because it is often difficult to identify and quantify different contaminants in wastewater, commonly used indicator tests of water quality are biological oxygen demand (BOD) and chemical oxygen demand (COD). Oxygen demand (OD) is a measure of whether or not wastewater will have a significant adverse effect upon aquatic life. The BOD test measures only OD of biodegradable pollutants, whereas the COD test additionally measures OD of nonbiodegradable oxidizable pollutants (Metcalf and Eddy, 2003). Other major water quality parameters are pH, conductivity, total suspended solids (TSS), total Kjeldahl nitrogen (TKN), ammonium (NH_4^+-N), nitrate (NO_3^--N), and total phosphorus (TP) (Metcalf and Eddy, 2003).

"Pollution" is the release of any contaminants (ie, material impurity) to the environment. It may represent a significant "fast hazard" (eg, highly toxic spill) or "slow hazard" (low-level but chronic/cumulative emissions) to coastal environments and communities; "hazard" being defined as a process or phenomenon that may by a certain probability constitute a damaging event (cf. Cochard et al., 2008). In pollution science, hazard identification involves assessing whether a substance or constituent may exhibit a particular adverse environmental or health hazard. Hazard exposure assessment involves identification of a potential receptor ecosystem (eg, a mangrove forest), evaluation of exposure pathways and routes (eg, estuarine creek systems and tidal flow patterns), and quantification of exposure (eg, yearly amount of BOD discharged into the river system) (Metcalf and Eddy, 2003).

In the case of chronical pollution exposure, the assessment of community and ecosystem "vulnerability" and "resilience" is commonly difficult and not nearly as clear as in the case of fast hazards (Islam and Tanaka, 2004; cf. Cochard et al., 2008); various aspects of complexity and approaches to pollution risk assessment in coastal environments are, for example, presented and discussed in Newman et al. (2001), Keller (2006), and Donohue et al. (2005). One first challenge for accurate pollution risk assessment is posed by identifying an actual cause-effect relationship for which risk is to be estimated (cf. Newman and Evans, 2001). A second challenge is to establish an understanding about the nature of the pollution hazard; that is, its various potential impacts within the spatial and temporal complexities of coastal ecosystems (cf. Roberts et al., 2001). Impacts may only occur and become obvious after a considerable lag time during which pollutants accumulate. After reaching a certain threshold, degradation may then occur in a cascading, sometimes catastrophic manner. The actual impact of a certain pollution exposure often becomes apparent only by the interaction with other hazards; for example, pollution by pesticides may be very afflictive to biota only under conditions of eutrophication (Islam and Tanaka, 2004). The development of meaningful and accurate indicators for pollution risk management is therefore a particularly intricate task. As demonstrated by Jones et al. (2001a), the composition of wastewater may differ between seemingly similar sources and exhibit significantly different ecological impacts. Various sets of ecological indicators have frequently been used as means to trace, assess, and model the effects of pollution constituents in the marine environment (see, for example, Luan and Debenay, 2005; Methratta and Link, 2006; Vassallo et al., 2006; Craig et al., 2003; Maurer et al., 1999; Macintosh et al., 2002; Hale et al., 2004).

12.2.2 Landborne Pollution Hazards in Thailand and Indonesia

Islam and Tanaka (2004) reviewed marine pollution hazards and possible environmental impacts in the seas around the world. FAO-UNEP (1980) provided some detailed early pollution data in the Gulf of Thailand, specifically with regard to mangroves. A summary of pollution problems in the Gulf of Thailand was provided by Cheevaporn and Menasveta (2003) that, however, also present old data only up until 1986. Some more recent data were available from the website of the Thai Pollution Control Department (PCD); most documents were, however, in the Thai language. An English report (PCD, 2004) indicated that, according to a water quality index assessed for Thailand's major 48 rivers, streams of "deteriorated" water quality decreased from 35% to 26% between 2002 and 2004, whereas rivers classified as being of "good" water quality also decreased from 40% to 23%, accompanied by an increase from 25% to 51% of rivers classified as being in "fair" condition. Notably, the water quality of Chao Phraya River, which flows through Bangkok, was highly deteriorated. Water quality measured on 240 stations in the marine environment showed a slight increase between 2002 and 2004 from 6% to 9% of "deteriorated" states of water quality, but a marked decrease of "excellent" water quality from 47% to only 3% with a total decrease from 83% to 46% of "good" water quality (PCD, 2004). None of the stations in the Inner Gulf measured values better than "deteriorated," with several parameters such as low dissolved oxygen (DO) (0.7–3.8 mg/L), and high TSS (114.4–914.8 mg/L), nitrate (100–1123 µg/L), ammonia (1028–1686 µg/L), phosphate (79–253 µg/L), iron (370–21,850 µg/L), and tributyltin (TBT) (34–43 ng/L) below water quality standards at many measuring stations. Overall assessments of water quality in the eastern and western Gulf regions were less serious, with stations in only a few areas reporting "deteriorated" quality measurements. The few records of "excellent" water quality were recorded from islands (Ko Phuket and Ko Phi in the Andaman Sea, and Ko Samui and Ko Pha Ngan in the western Gulf) and a few beaches in Prachuapkhirikan and Chumporn Provinces in the western Gulf region (PCD, 2004).

According to Cheevaporn and Menasveta (2003), untreated municipal and industrial wastewater was posing the most serious pollution problem along the Thai coastlines in 1986; about 29,033 tons of domestic and 5343 tons of industrial BOD was discharged annually into the Gulf from the central coastal basin, which includes Bangkok metropolis, representing about 95% of the total wastewater runoff in eastern Thailand. Chongprasith and Praekulvanich (2003) put figures of total estimated Thai BOD loads far higher at 274,898 tons/year, with a contribution of 37% from domestic sources, 35% from cultivation, 15% aquaculture, 11% livestock, and only 2% from industrial sources. As data sources were not indicated, the reliability of these percentage figures is unclear. Tonmanee and Kanchanakool (1999) quite in contrast emphasized the important contribution of industry before agriculture. In Bangkok in 1986 about 60–70% of domestic waste was estimated to be released to the Chao Phraya River and eventually into the Gulf without prior treatment (Cheevaporn and Menasveta, 2003). Application of agricultural fertilizer in Thailand increased from an average 25 kg/ha in 1983 to 47 kg/ha in 1990, whereby fertilizers contained 26 kg nitrogen, 14 kg phosphorus (P_2O_2), and 7 kg potassium (K_2O); this increase was three times less than in European countries during the same period, but application efficiency was often low and soil erosion added to nutrient losses (Tonmanee and Kanchanakool, 1999). Nutrient enrichment consequently often led to algal blooms in coastal zones in the Inner Gulf. According to Suvapepun (1991), nitrate measurements in the Inner Gulf increased from 3.0 µg/L in 1974 to 24.9 µg/L in 1989. In parallel, phytoplankton numbers doubled from 4453 cells/L in 1976 to 11,553 cells/L in 1987. Less dramatic increases of nitrate in watersheds were presented by Tonmanee and Kanchanakool (1999). The levels measured for heavy metal contaminants including arsenic, cadmium, chromium, copper, iron, mercury, manganese, lead, and zinc at the time did not seem to pose a major problem in the Gulf, despite an estimated discharge of over 51 tons per day in 1983 (Cheevaporn and Menasveta, 2003). According to Hungspreugs et al. (1989) and Cheevaporn and Menasveta (2003), uses of major pesticide pollutants were declining in the 1980s, and then not posing any serious problems. This, however, contrasts with statements by Tonmanee and Kanchanakool (1999) who listed it as an increasing and serious problem. A study by Boonyatumanond et al. (2006) concluded that overall contamination levels of Thai sediments with polycyclic aromatic hydrocarbons (PAHs) could be ranked as low to moderate as compared with ranges of a worldwide survey. PAH levels were highest in Bangkok, probably still posing a significant risk at certain locations. PAH contamination was probably attributable to various, mostly diffuse sources. Wattayakorn et al. (1998), in contrast, reported "chronic" pollution of the Inner Gulf of Thailand with dissolved and dispersed petroleum hydrocarbons of about $4 \mu g/L$ with occasional acute pollution events of $>40 \mu g/L$ during about 25% of the time. They also assessed water circulation and found that water movement was sluggish (mean currents of $<0.07 m/s$) and that the Inner Gulf was poorly flushed, with pollution-loaded brackish water trapped along the coastline during much of the time. Tabucanon (1991) referred to inland soil erosion problems as an important source of water pollution in the Gulf: an estimated 47 million cubic meters of sediments contributed to mangrove succession in some places; on the negative side, it affected water quality and smothered corals and seagrass beds offshore. Also, tin mining was listed as a significant contributor to sedimentation problems in the early 1980s, particularly along parts of the Andaman coastline (Hungspreugs et al., 1989). Tin mining has since ceased in most parts of this coastline.

The Indonesian Archipelago consists of a multitude of islands dividing the ocean into several sea basins; these include the Java, Flores, Banda, Celebes, Ceram, Molucca, and the South China Sea on the northern side, and the Arafura, Savu, and Timor Sea and the Indian Ocean Basin on the southern side. Overall assessments and quantifications of pollution hazards are much more difficult than in the case of the Gulf of Thailand. Virtually no pollution data are available for many of the more remote parts of the Archipelago. However, for example, an attempt was made by Morrison and Delaney (1996) to assess overall pollution in the remote Arafura and Timor Seas. Also, a comprehensive review of pollution was prepared for the South China Sea Basin by Morton and Blackmore (2001), although focusing more on continental countries and specifically Hong Kong. As in the case of Bangkok/Thailand, major centers of pollution in the Archipelago are highly populated areas, particularly metropolitan areas; for some areas in Java and Ambon, quite detailed data sets of coastal pollution are available from the literature (eg, Williams et al., 2000; Booij et al., 2001; Takarina et al., 2004; Evans et al., 1995; Dsikowitzky et al., 2011).

12.2.3 Contribution of Aquaculture to Pollution Along the Thai Coastlines

With the expansion of shrimp farm enterprises, aquaculture effluents have increasingly contributed to pollution in Southeast Asia as well as in other parts of the world (Paez-Osuna, 2001). In Thailand, shrimp harvests have shifted between 1987 and 1993 from predominantly wild-capture (105,000 metric tons in 1987 to 80,000 in 1993) to farm-raised (30,000 in 1987 to 200,000 in 1993) (FAO/NACA in Dierberg and Kiattisimkul, 1996). Equally, a shift from predominantly extensive shrimp farming to intensive farming practices could be recorded; production efficiency increased about 20-fold from 1977 (125 kg/ha/crop) to 1992 (2580 kg/ha/crop). Dierberg and Kiattisimkul (1996, p. 650) noted at the time that hardly any research in Thailand had focused on the environmental impacts of aquaculture and data for quantitative assessments of pollution were unavailable: "Reports are not circulated widely outside the region or the aquaculture community. Most of the reports rely on surveys, anecdotal information, and historical events upon which to base generalizations." Even now, comprehensive numbers are not readily accessible. In 1994 estimates by Briggs and Funge-Smith (in Dierberg and Kiattisimkul, 1996) indicated, however, that an approximate area of 40,000 ha of intensively operated shrimp farms in Thailand produced the waste equivalent of 3.1–3.4 million people for nitrogen and 4.6–7.3 million for phosphorus. This was about equivalent to an increase of coastal population by 50–100% without any sewage treatment. There has also been a marked increase in marine cage aquaculture from a production of about 1000 mt in 1988 to 3800 tons in 1996 (Islam, 2005). Being an essentially open system, cages are usually characterized by a high degree of direct interaction with the environment, and the large bulk of wastes frequently produced from cage systems dissipates freely into the environment (Islam, 2005).

Depending on stocking density, pond effluents discharged from intensively operated shrimp farms usually contain high concentrations of nutrients, suspended solids, oxygen-demanding substances, and chlorophyll *a*, whereby aquaculture feeds are normally the main source of wastage (Tacon et al., 1995). The daily exchange of pond water with outside water can be as much as 40% of semiintensive and intensive systems to remove excess nutrients and organic matter, and wastewater effluent into receiving water can be very significant (Dierberg and Kiattisimkul, 1996). Fifty-one to sixty-six percent of the total nitrogen (N) and 94% of TP inputs into shrimp ponds are either released in the discharge water or deposited on the pond bottom, as was indicated by nutrient budgets carried out on intensive shrimp ponds in Thailand (Funge-Smith and Briggs, 1998; Briggs & Funge-Smith in Dierberg and Kiattisimkul, 1996). Wastewater discharge is particularly afflictive to the environment during pond harvesting; nutrients, BOD, and TSS exported during harvest were in the range of 23–71% of total discharge during a 4-month pond operation period (Dierberg and Kiattisimkul, 1996). Another significant impact is the disposal of accumulated pond sediments after harvest. Improper disposal can contribute almost twofold to loadings of TKN and BOD, and even 9- to 10-fold for TP and 13- to 18-fold for TSS as compared to discharges during operation and harvest. Although there is legislation for proper disposal in Thailand, illegal disposal by pond flushing with high-pressure hoses appeared to be widely practiced there (Szuster, 2003). Another practice is to allow pond outlet structures to remain open over several tidal cycles. The BOD effluent produced by such sludge disposal practices is extremely high and can reach 1500 mg/L (Satapornvanit, 1993). On average 238–321 kg/ha, 455–668 kg/ha, and 196,000–215,000 kg/ha per operation cycle for TP, TKN, and TSS, respectively, was calculated for intensive shrimp farming in Thailand (Dierberg and Kiattisimkul, 1996). According to figures by Lin (2000), total annual amounts of wastes released from Thai shrimp farms amounted to 131,250 metric tons (mt) of organic matter, 8400 mt of N and 3150 mt of P annually. Lindberg and Nylander (2001) estimated that approximately 16 million mt of dry sediment was produced from the 40,000 ha of intensive ponds in operation in Thailand in 1994. Information on cage aquaculture nutrient budgets is still scarce, but a conceptual model by Islam (2005) estimated that for each ton of fish produced about 132.5–462.5 kg N and 25–80 kg P are released to the environment, putting 1996 Thai figures for total effluents at 504–1757 tons for N and 95–304 tons for P.

Other hazardous wastewater constituents emanating from shrimp ponds and other types of aquaculture are antibiotics, anesthetics, pesticides, and disinfectants (Szuster, 2003). Gräslund et al. (2003) found in a survey that shrimp farmers used on average 13 different chemicals and biological products to treat shrimp ponds. The most commonly used products were soil and water treatment products, pesticides, and disinfectants. Overall, the use of more than 290 different chemical and biological products was documented, many of which could have negative effects on the cultured shrimp, cause a risk for food safety, occupational health, or affect adjacent environments (Gräslund et al., 2003). The use of antibiotics, in particular, can affect microbial communities thereby affecting and changing microbiological processes, such as organic matter breakdown, within receiving waters. It can also lead to increased concentrations of resistant microbial strains associated with animal and human diseases (Szuster, 2003). While misuse of antibiotics is widespread in Thailand, its environmental impacts are still difficult to assess; persistence of antibiotic residues in the environment is influenced by a complex set of variables (Weston in Szuster, 2003). According to a study by Le et al. (2005) in Vietnam, incidence of resistance by bacteria collected from mangrove mud samples was relatively high, although the relation between concentration and antibiotics residues and the incidence of resistance was not clearly defined. Results also indicated that antibiotics degraded more rapidly in media with resistant bacteria.

12.2.4 Effects of Aquaculture and Pollution on Coastal Stability Along the Thai Coastlines

Production of operational ponds generally declines at a rate of 3–9% per production cycle. With an average lifetime of about 7 years, abandonment of the ponds is a common phenomenon (Dierberg and Kiattisimkul, 1996). As a consequence of frequently occurring acidification and release of heavy metals from the sediments, shrimp farms located in mangrove environments are generally more susceptible to aquaculture diseases than inland farms, and turnover rate of productivity is shorter (Kautsky et al., 2000; Eng et al., 1989). The expansion of shrimp farming has coincided with much of the mangrove destruction in Thailand; a total decrease in mangrove forest cover from 367,900 ha to 167,582 ha has been estimated between 1961 and 1997 (Cochard, 2011a). Not all the mangrove destruction was due to shrimp farming, but as the single largest impact it has been devastating in some regions; for example, in Pak Phanang District 85% of the destruction of a mangrove area of 69.4 km^2 was attributable to shrimp farming (Dierberg and Kiattisimkul, 1996). Destruction of mangrove wetlands in many areas led to coastal erosion and further increase of BOD loads and other pollution. Coastal protection and stabilization was significantly diminished and any intrinsic "wastewater treatment function" of mangroves was lost (see later discussion). According to a study by Thampanya et al. (2006), net annual land losses of 0.91 km^2 were measured between 1966 and 1998 along the western Gulf of Thailand coastline, where about 90% of mangroves had been destroyed.

Somewhat exemplary is the boom-and-bust evolution of shrimp farming over the last two decades along the coasts of the Chao Phraya River delta. After a phenomenal increase of shrimp farm production from about 8000 tons of shrimp in 1987 to 21,000 tons in 1989, production crashed as a result of viral disease pathogens, self-pollution, and general environmental degradation, and in 1992 production dropped to less than 15% of 1989 peak levels (Szuster, 2003). Approximately 19,900 ha of shrimp farmland along the upper Gulf of Thailand coastline and possibly 45,000 ha in the entire central region of Thailand were abandoned just in 1990–1991 (Potaros in Szuster, 2003). A move from coastal shrimp farming to inland low-salinity shrimp aquaculture has since taken place to maintain Thailand's production levels (Flaherty and Vandergeest, 1998). While some of the abandoned ponds along the coast have again been returned to more extensive shrimp farm use, much of it remains as wasteland. The destruction of mangrove forests, in particular, led to a fast rate of erosion of approximately 25 m of land loss per year between 1969 and 1987, whereby at Bang Khun Tien the coastline now follows the layout of abandoned former fish and shrimp ponds (Winterwerp et al., 2005; Fig. 12.1).

FIG. 12.1 Erosion processes along the coastline near the Chao Phraya River estuary south of Bangkok. *Top image:* aerial picture taken in 1999 of the coastline at Bang Khun Tien. After destruction of the mangrove belt under the expansion of intensive shrimp farming followed by abandonment of shrimp enterprises, the coastline is highly irregular and follows the layout of former shrimp ponds. *Picture source: Google Earth™. Middle left:* view away from the estuary toward the marina belonging to the royal Thai Naval Base (at *horizon*). The ripples in the algae-covered mud near to the rubble wall (see *arrow* a) illustrate the wave energy rebound from the wall and associated sediment scour in the mud flat. *Middle right:* old surviving *Avicennia marina* tree in the mud flat (see *arrow* b). The position of the tree, that is, its distance away from the other mangroves as well as its elevation above the mud flat (with roots forming a "sediment Island") also illustrate the erosion. Note the lack of any rejuvenation. *Bottom left:* detail of mud surface (see arrow c) near the mangrove. Note algal cover and the erosion dips in the mud. *Bottom right:* debris deposited at a remaining mangrove stand in front of the concrete wall near the marina. This location was designated for replantation (see wave breaker poles to the right), obviously without any success. *Photographs by R. Cochard, June 2006.*

III. LEARNING FROM THE FIELD CASES/ISSUES

To stop coastal erosion south of Bangkok, Winterwerp et al. (2005) proposed to restore the intertidal area by sacrificing some of the now-existing fish and shrimp ponds, to protect it from lateral wave transport of sediments by installing permeable groynes perpendicular to the coast, and finally to reinitiate sedimentation in the long-term by reestablishing a mangrove greenbelt 300–500 m wide. Such a scheme obviously would be very difficult and costly, and whether restoration will be possible under the given rate of pollution in the Gulf remains another question. Winterwerp et al. (2005) mentioned that virtually no rejuvenation took place in the remaining mangrove stands along this stretch of coastline without, however, discussing in detail the possible reasons for lack of seedling establishment and survival. So far, replanting schemes have been largely unsuccessful along this coastline (Winterwerp et al., 2005). Observations at the Royal Thai Naval Base on the shoreline east of the Chao Phraya estuary provided some clues. One major problem is certainly posed by a change of wave energy and lateral sedimentation processes along the stretches of coastline that were revetted by rubble mounds. Slow erosion occurs in the forelying tidal flats, but nearby *Avicennia marina* mangrove stands to the east are still experiencing insufficient sediment inputs from lateral drift, probably primarily due to a lack of sediment trapping closer to the estuary (Fig. 12.1). Wave energy is increased to a degree that probably does not allow further sedimentation and seedling establishment. Furthermore, debris and chemical pollutants transported from the Chao Phraya River contribute to the destruction and degradation of the mangrove front (Fig. 12.1). Near the frontal mangrove stands the fine sediment deposits are covered by dense layers of algae that are likely to negatively influence the deposition of moveable sediments. This probably also represents an additional hindrance to the burial of floating propagules and successful establishment of seedlings (Fig. 12.1). The only place where mangrove reforestation along this coastline was so far successful was inside abandoned shrimp ponds, which are protected from waves by rubble mounds, such as a prominent place east of the marina of the Royal Thai Naval Base that belongs now to the World Wildlife Fund Thailand, now called the Bang Pu Conservation Area (cf. Wickramasinghe et al., 2009).

12.3 POLLUTION IMPACTS ON TROPICAL COASTAL ECOSYSTEMS IN SOUTHEAST ASIA

12.3.1 Biodiversity and Biophysical Gradients in the Coastal Landscape

Southeast Asia is a center of marine biodiversity (Tomascik et al., 1997). The region contains more than 600 of the world's nearly 800 reef-building coral species (Scleractinia), and 23 of the 51 species of seagrasses (Burke et al., 2002). With a cover of more than 100,000 km² the region has about 34% of the world's coral reefs. These reefs are the most species rich, but they are also the most threatened (Burke et al., 2002). Less is known about the cover of and threats posed to seagrass beds of the region. Estimates of destruction are in the range of 20–40% along the coasts of Thailand and Indonesia (Green and Short, 2003). Southeast Asia is equally the center of mangrove diversity with 51 of the world's 70 true mangrove species present, and containing some 250 other plant species and more than 1000 marine invertebrate and vertebrate species. In addition, there are 177 bird and 36 mammal species associated with mangroves (Chou, 1994). Mangrove forests cover a total area of about 68,384 km² with some 2641 km² in Thailand and 42,550 km² in Indonesia; the country with by far the largest cover of mangroves in the world (almost a quarter of the world total of mangroves are found in Indonesia, possibly now diminished to some 25,000 km²) (Spalding et al., 2010; Cochard, 2011a). Furthermore, the region provides for about 40% of the world's fish catch (Thia-Eng, 1999).

While mangrove forests, coral reefs, and seagrass beds can occur in isolation, these systems often occur together and are linked by considerable physical as well as biological interactions. In many areas, the coastal profile moves from the intertidal mangrove mudflats to shallow water with seagrass beds to seaward-fringing coral reefs (Burke et al., 2002). Corals play a foundation and keystone role in many tropical coastal ecosystems, providing essential sheltering habitat for marine species, protecting shorelines, and thereby facilitating the growth of mangroves and seagrasses. On the other hand, mangroves and seagrasses bind soft sediments, thereby facilitating coral reef development in areas that may otherwise have too much silt for coral growth (Burke et al., 2002). Mangrove forests provide shelter and nutrients, and are therefore important feeding, spawning, and nursery grounds for fishes, crustaceans, and other marine life (Ikejima et al., 2003; Chong et al., 1990). Research has found that fish catch is higher in reefs that are close to mangrove areas, whereby reefs and mangroves are often linked by seagrass beds (Dorenbosch et al., 2004; Mumby et al., 2004). Some economically important species are actually dependent on seagrass beds for some stage of their life cycle (Chou, 1994). The value of the region's fisheries in coral reefs alone is estimated to be US$2.4 billion per year; in Indonesia, economic benefits from reefs including tourism are estimated to be US$1.6 billion per year (Burke et al., 2002). There are various economic valuations of mangrove resources (see Cochard, 2011a), but relatively few attempts have been made at valuations of seagrass beds (cf. Unsworth and Cullen-Unsworth, 2013).

At a higher landscape level the interactions of marine pollution, ecosystem destruction, and subsequent follow-on consequences, such as coastal erosion and biodiversity loss, and environmental and economic rebounds, are not yet well documented nor understood for the region of Southeast Asia. The most comprehensive data about water pollution effects on tropical coastal environmental processes and interactions from landscape catchment-to-reef down to patch scale ecology probably exist for the Great Barrier Reef Marine Park (GBRMP) in northeastern Australia (see summary of studies in Hutchings et al., 2005a). The GBRMP (particularly the northern region) could be considered to be still in a fairly pristine condition as compared to many other regions, such as Southeast Asia. Yet, a study by Wooldridge et al. (2006) concluded that the nutrient-enriching impact from river runoff in the GBRMP is now reaching reefs as far as 20–30 km offshore, whereas estimates of pre-European enrichment extent were only 1–2 km from river estuaries.

12.3.2 Vulnerability of Coral Reefs to Pollution Hazard Exposure

Based on available data and model inferences, human activities now threaten 88% of Southeast Asia's, and 85% of Indonesia's coral reefs. For about 50% of the reefs the level of threats is estimated to be high or very high (Burke et al., 2002). Data on pollution are often incomplete for coastal areas, but the state of forelying coral reefs often tells an indirect story (cf. maps provided by Burke et al., 2002, which show affected reefs primarily near population centers), even though the actual causes of coral reef mortality are commonly difficult to backtrace, and pollution effects and other disturbances are typically confounded (cf. Fabricius, 2005;

Hoeksema and Cleary, 2004). Burke et al. (2002) estimated that the highest threats were posed by overfishing and destructive fishing, threatening 65% and 51% of reefs, respectively. Sedimentation and pollution from deforestation and agricultural activities threatened about 20% of coral reefs, and coastal pollution and activities, such as sewage discharge, dredging, landfilling, mining of sand and coral, and coastal construction, threatened about 25% of reefs (Burke et al., 2002).

In a study in Indonesia (Java, Sulawesi, and Ambon) Edinger et al. (1998) concluded that reefs subjected to high land-based pollution (sewage, sedimentation, and/or industrial pollution; at chlorophyll *a* levels up to $1.24 \, mg/m^3$) showed reductions in coral species diversity of 30–50% at 3 m depth, and 40–60% at 10 m depth, relative to unpolluted controls in the region. This demonstrates how reef ecosystems increasingly simplify with increasing exposure to runoff (see also Dikou and van Woesik, 2006), compromising their ability to maintain essential ecosystem functions and remain inherently resilient to wave impacts (cf. Cochard et al., 2008). The death of key organisms on the reef or a shift from an autotrophic to a heterotrophic (suspension/detritus feeding) benthic community changes the dominant ecological process of carbonate deposition to erosion. Reef disintegration may be accelerated by increasing densities of organisms, such as urchins, sponges, and fishes, which contribute to the destruction of the reef and the formation of sand (Peters et al., 1997).

Fabricius (2005) provided a detailed review about the state of knowledge on the considerably complex effects of terrestrial runoff on coral reef ecology, also considering various stages in the coral reproduction and recruitment cycle and interacting organisms. The four main water quality parameters affecting corals initially are (1) increased dissolved inorganic nutrients, (2) enrichment with particulate organic matter, (3) light reduction from turbidity, and (4) increased sedimentation (Fabricius, 2005). Nutrient enrichment in typically nutrient-poor reefs can reduce coral calcification and fertilization rates and increase macroalgal abundances. As nutrients increase, nutrient cycling coral communities (in oligotrophic waters) generally shift to increasing proportions of macroalgae and further to heterotrophic filter feeders (Fabricius, 2005). Even slight increases of nutrients can lead to corals and surrounding substrates becoming overgrown with algae, which not only diminishes valuable solar energy for the corals, but also inhibits colonization of substrate by larval recruits. Sedimentation and algal growth may also indirectly lead to reef bioerosion by changing community composition of organisms that affect corals and coral communities (Hutchings et al., 2005b; Fabricius, 2005). Increasing bacterial or other microbial activity linked to organic matter may directly affect coral health (Fabricius, 2005). Algae-dominated reefs are commonly also poorer in fish diversity and biomass (Burke et al., 2002). Most importantly, turbidity by increased water sediment levels and/or suspended algal concentrations limit light availability and can completely smother corals. Under decreased irradiation, zooxanthellae may not get enough light to photosynthesize and feed corals, reducing growth or causing coral bleaching and death (Burke et al., 2002). Other pollution hazards are posed by heavy metals such as copper, zinc, and some hydrocarbons, which have been linked to reduced fertilization, fecundity, and growth of adult corals (Fabricius, 2005). In addition, mass coral bleaching, as a relatively recent (and not well understood) phenomenon within the last two decades poses major threats to coral reefs. It is a stress response commonly correlated to increased sea temperature levels (often attributed to global warming) which may, however, be particularly harmful in interaction with other chronic stresses, such as those posed by land-based pollution (Vega-Turber et al., 2014; Hughes et al., 2003).

12.3.3 Vulnerability of Seagrass Beds to Pollution

Like coral reefs, seagrass beds are an important permanent or transitory habitat for many fish and invertebrate populations. Seagrass beds also serve as feeding grounds for threatened marine species such as sea turtles and dugongs (*Dugong dugon*). Like mangrove forests, seagrass communities to a certain extent contribute to water filtering by trapping suspended matter and absorbing dissolved nutrients. Most importantly, seagrasses regulate sediment dynamics by reducing sediment resuspension and increasing stabilization with roots and rhizomes, and contributing to the formation of carbonate sediment particles (Gacia et al., 2003). Research indicates that seagrass beds may have a certain buffering role against tsunamis and other types of wave hazards (Chatenoux and Peduzzi, 2007). There are only a few data of impacts of pollution on seagrass meadows for the Southeast Asia region, but Waycott et al. (2005) and Schaffelke et al. (2005) provide useful summaries of the increasing body of scientific knowledge in the GBRMP.

Like corals, seagrasses can be negatively affected by chronically elevated turbidity, but the sensitivity to shading differs between seagrass species. Some species can survive for more than 30 days under full shading. Other light-dependent species are highly sensitive to irradiation flux changes (Waycott et al., 2005; Gacia et al., 2003, 2005; Terrados et al., 1998). In Hervey Bay, Australia, seagrass losses have been observed in a variety of habitats in an area of >1000 km^2 due to pulsed turbidity events from river discharges (Preen et al., 1995). In Southeast Asia, losses of mangroves on fringing coastlines, logged for aquaculture development or wood chipping, were specifically blamed for sediment pollution impacts and losses of seagrass beds (Fortes, 1991).

Seagrasses are generally nutrient limited and therefore they have the ability to act as a biosink for nutrients, sometimes containing high levels of nitrogen and phosphorus in their tissue (Waycott et al., 2005). Yet research in Indonesia by Erftemeijer et al. (1994) indicated that tissue absorption may be limited. No increase of biomass was observable under increased levels of nitrogen and phosphorus. Seagrass beds offshore from the city of Cairns in the Great Barrier Reef (GBR) region have increased in area between 1936 and 1994, which was attributed to an increase of net total nutrient levels over this period (Udy et al., 1999). Chronically elevated concentrations of dissolved nutrients can, however, also promote the proliferation of algal growth (eg, phytoplankton or algal epiphytes on seagrass leaves and stems), thus also reducing the amount of light reaching the seagrasses (Schaffelke et al., 2005). Concentrations of nutrients have apparently not yet reached critical levels for seagrasses in the GBR region. On the other hand, Neverauskas (1987) reported on the destruction of a South Australian seagrass meadow of almost 2 km^2 extent following several years of exposure to municipal sewage (MS), attributing seagrass death to increased growth of leaf epiphytes.

Sediment changes (alteration in grain size, composition, and porosity) represent another, possibly very significant threat to seagrass communities. Holmer et al. (2003), for example, reported on how organic load emanating from Philippine fish farms changed sediment biogeochemistry and benthic communities. Sediments became highly reduced (ie, oxygen depleted) due to the accumulation of sulfides in the sediment pore waters. Such substrate is unfavorable for growth and survival of seagrasses and associated organisms. Interaction between various factors (ie, sediments, nutrients, and other limiters of plant growth) are as yet not well understood and require further investigation and experimentation (Waycott et al., 2005).

12.3.4 Impacts of Pollution Hazards in the Water Column: Fisheries and Algal Blooms

Coral reef, seagrass, and mangrove ecosystems are relatively clearly defined in space. Biotic communities in the open water sphere, in contrast, are difficult to trace and monitor. Assessing pollution impacts in this system is complicated by high spatiotemporal variability. Similar to what has been found for seagrass meadows, slightly increased levels of organic matter and nutrients may have initial stimulating effects on pelagic food chains and productivity, with possibly some short-term positive effects for fish production. Short-term increases of fish abundance due to sewage outfall are commonly observed near the points of discharge (Islam and Tanaka, 2004). According to Lin (2000), some unsubstantiated reports had also claimed that the quantity and quality of coastal fisheries had improved at the onset of the shrimp farm boom. There is, however, a lack of robust data. The overall longer-term trends of demersal fisheries harvests in the Gulf of Thailand show a diametrically opposed picture: a steep decline from about 350 kg of fish caught per hour of fishing effort in 1960 to less than 10 kg/h in 1995 was recorded according to Islam and Tanaka (2004). Equally, trawlable biomass calculated at 680,000 tons for 1961 was estimated at only 56,000 tons in 1995 (ie, down to 8.2% from the 1961 level) by Kongprom et al. (2003), with similar trends observed in neighboring countries (Stobutzki et al., 2006). Islam and Tanaka (2004) referred to water pollution and habitat degradation as major causes, but overfishing was probably another important driving force of fish stock depletion in the seas: according to Stobutzki et al. (2006), time series of exploitation ratios for 17 species showed significantly increasing fishing pressure over time (see also Pauly and Chuengpadgee, 2003). According to Vidthayanon and Premcharoen (2002), fish species diversity in Thailand's estuaries had drastically decreased owing to habitat loss, overfishing, and pollution, and 75 of the total 607 species recorded since 1966 were threatened. Similar parallels between coastal fish resource degradation, mangrove habitat destruction, and pollution have also been documented from Vietnam by De Graaf and Xuan (1998).

The complex effects of an algal bloom on the nekton in an estuary in northeastern Australia were described by Pittman and Pittman (2005). Long-term increased nutrient input in marine waters, as well as depletion of consumer species by overfishing, can be highly detrimental to food chains due to significant changes of microbial communities, and chemical and biological chain reactions, also leading to decreased biodiversity (Islam and Tanaka, 2004; Lin, 2000; Pauly and Chuengpadgee, 2003). Particularly in relatively calm waters, certain species of dinoflaggellates are promoted, which can under hypereutrophic levels cause harmful algal blooms (HABs) or the so-called red tides (Valiela et al., 1997; Anderson et al., 2002). Such blooms can lead to depletion of DO and increases of toxic metabolites, leading to mass mortality of fish and other marine animals (Anderson et al., 2002). Secondary and tertiary consumer species, such as fish, seals, and even humans, can also be poisoned by mussels and other filter feeders that have been accumulating toxins in their tissue. Paralytic shellfish poisoning caused by ingestion of fish, shellfish, and other marine life during red tide blooms have resulted in fatalities throughout Southeast Asia. In Thailand, many incidents of poisoning have recurrently occurred as a result of the consumption of horseshoe crabs, *Carcinoscorpius rotundicauda* and *Tachypleus gigas* (Lehane, 2000). In the Gulf of Thailand, several red tides have been reported since the 1980s (Cheevaporn and Menasveta, 2003).

Using satellite imagery, Singhruk (2001) visualized chlorophyll *a* concentrations between September 1999 and February 2001, indicating the entrapment of excessive nutrient effluents particularly along the coastline of the Inner Gulf of Thailand. HABs are very difficult to forecast as their generation and dynamics are influenced by a multitude of interactive environmental factors. Algal blooms are not new phenomena, but it can be said with certainty that their appearance frequently correlates with extensive marine pollution, and that they have increased significantly over the last two decades (Anderson et al., 2002). Efforts to develop early warning systems for HABs are, for example, being undertaken in Hong Kong (Lee et al., 2005).

12.3.5 Vulnerability of Mangrove Forests to High Loads of Nutrient-Rich Organic Wastewater

According to Schaffelke et al. (2005) and Peters et al. (1997), mangroves appear to be overall fairly resistant to wastewater pollution. Serious detrimental effects on trees and large-scale pollution-related diebacks are relatively rarely observed, except in the case of oil spills and serious exposure to herbicides. Yet, degradation of mangroves due to pollution may not be apparent at first sight. According to the MS Swaminathan Research Foundation (MSSRF) (in IUCN-WWF, 2005), for example, observations in the extensive Pichavaram mangroves in India (~1500ha) after the Asian tsunami indicated that mangrove health had unexpectedly improved as trees were exhibiting a remarkable leaf flush. According to their interpretation of events, the waves had flushed out high levels of hydrogen sulfide that had built up in the sediments. This appears to be one of the very few reported positive effects of this otherwise devastating tsunami (see also Ranjan et al., 2011).

According to Schaffelke et al. (2005), excessive nutrients have so far not been directly linked to mangrove dieback and damage, but indirect effects have been reported. For example in South Australia, 250ha of *Avicennia marina* mangroves have died since 1956 when a MS outfall was put in place (Environment Protection Authority in Schaffelke et al., 2005). The dieback, which was still ongoing, was attributed to smothering by algae of the genus *Ulva*. The assumed effects were a retardation of the growth of mangrove seedlings and asphyxiation of aerial roots. Algal smothering may also have been the cause of tree dieback in a mangrove stand in Queensland (Laegsgaard and Morton in Schaffelke et al., 2005). In contrast, abnormal growth observed in a mangrove stand near Jeddah, Saudi Arabia, was possibly induced by direct effects of high loads of wastewater and other types of industrial pollution (Mandura, 1997).

Similar to the cases reported in the literature, mangrove smothering was apparent in a part of an *Avicennia marina* stand (~5ha), which over several years had been exposed to concentrated wastewater runoff emanating from a piggery puddle (~70pigs) at Bang Pu south of Bangkok (Figs. 12.3 and 12.4; pers. obs. R. Cochard; the site was investigated by Wickramasinghe et al., 2009, cf. discussion in Section 12.5.2.1). At the Bang Pu site, some direct and indirect effects on the mangrove rooting environment appear to be implicated, as soil consistence and the infauna were obviously severely affected by chronic long-term eutrophication, algal bloom, and probably temporarily toxic levels of ammonium (Figs. 12.3 and 12.4).

Algal blooms in shallow waters of estuaries and mangroves can significantly alter the sediment biogeochemistry by dominating oxygen profiles, intercepting nutrients regenerated from sediments, and uncoupling biogeochemical sedimentary cycles from those in the water column. Algal blooms can also change carbon pathways and dynamics by dominating microbial and consumer food webs, and affecting benthic fauna in terms of biomass, functional composition, as well as species richness (Valiela et al., 1997). Observations by Pittman and Pittman (2005) during an algal bloom event in an Australian tropical estuarine mangrove swamp may provide some clues to also explain the Bang Pu case. Ammonia, TKN, and reactive phosphorus and TP were substantially elevated during the algal bloom, whereas the decomposing bloom resulted in a buildup of toxic sulfides in the sediment, and highly reduced oxygen conditions occurred both in the sediment and near-bottom water (Pittman and Pittman, 2005). Such changes are frequently accompanied by mortality or emigration of infauna, including crabs and fish (Pittman and Pittman, 2005; Wannamaker and Rice, 2000; Eby and Crowder, 2004). High ammonia concentrations can cause damage to gill, liver, kidney, spleen, and thyroid tissue in fish, crustaceans, and mollusks. Concentrations of between 2.5 and 13 mg/L are considered harmful for fish, whereas maximum acceptable levels for crustaceans have been suggested in the range of between 4.1 and 7.7 mg/L for crustaceans (Pittman and Pittman, 2005). Low DO has been quantified to be lethal (>24 h exposure; LC_{50}, ie, >50% mortality) to fish and crabs at concentrations of 1.4–3.3 mg/L for larvae, 1.0–2.2 mg/L for postlarvae, and 0.5–1.6 mg/L for juveniles (Miller et al., 2002). Decreasing oxygen levels have, furthermore, been shown to increase ammonium toxicity for fish and crustaceans (Pittman and Pittman, 2005). In contrast to most demersal fish and crabs, species of mudskipper appear to be largely resistant to excessively high levels of ammonia (Ip et al., 2004, Fig. 12.4). In East Africa, mudskipper morphology and growth have, however, been significantly affected by pollution, although the implicated agents could not be identified specifically (Kruitwagen et al., 2006).

At the Bang Pu site, measurements indicated that oxygen levels were particularly low near the piggery puddle (0.6–0.8 mg/L), whereas they did not appear to be critical at some distance away (~4–6 mg/L) (cf. Wickramasinghe et al., 2009, including unpublished data). However, as in the case of the Australian study, levels of ammonia-N in residual water in the Bang Pu mangrove were overall higher (~5–10 mg/L) than the critical levels for crab and fish survival. The levels were clearly excessively high in the piggery puddle (200–560 mg/L; sludge 1530 mg/L) where many dozens of dead crabs (primarily *Episesarma* spp. and some *Metopograpsus* spp.) and even some dead mudskippers were observed (Figs. 12.3 and 12.4; Wickramasinghe et al., 2009, including unpublished data).

The feeding ecology of crabs and fishes surviving in the vicinity may also have been disrupted by algal bloom, excessive ammonia-enrichment, and associated sedimentary changes. Some crabs depend on sediment uptake for their nutrient requirements, and under natural, nutrient-limited conditions may actively contribute to N enrichment of their sediment environment by leaf consumption and fragmentation (cf. Skov and Hartnoll, 2002). In their function as leaf shredders crabs are also important agents of leaf carbon breakdown, playing a keystone role at the base of food chains in mangrove swamps and estuaries (Werry and Lee, 2005). Food webs that support fishes over mangrove mudflats are primarily based on transported organic matter and not microalgae (Melville and Connolly, 2005). At the Bang Pu site, no fishes were observed in residual water, except for air-breathing mudskippers.

Correspondingly, Pittman and Pittman (2005) stressed the wider ecological consequences of algal blooms on the mangrove utility as a fish nursery.

Overall, crab populations at Bang Pu appeared to be significantly affected in the mangrove forest near and downstream of the piggery puddle (Figs. 12.3 and 12.4). In the direct vicinity laterally (ie, parallel to the coastline) of the piggery losses of crabs due to high mortality (predominantly smaller *Episesarma* spp.) were apparently compensated by new immigration from nearby, less affected upstream mangrove stands, whereby mangrove stands near the piggery appeared to be nutritionally attractive (presumably due to increased nutrient levels in their leaves, cf. later discussion). In contrast, in a heavily affected area downstream of the piggery (where posttidal waters tended to be stagnant in puddles) hardly any crab holes were observed in structurally deteriorated, highly sulfidic sediments underneath mostly dead trees (>70% mortality). Crabs living in mangrove sediments play a keystone role in mangrove forests, as has been demonstrated by Smith et al. (1991) in a 12-month crab removal experiment in *Rhizophora*-dominated forests in North Queensland. After a few months, they found significant increases of soil sulfide (up to ×1.4) and ammonium (up to ×1.8) concentrations on crab removal plots as compared to controls. Cumulative forest growth, as measured by stipule fall and tree reproductive output, was significantly lowered in removal plots as compared to controls. Smith et al. (1991) noted aeration of the soil by crab burrows as a most important factor to explain their results, but the flushing function of crab burrows for excessive salt removal, as well as nutrient provision to roots, may be of similar importance (Ridd, 1996; Stieglitz et al., 2000). Mangrove roots are known to exclude salts, whereby osmotic gradients are created that may increase stress on plants (Passioura et al., 1992). Similar to those field studies, Penha-Lopez et al. (2010) provided a detailed experimental demonstration of the role of crabs and the influences of wastewater on biochemical processes in mangrove sediments.

All in all, a clear pattern of wastewater impact was recognizable at the Bang Pu site, even if the exact cause-effect relationships were not fully established. For example, whether tree dieback was caused by algal growth, or whether algae proliferation was only triggered by light penetration after tree death was not clear. From all the observations, a favored hypothesis was that excessive crab mortality was a key factor that led to stand dieback in a slow cascading manner; that is, sediment changes, tree stress, and gap creation followed by algal growth and associated biochemical changes, and eventually tree death and further sediment changes.

While mangrove mortality due to excessive wastewater has rarely been documented in the literature, these observations seem to be relevant, because villages and agricultural enterprises are typically located directly adjacent to mangrove forests, and wastewater effluents are normally discharged without any informed design that could minimize the impact or possibly even enrich some parts of the forest without any major harm. Observations at Bang Pu indicate, however, that forest regeneration may readily occur once excessive nutrient pollution is removed. Resprouts survived the impact, and apparent recovery of mangrove infauna and the restitution of biochemical soil processes probably facilitated forest regeneration after the closure of the piggery in 2006, as visible on aerial images (Figs. 12.3 and 12.4).

12.3.6 Vulnerability of Mangrove Forests to Other Types of Water Pollution

The potentially destructive effect of wastewater nutrient loads on mangrove stands has mainly gained publicity in their interaction with other toxic constituents, most notoriously with herbicides, as rates of uptake of nutrients and other wastewater constituents are stimulated in mangrove trees exposed to increased nutrient levels (Schaffelke et al., 2005). Mangroves appear to show relatively lower physiological resistance to herbicides than other vegetation types (Peters et al., 1997). According to Duke et al. (2005), more than 30 km^2 of mangroves have, for example, experienced serious dieback and degradation as a result of exposure to diuron and other agricultural herbicides contained in catchment runoff in the intensively cultivated Mackay region of Queensland, Australia. The authors stressed the implications of this observation with regard to knock-on effects such as erosion and sediment smothering of offshore ecosystems in the GBRMP.

Probably the largest ever destruction of mangrove forest by dispersion of synthetic chemicals was caused by the infamous application of the defoliating herbicide "Agent Orange" during the Second Indochina War: 18.9 million gallons were sprayed on agricultural land and forests (Peters et al., 1997). Mangrove forests were the most exposed vegetation types, not only receiving direct loads of air-sprayed herbicides, but also residual loads in runoff from catchments: in total about 124 km^2 or 41% of Vietnam's mangrove cover was destroyed (Peters et al., 1997). Long-term effects seem to be uncertain. Peters et al. (1997) stated that herbicide products are generally characterized by a relatively fast rate of breakdown, but the authors referred to estimates that put recovery time at between 20 and 100 years. In contrast, Kogo and Kogo (2004) noted that replanting of mangroves started shortly after the war, and almost 100 km^2 of mangrove forest have since been reforested.

Heavy metals may also be toxic to mangroves at high concentrations, particularly metals such as arsenic (As), lead (Pb), cadmium (Cd), and mercury (Hg), which are not essential for plant growth (in contrast to zinc (Zn), copper (Cu), nickel (Ni), and chromium (Cr); Schaffelke et al., 2005). However, levels of concentration that have been found to be affecting mangrove plants in laboratory tests (eg, MacFarlane and Burchett, 2001; Yim and Tam, 1999) are rarely encountered under natural conditions. Mangroves also appear to be quite resistant and are rarely affected despite the high efficiency of mangrove sediments to sequester heavy metals and to bind them as sulfides under anaerobic conditions (Peters et al., 1997; Agoramoorthy et al., 2008; Zhou et al., 2010; Lewis et al., 2011; Marchand et al., 2011). A study in India by Selvakumar et al. (1996) on the effects of heavy metals and pesticides also concluded that natural concentrations of heavy metals were too low to seriously affect the crab infauna. Nonetheless, chronic exposures of organisms to heavy metals may well influence their growth and cause shifts in species dominance, potentially affecting the stability and resilience of mangrove stands (eg, Kruitwagen et al., 2006; De Wolf and Rashid, 2008; Chaiyara et al., 2013).

With much of land erosion taking place in river catchments, sediment accretion with successive expansion of mangrove forests has been observed in estuaries throughout the Asia-Pacific tropical zones (Tabucanon, 1991; Walters, 2003; Schaffelke et al., 2005). On the other hand, excessive sediment accumulation has also led to tree mortality owing to root smothering. Ellison (1999) summarized the literature that documented impacts of sediment burial on mangroves, either in estuaries due to upstream erosion in catchments, or elsewhere due to other factors, such as coastal construction, aquaculture, or impacts of cyclonic storms (see also Vaiphasa et al., 2007). Natural sedimentation rates of mangroves are in a range normally lower than 0.5 cm/y, with a maximum of about 1 cm/y, whereby *Rhizophora* spp. with stilt roots appear to be slightly less sensitive than species with pneumatophores, such as *Avicennia* and *Sonneratia* spp. Also, tree death is more likely to occur if roots are being smothered by fine sediments, probably because aeration is lower than in rough sediments such as sand (Ellison, 1999). Terrados et al. (1997) studied the effect of sediment deposition on seedlings of *Rhizophora apiculata* and found that seedling mortality increased linearly with increasing sediment accretion at a rate of 3% per cm of deposited sediment; also, seedling growth decreased linearly.

Unfortunately, mangrove roots not only facilitate sedimentation but also provide excellent traps for oil slicks, whereby aerial roots are being smothered by oil films and trees die because of lack of oxygen and absorption of toxic elements (Peters et al., 1997). Death often does not take place immediately, but dieback of highly stressed trees occurs over time as a result of additional perturbation and stresses, which may include the death of crustaceans and other infauna (Malan et al., 1988). In an experiment by Dodge et al. (in Peters et al., 1997), it was, for example, found that 17% of trees died after 2 years and 47% only after 10 years. Anaerobic conditions and low redox potentials of mangrove sediments are unfavorable for biodegradation of hydrocarbons, and from studies in the Caribbean it has been found that oil can remain in sediments for decades (Ellison and Farnsworth, 1996). At least 20 years are required until mangrove rehabilitation may again be possible at heavily impacted sites. Often, the time period is considerably longer; in particular, impacts on some components of mangrove infauna may hardly be reversible in foreseeable time spans (Peters et al., 1997). Therefore, while large-scale oil spills are fortunately relatively rare, they still pose a considerable threat to coastal ecosystems, as impacts are persisting over a very long time period. In the Straits of Malacca, one of the world's busiest supertanker routes, tanker-derived smaller oil spills occur frequently, and domestic oil spills have also been increasing due to growing industrialization in the region (Zakaria et al., 2000). Indeed, spatial oil distribution maps derived from satellite imagery show good correlations of oil slicks with the major shipping routes of the region (Lu, 2003).

12.4 USES OF NATURAL AND CONSTRUCTED WETLANDS AS WASTEWATER FILTER

12.4.1 Industrial Wastewater Treatment in Thailand

In 2002, industrial reticulation and treatment of wastewater barely existed in Southeast Asia except in richer urban centers such as in Brunei, Singapore, and Malaysia (Mogg, 2002). In Thailand, government legislation required large buildings to have their own wastewater treatment systems, but Mogg in 2002 (p. 34) deemed that incentives for implementation were mostly absent and enforcement was feeble: "Wrangling between different government departments and municipal feuding makes enforcement of environmental laws problematical, and Thailand's PCD is a paper tiger without legislative teeth to enforce a clean-up." Conventional wastewater treatment processes are costly in construction and operation (Aramaki et al., 2006; Lønholdt et al., 2005). In a survey presented by Visvanathan and Cippe (2001), only 10.5% of Thai industries were involved with any water reuse projects. Of the other 89.5% of industries, 48% mentioned investment costs as the main reason for the nonadoption of wastewater reuse, 16% a lack of incentives for reuse, 10.5% noted that they were unaware of technologies, and 15% simply did not find it necessary or mentioned other reasons.

Tackling the water pollution problem emanating from diffuse sources, such as agriculture or unplanned domestic areas, is an even greater challenge in Thailand as well as in other countries of Southeast Asia (cf. Tonmanee and Kanchanakool, 1999; Schouw et al., 2003; Bandara, 2003; Maneepitak and Cochard, 2014). As for shrimp farming, only about 36% of farmers surveyed in the Hua Sai and Ranot districts in Thailand actually disposed of pond sediments in areas designated by the government (Dierberg and Kiattisimkul, 1996). New sources of industrial and agricultural pollution are emerging and additional loads arise from steady economic growth. Nonetheless, at least in Thailand there have been some tangible efforts to address the problems. According to the PCD (2004), about 30% of municipal wastewater was treated by 2004. With the construction of the Samut Prakarn and the Yannawa wastewater treatment plants in Bangkok, Thailand possesses the two largest treatment plants in the region servicing approximately 600,000 residents and 2300 factories, and about 500,000 residents, respectively (Lønholdt et al., 2005; Kirkwood, 2004). Nevertheless, for a majority of Southeast Asia's (including Thailand) population wastewater still flows off without industrial treatment.

12.4.2 Treatment of Land-Based Point and Diffuse Pollution Sources Using Wetlands

As an alternative to expensive industrial treatments, the use of wetlands (both constructed and natural) as low-cost and low-technology biological treatment systems for effluent purification has found increased recognition in many developing countries (Kivaisi, 2001; von Sperling, 1996; Brix, 1994; Denny, 1997; Koottatep et al., 2005; Shipin et al., 2005). Wetlands can be used to mitigate diffuse-source pollution. By dispersing waste over a large area they can absorb large volumes of nutrient-loaded water (Raisin and Mitchell, 1995; Knight et al., 2000; Kao et al., 2001). Much of the stormwater and other wastewater in Bangkok is being caught in an extensive canal system. Densely covered with floating vegetation such as water lettuce (*Pistia stratiotes*), water hyacinth (*Eichhornia crassipes*), and floating ferns (*Salvinia* spp.), the canals are managed for wastewater treatment. These systems are quite efficient if water flow can be controlled at slow rates (Banjongproo and Wett, 2002; Fig. 12.2). Other wetland areas around Bangkok primarily vegetated with cattail (*Typha angustifolia*) also receive considerable loads of wastewater (Fig. 12.2). These species are well known for their wastewater treatment capacity (eg, Kao et al., 2001; Koottatep and Polprasert, 1997).

Conventional wastewater treatment in the activated sludge process or trickling filter system is not only costly, it is often also quite unsatisfactory if not implemented thoroughly, particularly in terms of nutrient and heavy metal removal. This is because it relies solely on bacteria and the associated protozoa (Metcalf & Eddy, 2003; Breaux et al., 1995). In contrast, wastewater stabilization ponds are comparatively cheap, easily operated and more efficient to maintain, and their performance in achieving treatment goals has often been found satisfactory. In the tropics removal efficiencies of BOD, nitrogen, phosphorus, and indicator bacteria have been reported to be 75–90%, 30–50%, 20–60%, and 60–99% respectively; this at daily loading rates of 180–500 kg BOD (Kivaisi, 2001). In other wetlands planted with species such as *Phragmites*, *Zizania*, *Typha*, and/or bulrushes (*Scirpus*, *Cyperus*, *Schoenoplectus*) almost 80–100% of phosphorus, nitrogen, and COD removals were achieved (Watugala et al. in Wong et al., 1995; Koottatep et al., 2005). Breaux et al. (1995) estimated the wastewater treatment

FIG. 12.2   Freshwater plant species important for water purification at Bangkok. *Left*: Floating weeds such as water lettuce (*Pistia stratiotes*), water hyacinth (*Eichhornia crassipes*), and floating ferns (*Salvinia* spp.) in a canal at the Asian Institute of Technology. *Right*: A dense stand of cattail (*Typha* spp.) surrounding a wetland. *Photographs by R. Cochard and Wikipedia.org.*

value of a large wetland in Louisiana to be around $785–$1500 per acre. Capitalized cost savings for using wetlands instead of conventional systems were calculated to be $6231–$34,700 for two industrial case studies. In addition to wastewater treatment, wetlands may offer other benefits such as their utilization for swamp fisheries, biomass production, seasonal agriculture, water supply, public recreation, wildlife conservation, and scientific study (Kivaisi, 2001; Knight, 1997).

In wetland systems the combination of saturated soil, plants, and microorganisms provide both aerobic and anaerobic conditions under which pollutants are removed through a complex variety of biological, physical, and chemical processes. The vegetation itself is adapted to filter nutrients from the water, whereby nutrient uptake for most plants is via the roots and pore water. In an experiment in Thailand with *Typha angustifolia*, N uptake by plants was found to amount to 7.5 kg/ha/day with 8 weeks operation and a hydraulic retention time of 5 days. This was accompanied by a total N removal of 84–86%, including other absorption and reduction processes (Koottatep and Polprasert, 1997). In an engineered wastewater treatment swamp in Massachusetts various species of water plants accounted for about 6.5% of the total 68% nitrogen removal at a loading rate of 156 kg N/ha/day, and for 2.6% of a total 38% N removal at 52.3 kg/ha/day (Peterson and Teal, 1996). If the wetlands are not harvested or burned the vast majority of nutrients which had accumulated into plant tissue will eventually be returned to the water by decomposition processes (Brix, 1997). Optimal efficiency of nutrient removal by plant uptake therefore depends on harvest interval and intensity, on hydrology and water level, and on plant type (Gopal, 1999).

The main importance of macrophytes for water purification is more indirect. Plants are important as a structural medium: macrophyte vegetation attenuates water flow causing sedimentation of particulate matter (eg, organic carbon, which is responsible for a large part of BOD), and facilitating chemical processes such as adsorption, chelation, and precipitation, which are responsible for the major removal of phosphorus and heavy metals (Kivaisi, 2001; Brix, 1997; Gopal, 1999). Vegetation also stabilizes the soil surface, and wetland soils serve as absorption medium for pollutants (Brix, 1997). Most significantly vegetation stems, leaves, and roots provide a large area for biofilms; plant tissues are colonized by dense communities of algae, bacteria, and protozoa. These biofilms are responsible for the majority of microbial processing, which accounts for most of the total removal processes, including oxidation and reduction of carbon, nitrogen, and sulfur, depending upon the availability of oxygen (Brix, 1997; Gopal, 1999; Salvato et al., 2012). Microorganisms are not only pivotal for geochemical transformation of nutrients, they also exhibit a capability to remove toxic organic compounds (Kivaisi, 2001). Interactions between various chemical and microbiological processes are generally fairly complex. Several elements such as iron, aluminum, and manganese play an important role in the removal of phosphorus (Gopal, 1999). Anaerobic conditions that favor denitrification are not favorable for adsorption and precipitation of phosphorus. The addition of high BOD wastewater actually often entails secondary release of phosphorus accumulated in sediments (Gopal, 1999). Other potentially important roles of vegetation in the wastewater treatment process include the oxygen release from roots and the release of other substances, such as antibiotic chemicals (eg, *Schoenoplectus* spp.) and carbon (Brix, 1997).

The hydrology, vegetation, and soil at any given wetland site are the main factors influencing water quality treatment efficiency (Kivaisi, 2001; Gopal, 1999). Removals of wastewater constituents are to a significant degree a function of inlet pollutant concentrations, hydraulic loading rates, and hydraulic retention time. A system already saturated with nutrients has low potential to process the additional load (Knight et al., 2000; Gopal, 1999). Natural wetlands, especially in tropical regions, experience large water level changes to which the vegetation is adapted. High runoff and low retention periods (eg, during extreme rainfall) may scour sediments and release nutrients and organic matter into the open water (Gopal, 1999). Hydrology may determine the type of vegetation growing in wetlands; vice versa, vegetation growth and succession may change the hydrology and therefore the wastewater treatment function and utility. Different plant species offer different physical flow resistance and nutrient absorption capacities. Furthermore, hydrology is an important determinant of microbial activity and biogeochemical cycling of nutrients in the soil (Kivaisi, 2001). In summary, different wetland systems with different plant species, biomass, soil properties, and interactions between water-plant-soil would have different capacities for wastewater cleaning. To evaluate treatment efficiency, several indicators have been suggested: carbon and toxic organic compound removal may be assessed by measurement of soil and water OD, microbial biomass, soil reduction potential (E_h), and pH; nitrate removal may be assessed by measuring dissolved organic carbon (DOC) and microbial biomass; P retention can be described by the availability of reactive iron (Fe) and aluminum (Al) in acid soils and calcium (Ca) and magnesium (Mg) in alkaline soils (Reddy and D'Angelo, 1997).

The careful management of natural wetlands can often improve their water purification functions. To apply efficient hydraulic and biological manipulation to this end, the functions and values of any wetland first need to be assessed and prioritized (Denny, 1997). The concept of wetlands for pollution control must be holistic and realistic, to include biodiversity conservation and sustainable development. The utilization of natural wetlands, rehabilitation and construction of additional (semi-) natural wetlands, and construction of artificial wetlands specifically designed as wastewater treatment facilities may all have to be considered (Denny, 1997). Particularly integrated schemes, which adapt various strategies to the pollution types and source localities and account for pollution diffusion patterns within the landscape, may be the best recipe to optimize the treatment efficiencies. Intensively managed constructed wetlands may be set up strategically for the treatment of point sources that pose a clear hazard to the environment, while (semi-) natural wetlands may be managed for diffuse pollution mitigation. Wetland areas of high biodiversity value could, on the other hand, be protected from eutrophication by adapting a well-managed surrounding wetland greenbelt pollution buffer, or by the diversion of potentially hazardous water streams (eg, into vegetated stabilization ponds).

Despite the potential of wetlands for wastewater treatment, Gopal (1999) emphasized several significant constraints. Wetlands require a lot of land per unit volume of wastewater to be treated. Estimates of land requirements for treating per capita domestic waste range from 10 to 20 persons per hectare. In many developing countries, land and water resource availability as well as loading and composition of pollution in water is a major deterrent to install or use wetlands for wastewater treatment. In addition, little attention has so far been paid to water level fluctuations between dry and wet seasons. Yet, Gopal (1999) also pointed to the general need of restoration of destroyed or disturbed wetland systems in developing countries, be it to reestablish them as natural buffers and treatment systems for wastewaters from nonpoint sources, or as an amenity that harbors biodiversity and other resources.

12.4.3 Treatment of Aquaculture Effluents Using Constructed Wetlands

The main threat to aquaculture enterprises worldwide is posed by viral and bacterial diseases, whereby disease problems have led to the collapse of coastal industries in several Asian countries since the 1980s, including Thailand. Some symptom-driven immediate disease risk mitigation steps in production management commonly included decreasing the grow-out period in shrimp farming (implying harvests of smaller shrimps), close monitoring of production with immediate harvest if signs of disease appeared (normally producing lower-quality shrimps), and increased application of antibiotics and other chemicals (Kautsky et al., 2000). Yet, clearly this did not address the root of the problem. The spread of disease was frequently facilitated by inappropriate management of wastewater. In areas with high pond density, the emitted chemical and biological pollutants (including disease agents) were recirculated among farms with consequential self-pollution (Kautsky et al., 2000). There was a clear linkage between pollution and disease as, for example, the release of metals from acid sulfide soils and other alterations of water chemistry (eg, cadmium and magnesium), fluctuations of pH, salinity, and temperature, decreased oxygen levels, or interactions between any such parameters, may have an adverse influence on disease susceptibility by lowering the shrimps' immune response (Kautsky et al., 2000; Eng et al., 1989). Kautsky et al. (2000, p. 154) stressed the environmental context: "When problems appear in shrimp ponds or fish cages, people tend to look at what is going on inside the pond or cage, without realizing that the farm is an integral part of a much larger surrounding ecosystem. The surrounding ecosystems provide the feed, seed, clean water and other necessary natural resources and ecosystem services, including waste assimilation."

Hopkins et al. (1995) listed several management principles to reduce the aquaculture ponds' effluent impact on the environment and minimize self-pollution. These included the reduction or elimination of water exchange by careful management of stocking densities and feeding rates, and by taking advantage of in-pond digestion processes; removal of sludge deposits to a disposal site on high ground; use of polishing ponds to remove nutrients, solids, and BOD before water is discharged; and transferring water to other ponds or reservoirs in the harvest process, reusing water for subsequent crops. Water stored for reuse after harvesting develops "a community of natural forage items, an 'aquatic pasture,' which is consumed by the subsequent crop" (Hopkins et al., 1995, p. 157). The food web structure in such a "pasture" in secondary ponds of a shrimp farm in Thailand were examined by Yokoyama et al. (2002) using the stable carbon (C) and N isotope ratio technique. They found that shrimp feed was the main food source for mussels, but that other macrobenthic animals were mainly sustained by sediments and particulate organic matter in the pond.

Since the importance of the link of disease infestation with self-pollution problems became clear at the end of the 1980s, shrimp farm wastewater treatment by use of constructed or rehabilitated salt and brackish water wetlands, as a means to reduce the size of the ecological footprint, has increasingly gained attention. Such alternative (but land-requiring) systems have also gained further interest because vast areas of former shrimp ponds have been laying idle as a consequence of the collapse of industries: 40–45 km^2 of unused wasteland south of Bangkok, and up to 70% in many other regions (Stevenson, 1997). While some of the disused ponds have been reverted to more extensive uses, often sufficient surrounding land is available to consider the restoration of mangrove forests or other saline wetlands, specifically designed for wastewater treatment (Stevenson, 1997). Here, we shortly introduce some systems other than mangroves (these are discussed in Section 12.5).

Emphasizing that any pollution treatment system should be simple and attractive to shrimp farmers, if possible providing secondary income, Enander and Hasselstrom (1994) tested the combined use of a mussel (*Scapharca inaequivalis*) and of seaweed (*Gracilaria* sp.) in Malaysia. Their results were quite promising, with a reduction after 1 month of operation to only 17% of initial phosphate, 39% TP, 19% ammonium, 81% nitrate, and 28% TKN in wastewater. At the same time the increase in mussel biomass was 27%, whereas the growth of the seaweed was suboptimal, due to the growth of competing green algae, which may, however, be treated with an algal inhibitor (Enander and Hasselstrom, 1994). Jones et al. (2001b) experimented with a laboratory three-stage effluent treatment system in Australia, applying first a sedimentation and then an oyster filtration (*Saccostrea commercialis*), and finally a macroalgal absorption treatment (*Gracilaria edulis*; each treatment for 24 h). Such treatment overall reduced TSS to 12%, TKN to 28%, TP to 14%, ammonium to 76%, nitrate to 30%, phosphate to 35%, bacteria to 30%, and chlorophyll *a* to only 0.7%. The authors emphasized, however, the possible differences and likely reductions in efficiency, if treatment would be scaled up to farm size. In another experiment, using the same species, Jones et al. (2002) found that treatment efficiency could be even further improved by applying a recirculating flow regime. Neori et al. (2004) provided a review of developments in integrated aquaculture in the preceding decades, primarily focusing on work in the Mediterranean and Europe. They described examples of technically elaborate integrated polyculture designs that now allow production of up to 25 tons of fish, 50 tons of bivalves, and 30 tons of seaweeds annually on a one-hectare farm without any pollution to the surrounding environment. Also, the use of vegetated saline ponds has gained increased attention.

12.5 ARE MANGROVES EFFICIENT COASTAL WASTEWATER POLLUTION FILTERS?

12.5.1 Pollution and the Role of Mangrove Diversity, Structure, and Coastal Setting

Mangrove ecosystems—the most common types of natural coastal wetlands in the tropical zones—represent last potential "filter" systems that may trap landborne pollutants before reaching the open water bodies, with ensuing potential impacts on sensitive offshore ecosystems. Mangrove forests (mangal communities) consist of various species of specialized, mostly long-lived trees that are characterized by potentially high productivity and biomass. The coastal hydrological regime and marine water chemistry and biota differ in significant ways from freshwater systems, with implications for wastewater treatment function and efficiency. Compared to freshwater ponds, water pollutants are not as likely to be stabilized and confined within the mangrove system. This is because mangroves are exposed to a dynamic tidal regime, whereby intertidal areas are not permanently submerged in water. Ecological interactions are commonly complex and often sensitive to changes.

True mangroves worldwide comprise some 70 species (51 of these in Southeast Asia) in 20 genera belonging to 16 families. There is an additional number of mangrove associate species that may grow in mangals, particularly near the high-tide mark (HTM), as well as in other coastal forest types. In Thailand, 35 true mangrove and 52 mangrove associate species belonging to 41 families are found, and in Indonesia 45 true mangrove species occur (Spalding et al., 2010). As compared to other tropical forests, even the most luxuriant mangrove forests are structurally simple. Mangals normally lack an understorey of shrubs and ferns. There is also no dense herbal vegetation cover, but synonymous to the structural role of freshwater plants in vegetated ponds, the extensive shallow, laterally spreading tree root systems of cable roots and pneumatophores (*Avicennia* and *Sonneratia* spp.), prop and stilt roots (*Rhizophora* and *Kandelia* spp.), kneed roots (*Bruguiera, Ceriops, Aegiceras, Lumnitzera,* and some *Xylocarpus* spp.), or plank roots (*Heritiera* and some *Xylocarpus* spp.) attenuate water movement and may serve a similar, albeit probably somewhat less efficient filter function for sediment particles, nutrients, and other wastewater constituents.

Mangrove forests predominantly occur in areas of slow water movement, such as in estuaries, behind coral reefs, sand spits, or forelying islands. Lower filtering efficiency by roots (as compared to dense freshwater pond vegetation) may therefore be counterbalanced by slow water movement of a thin water layer spreading over an extensive intertidal flat area, often measuring many dozens of hectares. Mangrove greenbelts are often near monospecific, particularly in species-poor subtropical zones, or in disturbed or rehabilitated sites, such as along the coastline near to the mouth of the Chao Phraya River south of Bangkok, which is dominated by *Avicennia alba* and *A. marina* (cf. Fig. 12.4). Pristine stands close to the equator are, however, commonly characterized by a distinct species zonation ranging from near the mid-tidal mark (MTM), which is often dominated by species with pneumatophore roots such as *Sonneratia* and *Avicennia*, to the approximate mark reached only by the exceptional high-tide mark (EHTM), often dominated by brackish water species and mangrove associates, which can intergrade with rain forest. On the almost untouched Thai Andaman coastline in Ranong Province and in extensive mangrove forests in Sumatra, mangal zonation patterns are very rich, with zonational forests dominated by genera (in general sequence from MTM to EHTM) such as *Sonneratia, Avicennia, Rhizophora, Bruguiera, Ceriops, Aegiceras, Xylocarpus, Lumnitzera, Acanthus, Nypa, Heritiera, Excoecaria*, and *Phoenix* (Spalding et al., 2010; Tomascik et al., 1997). Mangal vegetational zonation reflects the different adaptations of tree species to wave actions, salinity, and other gradients. Gradients, for example, from dense systems of *Avicennia* and *Sonneratia* pneumatophores into robust stilt-rooted *Rhizophora* stands, also represent gradients of sedimentation and vegetation succession.

Even though in various studies mangroves have been compared to and viewed in similar ways as "wastewater stabilization ponds," such mangrove systems are neither anything near uniformly effective, nor insensitive and immune to high pollution impacts (cf. Sections 12.3.5 and 12.3.6). When assessing any pollution mitigation capacities of mangroves, the area of the entire mangrove forest, the absorption effectiveness of various species within their respective habitats, the structure of tidal channels within the mangroves, and the tidal regimes and water flows would have to be considered. Wastewater is commonly discharged into creeks that meet the sea in estuarine systems, often vegetated by mangroves. While some of the pollution load may be dispersed and filtered through mangrove forests via dissipation and tidal currents, much of the load may, however, be freely discharged into the sea via estuarine creeks (cf. Jennerjahn et al., 2009). Some fraction of wastewater constituents may then still be trapped by mangrove forests along the coast through the interaction of lateral currents and tidal movements. However, in the case of many mangrove forest systems, any pollution "buffer" function may not be effective to protect outer pollution-sensitive ecosystems such as coral reefs, especially if wastewater discharges into creek systems occur in high concentrations. Despite these "real-world" limitations, several studies have investigated the use of mangrove forests as "wastewater filter facilities."

12.5.2 Studies on Wastewater "Treatment" in Mangroves

12.5.2.1 Field Studies

A major longer-term experiment was set up in September 1991 at the Futian National Nature Reserve, Shenzen, southern China. At the study site the dominant tree species were *Aegiceras corniculatum, Kandelia candel*, and *Avicennia marina* (Wong et al., 1995). In the forest an elongated experimental site was set up 180 m long and 10 m broad and stretching from the landward HTM to the seaward edge of the forest. This site was sealed by a wall of metal plates from the rest of the forest. Sewage wastewater was collected from local premises at Shenzen, and a load of 20 m^3 wastewater was discharged at the landward side three times per week (Wong et al., 1995). Various parameters were sampled at frequent intervals and compared to a control site running parallel at about 150 m distance. Between September 1991 and October 1992, sewage loads of discharged water were relatively low with an average BOD of 55.9 mg/L and a TKN of 24.58 mg/L. Levels of heavy metals were also low compared to sewage from nearby Hong Kong (Wong et al., 1995). After 1 year no effect of the treatment was detectible. Between the experimental and control sites, there was neither any significant difference of nutrients in the soil nor in the plants, nor in pH and levels of heavy metals. Also, plant growth and biodiversity parameters were similar between experimental and the control site (Wong et al., 1995). To explain this Wong et al. (1995) mainly referred to the relatively low sewage concentrations and suggested that nutrient accumulation and changes may be subtle and gradual (see also Tam and Wong, 1995). A second experiment was performed between December 1994 and June 1996. Sewage concentrations during this period were significantly higher with an average BOD of 194.2 mg/L and a TKN of 38.5 mg/L (Wong et al., 1997). This second assessment of the same variables again revealed virtually no differences between the two sites: the only measurable impacts were increased nutrient concentrations in the sediments close to the sewage discharge point, which may be expected (Wong et al., 1997). There was also no difference in tree litter production and decomposition rates (Tam et al., 1998). The studies showed no detrimental impacts on the mangrove trees, but also no apparent positive effects of wastewater on mangrove tree growth were observed. Provided that the studies did not present any detailed data about the tidal hydrology of the site and nutrient removals via tidal action, the proposition by Wong et al. (1997) that mangrove intertidal wetlands are of "great potential" for natural wastewater treatment may appear somewhat venturesome.

Similar limitations may be noted for a study conducted south of Bangkok by Wickramasinghe et al. (2009). This publication is very valuable as it presents detailed data on soil and water quality parameters sampled at various locations near and surrounding a hypernutrified pond below a piggery stall (ie, manure and urine from ~70 pigs; Figs. 12.3 and 12.4). In particular, the data on microbial species found in water samples are interesting and unique for the region. Nonetheless, some conclusions of the paper may be considered with reservation, because the stated distinctions of various areas of the mangrove forest (eg, termed "hypernutrified treatment zone," "nitrified enhancement zone," "hydraulically inactive zone," etc.) arose partly from preconceived ideas rather than from empirically established baselines. As a matter of fact, movement patterns of tidal flows and pollutants transported therein (cf. Fig. 12.3; somewhat changeable with different tidal regimes; R. Cochard pers. obs.) were not described and discussed in detail in the paper, and the "zones" could hardly be considered separate such as in discrete, interlinked, constructed treatment ponds. Hence the overall efficiency of various "zones" in terms of pollution treatment remains partly a matter of speculation. In addition, impacts of pollutants on the mangrove forest were apparent in parts downstream of the piggery (Figs. 12.3 and 12.4; the site was visited and studied in 2006; cf. Section 12.3.5), but these were not specifically discussed in the publication. As stated in the paper, populations of crabs may have profited from mildly increased nutrients levels in certain parts of the mangrove forest. Minor negative impacts, or even positive impacts of wastewater applications on crab populations, have also been reported for other studies (eg, Amaral et al., 2009; Penha-Lopez et al., 2009; Cannicci et al., 2009). Trees upstream of the piggery (mainly to the east) may equally have profited from nutrients added through diffusion and tidal transport, similar to what has been reported in a study by Herteman et al. (2011). Downstream of the piggery there was, however, a large area where crabs were virtually absent, mostly coinciding with the area of dead trees (Fig. 12.3; cf. Section 12.3.5). Furthermore, many dead crabs were found after tidal floods in a temporally exposed area (location c, Figs. 12.3 and 12.4) where probably continuous in-migration took place from an intact mangrove stand to the north; pers. obs.).

In the case of many studies, the long-term implications of wastewater exposures remain largely unclear. Studies such as those conducted by Amaral et al. (2009) and those conducted at Shenzen in China were undertaken over relatively short periods of maximally several months and were using comparatively low levels of wastewater nutrients (as compared to Bang Pu and other locations; cf. also Mandura, 1997). Even under conditions where crabs would not be negatively affected by high water eutrophication, the behavior of crabs may

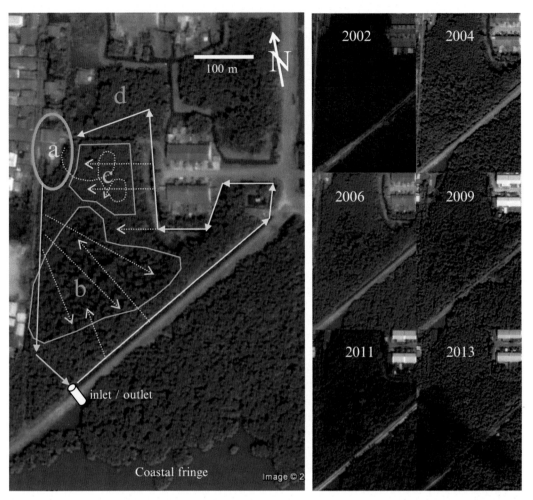

FIG. 12.3 The impact of piggery wastewater on an *Avicennia marina* forest stand near the Royal Thai Naval Base South of Bangkok. *Left image*: Map of the study site in 2006, showing the location of the piggery (~70 pigs) (a), and the location of wastewater effluent impact (b). The blue arrows indicate the main water movement during tidal influx (flood), starting from the inlet and then moving along the water channel parallel to the road. Water was entering the piggery site from the east. At location (c) piggery wastewater entered at lower concentrations during high tide through dissipation. This still caused mortality of many crabs (cf. Fig. 12.4), but after the tide crabs immigrated again from unaffected location (d), possibly due to attractive higher nutrient contents of trees and soils at location (c). The greenish-yellow arrows indicate the main movement of piggery sewage water during tidal outflux (ebb) toward the outlet and dispersing on the mangrove tidal flat (b). *Right images*: The images illustrate the forest dieback (location b) from 2002 until 2006, and then forest regeneration after the piggery was closed in 2006. *Picture source: Google Earth™.*

significantly change so as to potentially affect the mangrove ecosystem in the longer term. For example, Bartolini et al. (2009, 2011) found that the bioturbation activity of fiddler crabs decreased under increased availability of nutrients from wastewater. They speculated that this could potentially lead to "cryptic ecological degradation" of a mangrove system. Observations from Bang Pu indicate that chronically persistent pollution of high nutrient loads (especially

FIG. 12.4 The impact of piggery wastewater on an *Avicennia marina* forest stand near the Royal Thai Naval Base South of Bangkok (cf. Fig. 12.3). *Top left*: picture of the piggery (~70 pigs; a) and the piggery wastewater puddle emanating from it. *Top right*: Intact *Avicennia* forest east of the piggery (location c). *Inset top right*: During incoming tide sesarmid crabs were climbing on trees at location (c), possibly because of toxic ammonium concentrations in the water. *Middle left*: Dead sesarmid crab adjacent to the piggery puddle (a). *Middle right*: Mudskipper in the piggery puddle (a). He was surviving. *Bottom left*: Tree dieback at the center of the mangrove swamp (b). *Bottom right*: Detail of pneumathophore roots smothered by algae (b). In this area, soil consistency was severely deteriorated, making it virtually impossible to walk without sinking to knee depth. There was an eminent sulfuric smell indicating decomposition under anoxic conditions. However, note the many *Avicennia* shoots sprouting from surviving roots, ready for forest regeneration once pollution pressures have ceased (cf. Fig. 12.3). *Inset middle*: View from the inlet/outlet to the low-tide mark. Note the dense green algal cover indicating eutrophication. *Photographs by R. Cochard and O. Shipin, June 2006.*

with high concentrations of ammonium) can result in severe impacts on the mangrove forest within the duration of a few years. However, affected forests may also readily regenerate once the pollution pressure ceases. As can be seen in Fig. 12.3, the forest area opened up during 2002–2006, but was again closing with regenerating trees after the piggery was closed in 2006. Even under high pollution in 2006, many small *Avicennia* saplings managed to survive in the opened gaps (cf. Fig. 12.4), and the area was readily reclaimed once the pollution pressure had ceased.

12.5.2.2 Studies Conducted in Experimental Microcosms

Studies conducted in constructed experimental settings can provide important insights into the mechanisms and capacities of mangrove systems to filter and reduce water pollutants. Equally, however, it needs to be kept in mind that small-scale, enclosed experimental settings do not necessarily reflect the true conditions encountered in natural settings, especially with regard to the dynamics of tidal flood regimes and influences of extreme weather events.

Chu et al. (1998) demonstrated the potentially high capacity of mangrove swamps to retain wastewater constituents using an experimental laboratory approach. In each of 12 tanks ($1 \times 0.5 \times 0.3 \, m^3$) they transferred mangrove soil and transplanted 16 one-year-old *Kandelia candel* plants from a nearby Hong Kong mangrove site. In the tanks, containing 30 liters of seawater, the tidal hydrological pattern of the swamp was simulated by a computer-controlled mechanism. After 1 year of acclimatization, the 12 tanks were divided into 4 triplicated treatment groups: a control treatment with 1 L of seawater, a treatment with the equal amount of average MS (1MS treatment; 500 mg/L COD, 40 mg/L NH_4^+-N), with concentrated sewage five times the average (5MS), and 25 times the average (25MS). During 16 weeks treatments were applied three times a week with 2-weekly cycles of seawater replenishment. Except for organic N, removal efficiencies of wastewater constituents were demonstrated to be up to 98% for nutrients and 96% for metals in the 1MS and 5MS treatments. In the 25MS treatment absorption rates were still considerable with about 75% for nutrients; 92% for cadmium (Cd), chromium (Cr), and copper (Cu); and 88% for nickel (Ni) and zinc (Zn), albeit absorption capacities generally decreased with increasing saturation (Chu et al., 1998). While this study provides some data of water pollutant removal capacities, it did not determine the exact mechanisms of removal, nor quantify the different mechanisms' contribution.

Tam and Wong (1993) conducted a laboratory column leaching study, using local Hong Kong mangrove soil (73.7% sand, 11.5% clay, and 1.98% organic matter) and simulating the tidal regime in mangrove swamps. They found that ammonium, TP, and certain metals (total zinc (Zn), manganese (Mn), and cadmium (Cd)) were significantly concentrated in the sediment column after 54 daily applications of 100 mL concentrated (COD 800 mg/L) wastewater. In contrast, concentrations of these nutrients and heavy metals were either decreased or remained constant under the control treatment (seawater) or the treatment with diluted (COD 200 mg/L) and medium (COD 400 mg/L) wastewater. Increased levels of extractable N, P, and metals (Cu, Zn, and Mn) in wastewater-treated sediments also implied that nutrients and metals were held loosely within the sediments and were available for plants; extractable nutrient and metal levels were particularly high in the highest sediment layers (cf. also Tam and Wong, 1993, 1996a,b; Ye et al., 2001). Organic matter, total N and K concentrations, were not found to be elevated. With regard to TN, Tam and Wong (1993) argued that most N was probably in the form of ammonium and the addition was relatively small in comparison to inherently bound N. Additional experimenting confirmed that the soils exerted a high capacity to retain heavy metals, but showed that the role of mangrove plants in retaining metals depends on the plants' age and biomass production (Tam and Wong, 1996b, 1997). Furthermore, the efficiency of retention of pollutants was negatively affected by high salinity (Tam and Wong, 1999; Ye et al., 2001), and the growth of mangrove plants (*Bruguiera gymnorrhiza* in this case) was diminished under high applications of wastewater-borne heavy metals (Yim and Tam, 1999).

In experiments conducted during 6 months with intermittent subsurface flow mangrove microcosms removal percentages of DOC, ammonia-N, and TKN were over 70%, for inorganic-N removal over 47%, and for phosphorus over 86% (Wu et al., 2008a). Under elevated salinity levels, the removal percentages of respective pollutants were almost about 20% lower as compared to treatments with nonsaline, pure wastewater (Wu et al., 2008b). Lower removal percentages of various pollutants were reported by Boonsong et al. (2003) after a 1-week experimenting period. Similarly, in experiments by Tam et al. (2009) TKN and inorganic nitrogen (NH_4^+-, NO_2^--, and NO_3^--N) were completely removed during nine simulated tidal water cycles (ie, less than a week), with enhanced microbial transformation processes (nitrification and denitrification) under nutrient additions.

12.5.2.3 Studies on the Role of Mangrove Bacteria in Wastewater Treatment

Nitrogen decontamination capacities can differ considerably among freshwater treatment ponds stocked with different types of emergent plant species. This is mainly because the various plants differ in their effectiveness to provide microenvironments where denitrifying bacteria and other microorganisms can settle and filter the nutrients (Salvato et al., 2012). In mangrove environments, denitrification by bacteria primarily occurs in the upper layers of the soil surface (or at the surface of animal burrows), especially in sediments with low-bulk density, and also on mangrove roots and in relatively stagnant water in pools (Adame et al., 2012; Suárez-Abelenda et al., 2014; Li et al., 2011; Wickramasinghe et al., 2009). Macroinvertebrates, such as crabs and other species, are important to increase the soil surface by digging tunnels beneath the mangroves. On the other hand, smothering algal films may seal off pores in the sediments, clog animal burrows, and thus decrease water penetration and percolation of soils, and diminish the substrate-water interface where water purification processes can take place (cf. Fig. 12.4).

The study by Wickramasinghe et al. (2009) showed that various denitrifying microorganisms were present at high densities within and surrounding the highly nutrient-enriched piggery puddle (cf. Sections 12.3.5 and 12.5.2.1). As such, this permanent (>1 m deep) puddle could be considered as something like a treatment pond. However, the continuing inputs of piggery wastes exceeded any "treatment" functions by microbes within the pond, and each tidal flow transported high loads of nutrients into the mangrove forest, with consequences as described previously (Sections 12.3.5 and 12.5.2.1). In this case, pollution was damaging to the forest because it represented a permanent, chronic exposure of excessive nutrient loads.

In other cases, however, pollution may not have had such drastic effects, because wastewater concentrations were either considerably lower and/or because pulses of wastewater were only transient. Important soil organisms (eg, crabs) may either have remained largely unaffected, or were able to recolonize the exposed areas. Under such circumstances of low exposures, metabolism and productivity in the mangrove ecosystem may be boosted by additional inputs of nutrients, and treatment functions of mangroves may be somewhat effective (cf. Suárez-Abelenda et al., 2014; Adame et al., 2012; Chen et al., 2009). According to Tam (1998), the growth of bacteria and other microbes was stimulated by the experimental addition of household wastewater nutrients (N and P), whereby the retention of wastewater heavy metals (Cu, Zn, Cd) in sediments did not exhibit any inhibitory effects on growth. According to Chen et al. (2011), wastewater effluents from shrimp ponds were most effective to stimulate metabolism (as measured by gas emissions such as N_2O and CO_2), possibly because of the mixture of different nutrients available. In contrast, MS and—in particular—pure livestock wastewater were less effective. In a study by Keuskamp et al. (2013), it was shown that nitrogen addition in wastewater lead to increased decomposition of labile organic carbon, but nitrogen addition had an inhibitory effect on the mineralization of soil organic carbon. This may also lead to changes in mangrove soil composition and porosity, ultimately with effects on the sediments' wastewater cleaning utility. Such processes may also partly explain the impacts observed at Bang Pu (cf. Sections 12.3.5 and 12.5.2.1).

Some bacteria in mangrove sediments are also capable to remedy contamination of PAHs derived from oil slicks and urban runoff. Their cultivation has been suggested as a means to catalyze remediation processes in tropical coastal sediments affected by oil spills (Tam and Wong, 2008; McGenity, 2014; Wang et al., 2014).

While the effective wastewater cleaning functions of mangroves may be difficult to assess and quantify in reality, it is clear that the wastewater treatment functions of a tidal flat area can be expected to decrease dramatically if mangroves are cut down. This is mainly because the mangrove root environments will be lost and many soil macroorganisms will disappear or be replaced by other species as associated carbon and nutrient cycles also change. Microenvironments for microorganisms will correspondingly simplify. A decreasing diversity will be recorded, as was found in mangrove sediments in Tanzania (Sjöling et al., 2005; cf. also Li et al., 2011).

12.6 CONCLUDING REMARKS: POLLUTION, MANGROVES, AND RESPONSIBILITY

Marine pollution sometimes makes headlines in daily newspapers, for example when it surfaces in connection with other more immediately obvious tragedies such as in the case of Malaysia Airlines Flight MH 370, which mysteriously disappeared on March 8, 2014. Until today the airplane has not been found despite the most extensive search in aviation history. Instead, during its search thousands of pieces of other types of large marine debris have been detected floating in remote regions of the Indian Ocean (Alverson, 2014). It was a reminder to the public that pollution of the oceans with large solid debris is an increasing problem. The immense problems and changes in the marine environments brought about by persistent cryptic pollution with anthropogenic chemicals are, nonetheless, still largely beyond the sights and concerns of large sections of the public. Until now, it is only coastal communities dependent on fisheries or tourism (especially coral reef diving) that may see and bear significant costs of the failures to address the problem.

Wastewater can be discharged into open water bodies often without any effects becoming immediately obvious to the public. In Southeast Asia and elsewhere immense loads of nutrients (from fertilizers) and synthetic toxic chemicals (from pesticides) are sprayed every year on agricultural fields, altering the functioning of the agroecosystems where they are applied, and polluting downstream ecosystems where they are exported through rains and floodwaters (Maneepitak and Cochard, 2014; Cochard et al., 2014). It has been shown in many studies that natural wetlands, such as mangrove ecosystems, can play a considerable role in mitigating the effects of incident landborne pollution in the coastal environment. This "water cleaning" ecosystem service may also represent one of many other incentives (eg, coastal stabilization and shelter, fisheries) to preserve and replant mangroves along the coasts, particularly near large estuaries and population centers, such as Bangkok. "Water cleaning" services provided by natural mangroves should, however, not be relied upon as the sole strategy to solve the problems of anthropogenic coastal pollution. Natural mangrove ecosystems are not perfect "facilities" for water purification. They can never fully replace the important functions of industrial wastewater treatment plants (especially for point pollution sources) or the functions of wetlands constructed and maintained specifically for wastewater treatment (point and/or diffuse sources). In constructed wetlands (especially freshwater, but also saline) the plants and associated biota are generally better adapted to incurring pollution loads, and—importantly—the purification processes can be monitored and controlled. In natural mangrove wetlands, on the other hand, the hydrological processes (including tidal movements intermixing with rain flood waters, etc.) are complex and difficult to control. Furthermore, mangrove ecosystems tend to be more pollution-sensitive than freshwater wetlands. Hence, wastewaters need to be diluted and spread over a larger area to prevent harm to critical infauna and, consequently, to the mangrove trees.

Research undertaken with the aim to illuminate the wastewater cleaning capacities of mangrove ecosystems (including their stability and resilience) can help to support the conservation and restoration of mangroves. Especially in shrimp production areas where many farms may no longer be economically viable, abandoned shrimp ponds can be replanted with mangroves and can possibly be reverted to economic utility as multifunctional secondary shrimp wastewater cleaning facilities (cf. Chickering, 2014; Stevenson, 1997). In such a way, parts of the biological functions formerly provided by the mangrove forests may be reestablished, whereby the remaining shrimp farms may reach more ecologically and economically sustainable equilibria (cf. Matsui et al., 2014). In strategies to combat coastal pollution mangrove forests should, however, only be seen as one potential last element after many other elements. Mangroves should certainly not serve as an alibi for failures to introduce more comprehensive and effective strategies and policies to reduce pollution. Such strategies should first aim to reduce unnecessary uses of chemicals and their release into the environment, especially in the agricultural sector but also in other sectors. The public, and especially farmer communities, should be better educated about how to reduce and manage the use of potentially hazardous chemicals, including fertilizers, pesticides, and household wastes. As a second element in the chain, various strategies should be set in place to filter out the chemicals before they reach natural water bodies, whereby constructed wetlands and other vegetated buffer zones may play an important role, especially to reduce the inputs of excessive nutrient loads. People may be educated about all the significant values of mangrove forests and other natural wetlands, but they should not be led to believe that mangroves offer true alternatives to preventing and mitigating pollution at and near the source point.

Acknowledgments

This review is a by-product of the surveys and reviews on the 2004 Asian Tsunami, a project made possible through the financial support of the Center for International Agriculture (ZIL Grant No. 41-3219.4/00256) and the Institute of Integrative Biology, Swiss Federal Institute of Technology (ETHZ). Site visits of the Bang Pu area south of Bangkok were kindly supported by Dr. O. Shipin (Asian Institute of Technology, Thailand) and Dr. S. Wickramasinghe (Rajarata University, Sri Lanka).

References

Adame, M.F., Reef, R., Herrera-Silveira, J.A., Lovelock, C.E., 2012. Sensitivity of dissolved organic carbon exchange and sediment bacteria to water quality in mangrove forests. Hydrobiologia 691, 239–253.

Agoramoorthy, G., Chen, F., Hsu, M.J., 2008. Threat of heavy metal pollution in halophytic and mangrove plants of Tamil Nadu, India. Environ. Pollut. 155, 320–326.

Alverson, K., 2014. Environment: ocean pollution foils search for plane. Nature 509 (7500), 288.

Amaral, V., Penha-Lopes, G., Paula, J., 2009. Effects of vegetation and sewage load on mangrove crab condition using experimental mesocosms. Estuar. Coast. Shelf Sci. 84, 300–304.

Anderson, D.M., Glibert, P.M., Burkholder, J.M., 2002. Harmful algal blooms and eutrophication: nutrient sources, composition, and consequences. Estuar. Coast. Shelf Sci. 25 (4b), 704–726.

Aramaki, T., Galal, M., Hanaki, K., 2006. Estimation of reduced and increasing health risks by installation of urban wastewater systems. Water Sci. Technol. 53 (9), 247–252.

Bandara, N.J.G.J., 2003. Water and wastewater related issues in Sri Lanka. Water Sci. Technol. 47 (12), 305–312.

Banjongproo, P., Wett, B., 2002. Mass balance of wastewater loaded canal system: case study of Bangkok. Water Sci. Technol. 46 (11), 451–456.

Bartolini, F., Penha-Lopes, G., Limbu, S., Paula, J., Cannicci, S., 2009. Behavioural responses of the mangrove fiddler crabs (Uca annulipes and U. Inversa) to urban sewage loadings: results of a mesocosm approach. Mar. Pollut. Bull. 58, 1860–1867.

Bartolini, F., Cimò, F., Fusi, M., Dahdouh-Guebas, F., Penha-Lopes, G., Cannicci, S., 2011. The effect of sewage discharge on the ecosystem engineering activities of two East African fiddler crab species: consequences for mangrove ecosystem functioning. Mar. Environ. Res. 71, 53–61.

Booij, K., Hillebrand, T.J., Nolting, R.F., van Ooijen, J., 2001. Nutrients, trace metals and organic contaminants in Banten Bay, Indonesia. Mar. Pollut. Bull. 42 (11), 1187–1190.

Boonsong, K., Piyatiratitivorakul, S., Patanaponpaiboon, P., 2003. Potential use of mangrove plantation as constructed wetland for municipal wastewater treatment. Water Sci. Technol. 48 (5), 257–266.

Boonyatumanond, R., Wattayakorn, G., Togo, A., Takada, H., 2006. Distribution and origins of polycyclic aromatic hydrocarbons (PAHs) in riverine, estuarine, and marine sediments in Thailand. Mar. Pollut. Bull. 52 (8), 942–956.

Breaux, A., Farber, S., Day, J., 1995. Using natural coastal wetlands systems for wastewater treatment: an economic benefit analysis. J. Environ. Manage. 44, 285–291.

Brix, H., 1994. Use of constructed wetlands in water pollution control: historical development, present status, and future perspectives. Water Sci. Technol. 30 (8), 209–223.

Brix, H., 1997. Do macrophytes play a role in constructed treatment wetlands? Water Sci. Technol. 35 (5), 11–17.

Burke, L., Selig, L., Spalding, M., 2002. Research report: reefs at risk in South Asia. World Resources Institute, Washington, DC. http://marine.wri.org/pubs_description.cfm?PubID=3144.

Cannicci, S., Bartolini, F., Dahdouh-Guebas, F., Fratini, S., Litulo, C., Macia, A., Mrabu, E.J., Penha-Lopes, G., Paula, J., 2009. Effects of urban wastewater on crab and mollusk assemblages in equatorial and subtropical mangroves of East Africa. Estuar. Coast. Shelf Sci. 84, 305–317.

Chaiyara, R., Ngoendee, M., Kruatrachue, M., 2013. Accumulation of Cd, Cu, Pb, and Zn in water, sediments, and mangrove crabs (sesarma mederi) in the upper Gulf of Thailand. ScienceAsia 39, 376–383.

Chatenoux, B., Peduzzi, P., 2007. Impacts from the 2004 Indian Ocean Tsunami: analysing the potential protecting role of environmental features. Nat. Hazards 40 (2), 289–304.

Cheevaporn, V., Menasveta, P., 2003. Water pollution and habitat degradation in the Gulf of Thailand. Mar. Pollut. Bull. 47, 43–51.

Chen, Q.H., Tam, N.F.Y., Shin, P.K.S., Cheung, S.G., Xu, R.L., 2009. Ciliate communities in a constructed mangrove wetland for wastewater treatment. Mar. Pollut. Bull. 58, 711–719.

Chen, G.C., Tam, N.F.Y., Wong, Y.S., Ye, Y., 2011. Effect of wastewater discharge on greenhouse gas fluxes from mangrove soils. Atmos. Environ. 45, 1110–1115.

Chickering, M., 2014. As shrimp farms fail, mangroves make a comeback. Thomson Reuters Foundation. Accessed July 2014, http://www.trust.org/item/20140516104944-ervdm/.

Chong, H.T., Sasekumar, A., Leh, M.U.C., D'Cruz, R., 1990. The fish and prawn communities of a Malaysian coastal mangrove system, with comparisons to adjacent mudflats and inshore waters. Estuar. Coast. Shelf Sci. 31, 703–722.

Chongprasith, P., Praekulvanich, E., 2003. Coastal pollution management in Thailand. In: Bruen, M. (Ed.), Diffuse Pollution and Basin Management. Proceedings of the 7th International Specialised IWA Conference, Dublin, Ireland, 17th–22nd August 2003. ISBN: 1902277767, pp. 1477–1483. http://www.ucd.ie/dipcon/docs/theme14/theme14_21.PDF.

Chou, L.M., 1994. Marine environmental issues in Southeast Asia: state and development. Hydrobiologia 285, 139–150.

Chu, H.Y., Chen, N.C., Yeung, M.C., Tam, N.F.Y., Wong, Y.S., 1998. Tide-tank system simulating mangrove wetland for removal of nutrients and heavy metals from wastewater. Water Sci. Technol. 38 (1), 361–368.

Cochard, R., 2011a. The 2004 Tsunami in Aceh and Southern Thailand: coastal ecosystem services, damages and resilience. In: The Tsunami Threat—Research and Technology. Tech Open Access Publisher, Rijeka, Croatia, pp. 179–216. http://www.intechopen.com/articles/show/title/the-2004-tsunami-in-aceh-a-southern-thailand-coastal-ecosystem-services-damages-and-resilience (Chapter 10).

Cochard, R., 2011b. On the strengths and drawbacks of Tsunami-buffer forests. Proc. Natl. Acad. Sci. U. S. A. 108, 18571–18572.

Cochard, R., 2013. Natural hazards mitigation services of carbon-rich ecosystems. In: Lal, R., Lorenz, K., Hüttl, R.F., Schneider, B.U., von Braun, J. (Eds.), Ecosystem Services and Carbon Sequestration in the Biosphere. Springer, Heidelberg, pp. 221–293 (Chapter 11).

Cochard, R., Ranamukharachchi, S.L., Shivakoti, G., Shipin, O., Edwards, P.J., Seeland, K.T., 2008. The 2004 Tsunami in Aceh and Southern Thailand: a review on coastal ecosystems, wave hazards and vulnerability. Perspect. Plant Ecol. Evol. Syst. 10, 3–40.

Cochard, R., Maneepitak, S., Kumar, P., 2014. Aquatic faunal abundance and diversity in relation to synthetic and natural pesticide applications in rice fields of Central Thailand. Int. J. Biodivers. Sci. Ecosyst. Serv. Manage. 10 (2), 157–173.

Craig, D.L., Fallowfield, H.J., Cromar, N.J., 2003. Effectiveness of guideline faecal indicator organism values in estimation of exposure risk at recreational coastal sites. Water Sci. Technol. 47 (3), 191–198.

Dahdouh-Guebas, F., Jayatissa, L.P., di Nitto, D., Bosire, J.O., Lo Seen, D., Koedam, N., 2005. How effective were mangroves as a defence against the recent tsunami? Curr. Biol. 15 (12), R443–R447.

De Graaf, G.J., Xuan, T.T., 1998. Extensive shrimp farming, mangrove clearance and marine fisheries in the southern provinces of Vietnam. Mangroves Salt Marshes 2, 159–166.

De Wolf, H., Rashid, R., 2008. Heavy metal accumulation in Littoria Scabra along polluted and pristine mangrove areas of Tanzania. Environ. Pollut. 152, 636–643.

Denny, P., 1997. Implementation of constructed wetlands in developing countries. Water Sci. Technol. 35 (5), 27–34.

Dierberg, F.E., Kiattisimkul, W., 1996. Issues, impacts, and implications of shrimp aquaculture in Thailand. Environ. Manage. 20 (5), 649–666.

Dikou, A., van Woesik, R., 2006. Survival under chronic stress from sediment load: spatial patterns of hard coral communities in the southern Islands of Singapore. Mar. Pollut. Bull. 52 (1), 7–21.

Donohue, I., Styles, D., Coxon, C., Irvine, K., 2005. Importance of spatial and temporal patterns for assessment of risk of diffuse nutrient emissions to surface waters. J. Hydrol. 304, 183–192.

Dorenbosch, M., van Riel, M.C., Nagelkerken, I., van der Velde, G., 2004. The relationship of reef fish densities to the proximity of mangrove and seagrass nurseries. Estuar. Coast. Shelf Sci. 60, 37–48.

Dsikowitzky, L., Nordhaus, I., Jennerhahn, T.C., Khrycheva, P., Sivatharshan, Y., Yuwono, E., Schwarzbauer, J., 2011. Anthropogenic organic contaminants in water, sediments and benthic organisms of the mangrove-fringed Segara Anakan Lagoon, Java, Indonesia. Mar. Pollut. Bull. 62, 851–862.

III. LEARNING FROM THE FIELD CASES/ISSUES

Duke, N.C., Bell, A.M., Pederson, D.K., Roelfsema, C.M., Nash, S.B., 2005. Herbicides implicated as the cause of severe mangrove dieback in the Mackay region, NE Australia: consequences for marine plant habitats of the GBR World Heritage Area. Mar. Pollut. Bull. 51, 308–324.

Eby, L.A., Crowder, L.B., 2004. Effects of hypoxic disturbances on an estuarine nekton assemblage across multiple scales. Estuar. Coast. Shelf Sci. 27 (2), 342–351.

Edinger, E.N., Jompa, J., Limmon, G.V., Widjatmoko, W., Risk, M.J., 1998. Reef degradation and coral biodiversity in Indonesia: effects of land-based pollution, destructive fishing practices and changes over time. Mar. Pollut. Bull. 36 (8), 617–630.

Ellison, J.C., 1999. Impacts of sediment burial on mangroves. Mar. Pollut. Bull. 37, 420–426.

Ellison, A.M., Farnsworth, E.J., 1996. Anthropogenic disturbance of Carribbean mangrove ecosystems: past impacts, present trends and future predictions. Biotropica 28 (4), 549–565.

Enander, M., Hasselstrom, M., 1994. An experimental wastewater treatment system for a shrimp farm. Infofish Int. 4, 56–61.

Eng, C.T., Paw, J.N., Guarin, F.Y., 1989. The environmental impact of aquaculture and the effects of pollution on coastal aquaculture development in Southeast Asia. Mar. Pollut. Bull. 20 (7), 335–343.

Erftemeijer, P.L.A., Stapel, J., Smekens, M.J.E., Drossaert, W.M.E., 1994. The limited effect of in situ phosphorus and nitrogen additions to seagrass beds on carbonate and terrigenous sediments in South Sulawesi, Indonesia. J. Exp. Mar. Biol. Ecol. 182, 123–140.

Evans, S.M., Dawson, M., Day, J., Frid, C.L.J., Pattisina, L.A., Porter, J., 1995. Domestic waste and TBT pollution in coastal areas of Ambon Island (Eastern Indonesia). Mar. Pollut. Bull. 30 (2), 109–115.

Fabricius, K.E., 2005. Effects of terrestrial runoff on the ecology of corals and coral reefs: review and synthesis. Mar. Pollut. Bull. 50, 125–146.

FAO-UNEP, 1980. The present state of mangrove ecosystems in Southeast Asia and the impact of pollution. Report prepared by T. Piyakarnchana, Chulalongkorn University, Bangkok. Food and Agriculture Organisattion of the United Nations and United Nations Environment Programme, Rome, Italy. http://www.fao.org/docrep/field/003/AB751E/AB751E00.htm.

Fernando, H.J.S., Mendis, S.G., McCulley, J.L., Perera, K., 2005. Coral poaching worsens Tsunami destruction in Sri Lanka. Eos. Trans. 86, 301–304.

Flaherty, M., Vandergeest, P., 1998. "Low-salt" shrimp aquaculture in Thailand: goodbye coastline, hello Khon Kaen! Environ. Manage. 22 (6), 817–830.

Fortes, M.D., 1991. Seagrass-mangrove ecosystems management: a key to marine coastal conservation in the ASEAN region. Mar. Pollut. Bull. 32, 113–116.

Funge-Smith, S.J., Briggs, M.R.P., 1998. Nutrient budgets in intensive shrimp ponds: implications for sustainability. Aquaculture 164 (1–4), 117–133.

Gacia, E., Duarte, C.M., Marba, N., Terrados, J., Kennedy, H., Fortes, M., Tri, N.H., 2003. Sediment deposition and production in SE-Asia seagrass meadows. Estuar. Coast. Shelf Sci. 56, 909–919.

Gacia, E., Kennedy, H., Duarte, C.M., Terrados, J., Marba, N., Papadimitriou, S., Fortes, M., 2005. Light-dependence of the metabolic balance of a highly productive Philippine seagrass community. J. Exp. Mar. Biol. Ecol. 316, 55–67.

Gopal, B., 1999. Natural and constructed wetlands for wastewater treatement: potentials and problems. Water Sci. Technol. 40 (3), 27–35.

Gräslund, S., Holström, K., Wahlström, A., 2003. A field survey of chemicals and biological products used in shrimp farming. Mar. Pollut. Bull. 46, 81–90.

Green, E.P., Short, F.T. (Eds.), 2003. World Atlas of Seagrasses. University of California Press, Berkeley.

Hale, S.S., Paul, J.F., Heltshe, J.F., 2004. Watershed landscape indicators of estuarine benthic condition. Estuar. Coast. Shelf Sci. 27 (2), 283–295.

Herteman, M., Fromard, F., Lambs, L., 2011. Effects of pretreated domestic wastewater supplies on leaf pigment content, photosynthesis rate and growth of mangrove trees: a field study from Mayotte Island, SW Indian Ocean. Ecol. Eng. 37, 1283–1291.

Hoeksema, B.W., Cleary, D.F.R., 2004. The sudden death of a coral reef. Science 303, 1293.

Holmer, M., Duarte, C.M., Heilskov, A., Olesen, B., Terrados, J., 2003. Biogeochemical conditions in sediments enriched by organic matter from net-pen fish farms in the bolinao area, Philippines. Mar. Pollut. Bull. 46, 1470–1479.

Hopkins, J.S., Sandifer, P.A., Browdy, C.L., 1995. A review of water management regimes which abate the environmental impacts of shrimp farming. In: Browdy, C.L., Hopkins, J.S. (Eds.), Proceedings of the Special Session on Shrimp Farming. World Aquaculture Society, Baton Rouge, LA, pp. 157–166.

III. LEARNING FROM THE FIELD CASES/ISSUES

Hughes, T.P., Baird, A.H., Bellwood, D.R., Card, M., Connolly, S.R., Folke, C., Grosberg, R., Hoegh-Guldberg Jackson, J.B.C., Kleypas, J., Lough, J.M., Marshall, P., Nyström, M., Palumbi, S.R., Pandolfi, J.M., Rosen, B., Roughgarden, J., 2003. Climate change, human impacts, and the resilience of coral reefs. Science 301, 929–933.

Hungspreugs, M., Utoomprurkporn, W., Dharmvanij, S., Sompongchaiyakul, P., 1989. The present status of the aquatic environment in Thailand. Mar. Pollut. Bull. 20 (7), 327–332.

Hutchings, P., Haynes, D., Goudkamp, K., McCook, L., 2005a. Catchment to reef: water quality issues in the Great Barrier Reef Region—an overview of papers. Mar. Pollut. Bull. 51, 3–8.

Hutchings, P., Peyrot-Clausade, M., Osnorno, A., 2005b. Influence of land runoff on rates and agents of bioerosion of coral substrates. Mar. Pollut. Bull. 51, 438–447.

Ikejima, K., Tongnunui, P., Medej, T., Taniuchi, T., 2003. Juvenile and small fishes in a mangrove estuary in Trang Province, Thailand: seasonal and habitat differences. Estuar. Coast. Shelf Sci. 56, 447–457.

Ip, Y.K., Chew, S.F., Wilson, J.M., Randall, D.J., 2004. Defences against ammonia toxicity in tropical air-breathing fishes exposed to high concentrations of environmental ammonia: a review. J. Comp. Physiol. 174, 565–575.

Islam, Md.S., 2005. Nitrogen and phosphorus budget in coastal and marine cage aquaculture and impacts of effluent loading on ecosystem: review and analysis towards model development. Mar. Pollut. Bull. 50, 48–61.

Islam, Md.S., Tanaka, M., 2004. Impacts of pollution on coastal and marine ecosystems including coastal and marine fisheries and approach for management: a review and synthesis. Mar. Pollut. Bull. 48, 624–649.

Islam, Md.S., Wahab, Md.A., 2005. A review on the present status and management of mangrove wetland habitat resources in Bangladesh with emphasis on mangrove fisheries and aquaculture. Hydrobiologia 542, 165–190.

IUCN-WWF, 2005. News: the Indian Ocean Tsunami. Arborvitae (IUCN/WWF Forest Conservation Newsletter) March: 2–4. http://cmsdata.iucn.org/downloads/arborvitae27.pdf>.

Jennerjahn, T.C., Nasir, B., Pohlenga, I., 2009. Spatio-temporal variation of dissolved inorganic nutrients related to hydrodynamics and land use in the mangrove-fringed Segara Anakan Lagoon, Java, Indonesia. Reg. Environ. Change 9, 259–274.

Jones, A.B., Dennison, W.C., Preston, N.P., 2001a. Integrated treatment of shrimp effluent by sedimentation, oyster filtration and macroalgal absorption: a laboratory scale study. Aquaculture 193, 155–178.

Jones, A.B., O'Donohue, M.J., Udy, J., Dennison, W.C., 2001b. Assessing ecological impacts of shrimp and sewage effluent: biological indicators with standard water quality analyses. Estuar. Coast. Shelf Sci. 52, 91–109.

Jones, A.B., Preston, N.P., Dennison, W.C., 2002. The efficiency and conditions of oysters and macroalgae used as biological filters of shrimp pond effluent. Aquacult. Res. 33, 155–178.

Kao, C.M., Wang, J.Y., Lee, H.Y., Wen, C.K., 2001. Application of a constructed wetland for non-point source pollution control. Water Sci. Technol. 44, 585–590.

Kautsky, N., Rönnbäck, P., Tedengren, M., Troell, M., 2000. Ecosystem perspectives on management of disease in shrimp pond farming. Aquaculture 191, 145–161.

Keller, V., 2006. Risk assessment of "down-the-drain" chemicals: search for a suitable model. Sci. Total Environ. 360, 305–318.

Keuskamp, J.A., Schmitt, H., Laanbroek, H.J., Verhoeven, J.T.A., Hefting, M.M., 2013. Nutrient amendment does not increase mineralisation of sequestered carbon during incubation of a nitrogen limited mangrove soil. Soil Biol. Biochem. 57, 822–829.

Kirkwood, S., 2004. Yannawa wastewater treatment plant (Bangkok, Thailand): design, construction and operation. Water Sci. Technol. 50 (10), 221–228.

Kivaisi, A.K., 2001. The potential for constructed wetlands for wastewater treatment and reuse in developing countries: a review. Ecol. Eng. 16, 545–560.

Knight, R.L., 1997. Wildlife habitat and public use benefits of treatment wetlands. Water Sci. Technol. 35, 35–43.

Knight, R.L., Payne Jr., V.W.E., Borer, R.E., Clarke Jr., R.A., Pries, J.H., 2000. Constructed wetlands for livestock wastewater management. Ecol. Eng. 15, 41–55.

Kogo, M., Kogo, K., 2004. Towards sustainable use and management for mangrove conservation in Viet Nam. In: Vannucci, M. (Ed.), Mangrove Management and Conservation: Present and Future. United Nations University Press, Tokyo, pp. 233–248.

Kongprom, A., Khemakorn, P., Eiamsa-Ard, M., Supongpan, M., 2003. Status of demersal fishery resources in the Gulf of Thailand. In: Silvestre, G., Garces, L., Stobutzki, I., Luna, C., Ahmed, M., Valmonte Santos, R.A., Lachica-Alino, L., Munro, P., Christensen, V., Pauly, D. (Eds.), Assessment, Management and Future Directions for Coastal Fisheries in Asian Countries. World Fish Center Conference Proceedings 67, pp. 137–152.

III. LEARNING FROM THE FIELD CASES/ISSUES

Koottatep, T., Polprasert, C., 1997. Role of plant uptake on nitrogen removal in constructed wetlands located in the tropics. Water Sci. Technol. 36 (12), 1–8.

Koottatep, T., Surinkul, N., Polprasert, C., Kamal, A.S.M., Koné, D., Montangero, A., Heinss, U., Strauss, M., 2005. Treatment of septage in constructed wetlands in tropical climate: lessons learnt from seven years of operation. Water Sci. Technol. 51 (9), 119–126.

Kruitwagen, G., Hecht, T., Pratap, H.B., Wendelaar Bonga, S.E., 2006. Changes in morphology and growth of the mudskipper (Periophthalmus argentilineatus) associated with coastal pollution. Mar. Biol. 149, 201–211.

Le, T.X., Munekage, Y., Kato, S., 2005. Antibiotic resistance in bacteria from shrimp farming in mangrove areas. Sci. Total Environ. 349, 95–105.

Lee, J.H.W., Hodgkiss, I.J., Wong, K.T.M., Lam, I.H.Y., 2005. Real time observations of coastal algal blooms by early warning system. Estuar. Coast. Shelf Sci. 65, 172–190.

Lehane, L., 2000. Paralytic shellfish poisoning: a review. National Office of Animal and Plant Health, Agriculture, Fisheries and Forestry, Canberra. http://www.affa.gov.au/corporate_docs/publications/pdf/animalplanthealth/chief_vet/psp.pdf.

Lewis, M., Pryor, R., Wilking, L., 2011. Fate and effects of anthropogenic chemicals in mangrove ecosystems: a review. Environ. Pollut. 159, 2328–2346.

Li, M., Cao, H., Hong, Y., Gu, J.D., 2011. Spatial distribution and abundances of ammonia-oxidizing archaea (AOA) and ammonia-oxidizing bacteria (AOB) in mangrove sediments. Appl. Microbiol. Biotechnol. 89, 1243–1254.

Lin, C.K., 2000. Development of shrimp farming and environmental sustainability in Thailand. Suisanzoshoku 48 (2), 267–272.

Lindberg, T., Nylander, A., 2001. Strategic environmental assessment on shrimp farms in Southeast of Thailand. Master Thesis, Uppsala University, Sweden.

Lønholdt, J., Elberg Jørgensen, P., O'Hearn, D., 2005. Setting-up a cost recovery system for the largest wastewater treatment plant in South-East Asia. Water Sci. Technol. 52 (12), 123–132.

Lu, J., 2003. Marine oil spill detection, statistics and mapping with ERS SAR imagery in south-east Asia. Int. J. Remote Sens. 24 (15), 3013–3032.

Luan, B.T., Debenay, J.-P., 2005. Foraminifera, environmental bioindicators in the highly impacted environments of the Mekong Delta. Hydrobiologia 548, 75–83.

MacFarlane, G.R., Burchett, M.D., 2001. Photosynthetic pigments and peroxidase activity as indicators of heavy metal stress in the grey mangrove, Avicennia marina (Forsk.) Vierh. Mar. Pollut. Bull. 42, 233–240.

Macintosh, D.J., Ashton, E.C., Havanon, S., 2002. Mangrove rehabilitation and intertidal biodiversity: a study in the Ranong mangrove ecosystem, Thailand. Estuar. Coast. Shelf Sci. 55, 331–345.

Malan, D.E., Ersamus, T., Baird, D., 1988. Aspects of Sesarma catenata (Grapsidae, Crustacea) burrows and its implications in the event of an oil spill. Estuar. Coast. Shelf Sci. 26, 95–104.

Mallin, M.A., Posey, M.H., Shank, G.C., McIver, M.R., Ensign, S.H., Alphin, T.D., 1999. Hurricane effects on water quality and benthos in the cape fear watershed: natural and anthropogenic impacts. Ecol. Appl. 9 (1), 350–362.

Mandura, A.S., 1997. A mangrove stand under sewage pollution stress: Red Sea. Mangroves Salt Marshes 1, 255–262.

Maneepitak, S., Cochard, R., 2014. Uses, toxicity levels, and environmental impacts of synthetic and natural pesticides in rice fields—a survey from central Thailand. Int. J. Biodivers. Sci. Ecosyst. Serv. Manage. 10 (2), 144–156.

Marchand, C., Lallier-Vergès, E., Allenbach, M., 2011. Redox conditions and heavy metals distribution in mangrove forests receiving effluents from shrimp farms (Teremba Bay, New Caledonia). J. Soils Sediments 11, 529–541.

Marris, E., 2005. Tsunami damage was enhanced by coral theft. Nature 436 (25), 1071.

Matsui, N., Songsangjinda, P., Wodehouse, D., 2014. Longevity of simultaneous operation of aquaculture and mangrove forestry as explained in terms of water and sediment qualities. Wetl. Ecol. Manage. 22 (3), 215–225.

Maurer, D., Nguyen, H., Robertson, G., Gerlinger, T., 1999. The infaunal trophic index (ITI): its suitability for marine environmental monitoring. Ecol. Appl. 9 (2), 699–713.

McGenity, T.J., 2014. Hydrocarbon biodegradation in intertidal wetland sediments. Curr. Opin. Biotechnol. 27, 46–54.

Methratta, E.T., Link, J.S., 2006. Evaluation of quantitative indicators for marine fish communities. Ecol. Indic. 6, 575–588.

Melville, A.J., Connolly, R.M., 2005. Food webs supporting fish over subtropical mudflats are based on transported organic matter not in situ microalgae. Mar. Biol. 148, 363–371.

Metcalf & Eddy, 2003. Wastewater Engineering. Treatment and Reuse. Metcalf & Eddy, Inc., fifth ed., revised by G. Tchobanoglous, F.L. Burton and H.D. Stensel. McGraw-Hill Companies, Inc., New York.

Miller, D.C., Poucher, S.L., Coiro, L., 2002. Determination of lethal dissolved oxygen levels for selected marine and estuarine fishes, crustaceans, and a bivalve. Mar. Biol. 140, 287–296.

Mogg R., 2002. Filtration and separation markets in Southeast Asia: Thailand focus. Filtration and Separation September 2002: 32–34. www.filtsep.com.

Morrison, R.J., Delaney, J.R., 1996. Marine pollution in the Arafura and Timor Seas. Mar. Pollut. Bull. 32 (4), 327–334.

Morton, B., Blackmore, G., 2001. South China Sea. Mar. Pollut. Bull. 42 (12), 1236–1263.

Mumby, P.J., Edwards, A.J., Arias-Gonzalez, J.E., Lindeman, K.C., Blackwell, P.G., Gall, A., Gorczynska, M.I., Harborne, A.R., Pescod, C.L., Renken, H., Wabnitz, C.C.C., Llewellyn, G., 2004. Mangroves enhance the biomass of coral reef fish communities in the Carribbean. Nature 427 (5), 533–536.

Neori, A., Chopin, T., Troell, M., Buschmann, A.H., Kraemer, G.P., Halling, C., Shpigel, M., Yarish, C., 2004. Integrated aquaculture: rationale, evolution and state of the art emphasising seaweed biofiltration in modern mariculture. Aquaculture 231, 361–391.

Neverauskas, V.P., 1987. Monitoring seagrass beds around a sewage outfall in South Australia. Mar. Pollut. Bull. 18 (4), 158–164.

Newman, M.C., Evans, D.A., 2001. Enhancing belief during causality assessments: cognitive idols or Bayes's theorem. In: Newman, M.C., Roberts, M.H., Hale, R.C. (Eds.), Coastal and Estuarine Risk Assessment. Lewis Publishers, New York.

Newman, M.C., Roberts, M.H., Hale, R.C. (Eds.), 2001. Coastal and Estuarine Risk Assessment. Lewis Publishers, New York.

Paez-Osuna, F., 2001. The environmental impact of shrimp aquaculture: causes, effects, and mitigating alternatives. Environ. Manage. 28 (1), 131–140.

Passioura, J.B., Ball, M.C., Knight, J.H., 1992. Mangroves may salinize the soil and in doing so limit their transpiration rate. Funct. Ecol. 6 (4), 476–481.

Pauly, D., Chuengpadgee, R., 2003. Development of fisheries in the Gulf of Thailand large marine ecosystem: analysis of an unplanned experiment. In: Hempel, G., Sherman, K. (Eds.), Large Marine Ecosystems of the World. Elsevier, Netherlands.

PCD, 2004. Thailand state of pollution report. Pollution Control Department, Ministry of Natural Resources and Environment,Bangkok.http://www.pcd.go.th/public/Publications/en_print_report.cfm?task=pcdreport2547&gt.

Penha-Lopez, G., Torres, P., Narciso, L., Cannicci, S., Paula, J., 2009. Comparison of fecundity, embryo loss and fatty acid composition of mangrove crab species in sewage contaminated and pristine mangrove habitats in Mozambique. J. Exp. Mar. Biol. Ecol. 381, 25–32.

Penha-Lopez, G., Kristensen, E., Flindt, M., Mangion, P., Bouillon, S., Paula, J., 2010. The role of biogenic structures on the biogeochemical functioning of mangrove constructed wetlands sediments—a mesocom approach. Mar. Pollut. Bull. 60, 560–572.

Peters, E.C., Gassman, N.J., Firman, J.C., Richmond, R.H., Power, E.A., 1997. Ecotoxicology of tropical marine ecosystems. Environ. Toxicol. Chem. 16 (1), 12–40.

Peterson, S.B., Teal, J.M., 1996. The role of plants in ecologically engineered wastewater treatment systems. Ecol. Eng. 6, 137–148.

Pittman, S.J., Pittman, K.M., 2005. Short-term consequences of a benthic cyanobacterial bloom (Lyngbya majuscula Gomont) for fish and penaeid prawns in Moreton Bay (Queensland, Australia). Estuar. Coast. Shelf Sci. 63, 619–632.

Preen, A.R., Lee Long, W.J., Coles, R.G., 1995. Flood and cyclone related loss, and partial recovery, of more than 1000 km^2 of seagrass in Hervey Bay, Queensland, Australia. Aquat. Bot. 52 (1–2), 3–17.

Raisin, G.W., Mitchell, D.S., 1995. The use of wetlands for the control of non-point source pollution. Water Sci. Technol. 32 (3), 177–186.

Ranjan, R.K., Ramanathan, A.L., Chauhan, R., Singh, G., 2011. Phosphorus fractionation in sediments of the Pichavaram mangrove ecosystem, south-eastern coast of India. Environ. Earth Sci. 62, 1779–1787.

Reddy, K.R., D'Angelo, E.M., 1997. Biogeochemical indicators to evaluate pollutant removal efficiency in constructed wetlands. Water Sci. Technol. 35 (5), 1–10.

Ridd, P.V., 1996. Flow through animal burrows in mangrove swamps. Estuar. Coast. Shelf Sci. 43, 617–625.

Roberts, M.H., Newman, M.C., Hale, R.C., 2001. Overview of ecological risk assessment in coastal and estuarine environments. In: Newman, M.C., Roberts, M.H., Hale, R.C. (Eds.), Coastal and Estuarine Risk Assessment. Lewis Publishers, New York.

Salvato, M., Borin, M., Doni, S., Macci, C., Caccanti, B., Marinari, S., Masciandaro, G., 2012. Wetland plants, microorganisms and enzymatic activities interrelations in treating N polluted water. Ecol. Eng. 47, 36–43.

Satapornvanit, K., 1993. The environmental impact of shrimp farm effluent. M.Sc. Thesis No. AE.93.30, Asian Institute of Technology, Bangkok.

Schaffelke, B., Mellors, J., Duke, N.C., 2005. Water quality in the Great Barrier Reef region: responses of mangrove, seagrass and macroalgal communities. Mar. Pollut. Bull. 51, 279–296.

Schouw, N.L., Bregnhøj, H., Mosbaek, H., Tjell, J.C., 2003. Technical, economic and environmental feasibility of recycling nutrients in waste in Southern Thailand. Waste Manage. Res. 21, 191–206.

Selvakumar, S., Khan, S.A., Kamaraguru, A.K., 1996. Acute toxicity of some heavy metals, pesticides and water soluble fractions of diesel oil to the larvae of some brachyuran crabs. J. Environ. Biol. 17 (3), 221–226.

Shipin, O., Koottatep, T., Kanh, N.T.T., Polprasert, C., 2005. Integrated natural treatment systems for developing communities: low-tech N-removal through the fluctuating microbial pathways. Water Sci. Technol. 51 (12), 299–306.

Singhruk, P., 2001. Circulation features in the Gulf of Thailand inferred from SeaWiFS data. In: Paper presented at the 22nd Asian Conference on Remote Sensing, 5–9 November 2001, Singapore. Centre for Remote Sensing, Imaging and Processing (CRISP), National University of Singapore, Singapore Institute of Surveyors and Valuers (SISV), Asian Association on Remote Sensing (AARS).

Sjöling, S., Mohammed, S.M., Lyimo, T.J., Kyaruzi, J.J., 2005. Benthic bacterial diversity and nutrient processes in mangroves: impact of deforestation. Estuar. Coast. Shelf Sci. 63, 397–406.

Skov, M.W., Hartnoll, R.G., 2002. Paradoxical selective feeding on a low-nutrient diet: why do mangrove crabs eat leaves? Oecologia 131, 1–7.

Smith III, T.J., Boto, K.G., Frusher, S.D., Giddins, R.L., 1991. Keystone species and mangrove forest dynamics: the influence of burrowing by crabs on soil nutrient status and forest productivity. Estuar. Coast. Shelf Sci. 33, 419–432.

Spalding, M.D., Kainuma, M., Collins, L. (Eds.), 2010. World Mangrove Atlas. Earthscan Ltd., London.

Stevenson, N.J., 1997. Disused shrimp ponds: options for redevelopment of mangrove. Coast. Manag. 25 (4), 423–425.

Stieglitz, T., Ridd, P., Müller, P., 2000. Passive irrigation and functional morphology of crustacean burrows in a tropical mangrove swamp. Hydrobiologia 421, 69–76.

Stobutzki, I.C., Silvestre, G.T., Abu, Talib A., Krongprom, A., Supongpan, M., Khemakorn, P., Armada, P., Garces, L.R., 2006. Decline of demersal fisheries resources in three developing Asian countries. Fish. Res. 78, 130–142.

Suárez-Abelenda, M., Ferreira, T.O., Camps-Arbestain, M., Rivera-Monroy, V.H., Macías, F., Nóbrega, G.N., Otero, X.L., 2014. The effect of nutrient-rich effluents from farming on mangrove soil carbon storage and geochemistry under semi-arid climate conditions in northern Brazil. Geoderma 213, 551–559.

Suvapepun, S., 1991. Long term ecological changes in the Gulf of Thailand. Mar. Pollut. Bull. 23, 213–217.

Szuster, B.W., 2003. Shrimp farming in Thailand's Chao Phraya river delta: boom, bust and echo. International Water Management Institute, Colombo, Sri Lanka. http://www.iwmi.cgiar.org/assessment/files/word/ProjectDocuments/ChaoPhraya/szuster.pdf&gt.

Tabucanon, M.S., 1991. State of coastal resource management strategy in Thailand. Mar. Pollut. Bull. 23, 579–586.

Tacon, A.G.J., Phillips, M.J., Barg, U.C., 1995. Aquaculture feeds and the environment: the Asian experience. Water Sci. Technol. 31 (10), 41–59.

Takarina, N.D., Browne, D.R., Risk, M.J., 2004. Speciation of heavy metals in coastal sediments of Semarang, Indonesia. Mar. Pollut. Bull. 49, 854–874.

Tam, N.F.Y., 1998. Effects of wastewater discharge on microbial populations and enzyme activities in mangrove soils. Environ. Pollut. 102, 233–242.

Tam, N.F.Y., Wong, Y.S., 1993. Retention of nutrients and heavy metals in mangrove sediments receiving wastewater of different strengths. Environ. Technol. 14, 719–729.

Tam, N.F.Y., Wong, Y.S., 1995. Mangrove soils as sinks for wastewater-borne pollutants. Hydrobiologia 295, 231–241.

Tam, N.F.Y., Wong, Y.S., 1996a. Retention of wastewater-borne nitrogen and phosphorus in mangrove soils. Environ. Technol. 17 (8), 851–859.

Tam, N.F.Y., Wong, Y.S., 1996b. Retention and distribution of heavy metals in mangrove soils receiving wastewater. Environ. Pollut. 94, 283–291.

Tam, N.F.Y., Wong, Y.S., 1997. Accumulation and distribution of heavy metals in a simulated mangrove system treated with sewage. Hydrobiologia 352, 67–75.

Tam, N.F.Y., Wong, Y.S., 1999. Mangrove soils in removing pollutants from municipal wastewater of different salinities. J. Environ. Qual. 28, 556–564.

Tam, N.F.Y., Wong, Y.S., 2008. Effectiveness of bacterial inoculum and mangrove plants on remediation of sediment contaminated with polycyclic aromatic hydrocarbons. Mar. Pollut. Bull. 57, 716–726.

III. LEARNING FROM THE FIELD CASES/ISSUES

Tam, N.F.Y., Wong, Y.S., Lan, C.Y., Wang, L.N., 1998. Litter production and decomposition in a subtropical mangrove swamp receiveing wastewater. J. Exp. Mar.Biol. Ecol. 226, 1–18.

Tam, N.F.Y., Wong, A.H.Y., Wong, M.H., Wong, Y.S., 2009. Mass balance of nitrogen in constructed mangrove wetlands receiving ammonium-rich wastewater: effects of tidal regime and carbon supply. Ecol. Eng. 35, 453–462.

Terrados, J., Thampanya, U., Srichai, N., Kheowvongsri, P., Geertz-Hansen, O., Boromthanarath, S., Panapitukkul, N., Duarte, C.M., 1997. The effect of increased sediment accretion on the survival and growth of *Rhizophora apiculata* seedlings. Estuar. Coast. Shelf Sci. 45, 697–701.

Terrados, J., Duarte, C.M., Fortes, M.D., Borum, J., Agawin, N.S.R., Bach, S., Thampanya, U., Kamp-Nielsen, L., Kenworthy, W.J., Geertz-Hansen, O., Vermaat, J., 1998. Changes in community structure and biomass of seagrass communities along gradients of siltation in SE Asia. Estuar. Coast. Shelf Sci. 46, 757–768.

Thampanya, U., Vermaat, J.E., Sinsakul, S., Panapitukkul, N., 2006. Coastal erosion and mangrove progradation of Southern Thailand. Estuar. Coast. Shelf Sci. 68, 75–85.

Thia-Eng, C., 1999. Marine pollution prevention and management in the East Asian Seas: a paradigm shift in concept, approach and methodology. Mar. Pollut. Bull. 39, 80–88.

Tomascik, T., Mah, A.J., Nontji, A., Moosa, M.K., 1997. The ecology of the Indonesian Seas. Periplus Editions (HK) Ltd., Singapore.

Tonmanee, N., Kanchanakool, N., 1999. Agricultural diffuse pollution in Thailand. Water Sci. Technol. 39 (3), 61–66.

Udy, J.W., Dennison, W.C., Long, W.J.L., McKenzie, L.J., 1999. Responses of seagrass to nutrients in the Great Barrier Reef, Australia. Mar. Ecol. Prog. Ser. 185, 257–271.

UNEP-WCMC, 2006. In the frontline: shoreline protection and other ecosystem services from mangroves and coral reefs. UNEP-WCMC, Cambridge. http://www.unep.org/pdf/infrontline_06.pdf>

Unsworth, R.K.F., Cullen-Unsworth, L.C., 2013. Biodiversity, ecosystem services, and the conservation of seagrass meadows. In: Maslo, B., Lockwood, J.L. (Eds.), Coastal Conservation. Cambridge University Press, UK.

Vaiphasa, C., de Boer, W.F., Skidmore, A.K., Panitchart, S., Vaiphasa, T., Bamrongrugsa, N., Santitamnont, P., 2007. Impact of solid shrimp pond waste materials on mangrove growth and mortality: a case study from Pak Phanang, Thailand. Hydrobiologia 591, 47–57.

Valiela, I., McClelland, J., Hauxwell, J., Behr, P.J., Hersh, D., Foreman, K., 1997. Macroalgal blooms in shallow estuaries: controls and ecophysiological and ecosystem consequences. Limnol. Oceanogr. 42 (5, part 2), 1105–1118.

Vassallo, P., Fabiano, M., Vezzulli, L., Sandulli, R., Marques, J.C., Jørgensen, S.E., 2006. Assessing the health of coastal marine ecosystems: a holistic approach based on sediment micro and meio-benthic measures. Ecol. Indic. 6, 525–542.

Vega-Turber, R.L., Burkepile, D.E., Fuchs, C., Shantz, A.A., McMinds, C., Zaneveld, J.R., 2014. Chronic nutrient enrichment increases prevalence and severity of coral disease and bleaching. Glob. Chang. Biol. 20 (2), 544–554.

Vidthayanon, C., Premcharoen, S., 2002. The status of estuarine fish diversity in Thailand. Mar. Freshw. Res. 53 (2), 471–478.

Visvanathan, C., Cippe, A., 2001. Strategies for development of industrial wastewater reuse in Thailand. Water Sci. Technol. 43 (10), 59–66.

von Sperling, M., 1996. Comparison among the most frequently used systems of wastewater treatment in developing countries. Water Sci. Technol. 33, 59–72.

Walters, B.B., 2003. People and mangroves in the Philippines: fifty years of coastal environmental change. Environ. Conserv. 30 (2), 293–303.

Wang, Y., Wu, Y., Pi, N., Tam, N.F.Y., 2014. Investigation of microbial community structure in constructed mangrove microcosms receiving wastewater-borne polycyclic aromatic hydrocarbons (PAHs) and polybrominated diphenyl ethers (PBDEs). Environ. Pollut. 187, 136–144.

Wannamaker, C.M., Rice, J.A., 2000. Effects of hypoxia on movements and behaviour of selected estuarine organisms from the southeastern United States. J. Exp. Mar. Biol. Ecol. 249, 145–163.

Wattayakorn, G., King, B., Wolanski, E., Suthanaruk, P., 1998. Seasonal dispersion of petroleum contaminants in the Gulf of Thailand. Cont. Shelf Res. 18, 641–659.

Waycott, M., Longstaff, B.J., Mellors, J., 2005. Seagrass population dynamics and water quality in the Great Barrier Reef region: a review and future research directions. Mar. Pollut. Bull. 51, 343–350.

Werry, J., Lee, S.Y., 2005. Grapsid crabs mediate link between mangrove litter production and estuarine planktonic food chains. Mar. Ecol. Prog. Ser. 293, 165–176.

Wickramasinghe, S., Borin, M., Kotagama, S.W., Cochard, R., Anceno, A.J., Shipin, O.V., 2009. Multi-functional pollution mitigation in a rehabilitated mangrove conservation area. Ecol. Eng. 35, 898–907.

Wikipedia, 2014. Accessed in July 2014, http://en.wikipedia.org/wiki/Waste.

III. LEARNING FROM THE FIELD CASES/ISSUES

Williams, T.M., Rees, J.G., Setiapermana, D., 2000. Metals and trace organic compounds in sediments and waters of Jakarta Bay and the Pulau Seribu complex, Indonesia. Mar. Pollut. Bull. 40 (3), 277–285.

Winterwerp, J.C., Borst, W.G., de Vries, M.B., 2005. Pilot study on the erosion and rehabilitation of a mangrove mud coast. J.Coast. Res. 21 (2), 223–230.

Wong, Y.S., Lan, C.Y., Chen, G.Z., Li, S.H., Chen, X.R., Liu, Z.P., Tam, N.F.Y., 1995. Effect of wastewater discharge on nutrient contamination of mangrove soils and plants. Hydrobiologia 295, 243–254.

Wong, Y.S., Tam, N.F.Y., Lan, C.Y., 1997. Mangrove wetlands as wastewater treatment facility: a field trial. Hydrobiologia 352, 49–59.

Wooldridge, S., Brodie, J., Furnas, M., 2006. Exposure of inner shelf reefs to nutrient enriched runoff entering the great barrier reef lagoon: post-European changes and the design of water quality targets. Mar. Pollut. Bull. 52 (11), 1467–1479.

Wu, Y., Chung, N.F.Y., Pi, N., Wong, M.H., 2008a. Constructed mangrove wetland as secondary treatment system for municipal wastewater. Ecol. Eng. 34, 137–146.

Wu, Y., Tam, N.F.Y., Wong, M.H., 2008b. Effects of salinity on treatment of municipal wastewater by constructed mangrove wetland microcosms. Mar. Pollut. Bull. 57, 727–734.

Ye, Y., Tam, N.F.Y., Wong, Y.S., 2001. Livestock wastewater treatment by a mangrove pot-cultivation system and the effect of salinity on the nutrient removal efficiency. Mar. Pollut. Bull. 42 (6), 513–521.

Yim, M.W., Tam, N.F.Y., 1999. Effects of wastewater-borne heavy metals on mangrove plants and soil microbial activities. Mar. Pollut. Bull. 39, 179–186.

Yokoyama, H., Higano, J., Adachi, K., Ishihi, Y., Yamada, Y., 2002. Evaluation of shrimp polyculture system in Thailand based on stable carbon and nitrogen isotope ratios. Fish. Sci. 68 (4), 745–750.

Zakaria, M.P., Horinouchi, A., Tsutsumi, S., Takada, H., Tanabe, S., Ismail, A., 2000. Oil pollution in the straits of Malacca, Malaysia: application of molecular markers for source identification. Environ. Sci. Tech. 34 (7), 1189–1196.

Zhou, Y., Zhao, B., Peng, Y., Chen, G., 2010. Influence of mangrove reforestation on heavy metal accumulation and speciation in intertidal sediments. Mar. Pollut. Bull. 60, 1319–1324.

Scaling the Costs of Natural Ecosystem Degradation and Biodiversity Losses in Aceh Province, Sumatra

R. Cochard[*],[†]

[*]Institute of Integrative Biology, Swiss Federal Institute of Technology, Zurich, Switzerland
[†]Asian Institute of Technology, Klong Luang, Pathumthani, Thailand

Abbreviations

CES	cultural ecosystem services
EIA	environmental impact assessment
ES	ecosystem services
HES	habitat ecosystem services
GAM	Gerakan Aceh Merdeka (Free Aceh Movement)
GIS	geographical information system
IUCN	International Union for Conservation of Nature
NGO	nongovernment organization
NTFP	nontimber forest product
PES	provisioning ecosystem services
REDD	Reducing Emissions from Deforestation and Forest Degradation (UN program)
RES	regulating ecosystem services
SES	supporting ecosystem services
TEV	total economic value
TNI	Tentara Nasional Indonesia (Military of Indonesia)
UNDP	United Nations Development Program
UNFCCC	United Nations Framework Convention on Climate Change
WTP	willingness to pay

13.1 INTRODUCTION: THE ENDANGERED NATURAL HERITAGE OF ACEH PROVINCE

Partly due to its unique and isolated geography and history, Aceh (Nanggroe Aceh Darussalam) is still the richest province on the island of Sumatra in terms of near-pristine tropical ecosystems. The province harbors vast tropical lowland, upland, and mountain forests of which a significant proportion is under formal protection within Leuser National Park and within the extended Leuser Ecosystem Conservation Area, as well as in other minor protected areas. These forests represent one of the last and most important refuges for highly endangered Southeast Asian megafauna, such as the Sumatran rhinoceros, tiger, elephant, and orangutan, as well as other rare, endangered, and/or endemic animal and plant species. The coasts and marine environments of Aceh equally harbor many natural treasures, such as species-rich coral reefs, seagrass beds, mangroves, and beach forests. The reefs of Weh Island on the northwestern tip of Sumatra are world-famous for scuba diving, whereby whale sharks and manta rays are occasionally sighted. The Banyak Islands archipelago in the south, with its extensive seagrass beds, harbor some of the last habitats and breeding sites of dugongs (sea cows) and sea turtles around Sumatra (ver Berkmoes et al., 2013). Coastal ecosystems and beach vegetation are important to protect the extensive coastlines against the strong waves of the Indian Ocean; in particular, they stabilize (at least to a certain degree) the land during storms and coastal flooding, such as during the 2004 Asian tsunami (Cochard, 2011a, 2013). The southern lowlands of Aceh also harbor extensive tropical peat swamp forests. These ecosystems are not only rich in unique fauna and flora, they also represent formidable carbon stores and may therefore play a significant role for the mitigation of global warming (Cochard, 2011b).

Due to large-scale commercial logging and agroindustrial development, vast areas of Sumatra have been deforested during the last decades. From 1990 to 2010 around 40% of primary forests (more than 7 million hectares), and 33% of degraded forest areas were entirely lost in Sumatra, with much deforestation occurring already before 1990 (Margono et al., 2012; Whitten et al., 1997). Most of the deforestation took place in lowland forests in South and Central Sumatra. In comparison, the forests in Aceh overall fared better, at least until recently (Fig. 13.1). The difference is partly attributable to the open conflict that raged in Aceh since the 1970s until 2004. The conflict ceased in the aftermath of a huge tsunami that flooded the coastlines in December 2004, causing immense disaster in many coastal towns. Aceh Province has since opened a new chapter in its history, with promises of peace and development, but also newly arising threats to its fragile natural heritage. Here in this chapter a brief outline of Aceh's distinct history is provided (Section 13.2), followed by a review of the main natural assets at stake (Section 13.3), and a discussion of their socioeconomic values in terms of providing various (ecosystem) services to humans, as well as descriptions of the threats occurring from excessive and/or ill-advised exploitation (Section 13.4). Finally, some special natural resource management issues are presented and discussed (Section 13.5): (1) the role of coastal ecosystems in mitigating coastal hazards, and the ecosystems' post-tsunami restoration or (natural) regeneration, (2) the destruction of peat swamp forests for expansion of oil palm plantations and efforts for conservation, and (3) excessive exploitation of forest biodiversity, especially the increasing threats to the most iconic forest animal species (ie, the Sumatran rhinoceros, tiger, elephant, and orangutan).

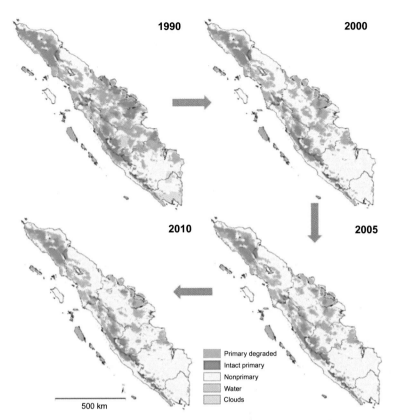

FIG. 13.1   Deforestation in Sumatra from 1990 until 2010. Lowland forests were most affected, and particularly in central and southern Sumatra. *Source: Margono B.A., Turubanova S., Zhuravleva I., Potapov P., Tyukavina A., Baccini A., Goetz S., Hansen M.C., 2012. Mapping and monitoring deforestation and forest degradation in Sumatra (Indonesia) using Landsat time series data sets from 1990 to 2010. Environ. Res. Lett. 7, 034010. doi:10.1088/1748-9326/7/3/034010, reprinted with kind permission.*

13.2   HISTORICAL PERSPECTIVES ON ENVIRONMENTAL ISSUES

13.2.1   Aceh: From Sultanate to Troubled Backwater

Owing to its specific geography, history, and culture Aceh Province has always occupied a special place within Indonesia. Aceh is situated on the northwestern tip of Sumatra, remote from the capital Jakarta. Aceh was one of the first places in Southeast Asia where Islam gained a foothold in the 13th century AD. The Sultanate of Aceh, which emerged in the 16th century AD, expanded south along the coasts of Sumatra to Padang and Riau, and also occupied lands on the Malay Peninsula. It established diplomatic connections to faraway Muslim powers, such as the Ottoman Empire. In the 19th century it still held a relatively strong position in the region due to maritime trading and lucrative production and exporting of pepper (Ricklefs, 2001). Following skirmishes triggered by pirates in the Straits of Malacca, the Dutch colonial army invaded Aceh in 1873. The following war of resistance lasted for decades, until the area was eventually pacified using heavy military might at considerable costs to the Dutch colonial government (Ricklefs, 2001).

In 1950, after Indonesia had attained independence, Aceh was politically dismantled and incorporated into the Province of North Sumatra. This eventually resulted in a first Acehnese rebellion lasting from 1953 to 1959, after which Aceh was reinstated as a province, including the granting of various rights of autonomy in religious affairs (Aspinall, 2009). During the 1970s secessionist tendencies again emerged as a result of disaffection toward several central government policies, including the transmigration program and oil exploitation in the province. Armed resistance formed in the "Free Aceh Movement" (Gerakan Aceh Merdeka, GAM), and open conflict with the Indonesian military (Tentara Nasional Indonesia, TNI), including various human rights abuses from both sides, reached a height during the 1990s. In 1999 the conflict was spreading even into remote areas in hitherto relatively calm southern Aceh (McCarthy, 2006). Just before the tsunami in 2004 a major military offensive against GAM was under way (Aspinall, 2009). The conflict was ended in 2005 as a result of events following the 2004 Asian tsunami, and aided by international mediation (Kelman, 2012). Under the peace agreement signed in 2005 Aceh was granted special autonomy and government troops were withdrawn in exchange for the disarmament of GAM. Local elections were held in 2006, whereby parties associated with the former GAM won the elections (Aspinall, 2009).

The conflict had resulted in an estimated 15,000 casualties with many more people injured, displaced, and traumatized (BBC, 2005; Aspinall, 2009). The environment may not have suffered to such a degree as in other conflicts (cf. Mohibbi and Cochard, 2014). Nonetheless, even if the conflict did not affect Acehnese society everywhere at a similar level, socioeconomic development was hampered in the entire province. This was also true in the realms of environmental research and management. For example, in a report on the 2004 Asian tsunami UNEP (2005) discussed the legal and institutional requirements of environmental impact assessments (EIAs), and the practicality of conducting EIAs in the context of such an extensive disaster situation. To the observant reader, however, the report essentially revealed that hardly any environmental reports and data were available in Aceh from before 2005. A look at recently updated research databases provide similar insights. Perhaps with the exception of the Leuser National Park in the south, there are very few published scientific studies on ecological and environmental issues in Aceh from before 2005. Correspondingly, major ecological works on Sumatra (eg, Whitten et al., 1997; Laumonier, 1997; Tomascik et al., 1997) mainly focused on the provinces to the south. To the world Aceh was a troubled and otherwise largely forgotten backwater, until the disaster in 2004 violently brought it to international attention.

13.2.2 The 2004 Asian Tsunami: A Flood of Disaster and Transformation

Triggered by a massive earthquake, the Asian tsunami of Dec. 26, 2004, caught many communities along the fringes of the Indian Ocean completely off guard. Most communities did not remember a similar event, and in Aceh the enduring conflict probably contributed to failure of tsunami warning and communication, and the ensuing calamities and chaos. In contrast, owing to an old native oral tradition, the indigenous people of Simeulue Island off the coast of Aceh understood the warning signs of the earthquake, and mostly managed to escape the tsunami flood to higher ground (McAdoo et al., 2006). Coastal towns of Aceh Province, including Banda Aceh, Meulaboh, and Calang, and hundreds of villages were largely or completely devastated. Overall estimated casualties were in the range of 170,000 to over 220,000 people (ie, 4–6% of a population of ~4.3 million), with many more injured and/ or rendered homeless (Cochard et al., 2008).

Global aid response was relatively swift and of unprecedented scale. While attention of the international media and public was initially set on tourist centers such as the coasts of Thailand, Sri Lanka, or the Maldives, reporting and aid responsiveness soon turned to Aceh, which was by far the worst-hit region. The devastation brought by the tsunami event put pressure on the two conflict parties for a cease-fire, to allow and facilitate national and international assistance. Before the tsunami only a few international parties were interested in resolving the conflict. With posttsunami aid contributions amounting to several hundred million U.S. dollars, donor countries now had an interest in conflict mediation to allow aid to effectively reach the affected communities (Gaillard et al., 2008).

Following the tsunami, emergency aid and temporary housing, and the reestablishment of critical service infrastructures (eg, medical and sanitary facilities, coastal roads, landing stages, etc.) were regionally and internationally coordinated. After a few months the focus started to shift toward the rebuilding of housing and the restoration of livelihoods (eg, fishing fleets, shrimp ponds, rice fields, etc.). Reconstruction was conducted with the proclaimed vision to "build back better," but activities conducted by governmental and nongovernmental aid agencies were often influenced by the agencies' culture, the available funds, and time constraints, with variable efficiencies and outcomes. In some cases, planning was conducted in a rash top-down approach, and the donor funds were not used efficiently. For example, some highly questionable investments occurred in the case of coastal protection measures. In efforts to stabilize the coastlines, concrete buffers were put in place against coastal erosion on beaches. The usefulness of such measures was criticized on general principles, but on accreting beaches these buffers were evidently without any value (Cochard, 2011a). Similarly, efforts by environmental nongovernmental organizations (NGOs) to stabilize the coasts through the planting of mangroves and other coastal vegetation met with substantial problems. As the environmental conditions had changed considerably (eg, through land subsidence, tsunamite sediment deposits, and/or altered ecological parameters), most large-scale mangrove plantation schemes near Banda Aceh were condemned to failure (Cochard, 2011a; cf. Section 13.5.1).

While reconstruction was itself rather "imperfect" in several ways, relatively few efforts were as yet made to foster a culture of real transparency, and to effectively learn from the mistakes made during the reconstruction phase. "Building back better" was a slogan permeating the academic literature. However, it could be stated that comparatively little was invested to improve the education of students in schools and at the university, and to teach the communities the management tools and strategies, so that they would eventually be able to manage their environment (and the associated natural risks) in self-responsible and sustainably gainful ways. If some of these persisting weaknesses can be genuinely addressed in the future (cf. Shah and Cardozo, 2014), there may be prospects for progress through learning from such past "imperfect" developments and associated failures. The 2004 Asian tsunami was a tremendous catastrophe for the affected communities. If the province will, however, continue to develop peacefully and thoughtfully, some useful lessons may be extricated from the tragedy; the event may emerge as a turning point to a more knowledge-based "culture." Future generations will continue to face the haphazard threats of nature at the convergence of tectonic plates. Everything is interconnected, and when such a disaster hits again, communities will have to be prepared and the crucial emergency resource reserves will have to be there.

Aceh still harbors the most extensive and intact rainforest ecosystem in Southeast Asia (on the Sundaic side of the Wallace line), as well as many near-pristine coastal ecosystems. The population density of Aceh (81 persons per km^2; total population of ~4.7 million people in 2014) is lower than the average of Indonesia (125 persons per km^2, with 1117 persons per km^2 on densely populated Java) (Wikipedia, 2014). Hence, given its relatively low population density one may assume that the province's land resources would be more than sufficient to support a healthy and prospering population, provided that these resources are efficiently managed. Equally, one may assume that the living forests with all their unique and rare wildlife represent an ever more precious asset that need not be squandered carelessly. Currently it does not seem certain, however, that the province is on a good path for long-term sustainable development. On the contrary, many resources are increasingly exploited in unnecessarily destructive ways to the environment.

13.3 A BRIEF REVIEW OF ACEH'S NATURAL RICHES: ECOSYSTEMS AND BIODIVERSITY

13.3.1 Terrestrial Forest Ecosystems

The earthquake that triggered the 2004 Asian tsunami had its epicenter below the sea between Simeulue Island and the mainland of Sumatra, at a depth of around 30 km. The location was at the center of the Sunda megathrust, a geologic subduction fault line that stretches from Bali up to Myanmar. According to new research, in northern Sumatra subduction ruptures (which often result in large and measurable seismic events with the potential to trigger tsunamis, and associated land subsidence and uplift) occur at intervals of around 230 years on average (Nalbant et al., 2005; Natawidjaja et al., 2006). The geological dynamics of northern Sumatra represent considerable hazards for the resident communities (eg, due to earthquakes, tsunamis, landslides, and in some locations volcanic eruptions). In geologic and evolutionary times these dynamics, however, also contributed to creating the high mountains, the rugged terrain, and diverse habitats of the region, which are one of the reasons for the high natural biodiversity. Of the 51 mountains in the Barisan Range in Sumatra that are higher than 2000 m, 12 are found in Aceh, with the highest peak, Gunung Leuser, at 3404 m above sea level (Fig. 13.2; Barber et al., 2005). Around 80% of the territory of Aceh is situated above 100 m a.s.l., and most of the hilly or mountainous ranges are covered with tropical forests (de Koninck et al., 2012). Until recent decades well over half of the province was occupied by two major forest ecosystem complexes. The Ulu Masen Ecosystem (covering around 7000 km^2) is situated on the northwestern mountain ranges stretching from close to the northern tip of Aceh to around Meulaboh city in the south. Farther in the south lies the larger Leuser Ecosystem (covering around 26,000 km^2) of which a part stretches into the province of North Sumatra, and also covers some peat swamp ecosystems along the southwestern lowlands (Fig. 13.2; de Koninck et al., 2012).

The high variability in the terrain, creating patchy habitats with different climate and soil conditions and diverse levels of disturbances, promoted speciation—especially under climatic conditions that were generally stable and mostly wet-tropical (Cochard, 2011b; Gurevitch et al., 2006; Gaston, 2000). The lush lowland rain forests are characterized by a rich structure of

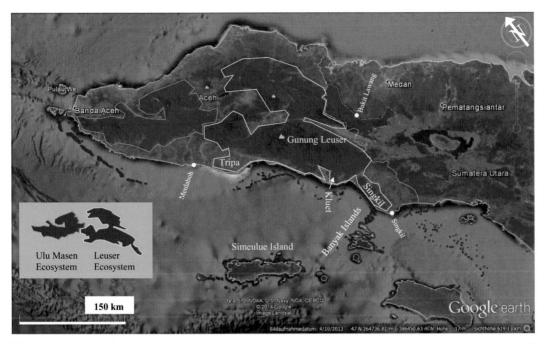

FIG. 13.2 Physical map of Aceh Province, with the major forest areas (Leuser and Ulu Masen Ecosystems) indicated. *Source: Google Earth™.*

different tree and canopy layers, with emergent trees over 50 m high. These forests represent diverse habitats, with many ecological niches in space, especially for birds and tree-climbing animals (eg, the orangutan and other primates). Richness of forest tree species as well as vertebrate species (especially birds, amphibians, and the larger mammals) typically increases in moister areas (Currie, 1991; McArthur and McArthur, 1961). Due to higher orographic rainfalls, some moist mid-altitude forests may be more species-rich than adjacent forests in the lowlands. In particular, some mist forests are characterized by a high diversity of epiphytes (including various vascular plants, mosses, and lichens) and an associated richness of insect species.

In 1982, MacKinnon (in Whitten et al., 1997) recorded about 55,320 km² of forest lands in Aceh, of which 800 km² were montane moss forests and 15,900 km² were other montane forests on mineral soils (ie, volcanic soils, limestone, alluvium, and other rocks/soils), 19,200 km² were "wet" lowland forests on mineral soils, 8700 km² were "moist" lowland forests on mineral soils, 4300 km² were "dry" lowland forests on mineral soils, and 4700 km² were freshwater swamp forests on alluvial or on peat soils. Less than 30 years later the total area of forests had apparently been reduced by almost 40% to 33,357 km², whereby lowland forests (in particular freshwater forests and partly peat swamp forests) were most affected by deforestation (Blackett and Irianto, 2007). Despite stated intentions by politicians to halt deforestation and taking a path of developing a "Green Aceh" (cf. van Beukering et al., 2009), uncontrolled deforestation is presently still ongoing in various parts of the province, even within some protected areas (Fig. 13.3; Perdani, 2013; Butler, 2013a). According to Hansen et al. (in Meijaard, 2014a) forest loss in Aceh has actually been increasing between 2004 and 2012.

FIG. 13.3 Impacts in the rainforest blocks of Central Aceh (Leuser and Ulu Masen Ecosystems) and in the Tripa swamps. (A/B) Views of primary forest in the most western parts of the Leuser Ecosystem. (C) A river in the eastern Ulu Masen Ecosystem. (D) Deforestation on hillsides in the Ulu Masen Ecosystem. (E/F) Paddy fields situated within the Ulu Masen Ecosystem. (G) Deforestation, drainage, and planting of oil palm trees in the Tripa peat swamp forests east of Meulaboh. (H) Deforestation of a relict peat swamp forest west of Meulaboh. *Photographs by R. Cochard, August 2006.*

13.3.2 Coastal Ecosystems

Even though many millions of U.S. dollars were invested in posttsunami reconstruction in Aceh, the surveying of coastal ecosystems was not a high priority, and the databases remain, until this date, relatively preliminary and in some respects confusing (cf. Cochard, 2011a). The southern and western coastlines of Aceh are exposed to the strong waves of the open Indian Ocean, and, hence, the coastlines are mostly long stretches of carbonate sandy beaches that (at least before the 2004 tsunami) were mostly covered with beach vegetation dominated by dune-colonizing creepers (eg, *Ipomea pes-caprae*) and trees such as *Casuarina equisetifolia* and *Barringtonia* spp., or cultivated *Cocos nucifera* (Cochard et al., 2008). Along these coastal stretches mangroves were only found in some estuarine areas (behind beach sand bars) where relatively large rivers meet the sea. The area covered by mangroves was minor along the southern coastlines, but much of the shrimp farming area near the town of Banda Aceh was probably initially occupied by vast mangrove forests. From field observations and reports it appeared that virtually all of the remaining mangroves were destroyed by the 2004 tsunami, and limited regeneration occurred only in a few locations. Today, very extensive mangrove stands can, however, still be found along the northeastern coastlines that are fringing the Strait of Malacca and that are fed with sediments from large rivers. These mangroves were not affected by the tsunami (Cochard, 2011a). There are 17 species of true mangroves found in mangals in Sumatra (Whitten et al., 1997).

Some magnificent coral reefs are found around the islands in the north (Weh and Aceh Islands) and in the south (Simeulue and Banyak Islands), as well as along northern and western coastal stretches of the mainland. Not much scientific information is as yet available on the coral reefs of Aceh, although some basic surveys have been conducted around Weh Island (Baird et al., 2005; Campbell et al., 2012) and around the southern islands (Herdiana et al., 2008). Around 360 species of coral reef fishes were recorded by Herdiana et al. (2008). While most of the reefs around the northern islands were relatively little affected by the 2004 tsunami, several reefs around the southern islands (mostly Simeulue Island) were destroyed due to tectonic uplifting (Baird et al., 2005; Foster et al., 2006). Even though there are significant seagrass areas around the northern and southern islands, with small populations of dugong (*Dugong dugon*), there are apparently no clear data either on their extent as well as on the impacts of the 2004 tsunami (Cochard, 2011a).

13.4 AN ECONOMIC APPRAISAL OF ACEH'S NATURAL RICHES

13.4.1 Valuation of Terrestrial Forest Ecosystem Services

13.4.1.1 Ecosystem Services and Their Valuation

Economic values of ecosystems are still mostly seen in the natural products that can be extracted, processed, and sold for profit on a market. If viewed from a local stakeholder perspective there are, however, many other valuable goods and services provided by ecosystems; their degradation can cause substantial costs, diminish the options for livelihoods, and, thus, increase poverty. Hence, more holistic assessments of ecosystem values are needed. Valuation of ecosystem services (ES) is a way to reveal the true overall economic value of ecosystems.

In the case of Aceh, van Beukering et al. (2003, 2009) attempted to assess the total economic value (TEV) of the Leuser ecosystem, respective of all of the remaining forest areas in Aceh Province. The assessments were based on various environmental and economic assumptions and used computer modeling.

Traditional economic decision making is mostly based on parameters such as capital investments and expected returns. When considering ecosystems, inherent values are often not marketable and readily observable in an economic way, and valuable ecosystem services are often taken for granted. The costs of ecosystem degradation are often borne by the parts of society that do not gain any profit from the extraction of specific ecosystem resources. For example, landslides and floods as a consequence of deforestation will often affect local communities which may not profit from logging operations (Cochard, 2013). Using modeling, the TEV of different forest resource-use scenarios can be compared, and insights can be gained regarding the cost-benefit effects of various regimes of resource uses (van Beukering et al., 2003, 2009). Assessing the TEV of ecosystems can thus be very useful for decision makers, granted that these decision makers have an inherent will, and the power, to lead their constituency on a path of long-term sustainable and equitable development and welfare.

13.4.1.2 Supporting Ecosystem Services of Forests

Supporting ecosystem services (SES) of forests are those ES that are necessary for the production of all other ecosystem services. SES may include inherent ecological functions such as the recycling of plant nutrients, the formation of soils and other types of habitats, respectively their essential biotic and physical constituents. Van Beukering et al. (2003, 2009) did not specifically consider any of these services in detail. In regards to economic modeling of various use scenarios the important questions essentially relate to assessments of ecosystem stability under different impact regimes, and such assessments may be very complex (cf. Cochard, 2011b). For example, the effects of selective logging on forest biodiversity and, subsequently, forest integrity and stability may be manifold and may significantly differ among different types of forests. On steep slopes soil resources may be disproportionately affected, leading to relatively fast soil erosion and further degradation of the forests. Due to changes in microclimates and losses of key species forest structure may, however, also collapse in damaged forests on flatlands (cf. Laurance, 1997; Burivalova et al., 2014). Furthermore, the extirpation of key species, such as top predators (eg, tigers), may lead to ecological chain reactions (eg, disproportionate increase of certain herbivores) affecting forest composition (Cochard, 2011b). The degradation of forests may also affect the ES of other ecosystems in distant locations. For example, increased erosion of forests soils may lead to the siltation of rivers (which may increase the risks of river flooding) and it may smother offshore ecosystems such as coral reefs and sea grass beds (which may lead to increased coastal erosion and a decline of fishery output) (Cochard, 2013). Hence, if decision makers could be brought to recognize the overall importance of SES, this could easily convince them to stop many business operations from resulting in economically damaging environmental degradation. SES encompass habitat ecosystem services (HES) as specified by Kumar (2010). The term "habitat" may, however, be problematic in the sense that services may not be provided by the "habitat" but actually by the populations and communities of organisms inhabiting the habitat.

13.4.1.3 Provisioning Ecosystem Services of Forests

Forests provide various products which are (directly or indirectly) of economic value to local communities or to other stakeholders. Such products include water resources (for drinking, irrigation, aquaculture, and the production of electricity), foods (wild crops and fruits or game animals), raw materials (wood for fuel and lumber, fodder, fertilizers, or animal skins) and minerals (various mining products), medicinal products (from plant saps and other sources, including synthetically extracted pharmaceutical chemicals), ornamental resources (plant flowers, pet animals, shells, etc.), energy (primarily from biomass fuels and hydropower), and genetic resources (eg, for the genetic enrichment of domesticated species). In the following the forest provisioning ecosystem services (PES) most relevant in Aceh are described.

WATER SUPPLIES

Water supplies of rain forests may be considered in terms of PES, but also in terms of regulating ecosystem services (RES), because extreme events relating to water cycles, such as floods or droughts, may exert potentially catastrophic damaging effects (cf. section "Natural Hazards Mitigation: Floods and Landslides"). Rain forests serve as an important water store, whereby the main storage facility is provided by the forests' soils. Maintaining the tree cover is, however, essential to safeguard the soils and protect them from erosion. The trees' canopies attenuate the impact and spread out the timing of heavy downpours. On the other hand, the trees' roots stabilize the soil substratum and facilitate water percolation into the soils (Cochard, 2013). If the tree cover is removed, soils are often initially covered by grass and herb vegetation. However, because overall transpiration is significantly decreased, the soils' saturation with water will remain high; in combination with the deterioration of important physical reinforcements (from tree roots) the risks of landslides (on steep slopes) and catastrophic soil erosion will increase. In addition, plant nutrients may be washed out, diminishing the protective cover of grasses or herbs and thus accelerating soil erosion (Cochard, 2013). Once forests and soil resources are degraded to a certain level the hydrological cycles of entire catchments may alter. Water flows may no longer be relatively constant, but flows from intensive rains will be released faster (potentially leading to flooding) whereas rivers may run dry during spells without rainfalls. In addition, due to potential changes in local climatic conditions incurred by the loss of forests, less water may be generated and kept within the extensive forest ecosystems. The effects of deforestation on dry season groundwater tables can, however, be rather complex and variable (cf. literature cited in van Beukering et al., 2003, 2009; Cochard, 2013).

According to the Leuser Management Unit (in van Beukering et al., 2009) approximately 50% of the streams in Aceh have now less than 50% of the water flow in springtime as compared to 10 years ago, and 20% of the flows are completely dry throughout the year. Changes in the rivers' water delivery brought about by forest losses may lead to impacts on the agricultural and fisheries (mostly freshwater aquaculture) sector, and on various other industries and household supplies. The impacts may be direct in the sense that various economic activities (especially in agriculture/aquaculture) may be hindered or reduced. There may, however, also be indirect impacts. For example, the water qualities may decline if water has to be kept in storage tanks, or if it has to be reused in various ways (eg, in aquaculture systems). This may lead to an increased incidence of health issues for humans as well as for domestic

TABLE 13.1   Values of Ecosystem Services (in Millions of U.S. Dollars per Year) as Estimated Through Projective Modeling Over a 30-Year Period From 2008 to 2038 by van Beukering et al. (2009)

Forest Ecosystem Services	Valuation Sources or Methods	Conservation	Deforestation	Value Difference
		Million US$ year^{-1}		
Provisionary ecosystem services (PES)				
Water supply (housing, industries)	Water prices	138	45	67%
Fisheries	Market values	128	98	23%
Hydro-electricity	Energy prices	1.4	0.7	50%
Agriculture	Market/various	206	166	19%
Nontimber forest products	Market/various	21	9	57%
Timber	Timber prices	−10?	149	−100%
Subtotal PES		484	467	4%
Regulating ecosystem services (RES)				
Carbon conservation/sequestration	Carbon price REDD	56	0	100%
Fire prevention	Various/studies	12	9	25%
Landslide prevention	No reliable data	?	?	?
Flood prevention	Damage data/var.	105	90	14%
Subtotal RES		173	99	43%
Cultural ecosystem services (CES)				
Tourism	WTP/CVM	7	1	86%
Biodiversity conservation	WTP/various	30	6	80%
Subtotal CES		37	7	82%
Total		694	573	17%

? Value unknown

The values were assessed in a conservation scenario and compared to the values assessed in a deforestation scenario. The "value difference" indicates the decrease of value (in percentage) from the conservation to the deforestation scenario.

animals or aquaculture crops (fishes or shrimps) (van Beukering et al., 2009). Van Beukering et al. (2003) estimated an annual increase of water prices under the deforestation scenario at Leuser National Park. Regarding water uses of households and industries, van Beukering et al. (2009) estimated an annual economic value of water supply PES of US$138 million for the forest conservation scenario, and US$45 million for the deforestation scenario (within a 30-year time frame from 2008 to 2038; Table 13.1). These estimates were based on current water prices (in 2008) and various assumptions that were drawn from other literature.

FISHERIES AND AQUACULTURE

Fisheries and aquaculture provide an important income to Acehnese communities, totalling around US$128 million in 2008 (van Beukering et al., 2009). Marine fisheries provide by far the largest share (123,673 tons in 2008, or 65.4% of fisheries). These types of fisheries would, however, likely be minimally affected by the effects of terrestrial deforestation (exceptions are mangroves and possibly coastal swamp forests; cf. Cochard, 2011a). In contrast, freshwater fisheries (33,396 tons, 17.7% share) as well as aquaculture using freshwater (12,380 tons, 6.5% share) and brackish water (19,596 tons, 10.4% share) could be directly affected through declining quantities and qualities (especially increased turbidity through erosion) of freshwater resources provided from forests. There are, however, still many "unknowns" as no detailed studies have so far been conducted. Over 114 fish species have been recorded in Acehnese inland waters (Muchlisin and Siti Azizah, 2010), but no detailed data exist on the fishes' ecology and sensitivity to disturbances. Regarding overall fisheries, van Beukering et al. (2009) estimated an annual economic value of water supply/quality PES of US$128 million for the forest conservation scenario, and US$98 million for the deforestation scenario (Table 13.1).

HYDROELECTRICITY

According to van Beukering et al. (2009), over 700,000 households and industries were served electricity from hydropowered plants in 2008. This amounted to around 839 GWH sold. Increased soil erosion does reportedly already lead to various problems for hydropower. Sediments accumulate in the reservoirs, diminishing their water storage capacity. The sediments can also damage the turbines, incurring higher costs of repairs and needs for replacements. Furthermore, decreased water supplies may lead to shortcuts. Van Beukering et al. (2009) estimated an annual economic value of water supply/quality PES of US$1.4 million for the forest conservation scenario, and US$0.7 million for the deforestation scenario (Table 13.1).

AGRICULTURE AND PLANTATIONS

In Aceh 0.4 megatons of rice, 1.7 Mt of fruits and vegetables, and 1.0 Mt of crude palm oils were produced in 2006 (BPS in van Beukering et al., 2009). Agricultural productivity, which is concentrated mostly in the lowlands of Aceh (coastal plains and some valleys inland), depends on the provision of sufficient water for irrigation, on pollinators, and partly on beneficial insects that can keep pest species in check (cf. Settle et al., 1996). Van Beukering et al. (2009) made a direct connection of these services to the impacts of deforestation. While crops may be affected through shortages in irrigation waters due to similar reasons as previously outlined, it may, however, be questioned whether deforestation will have a significant negative impact on crop pollination and pest control. Many important crop plants, such as rice and oil palms, are wind-pollinated and thus not directly dependent on insects for pollination. For other crops, such as tree fruits, sufficient pollinators may still be available through the management of residual forest patches or hedges. Similarly, the overall value of the forests in regard to pest management is not clear. Many specific pests of rice, for example, are unlikely to be controlled by insects that are dependent on forests. In contrast, crops may also be damaged from animals (eg, rats and birds) nesting in and foraging out of the forests (cf. Rijal and Cochard, 2015).

Van Beukering et al. (2003, 2009) also considered that deforestation for farming may eventually lead to losses of soil nutrients and, in some cases, disintegration of soil structure and erosion. Their computer models actually captured an initial increase in economic benefits from crops (especially for vegetables) under the deforestation scenario, as cropping land would expand. After only around 12 years, however, the benefits would turn into costs compared to the forest conservation scenario. The link between deforestation and soil losses (and hence damages to crops) is, however, not straightforward. Major soil degradation may be mostly expected in steep areas (cf. Cochard, 2013). Major problems of land degradation can also be expected in areas with peat soils, which are mostly used for oil palm production (cf. Section 13.5.2.2.; Fig. 13.3G). In other areas, however, soil uses may be potentially sustainable, especially if planted with suitable crops and if organically managed. As most of the currently forested areas are in steep terrain or in peat swamps, the models by van Beukering et al. (2003, 2009) are, nonetheless, probably fairly realistic. Crop failures may also occur due to extreme events such as flooding or droughts; these are partly treated under regulating forest ecosystem services (RES, section "Natural Hazards Mitigation: Floods and Landslides"). Keeping in mind the limitations of their estimates, van Beukering et al. (2009) arrived at an estimation of annual economic value of forest PES and RES to agriculture of US$206 million for the forest conservation scenario, and US$166 million for the deforestation scenario (Table 13.1).

NONTIMBER FOREST PRODUCTS

Forests provide various products other than timber (nontimber forest products, NTFPs). Many of these products, such as wild foods or medicinal plants, are of direct economic value for local communities. Some products are traded, and approximate market prices can be established. Other products are predominantly exchanged in barter trade, or form part of illicit trading with no overt market prices available. Important low- to medium-value plant NTFPs in Aceh include cotton from cotton trees, various resins (damar), bamboo, rattan (a climbing palm, offshoots and leaves), palm leaves and nuts (eg, from rumbia, nipah or sagu palms), gum benzoin, nutmeg, aromatic oil, candlenut, cinnamon, palm sugar, and various other minor plant and mushroom products used for foods, for medicinal purposes, or for decoration and crafting. High-value products that may be traded for income include honey, vanilla, and edible bird nests (saliva from the nests of swiftlets) (van Beukering et al., 2003, 2009). Elliott and Brimacombe (1987) listed 158 species from Leuser National Park that were used by local communities for medicinal purposes. A study by Afolabi et al. (2008) indicated that communities in Central Aceh gained as much as 25% of their livelihoods from NTFPs. They also recorded an increasing shortage of various NTFP products, especially products derived from palms (including rattan), bamboo, and some tree species. In the models by van Beukering et al. (2003, 2009), exploitation of NTFPs increases first under the deforestation scenario as additional extraction in frontier areas becomes possible, but it soon decreases as the forest resources diminish overall. They estimated an annual economic value of NTFP PES of US$21 million for the forest conservation scenario, and US$9 million for the deforestation scenario (Table 13.1).

TIMBER

In Aceh the lowland forests on mineral soils are particularly rich in tree species whose timber is highly priced. Important timber species include dipterocarps (eg, *Shorea, Parashorea, Dipterocarpus,* and *Dryobalanops* spp.), leguminous trees (eg, *Koompassia excelsa* and *K. malaccensis*), and various trees from other plant families (Blackett and Irianto, 2007). In upland and montane forests dipterocarps are increasingly replaced by members of the oak and laurel families, whereby various species have economic potential as timber. In coastal peat and freshwater swamp forests, ramin (*Gonystylus bancanus*, family Thymelaeaceae) is by far the most important timber species (Blackett and Irianto, 2007).

In the past the lowland forests on mineral soils as well as freshwater swamp forests have declined the most because of logging and subsequent conversion to agriculture or plantation crops, especially palm oil and rubber trees. According to Whitten et al. (1997) lowland forests could produce >100 m³ of high-quality logs per hectare. Nowadays, many of the remaining forests in the lowlands are, however, degraded and less productive. Logging has thus increasingly turned to marginal peat swamp forests (cf. Section 13.5.2.2) and upland forests, even though extraction in these types of forests is limited by law. In the case of mountain forests, for example, logging operations are only permissible to an elevation of 2000 m a.s.l., and on slopes of less than 40% inclination, respectively less than 15% if the risk of soil erosion is considered high (Blackett and Irianto, 2007). In 2001 forest concessions were closed in an effort to control the exploitation of timber, but logging resumed again in the wake of the 2004 Asian tsunami, as much timber was needed for reconstruction. Another moratorium on logging was imposed in 2007 pending the preparation of a new forestry master plan (van Beukering et al., 2009). In practice, however, in addition to logging in lowland forests many upland forests are already exploited for timber, often in areas that are susceptible to soil erosion and landsliding, with corresponding consequences (cf. Fig. 13.3).

If the forests were to be completely conserved, a significant opportunity cost would be incurred in terms of nonused timber resources. Van Beukering et al. (2009) estimated the potential economic gains obtained if the forests were to be completely logged over the coming 30 years at around US$149 million annually (Table 13.1), whereby profits would be highest (>US$200 million) during the first 15 years and would then continually decrease (<US$100 million during the last 10 years). Timber prices may be expected to further rise in the future as rain forest timber may become increasingly overexploited within Indonesia and in other countries of Southeast Asia. This may also increase the value of timber trees in Aceh, heightening the pressures to exploit the remaining virgin forests. If logging would and could indeed be completely closed, then timber would have to be imported to Aceh at costs of possibly more than $10 million annually (Table 13.1; van Beukering et al., 2009).

13.4.1.4 Regulating Ecosystem Services of Forests

If ecosystems, such as tropical forests, are disturbed various processes maintained by these ecosystems may fall out of balance and change in dramatic and/or damaging ways, and/or may directly become more hazardous to humans and their livelihood assets. RES by forests therefore represent the benefits obtained from the regulation of ecosystem processes. This comprises water cycling and climate regulation, including the sequestration of carbon (thus serving to mitigate global warming). Carbon sequestration may also be viewed from the aspect of organic waste decomposition as an important RES. Forest ecosystems can also purify the air and the water from toxic or otherwise hazardous substances. Various pest species and diseases may be regulated by providing a genetically diverse pool of control agents, and forests may provide the pool of pollinators needed in agriculture (cf. section "Agriculture and Plantations"). Furthermore, tree roots will stabilize the soils and prevent soil erosion, and natural hazards such as landslides, floods, droughts, and fires may equally be diminished and/or mitigated via the presence of forests. In the following the forest RES relevant in Aceh are described.

CARBON SEQUESTRATION, CARBON STORAGE, AND FIRE PREVENTION

Processes of carbon fixation via plant photosynthesis are important to reduce the concentration of the greenhouse gas carbon dioxide (CO_2) in the atmosphere (ie, carbon sequestration from the atmosphere). Primary forests may not necessarily sequester carbon, as carbon may be lost in equal amounts from the ecosystem via respiration; but many types of forests, especially swamp forests (cf. Section 13.5.2.1), actually still do sequester substantial amounts of carbon (Jandl et al., 2013). Crucially, however, substantial emissions of CO_2 (as well as other greenhouse gases, eg, methane) occur when forests are converted to other land uses. It was estimated that land-use changes amount to about 20% of all CO_2 emissions (Houghton et al., 2001). These emissions thus add to the problem of global warming—a problem whose economic costs scientists and economists have been trying to estimate. If timber is used for construction materials some carbon may still be bound in buildings. Nonetheless, in logged forests carbon will generally be lost in vast amounts from decaying plant residues and eroding soils. The fastest and most "catastrophic" loss of carbon occurs through forest fires (often lit after logging) and, in the case of peat swamp forests, the peat fires (cf. Section 13.5.2.2). In those cases there will not only be emissions of greenhouse gases, but air pollution through soot may result in additional hazards to human and environmental health, with potentially significant economic costs arising.

Under the United Nations Framework Convention on Climate Change (UNFCCC), efforts have been undertaken since 2005 to establish an international market to trade carbon certificates, whereby carbon stored in forests may be quantified, economically valued, and sold to polluting industries. The program for Reducing Emissions from Deforestation and Forest Degradation (REDD) was developed with the objectives to mitigate climate change through the reduction of CO_2 emissions and to conserve and wisely manage the remaining forest areas. In Indonesia carbon stocks of virgin rainforests are in the range of between 180 and 390 tons per hectare (but up to about 2700 t ha^{-1} in peat swamp forests, cf. Section 13.5.2.1), whereby the average price for one ton of offset CO_2 was approximated at US$3.10 (van Beukering et al., 2009). Under the assumption that a functioning, efficient, and sustainable carbon market will indeed be set in place (cf. Angelsen et al., 2014), van Beukering et al. (2009) thus estimated potential annual benefits of US$56 million under the conservation scenario. Under the deforestation scenario obviously no benefits could be obtained from carbon credits (Table 13.1).

REDD carbon certification programs will only be set up for Acehnese forests if it can be convincingly demonstrated that the certified forests are maintained intact. Especially if forests have been degraded through extractive logging or agricultural encroachment the incidence of bush fires may pose an increasingly significant threat; virgin rain forests are much more resistant to penetration by fires. Under exceptional circumstances, such as for example during the drought resulting from the 1997 El Niño event, rain forests may, however, also be susceptible to fires (Page et al., 2002). Until 1997 such extreme conditions, with fire events in rain forests, were presumed to occur only around every 50 years (van Beukering et al., 2003). The recurrence intervals may become shorter, however, under global warming, especially if regional climates become warmer and drier due to decreased overall vegetation cover (Cochard, 2013). In fact, several significant events of forest and peat fires leading to regional smog alerts have occurred since 1997 (eg, in 2006, 2010, and just recently in 2013 and 2014) (BBC, 2014; Guardian Environment Network, 2014).

Fires destroying intact forests and peat soils obviously also severely affect all the ecosystem services provided by these forests (ie, including destruction of timber and NTFPs, carbon emission, resultant erosion, and natural hazards, etc.; Tacconi, 2003). The fires may, however, also affect ecosystems ex situ via pollution impacts of the soot. According to Abram et al. (2003; but challenged by Hoeksema and Cleary, 2004) iron fertilization in ocean waters induced by soot from the 1997 wildfires may have led to unusual red tides (harmful algal blooms) in the same year, leading to high mortality of coral reefs in the Mentawai Islands south of Sumatra. Even though research is still insufficient, large extended smog events will certainly also have various impacts on local and regional climates and ecosystems (cf. Nichol, 1997; Sambas, 2003; Page et al., 2002; Tacconi, 2003). This could lead to a chain of ecosystem degradation (cf. Section 13.4.1.2). Economic impacts may be more easily measured in regard to impacts in the anthroposphere. The vast forest fires that occurred in Sumatra and Borneo in 1997 exposed some 20 million people across Southeast Asia to harmful soot over several months (Tacconi, 2003). According to various estimates, over 100,000 people may have been killed annually from smog-induced ailments during the most recent hazes (Greenpeace, 2014). In addition to health issues, other damages are economically measurable in terms of declining tourism, hazards to transportation and diminished traffic (including air traffic), and under extreme conditions the intermittent closing of schools (BBC, 2014; Guardian Environment Network, 2014; Greenpeace, 2014; van Beukering et al., 2009).

Overall economic damages in Southeast Asia during the 1997 fires have been estimated at between US$3.5 and US$9.3 million (Barber and Schweithelm, 2000; Tacconi, 2003). Based on such figures, van Beukering et al. (2009) estimated amounts of potential fire damages, that is, maximum damages that may occur under a worst-case fires scenario (under various baselines and assumptions, eg, fire events in grasslands occur about every 7 years, in primary forest about every 50 years). Considered in the reverse, "avoided fire damages" have an economic value. According to the models, fire damages first increase under deforestation because more fires with high intensities occur initially. This levels out, however, as most forest areas are degraded. The average "avoided fire damage" under the deforestation scenario was thus calculated at around US$9.4 million annually. In contrast, under the conservation scenario the "avoided fire damage" remained constant (ie, a slight constant increase due to economic growth) with an annual average of US$12 million over the 30-year period (Table 13.1; van Beukering et al., 2009).

NATURAL HAZARDS MITIGATION: FLOODS AND LANDSLIDES

Northern Sumatra is often exposed to tectonic hazards (eg, earthquakes), and weather-related hazards (eg, severe monsoon rainfalls and resultant flooding). The earthquake that triggered the disastrous tsunami in 2004 was indeed the third strongest earthquake ever recorded globally, and the region is currently still subject to considerable seismic activity with frequent earthquakes of lesser strength occurring (Cochard, 2011a; Jakarta Post, 2013). Earthquakes may cause direct destruction to housing, and hence threaten human lives, but in steep terrain they may also trigger landslides or rock falls. The risk of landslides is, however, especially high if soils on steep slopes are saturated with water. Most landsliding thus occurs during extensive monsoon rainfalls or tropical cyclones (Cochard, 2013; Sidle et al., 2006). Until now, tropical cyclones have rarely crossed over northern Sumatra, but with global warming cyclones may become more frequent in the region (Cochard et al., 2008; Cochard, 2013).

Heavy rainfalls may not only trigger landslides but may also lead to local flash floods in mountain valleys, and extensive "carpet" flooding in the lowlands. It appears that landslides and floods have become increasingly common in Aceh during the last decades, as can be seen from the Reliefweb database. For example, over 90,000 people were displaced during flooding in northern Aceh in 2014, over 15,000 people in southern Aceh in 2009, and more than 200,000 were displaced in 2006 (Reliefweb, 2014a,b,c).

While there is still considerable debate among academics, there is indeed increasing scientific evidence that tropical forests play an important role in the mitigation of both landslides and floods (Cochard, 2013; Ogden et al., 2013). Trees buffer the impact of heavy rainfalls. Much of the water is caught in the trees' canopies and may evaporate during intermittent spells of sunshine. The trees' roots provide a reinforcement of the soils and facilitate percolation of rainfall waters, while preventing strong surface water flows with associated soil erosion. Hence the water is spread in space and partly kept back within the forest soil "sponge." This decreases the chances of fast accumulating river water leading to flash floods in steep terrain, or "carpet" flooding within lowland basins. Furthermore, tropical trees are characterized by high transpiration rates and thus act to decrease the soil pore water pressure. This provides absorption space for water in another rainfall event, and on steep slopes it diminishes the risk of landsliding. Surface landslides are also prevented by a dense network of tree roots.

If the forests are cut, rainfalls will impact directly on the unprotected and loose soils. The associated soil erosion and landsliding will not only preclude fast and effective rain forest regeneration, but it will significantly diminish the water absorption capacity during future rainfalls. Hence economic losses through disaster events will increase, with diminishing chances for remediation (Cochard, 2013). While the impacts of landslides can potentially be considerable in densely inhabited areas on steep terrain (cf. Cochard, 2013), it is difficult to quantify the overall economic damages caused by landslides in Aceh. In any case, it may be expected that the overall direct damages would be relatively minor as compared to floods.

Based on past data of damages, van Beukering et al. (2009) estimated amounts of potential flooding damages. Considered in the reverse, "avoided flood damages" (flood prevention) have an economic value. According to the models the values of flood prevention increased under both scenarios due to economic growth, thus increasing risk exposure of valuable assets. In the conservation scenario the increase in the value of flood prevention was, however, significantly steeper (average of US$105 million) than in the deforestation scenario (average of US$90 million) (Table 13.1; van Beukering et al., 2009).

13.4.1.5 Cultural Ecosystem Services of Forests

Ecosystems provide many nonmaterial (or indirect material) benefits to people. They provide space and environments for spiritual inspiration and recreation, reflection, and aesthetic experiences. Cultural ecosystem services (CES) thus include values related to cultural experiences and transmission (eg, motifs for legends, stories, paintings, national symbols, folklore, etc.), spiritual or historical experiences (religious inspiration and endowment, historical identification, etc.), recreational experiences (ecotourism, bird-watching, outdoor sports, other recreation, etc.), and science and education (scientific discovery and reference, intellectual development and school excursions, inspiration for technical innovation, etc.). The values of forest ecosystems for the culture of the Acehnese people are difficult to quantify in economic terms in the virtual absence of any studies. Such cultural values are certainly not unimportant, however, as Acehnese culture is strongly linked to its diverse natural environment. According to De Koninck et al. (2012), the name "Aceh" may actually be derived from the local name of a valuable and revered forest tree species. In the following, two CES are described that are most relevant in Aceh and whose values can be approximated in relatively straightforward economic terms.

TOURISM

Local and international tourism can provide important incentives for natural conservation, especially if local communities can gain tangible benefits from tourism. At the same time, unplanned development of tourism may also entail certain threats to the forests, especially if large facilities are constructed and/or dense networks of tracks and roads are expanded into prime conservation zones. Small-scale ecotourism can be of great benefit to local communities, and it can be one of the best forms of nonconsumptive uses of natural areas, if well planned and integrated into a broader conservation strategy (cf. Gössling, 1999; Stem et al., 2003; Ormsby and Mannle, 2006). Ecotourism may also be linked to education and research programs, involving both the tourists (who may actually be interested university students from foreign countries) and the locals (who may find new opportunities to revive old knowledge on the uses of natural resources) (cf. Brightsmith et al., 2008; Mbaiwa et al., 2011). Ecotourism is expected to continue growing, and may within a few years amount to revenues of more than US$470 billion annually, or 25% of the international travel market (International Ecotourism Society in van Beukering et al., 2009). On the other hand, tourism in the upper segments, if well planned and rigidly managed according to principles of sustainability, may generate important revenues for effective park management (eg, patrolling to prevent poaching and other incursion into the park). For example, various famous conservation areas in Africa, such as the Okavango Delta in Botswana or the Serengeti National Park in Tanzania maintain resorts and services for well-paying international tourists, which contribute to the national economies but also entail various social equity issues (cf. Thirgood et al., 2008; Mbaiwa, 2011).

During the 1990s up to >20,000 visitors were recorded at the eastern Bukit Lawang entrance to Leuser National Park, but in 1998 the visits decreased to less than 8000 due to the rise in political unrest in Aceh (van Beukering et al., 2003). Bukit Lawang is actually situated in the North Sumatra Province. The main attractions of the park are, however, situated in Aceh Province at the center (eg, treks along the Mamas River, or to Gunung Kemiri and Gunung Leuser; river rafting in the Alas River, etc.) and in the west (eg, treks through the peat swamp forests of Kluet, with highest population densities of orangutans) (cf. ver Berkmoes et al., 2013). Hence, now that peace has returned to the province tourism may be developed to become a substantial source of revenue for communities around Leuser as well as in many other spectacular forest areas of Aceh (van Schaik in van Beukering et al., 2003). According to van Beukering et al. (2009), there is evidence that tourism is starting to bloom in Aceh, with increasingly diverse offers by local travel agents in regards to ecotourism, trekking tours, and wildlife viewing. Under the conservation scenario, their model projections therefore depict a marked increase of annual revenues from US$3 million in 2008 to US$20 million in 2038 (average of US$9.0 million). In contrast, if the forests would be cleared, tourism would be expected to decrease to only US$0.3 million in 2038 (average of US$1.2 million; Table 13.1).

BIODIVERSITY

The term "biodiversity" was introduced and advocated by conservation biologists during the 1980s. Technically, biodiversity can be considered in various groupings and at different levels of organisms' variation (from genetic diversity to the diversity of ecological populations), but ecologists commonly equate it with "species diversity" (either described by simple "richness" of [taxonomic] species or by several "biodiversity indicators" such as the Shannon-Weiner index or the Simpson index) (Cochard, 2011b). Biodiversity is important to maintain the functioning of ecosystems. Each and every species may be expected to play a certain function of greater or of minor importance within species communities. Species extinctions (either complete or just functionally), and hence losses of "biodiversity," often lead to changes in ecosystems with sometimes dramatic effects. For example, the extinction of species such as the Sumatran rhinoceros may also lead to the decline (and potential extinction) of various fruit tree species that depend on rhinos for seed dispersal. In contrast, the extinction of the Sumatran tiger may lead to unchecked population increases of certain ungulate species that may, in turn, alter the ecosystem via excessive browsing of saplings of various tree species. Innumerable examples could be listed of important ecological roles played by any other species or groups of organisms, some of which may be microscopic in size but nonetheless very powerful in numbers (Cochard, 2011b). Biodiversity (respectively the species that form its basis) thus has an economic value as it provides essential SES (cf. Section 13.4.1.2). Especially in the case of such complex ecosystems as tropical rain forests; this value is, however, currently still virtually impossible to quantify.

Biodiversity also provides important CES. In this context "biodiversity" may not be a technically accurate term, because human perceptions and value systems tend to be biased in various ways. For example, species such as the tiger and the rhinoceros are certainly much more recognized and valued by the public than other potentially important ecosystem components such as certain species of termites or spiders. This is why conservation agencies often use large, characteristic mammalian species as "flagships" to promote the protection of highly biodiverse ecosystems. On the other hand, in the public and media "biodiversity" is nowadays seen as a synonym for "natural richness" per se. This somewhat more abstract "cultural" conception of biodiversity (respectively its inherent "flagship" species) does have an economic value if somebody is willing to pay for its conservation or promotion.

Biodiversity is also seen as a pool of genetic and molecular resources that can potentially be tapped for economic benefit by industries such as pharmaceutical firms; for agricultural breeding; as a genetic source of domestic, zoo, and pet animals or plants of botanical gardens; or as a source of scientific wonder and technical inspiration. While this view on biodiversity may ultimately be a utilitarian view that essentially seeks to obtain PES (eg, NTFPs or derived technical products), it does not necessarily need to be extractive in a detrimental sense. Scientific research tends to be seen as a continuous path of progress, whereby many new insights may be gained, and new biological products may be obtained in the future via observation and further discovery in "pristine" nature. This scientific view therefore values the existence and conservation of biodiversity as a repository of almost limitless information and as yet undiscovered biological treasures. In a certain sense it transcends and also taps into "cultural" views, traditions, and knowledge. Bioprospecting for pharmaceutically useful plant products, for example, often takes guidance from ethnobotanical studies conducted in native rain forest cultures (with questions sometimes arising in regards to "intellectual property"; cf. Soejarto et al., 2005). Hence, losses of biodiversity represent a degradation of this repository and a decrease of ecosystem services for scientific prospecting.

To weigh the values of biodiversity, van Beukering et al. (2009) considered the values of biodiversity in terms of potential benefits for bioprospecting by pharmaceutical firms, in terms of future funding for conservation programs from governmental and nongovernmental sources, and in terms of tourists' willingness to pay (WTP) for the conservation of biodiversity. Under a conservation scenario the biodiversity value of the area can be expected to significantly increase in coming decades, because most remaining forests in Southeast Asia continue to be degraded and destroyed, with foreseeable losses of species and diversity. Under the conservation scenario model projections by van Beukering et al. (2009) indeed depict a marked increase of values from annually US$10 million in 2008 to US$108 million in 2038 (average of US$39.6 million), whereby the steep increase is mainly due to an expected growth in the pharmaceutical potential of the forest. In contrast, if the forests would be cleared, van Beukering et al. (2009) would predict a decrease to only US$3 million in 2038 (average of US$6.2 million; Table 13.1).

13.4.1.6 Summary of Forest Values and Discussion

The analysis by van Beukering et al. (2009) showed that deforestation would generate higher socio-economic benefits during around 12 years from 2008 until 2020 (at more or less constant benefits of around US$750 million) than forest conservation (at benefits steadily increasing from US$500 to 700 million). This would be due to high benefits from timber and NTFP extraction, and still-limited negative effects in terms of SES, RES and CES. After 2020, however, the severe damages to the forests and resultant declines of all ecosystem services would weigh in and annual benefits would steeply decline to less than US$300 million after another 16 years (until 2036). In contrast, under the conservation scenario the annual benefits provided by the forests would have increased to over US$800 million.

The assessment is dependent on the discount rate that is used in the analysis. The values shown in Table 13.1 are derived from an assessment with a zero discount rate. From an ecological viewpoint a zero discount rate makes most sense, as the damages induced to the ecosystems will in most cases be very long-term, and in some cases (eg, the extinction of important species) forever. Within the time frame of 30 years considered by van Beukering et al. (2009) the conservation scenario would provide benefits that would be around 17% higher (ie, US$694 million annually) as compared to the deforestation scenario (ie, US$573 million annually). The difference would, however, continuously increase with time, and—predictably—later generations will have to deal with increasingly serious environmental problems as well as declining incomes and livelihood options. Some economists may, nonetheless, argue that high discount rates should be applied as revenues generated from the forests can help in "development" and in overcoming "capital scarcity" (see Table 13.2 for the effect of applying different discount rates). Separate from the issues of serious long-term environmental degradation (and associated costs) it can, however, be questioned whether such revenues would and could be used efficiently and invested sustainably under the prevailing sociopolitical conditions (cf. Section 13.6). It seems more likely that the socioeconomic benefits for the communities of Aceh would slowly evaporate along with the disappearance of the forests.

TABLE 13.2   Changes in the "Value Difference" in Relation to Which Annual Discount Rates Are Used in the Assessment by van Beukering et al. (2009)

Scenario		Annual Discount Rates in Percentage			
		0%	5%	10%	15%
Conservation	US$ mil.	694	358	220	155
Deforestation	US$ mil.	573	334	226	169
Value difference	%	17%	7%	−3%	−9%

The studies by van Beukering et al. (2003, 2009) may be criticized in various respects. Even though the theoretical framework and modeling was rather comprehensive, van Beukering et al. (2003) acknowledged that various assumptions were based on relatively weak data baselines and insufficient understanding of potential ecosystem changes. They stated that their models assumed that there were no sharp thresholds, discontinuities or irreversibilities in the ecosystem response function; however, noting (p. 51) that "the crude estimate we have assembled is a useful starting point for further research." Given the assets at stake, evidently further studies should follow to obtain more detailed information about the forests' extent in relation to terrain, and to gain a better understanding of the forests' functions and ecological stability under the prevailing conditions. Baseline economic models should then be revised and refined according to new research findings. Further uncertainties in the models include other aspects (some of which were examined by van Beukering et al. (2003, 2009) through sensitivity analyses), such as the growth and development of the population of Aceh, economic changes and volatility at various levels, and political issues inside of Aceh and Indonesia (eg, concerning the stability and consolidation of the complex democratic processes, including the mediation of past and enduring grievances, cf. Section 13.2), as well as outside of Aceh (eg, the question whether or not markets for carbon trading will be maintained and well managed).

13.4.2 Valuation of Coastal Ecosystems

13.4.2.1 Coastal Ecosystem Services and Their Valuation

In contrast to the forest resources, an attempt at valuation of coastal ecosystems has so far not been undertaken in Aceh. Some available data and literature were compiled, and various aspects of coastal ecosystem functions and services were discussed by Cochard (2011a). Mangrove resources are probably most important along the northern coastlines where extensive mangrove stands can still be found. Even the smaller, confined mangrove areas may, however, be highly valuable for communities along the southern coasts and in the tambak (shrimp and fish farm production) areas around Banda Aceh (cf. Griffin et al., 2013; Cochard, 2011a). Coral reefs are mostly found around the islands where they provide various valuable services to communities. There are many uncertainties and data limitations regarding the valuation (TEV) of coastal ecosystems, especially in Aceh, but comprehensive studies were conducted in other regions of Southeast Asia on mangroves, estuarine systems, and coral reefs (see Barbier et al., 2011; Barbier, 2012; Brander et al., 2012; Sarkis et al., 2013; Laurans et al., 2013; Pet-Soede et al., 1999). In the following, major coastal ecosystem services are briefly summarized.

13.4.2.2 Supporting Ecosystem Services of Coastal Ecosystems

There exist various biophysical interlinkages among coastal ecosystems (cf. Cochard, 2011a). In some areas mangroves may be protected from the impact of sea waves through the presence of frontal coral reef flats, or the presence of sand bars that are stabilized by beach vegetation. On the other hand, mangroves act as filter systems that trap sediment loads, thus protecting some coral reefs and/or sea grass beds from excessive smothering with sediments. Furthermore, various organisms, such as fishes and crustaceans, may move between different coastal habitats during various life stages. Some of these organisms may be important to maintain the ecological balance in one or the other of the habitats. Hence the destruction or partial degradation of one ecosystem (by humans or natural disasters) may also lead to impacts and changes in other nearby ecosystems. The tsunami-affected coastlines of Aceh would have provided valuable scientific benchmark sites to obtain a better understanding of coastal SES in Aceh and elsewhere, but few sufficiently detailed surveys have been conducted that may be useful in this regard (cf. Cochard, 2011a; Griffin et al., 2013).

13.4.2.3 Provisioning Ecosystem Services of Coastal Ecosystems

Coastal ecosystems provide various products which are (directly or indirectly) of economic value for local communities or for other stakeholders. Such products include foods (fishes and other seafood), raw materials (timber from mangroves, and leaves for roof thatch from *Nypa* palms and *Pandanus* spp.), medicinal products (from mangrove plant saps and other sources, including synthetically extracted pharmaceutical chemicals from various marine life forms), ornamental resources (shells, pearls, etc.), energy (biomass fuels and alcohol from *Nypa* palms), and genetic resources. The value of Southeast Asia's fisheries in coral reefs alone is estimated to be around US$2.4 billion per year (Burke et al., 2002). Estimations of the value of mangroves have ranged from around US$3000 up to US$60,000 per hectare per year (cf. literature in Cochard, 2011a).

13.4.2.4 Regulating Ecosystem Services of Coastal Ecosystems

Mangrove ecosystems bind large amounts of carbon in situ and through export to other coastal waters (Alongi, 2009), and significant amounts of carbon are also sequestered in extensive sea grass beds (Fourqurean et al., 2012). Coastal ecosystems provide a certain degree of protection against the impacts of sea wave hazards such as large wind-driven waves, storm surges, and violent flooding through tsunamis (Cochard et al., 2008; Cochard, 2011c; Laso Bayas et al., 2011; Das and Vincent, 2009; Barbier and Enchelmeyer, 2014). Most importantly, coastal ecosystems provide relative stability to coastlines even under heavy wave impacts (Thampanya et al., 2006; Cochard, 2013). Mangroves also act as sediment filters, and to some degree as filters for other pollutants (eg, household wastewater nutrients, heavy metals, etc.), and in certain places they are thus important to protect sea grass beds and coral reefs from excessive pollution impacts (Cochard, 2016; Wickramasinghe et al., 2009).

13.4.2.5 Cultural Ecosystem Services of Coastal Ecosystems

Like forests, coastal ecosystems provide many nonmaterial benefits to people. Especially the species-rich coral reefs of Aceh, and the surrounding waters provide magnificent realms for well-paying international dive tourism. On Weh Island, communities have become involved in artificial coral reef restoration, partly because of incentives from tourism and engaged scientists from overseas (Fadli et al., 2012). The scenic beaches around Calang and other places along the southern coastline have been a travelers' insiders' tip for a long time (ver Berkmoes et al., 2013). Various marine resources around Aceh may also be highly attractive for pharmaceutical bioprospecting.

13.5 ENVIRONMENTAL DEGRADATION: IMPORTANT FIELDS FOR ACTIVE ENGAGEMENT IN RESEARCH AND MANAGEMENT

13.5.1 Restoring and Managing Coastal Ecosystems: Learning from Nature

Following the 2004 Asian tsunami, large amounts of funds were made available for the rehabilitation of coastal communities. This included funding to agencies concerned with the ecological restoration of coastal ecosystems such as mangroves and coral reefs. Partly because of the difficult political situation in Aceh, and partly because of other challenges brought about by the tsunami catastrophe, only a few academic studies concerned with ecology, ecosystem, and socioeconomic dynamics were initiated and conducted in Aceh (cf. Buranakul et al., 2005). Hence, until today the distribution, state, role, and the utility of coastal ecosystems are still insufficiently documented and understood. Despite this, environmental agencies have taken action to restore degraded areas even in the absence of proper surveying and planning. Unfortunately, the outcome was often wasteful, as mangrove restoration resulted in failure (cf. Wibisono and Suryadiputra, 2006). According to Griffin et al. (2013), 68% of mangroves did not regenerate in the areas where they existed before the tsunami.

Some potentially relevant research questions in regard to the failures of mangrove plantations were outlined by Cochard (2011a). Coseismic land subsidence was probably a major factor precluding mangrove restoration at many locations along the southwestern coastlines. Other factors may have been increased wave exposures due to the demise of sand bars and vegetation wind barriers, mangrove peat collapse after the destruction of the vegetation, local extinction of key faunal components in mangroves (eg, burrowing crab species), and marine sediment deposition, changing the nature of the substratum. Before large-scale replantation schemes were put in place, there should have been detailed ecological feasibility studies, including experiments and mapping of actual and potentially suitable areas. These studies should have included the study of natural regeneration processes. Such processes were apparently nontrivial and appeared to proceed along a successional sequence (ie, as hypothesized; cf. Cochard, 2011a): (1) pioneer colonization by mangrove ferns and palms, (2) facilitation of seedling establishment of other true mangrove species, and (3) subsequent steps to the original, closed mangrove forest. Even if certain opportunities for insightful studies may already have been missed, ecological studies could still provide very useful information regarding the reestablishment and long-term sustainable management of mangroves and other coastal resources. Interesting and relevant questions involve the uses of coastal resources by local communities and the utility of ecosystems (eg, mangroves) in regard to the productivity of key resource components (eg, fisheries). Several collaborative studies in recent years provide useful baselines (eg, Fadli et al., 2012; Campbell et al., 2012; Griffin et al., 2013), but more efforts in education and research in coastal ecology and resources management should be undertaken.

13.5.2 Saving and Restoring the Remaining Peat Swamp Forests

13.5.2.1 *The Important Functions and Values of Aceh's Peat Swamp Forest Ecosystems*

The plains between the towns of Meulaboh and Singkil in western Aceh Province still comprise some of the best stands of coastal lowland forest remaining in Sumatra (Fig. 13.2). In particular the remaining peat swamp and freshwater swamp forests of Tripa (ca. 150 km² of remaining primary forest, not formally protected), Kluet (ca. 180 km² largely protected within the Leuser National Park) and Singkil (ca. 1000 km² protected within the Rawa Singkil Wildlife Preserve), as well as a few remaining forests close to Meulaboh (cover unknown) are of high conservation value, for several reasons.

First, the peat swamp forests are important for the mitigation of inland floods and coastal disasters. The forests act as overflow areas in flooding periods. Peat swamps have a great potential to absorb and store floodwaters, slowly releasing the caught waters in the dry periods and thus maintaining a steady base flow in rivers (Cochard, 2013). In some rural areas, peat swamps may be the only freshwater source for several months. In the lowlands of southeastern Sumatra, previously unknown droughts are directly or indirectly traceable to regional wetland destruction since the 1990s (Wösten et al., 2006). The forests are also important for coastal protection as they mitigate against the risks of salt water intrusion, land subsidence and relative sea level rise, and coastal flooding (eg, from storm surges or tsunamis) (Cochard, 2011a; Cochard et al., 2008).

Second, the peat swamp forests act as a major carbon repository and thus play an important role for mitigating global climate change. In conditions characterized by high rainfall, low drainage, permanent waterlogging, and substrate acidification, plant materials accumulate faster than they decay (Brady, 1997). In Southeast Asia peat soils developed in coastal floodplains as early as 30,000 BP (Page et al., 2004), and peat soils up to >20 m deep now constitute one of the largest carbon stores per unit area of any ecosystem worldwide (Phillips, 1998). Tropical peat swamp forests may on average comprise about 200 tons of carbon per hectare in the standing tree biomass and about 2500 t C ha^{-1} in the peat soil (for a median depth of 5 m); this is about nine times as much carbon as is stored in a tropical rain forest standing on mineral soils (Diemont et al., 1997). The average estimated rate of carbon accumulation in pristine peat swamps in Indonesia is between 0.4 and 1.9 tons per hectare annually (Immirzi and Maltby in Rieley et al., 1997; Sorensen, 1993).

Third, peat swamp forests harbor a great biodiversity, much of which remains undescribed. Vegetation mosaics and the catena from mangrove forests to "kerapah climax" forest communities on deep peats have been described in several regions (see Bruenig, 1990; Laumonier, 1997; Page et al., 1999), but no detailed study is known from Acehnese swamps, which are probably floristically quite distinct (Prof. C. van Schaik, personal communication). Swamp forests occupy a specific place in Southeast Asian landscapes and represent a unique and diverse biome with highly specialized, often endemic plants: over 1700 different species were identified in each of two surveys in Borneo and the Malay Peninsula (Yamada, 1997), and over 300 peat swamp tree species have been recorded in eastern Sumatra (Giesen, 2004).

Fourth, peat swamp forests provide various products for livelihoods of local communities and products of marketable value. A large number of valuable timber species (especially ramin, *Gonystylus* spp.) is often found in peat swamp forests, as well as other plant products for food, medicinal uses, and construction (Giesen, 2004; Kuniyasu, 2002; Smulders et al.,

2008). The swamps are also characterized by a high diversity of edible fishes (Rijksen et al., 1997; Palis, 2000). Socioeconomic studies indicate that people of local communities may depend for over 80% on the peat swamp forest resources for their livelihoods (Wösten et al., 2006; Kuniyasu, 2002; Ramakrishna, 2005; Limin and Putir, 2000).

Fifth, the peat swamp forests in the lowlands of western Aceh provide one of the last retreats for some critically endangered Asian mammal and bird species. Possibly more than 40% of the remaining populations of Sumatran orangutan (*Pongo abelii* Lesson; ~7500 animals) live in the coastal swamp forests of Aceh. The highest densities of orangutan were recorded in the Kluet swamps, which are part of the West Leuser lowlands (ie, largest contiguous population, ~2500 animals). The second largest population occurs entirely within the Singkil swamps (~1500 animals). Probably similar population numbers were found at Tripa before the deforestation and degradation of large parts of these swamps. Last estimates at Tripa were at <300 animals with a still decreasing population (van Schaik et al., 2001; Wich et al., 2003; UNEP, 2007). Other highly endangered animals found in the swamp forests include the Sumatran tiger (~250 animals in Sumatra), Asian tapir, otter civet, siamang, Storm's stork, masked finfoot, white-winged wood duck, several hornbills, and the freshwater crocodile (Veron et al., 2006; Lambert and Collar, 2002; Sunarto et al., 2013; Dudgeon, 2000; Rijksen et al., 1997).

13.5.2.2 *Peat Going Up in Smoke for Palm Oil: A Wasteful and Obscure "Business"*

If peat swamp forestlands are cleared they become a net source of carbon, with oxidation being accelerated by active water drainage and fires. The massive peat and forest fires in 1997 in Borneo and Sumatra released an estimated 0.81–2.57 gigatons of carbon, which was equivalent to 13–40% of global emissions from fossil fuels in that year (Page et al., 2002). Overall emissions were likely to be in the upper range. A study in Malaysia by Wösten et al. (1997) found that on average 7.2 tons of carbon were released per hectare annually from drained peatlands even in the absence of fire. This lead to continued land subsidence of 2 cm annually. In addition to carbon release, the destruction of peatlands leads to significantly higher emissions of other greenhouse gases, such as methane (CH_4) and nitrous oxide (N_2O) (Inubushi et al., 2003; Takakai et al., 2006; Hadi et al., 2005). Also, the local climate is affected, as forests act as wind break and absorb heat, which in turn attracts more rains (Wösten et al., 2006; Rijksen et al., 1997; Cochard, 2013).

All the swamp forests in the coastal lowlands of western Aceh have been affected by illegal selective logging, but the Kluet swamps still appear to be the most intact. The Singkil swamps are increasingly under pressure. According to estimates by the Leuser Foundation, about 2% (2000 ha) of the protected area has been completely logged and partly converted to oil palm plantations. Damages to the forests are rapidly increasing as new roads are being built through the forests, facilitating illegal logging (Bell, 2014a; Rijksen et al., 1997). It has been estimated that damages from wood extraction has already occurred in around 40% of the forest area. A problem lies with contradictory definitions of the borders of the protected area, and with capacities for the enforcement of protection (Bell, 2014a). Another problem, however, may be that natural areas are still viewed by powerful agents as a frontier for economic development (cf. McCarthy, 2006; cf. Section 13.6). If an area is logged and converted to agricultural use, it is (under the current circumstances) virtually impossible to reclaim the area and restore it to natural vegetation, which means that economic exploitation may be sanctioned de facto.

The Tripa swamps that were originally not protected (listed as "other uses [non-forest] area" Tata et al., 2014) are by far the worst affected: since the beginning of the 1990s about 70% of swamp forest lands have been logged, partially drained, and planted with oil palms by several palm estates of partly foreign ownership. With continuing peat oxidation and subsidence, lowering of the water level, and soil nutrient losses, the palm plantations at Tripa are becoming increasingly less viable (Rijksen et al., 1997). Peat swamp areas are generally not suitable for conversion into croplands, especially where the peat is deeper than 1–2 m (Rieley et al., 1997; Sorensen, 1993; Phillips, 1998). Most peat soils at Tripa appeared to be deeper than 5 m (Fig. 13.3G), and may be deeper than 8 m in many parts (Rijksen et al., 1997). For these reasons, only about half of the cleared areas had actually been planted with palms until 2006. As the Acehnese civil war reignited in the post–Suharto period, forest clearance stopped, allowing many unused logged areas to be recolonized by secondary bush and forest vegetation (Tata et al., 2014). In the wake of the 2005 Aceh peace accord, renewed exploitation occurred as a result of a lack of planning, coordination, and legal controls. Hence, pressures on the swamp ecosystems increased again (Fig. 13.3H).

13.5.2.3 Saving Tripa: Points for Action

As stated by Rijksen et al. (1997, p. 361), the swamps may, if converted to croplands, become "a deplorable sink of investments" rather than being a "carbon sink." In contrast, mechanisms of carbon trading are gaining interest and may provide strong economic incentives for peat conservation in the future (Tata et al., 2014). Under certain management regimes even the extraction of wood and peat resources can be conducted in sustainable ways - much in contrast to complete forest conversion to croplands (cf. Hahn-Schilling, 1994; Giesen, 2004; Prasodjo and Mukarwoto, 1997). The fewer tropical forests remain, the more such resources would be valued. The very particular value of the coastal swamp forests in southwestern Aceh, however, mainly lies by their "being an integral part of the unsurpassed Leuser Ecosystem, harboring a unique biodiversity and providing essential habitat, while supporting the living conditions of the surrounding population" (Rijksen et al., 1997, p. 361). This functional ecosystem integrity cannot be restored, once it is lost, and it will be increasingly valued by local communities as well as the world community at large.

Several environmental agencies, as well as local communities, have been struggling to achieve better protection of the remaining intact forests at Tripa, despite the palm oil concessions that were previously granted. In May 2011 parts of Tripa were included in a national moratorium on the issuance of new logging permits, extended until 2015 (Tata et al., 2014). Nonetheless, a logging permit was signed for a specific tract by the governor of Aceh. This became a test case of how different authorities and government levels in Indonesia interact with regard to forest resource issues. This specific legal case was finally won for continued conservation, albeit with mixed assessments (Tata et al., 2014; Butler, 2014a,b). Whether or not the remaining forests at Tripa will be saved for good, and degraded parts may even be restored, will probably be decided after 2015.

Proposals for integrated conservation and sustainable development have, for example, been prepared by the PanEco Foundation and its Indonesian partner Yayasan Ekosistem Lestari (Foundation for a Sustainable Ecosystem) (YEL). Such proposals included the following points (cf. www.paneco.ch): (1) the development and/or improvement of legal frameworks for protection and management, including delineation of conservation boundaries and use zones, and the strengthening of legal enforcements; (2) preparation of detailed economic assessments of the value and potential uses of the ecosystems, including consideration of implementing schemes for issuing carbon certificates, marketing of traditionally used plant and fishery resources, and development of ecotourism; (3) promotion of alternative economic schemes outside of peat swamp areas, such as horticulture gardens, *Spirulina* algae culture propagation, roof tile manufactures, and so forth; (4) rehabilitation of degraded fallow land on mineral soils as compensation for relinquished palm oil concessions in peatlands; and (5) implementation of active schemes of swamp rehabilitation, that is, reinstating original soil water levels and replanting of cleared sites with appropriate forest tree seedlings.

There exist little scientific data at present, and no systematic surveys of these ecosystems have ever been conducted (cf. Rijksen et al., 1997; a vegetation survey was conducted at the Kluet swamps; Purwaningsih and Yusuf, 2000). There is thus a vital need to collect baseline information on ecosystem status and functioning as an essential basis for conservation planning and management. Specifically, spatial baseline data is needed to (1) prioritize efforts of conservation (eg, location of prime habitats with endangered animals); (2) set up effective, sustainable, and trustworthy new economic schemes (eg, carbon trading); (3) identify suitable land for both rehabilitation of swamp forests, and for improved commercial development (eg, palm plantations); and (4) implement reforestation schemes on an informed basis in ecologically, economically, and socially sound ways. Scientific surveys and experiments also need to be conducted to establish an ecologically sensible and economically viable protocol for forest restoration (cf. Cochard, 2011a).

13.5.3 Forest Conservation under Encroachment: Securing the Survival of Rare Species

13.5.3.1 Notes on the State of Forest Degradation (Forests on Mineral Soils)

The Leuser and the Ulu Masen ecosystems form one of the last remaining great wildernesses in Southeast Asia. The forests of Aceh, however, are increasingly under pressure from encroachment through illegal logging, mining, and wildlife poaching. Before 2004 deforestation in Aceh was minor as compared to other parts of Indonesia, but almost $4000\,km^2$ of primary forests were lost between 2001 and 2013 (Harvenist, 2014). In a report by the United Nations Development Program (UNDP) the Province of Aceh was alleged to be the worst performer when it came to protecting Indonesia's remaining forests. Logging companies often engage in illegal logging operations (and associated bribing and corruption) to evade high charges and administrative hurdles for concession permits (The Jakarta Post, 2013). Interpretations of overall forestry developments, however, seem to differ (with shifting alliances) between communities and between provincial and national governmental agencies, making it difficult to reach consensus among stakeholders for strong legally based action (cf. McCarthy, 2006; cf. Section 13.6).

Of greatest concern to environmental agencies, NGOs, and engaged scientists is the planned delisting of large tracts of forests as "nonforest" areas, followed by logging operations and partial conversion of forests to other uses (The Jakarta Post, 2013; Butler, 2013a,b; Meijaard, 2014a). These plans were made without any proper environmental impact assessment and spatial planning. Many scientists fear that even under selective logging forest integrity and biodiversity will be significantly affected, especially in relatively steep terrain (cf. Burivalova et al., 2014). Other major concerns include the construction of roads crossing through major untouched forest blocks, which is often followed by forest exploitation (selective logging and poaching), settlement, and small-scale forest conversion to croplands (Rini, 2012; Butler, 2013b). Furthermore, the issuing of permits to companies for mining in remote forest areas has been criticized. Mining is often conducted without any environmental safeguards; for example, using mercury and cyanide, which has caused fish kills in several rivers (Bell, 2014b,c).

Current politics and many developments therefore indicate that a scenario of complete forest conservation in Aceh (cf. van Beukering et al., 2009) will be difficult to achieve in the near future. Irrespective of whether or not a turnaround could indeed be achieved in the coming years, it is in any case important to ensure that at the current time key species are not completely lost from the ecosystem. Some of the most threatened species are, however, also some the most iconic species that are a major pull for tourism in the area. As can be read in tourist guides for Indonesia (eg, ver Berkmoes et al., 2013) the Leuser ecosystem is the only place in Southeast Asia where orangutans, tigers, rhinos, and elephants can all still be found in one area. Whether this will remain so for future generations may, however, be decided in the coming years. As in many other regions of Indonesia (cf. Meijaard, 2014b), wildlife poaching is currently rampant in and around the Leuser National Park as the area is increasingly gaining the attention of international animal traffickers. During 2013–2014, park patrols have dismantled almost 300 makeshift traps targeted at large wildlife such as tigers, elephants, and rhinoceroses; and the situation was getting worse (Bell, 2014d). It is not only the large and iconic species that are edging toward extinction. Along with the large mammalian species many smaller species, such as birds and reptiles (eg, for pet trade and as food), are increasingly threatened (cf. Nijman, 2010). Here, a short review of the conservation status of some of the "flagship" species is provided.

13.5.3.2 Poached for Their Horn to the Brink of Extinction: Sumatran Rhinoceroses

The most threatened species is by far the Sumatran rhinoceros (*Dicerorhinus sumatrensis sumatrensis* Fischer). It is classified by the International Union for Conservation of Nature (IUCN) as "critically endangered." Originally found in rain forests throughout Southeast Asia, including parts of India and southern China, there now remain only very small populations on the Malaysian mainland (possibly already extinct) and in Sumatra (~150 individuals). The subspecies in Borneo (*D. s. harrisoni*) went extinct in the wild in 2015, and there currently remain only three individuals surviving in captivity (Hance, 2015). Estimates of the total population range from less than 100 to over 200 individuals, but true numbers are now more likely to be in the lower ranges. The current status of the species in Leuser National Park is obscure (for lack of data and/or restrictions for data publication), but recent literature still puts the largest remaining population (60–80 individuals) within the Leuser ecosystem (Zafir et al., 2011). The species that may roam from lowland swamp forests to montane forests (up to 2000 m a.s.l.) is threatened by habitat loss, but the main threat is posed from poaching, especially for its horn, which is highly priced in Chinese medicine. Some experts believe that the species may now be more endangered than the Javanese rhinoceros, of which only a few dozen remain in Ujung Kulon National Park, Java. The reason is that the remaining populations are highly fragmented and small, so that there are high risks of inbreeding depression, even if poaching threats could be sufficiently mitigated (van Strien et al., 2008). While there are a few individuals in captivity, breeding in captivity has in the past rarely been successful. In any case, significant conservation efforts will need to focus on the species, so that it will not be lost entirely (Leader-Williams, 2013).

13.5.3.3 Poached for Their Bones: Sumatran Tigers

The Sumatran tiger (*Panthera tigris sumatrae* Pocock, only found in Sumatra) is the smallest subspecies of the tiger, and the last representative of the Sundanese tigers (the Javanese and Bali tigers have already become extinct). Like the rhinoceros, the Sumatran tiger occurs from lowland forests up to mountain forests (up to over 3000 m a.s.l.), and like the rhinoceros it is classified as "critically endangered" by IUCN. The case of the tigers is less dramatic, however, since around 200 Sumatran tigers are kept in zoos around the world, and the species has been successfully breeding in captivity. In the wild, nonetheless, the longer-term survival of the Sumatran tiger is uncertain. The population has been declining from an estimated ~1000 individuals in 1978 to between 400 and 700 individuals in 2008 (Borner, 1978; Linkie et al., 2008). The last estimates from 1992 suggest that possibly the highest numbers of tigers may still exist in the Leuser National Park (110–180 individuals) together with the Kerinci Seblat National Park in the south (145 individuals) (Linkie et al., 2008). A recent survey indeed indicated that the entire combined Leuser and Ulu Masen ecosystems covers the highest density of tigers in Sumatra; the region was thus upgraded as a "global priority" region for wild tiger conservation (Wibisono et al., 2011). Previously the main threat for the species' decline has been habitat destruction and fragmentation, and isolation of tiger populations. Poaching (tiger parts mostly for Chinese medicines) and human-wildlife conflicts are, however, becoming an increasingly important threat. According to Shepherd and Magnus (2004), at least 51 tigers were killed annually during 1998–2002, with 76% for trade and 15% due to human-tiger conflicts. In surveys around the island, parts of at least 23 tigers were found for sale (Ng and Nemora, 2007).

13.5.3.4 Poached for Ivory: Sumatran Elephants

The Sumatran elephant (*Elephas maximus sumatranae* Temminck, only found in Sumatra) is the smallest subspecies of the Asian elephant. In 2011, the Sumatran elephant was classified as "critically endangered" (A2c) by IUCN as the population has declined by at least 80% over the last three generations (~75 years). Estimates in 2007 were in the range of 2400–2800 elephants surviving in Sumatra (Soehartono et al. in Gopala et al., 2011). Sumatran elephants mostly occur in lowland rain forest areas (on mineral soils) below 300 m a.s.l., but they may be found up to 1600 m a.s.l. (Rood et al., 2010). The elephants are threatened by habitat loss, degradation, and fragmentation, and by poaching for ivory. Around 70% of potential elephant habitat in Sumatra has been lost during the last 25 years, and many of the remaining forest blocks are smaller than the minimum area (around 250 km²) required to support viable elephant populations (Gopala et al., 2011). Populations within Leuser National Park (which covers mostly upland and peat swamp forests) are minor, but some of the largest remaining elephant populations in Sumatra (recent estimates ~460 individuals; Tribun Network, 2014) are found in the northern Leuser ecosystem and in the Ulu Masen Ranges (Gopala et al., 2011). The elephants are mostly found in the valleys and along forest edges. In areas that are being converted to agriculture this often leads to conflict with humans (Rood et al., 2010). These problems (sometimes leading to targeted poaching of elephants) are increasing in Aceh, but they have not yet led to such critical levels as are observed in other provinces such as Lampung Province in southern Sumatra (cf. Hedges et al., 2005). If appropriate steps for conservation are taken in due time, Aceh may be the one province in Sumatra that could maintain stable and healthy populations of elephants.

13.5.3.5 Losing Habitat and Losing Connection: Sumatran Orangutans

The Sumatran orangutan (*Pongo abelii* Lesson, only found in northern Sumatra) is certainly the most iconic wild animal species in Aceh, if not in all of Indonesia (together with the Bornean orangutan). The word orangutan means "the man of the forest" in Indonesian, and similarities in appearance make the species particularly amiable to humans. It is one of the main reasons why thousands of tourists visit the Gunung Leuser National Park each year, with a good chance to observe orangutans in places such as Bukit Lawang, the Mamas valley, or in the peat swamp forests of Kluet and Singkil. Despite their attractiveness, however, these great apes have become rare in many parts of their former range, and are now listed as "critically endangered" by IUCN. The orangutans mostly occur in lowland rain forest areas below 1000 m a.s.l., whereby their density is particularly high in freshwater swamp and peat swamp forest areas (Singleton et al., 2008; Singleton and van Schaik, 2000). Until the 19th century orangutans were also found in southern Sumatra, but only around 6600 individuals now survive in the province of Aceh and in bordering areas of North Sumatra Province (Nater et al., 2013). The majority are found in the Leuser ecosystem (~79%, thereof ~31% in peat swamp forests, and ~25% within Leuser National Park), and around 11% in the Ulu Masen ecosystem and other forests of Aceh (Singleton et al., 2008). The species is seriously threatened by logging and habitat fragmentation. In contrast to the Bornean orangutan (no predatory tigers are found in Borneo) the Sumatran orangutan is entirely arboreal, and hence orangutan population densities typically plummet (by up to 60%) even under limited selective logging (Hardus et al., 2012; van Schaik and Rao, 1997). Various populations outside of protected areas, especially in the preferred peat swamp forest habitats (cf. Section 13.5.2), have been significantly reduced by wholesale deforestation. Road construction through orangutan habitat is threatening to bisect populations, and animals are hunted (or caught and traded as cage animals) primarily as a result human-wildlife conflicts under the expansion of agricultural areas (orangutans are often perceived as pests of fruit crops) (Singleton et al., 2008). Isolated populations are in danger of becoming extinct due to constrained possibilities for migration and the associated lack of food resources, and due to potential inbreeding depression (Nater et al., 2013; Buij et al., 2002). Efficient protection of the remaining forest habitats and restoration of wildlife corridors will therefore be crucial for the longer-term survival of the species. In addition, efforts need to be undertaken to improve antipoaching patrols and to resolve human-wildlife conflicts in forest border areas (cf. Campbell-Smith et al., 2012; Marchal and Hill, 2009).

13.5.3.6 Approaches and Technical Means for Improved Protection of Endangered Species

As already outlined (Section 13.5.3.1), there is considerable concern that new spatial land management plans will lead to further accelerated degradation of the Leuser and Ulu Masen ecosystems (cf. Meijaard, 2014a). Certainly, clear legislative frameworks for forest protection (in the optimal case sufficiently supported and understood by local communities) would be fundamental for effective and longer-term sustainable conservation. In the past, protection was often hampered due to a shortage of resources combined with the vastness and complexity of the area to be surveyed. New technical advancements could, however, make knowledge-based spatial planning, patrolling, and monitoring significantly more efficient and cost-effective. Already now fairly high-quality geographical information system (GIS)-based data on terrain and forest cover are available to managers and planners of the government of Aceh agencies and to scientists. Data files include information on watershed forest areas, environmental risk, soil types, geological hazards, human population centers, rainfall and—to some measure—the distribution of Aceh's wildlife (Butler, 2013b). Deforestation, respectively severe forest degradation, can now be easily tracked within a few days through high-resolution satellite imagery (cf. Margono et al., 2012), and (if the political will/consensus exists) countermeasures can be taken. For higher-resolution, closer-up surveillance so-called conservation drones have recently been developed. These are relatively cheap and easy to handle, and they can be used to monitor several square kilometers of terrain. Some of these drones have been tested in the Leuser ecosystem, whereby wildlife such as orangutan and elephants could be readily detected (Koh and Wich, 2012). It may be seen how useful such drones could be to counteract poaching. To survey and monitor more cryptic wildlife such as tigers and rhinos, camera traps have been used with some success (cf. Sunarto et al., 2013). Camera traps have also been used to assess the overall impact of poaching in nonprotected versus protected areas (cf. Meijaard, 2014b). Regarding the poaching of timber, it is now technically possible to identify from which species a certain type of wood was derived (cf. Tsumura et al., 2011; Smulders et al., 2008). Timber traders therefore have the means that could allow them to identify and thus boycott illegitimate wood products. Strict applications of such monitoring could help them gain market advantages through maintaining a credible "clean" trading image.

13.6 CLOSING REMARKS: NATURAL ASSETS FOR FUTURE GENERATIONS

In this chapter, past and present environmental issues and developments were described and discussed. Referring largely to the work of van Beukering et al. (2003, 2009), it was outlined that continued intensive logging of the natural forest assets in Aceh will likely be significantly more costly overall to future generations in Aceh, than if a path of "forest conservation" is being taken. Some researchers and decision makers may still view the studies by van Beukering et al. (2003, 2009) critically because of various limitations. The notion that several models are based on a relatively weak database, should however be seen as a motivation to improve this database and refine the models, rather than as any indication a priori that their overall findings are unsound. The overall findings are indeed unlikely to be substantially different even under an improved database. It can be said with high certainty that logging the remaining natural forests will not provide any longer-term net benefits to the people of Aceh. In contrary, it will deprive future generations of valuable development options and bequeath them an environment that will be less productive and more prone to climatic capriciousness. Hence, future generations will also be more threatened by agricultural risks (pests, diseases, droughts) and natural disasters. Such a conclusion is also reflected in several of this author's previously published reviews and surveys, concerned with the roles and functions of tropical forests in Aceh or in similar regions (Cochard, 2011a,b, 2013; Cochard et al., 2008).

This chapter primarily focused on the environmental and resultant socioeconomic effects of deforestation versus forest conservation. Both scenarios ("deforestation" and "conservation") essentially represent the two opposite poles along a continuous gradient of impact levels (from near 0% to near 100% deforestation). The pathways that will be valid options for the Acehnese people will probably largely be determined by sociopolitical and socioeconomic factors and constellations. Describing and discussing the societal changes and politics and the associated "political ecology" of Aceh could be the theme of additional extensive chapters. Probably the most notable and detailed study on the "political ecology" of Aceh's "rain forest frontier" (focusing on areas in southern Aceh adjacent to Leuser National Park) was conducted between 1996 and 1999 by McCarthy (2005, 2006). The study began when the autocratic government under President Suharto was still in power. McCarthy (2006, 2005) outlined the direct and indirect effects of the Asian economic crisis of 1997, which at the national level ultimately contributed to the end of the Suharto era and the transition to a democratic regime in Indonesia. At the study sites in Aceh the 1997 crisis led to volatile fluctuations in the prices of essential commodities (especially cash crops such as nilam/patchouli, *Pogostemon cablin*, a herb used for scent, and pala/nutmeg, *Myristica fragrans*) upon which the villagers depended for income. With the collapse of prices (combined with damages of insect pests to crop plants in those years) many villagers turned to the timber of the adjacent forests to compensate for some of the lost income. This coalesced with a changing political situation that affected the capacity of the state to enforce existing resource management laws. Conditions and the impacts on the forests were largely determined by an interplay of local economic interests (mostly by powerful members of village communities) aided by the advent of timber entrepreneurs, various (sometimes conflicting) interests and power constellations of personnel within the forestry department, and the presence or absence of conservationist actors in a certain area (McCarthy, 2006).

To this author's knowledge no study comparable to that of McCarthy (2005, 2006) in terms of scope and detail has since been conducted in Aceh. It is clear, however, that certain parameters have changed markedly during the last decade. Between 1999 and 2004, logging rates decreased again in several parts of Aceh because of the heightening conflict (cf. Tata et al., 2014). In the years following the 2004 tsunami, however, rates of logging picked up again, partly because of high timber demands for house reconstruction, and partly because of improved business conditions after the cessation of hostilities (cf. Hansen in Meijaard, 2014a; De Koninck et al., 2012). Aceh has since become more integrated in the world economy as international public and private actors have increasingly made investments and directed business operations in the province (cf. Phelps et al., 2011). While this is not necessarily undesirable overall, new problems of resource exploitation have emerged, whereas the conditions and constellations described by McCarthy (2006) may not have changed fundamentally at various administrative levels and for the communities. For example, according to Bell (2014e), several large palm oil companies in Indonesia now follow a strategy to buy oil palm fruits at attractive rates from small holders and local villages who agree to clear forest and plant palm crops in nearby reserves. In this way they provide incentives to poor farmers to encroach on protected forestlands while as a company circumventing existing laws and feigning a clean, green image. In Aceh, peat swamp areas especially have been under pressure from such encroachment (cf. Section 13.5.1). If environmental degradation goes apace through "business as usual," conditions in and around Leuser National Park and in other protected areas may ultimately become similarly intricate as in other parts of Sumatra. In Bukit Barisan Selatan National Park in southern Sumatra, for example, poor landless migrants and local opportunists took advantage of weak law enforcement and settled within the national park. Local politicians now try to build a constituency by backing illegal activities within the park (Levang et al., 2012). Similar constellations have been documented in West Sumatra (Yonariza and Webb, 2007). Despite some disquieting recent signs (cf. Meijaard, 2014a) it may be hoped that decision makers and the Acehnese people will be wise enough to foresee such vicious circles in natural resource mismanagement, and take the necessary countermeasures in terms of land resource spatial planning, community education and partnerships for development, and the capacitation of lawfulness.

In a very personal essay Siapno (2009) described various complexities of the recent Acehnese secessionist conflict from the perspectives of internally displaced persons, especially women. Regarding narratives on "terror and displacement" she noted (p. 50) that "the literature [] is primarily written by government officials, human rights 'experts', rebel forces, scholars (especially anthropologists), or journalists - all of whom, even with good intentions, tend to give very little agency or subjectivity to the displaced persons themselves." She added that she was "so tired [] of articles on 'Living through Terror' that are just about victims being victimized. Victims, at one point in their lives, become more visionary, and get wise. They begin to focus on 'emotional healing' … and begin to train every day to align their body with their mind and soul and spirit, and the universe." While some wounds may remain, history and past events may not necessarily stipulate the future. With peace in place, broader education, better health services, new developments, and improved economic welfare, the Acehnese society may be expected to transform, even reinvent itself on new as well as on old foundations. The wounds of the conflict and of the tsunami disaster will eventually be healed as coming generations will hopefully make their living in peace and improved security. Within this outlook it will also be vital not to create new misadventures by thoughtlessly expending the natural assets of Aceh. Similar to the small country of Costa Rica in Central America, Aceh (as well as other parts of Sumatra) could be a peaceful and self-confident place, with a reputation of environmental stewardship, and unique natural treasures within the region of Southeast Asia.

References

Abram, N.J., Gagan, M.K., McCulloch, M.T., Chappell, J., Hantoro, W.S., 2003. Coral reef death during the 1997 Indian Ocean dipole linked to Indonesian wildfires. Science 301, 952–955.

Afolabi, O.R., Razafinjatovo, C., Dewiyanti, I., 2008. Socio-economic and ecological assessment of forest products harvesting in Central Aceh and Bener Meriah District, Nanggroe Aceh Darussalam, Indonesia. In: The Annual Conference on Tropical and Subtropical Agricultural and Natural Resource Management (TROPENTAG), October 7–9 2008, Hohenheim, Germany. http://www.tropentag.de/2008/abstracts/links/Afolabi_cLRhyrQi.php.

Alongi, D.M., 2009. The Energetics of Mangrove Forests. Springer Verlag, Berlin.

Angelsen, A., Gierløff, C.W., Mendoza Beltran, A., den Elsen, M., 2014. REDD credits in a global carbon market. Options and impacts. TemaNord 2014: 541. Nordic Council of Ministers, Copenhagen.

Aspinall, E., 2009. Islam and Nation: Separatist Rebellion in Aceh, Indonesia. National University of Singapore Press, Singapore.

Baird, A.H., Campbell, S.J., Anggoro, A.W., Ardiwijaya, R.L., Fadli, N., Herdiana, Y., Kartawijaya, T., Mahyiddin, D., Mukminin, A., Pardede, S.T., Pratchett, M.S., Rudi, E., Siregar, A.M., 2005. Acehnese reefs in the wake of the Asian Tsunami. Curr. Biol. 15, 1926–1930.

Barber, C.V., Schweithelm, J., 2000. Trial by Fire: Forest Fires and Forestry Policy in Indonesia's Era of Crisis and Reform. World Resources Institute, Washington, DC.

Barber, A.J., Crow, M.J., Milsom, J.S., 2005. Sumatra: Geology, Resources and Tectonic Evolution. Geological Society Memoir No. 31, Geological Society Publishing House, Bath, UK.

Barbier, E.B., 2012. Progress and challenges in valuing coastal and marine ecosystem services. Rev. Environ. Econ. Policy 6, 1–19.

Barbier, E.B., Enchelmeyer, B.S., 2014. Valuing the storm surge protection service of US Gulf Coast wetlands. J. Environ. Econ. Policy 3, 167–185.

Barbier, E.B., Hacker, S.D., Kennedy, C., Koch, E.W., Stier, A.C., Siliman, B.R., 2011. The value of estuarine and coastal ecosystem services. Ecol. Monogr. 81, 169–193.

BBC, 2005. Indonesia agrees Aceh peace deal. http://news.bbc.co.uk/2/hi/asia-pacific/4690293.stm (accessed September 2014).

BBC, 2014. Singapore Haze Hits Record High From Indonesia Fires. http://www.bbc.com/news/world-asia-22998592 (accessed October 2014).

Bell, L., 2014a. Aceh's largest peat swamp at risk from palm oil. Mongabay. http://news.mongabay.com/2014/0810-lbell-singkil-peat-swamp.html (accessed September 2014).

Bell, L., 2014b. Aceh backtracking on mining moratorium, continue to issue permits. Mongabay. http://news.mongabay.com/2014/0813-lbell-aceh-mining moratorium.html (accessed September 2014).

Bell, L., 2014c. Under pressure over pollution complaints, Aceh calls for closure of gold mines. Mongabay. http://news.mongabay.com/2014/0822-lbell-aceh-illegal-gold-mines.html (accessed September 2014).

Bell, L., 2014d. Poachers target elephants, tigers in Sumatran park. Mongabay. http://news.mongabay.com/2014/0731-sumatra-poaching-lbell.html (accessed September 2014).

Bell, L., 2014e. Companies hire local communities to evade palm oil restrictions in Indonesia. Mongabay. http://news.mongabay.com/2014/1004-lbell-small-palm-oil.html (accessed September 2014).

Blackett, H., Irianto, N., 2007. Forest resources and forest industries in Aceh. FAO forestry program for early rehabilitation in Asian tsunami-affected countries (Contract: OSRO/GLO/502/FIN). Food and Agriculture Organisation of the United Nations, Rome.

Borner, M., 1978. Status and conservation of the Sumatran tiger. Carnivore 1, 97–102.

Brady, M.A., 1997. Effects of vegetation changes on organic matter dynamics in three coastal peat deposits in Sumatra, Indonesia. In: Rieley, J.O., Page, S.E. (Eds.), Biodiversity and Sustainability of Tropical Peatlands. Samara Publishing Limited, Cardigan, UK, pp. 113–134.

Brander, L.M., Wagtendonk, A.J., Hussain, S.S., McVittie, A., Verburg, P.H., de Groot, R.S., van der Ploeg, S., 2012. Ecosystem service values for mangroves in Southeast Asia: a meta-analysis and value transfer application. Ecosyst. Services 1, 62–69.

Brightsmith, D.J., Stronza, A., Holle, K., 2008. Ecotourism, conservation biology, and volunteer tourism: a mutually beneficial triumvirate. Biol. Conserv. 141, 2832–2842.

Bruenig, E.F., 1990. Oligotrophic forested wetlands in Borneo. In: Lugo, A.E., Brinson, M., Brown, S. (Eds.), Forested Wetlands. Ecosystems of the World 15. Elsevier Science Publishers, Amsterdam, pp. 299–334.

Buij, R., Wich, S.A., Lubis, A.H., Sterck, E.H.M., 2002. Seasonal movements in the Sumatran orang-utan (*Pongo pygmaeus abelii*) and consequences for conservation. Biol. Conserv. 107, 83–87.

Buranakul, S., Grundy-Warr, C., Horton, B., Law, L., Rigg, J., Tan-Mullins, M., 2005. The tsunami, academics and academic research. Singap. J. Trop. Geogr. 26, 244–248.

Burivalova, Z., Şekercioğlu, C.H., Koh, L.P., 2014. Thresholds of logging intensity to maintain tropical forest biodiversity. Curr. Biol. 24, 1893–1898.

Burke, L., Selig, L., Spalding, M., 2002. Research report: Reefs at risk in South Asia. World Resources Institute, Washington, DC. http://marine.wri.org/pubs_description.cfm?PubID=3144.

Butler, R.A., 2013a. Indonesian governor proposes opening protected areas to logging. Mongabay. http://news.mongabay.com/2013/0212-aceh-spatial-plan.html (accessed September 2014).

Butler, R.A., 2013b. Conservation scientists: Aceh's spatial plan a risk to forests, wildlife, and people. Mongabay. http://news.mongabay.com/2013/0322-atbc-aceh-declaration.html (accessed September 2014).

Butler, R.A., 2014a. In precedent-setting case, palm oil company fined $30M for destroying orangutan forest. Mongabay. http://news.mongabay.com/2014/0109-aceh-tripa-court-decision.html (accessed September 2014).

Butler, R.A., 2014b. Environmentalists lament light sentence in Tripa peatland destruction case. Mongabay. http://news.mongabay.com/2014/0509-tripa-punishment-subianto.html (accessed September 2014).

III. LEARNING FROM THE FIELD CASES/ISSUES

Campbell, S.J., Cinner, J.E., Ardiwijaya, R.L., Pardede, S., Kartawijaya, T., Mukmunin, A., Herdiana, Y., Hoey, A.S., Pratchett, M.S., Baird, A.H., 2012. Avoiding conflicts and protecting coral reefs: customary management benefits marine habitats and fish biomass. Oryx 46, 486–494.

Campbell-Smith, G., Sembiring, R., Linkie, M., 2012. Evaluating the effectiveness of human-orangutan conflict mitigation strategies in Sumatra. J. Appl. Ecol. 49, 367–375.

Cochard, R., 2011a. The 2004 tsunami in Aceh and Southern Thailand: coastal ecosystem services, damages and resilience. In: The Tsunami Threat: Research and Technology. InTech Open Access Publisher, Rijeka, Croatia, pp. 179–216. (Chapter 10). http://www.intechopen.com/articles/show/title/the-2004-tsunami-in-aceh-a-southern-thailand-coastal-ecosystem-services-damages-and-resilience.

Cochard, R., 2011b. Consequences of deforestation and climate change on biodiversity. In: Trisurat, Y., Shrestha, R., Alkemade, R. (Eds.), Land use, Climate Change and Biodiversity Modeling: Perspectives and Applications. IGI Global, Hershey, pp. 24–51.

Cochard, R., 2011c. On the strengths and drawbacks of tsunami-buffer forests. Proc. Natl. Acad. Sci. U. S. A. 108, 18571–18572.

Cochard, R., 2013. Natural hazards mitigation services of carbon-rich ecosystems. In: Lal, R., Lorenz, K., Hüttl, R.F., Schneider, B.U., von Braun, J. (Eds.), Ecosystem Services and Carbon Sequestration in the Biosphere. Springer, Heidelberg, pp. 221–293.

Cochard, R., 2016. Coastal water pollution and its potential mitigation by vegetated wetlands – an overview of issues in Southeast Asia. In: Shivakoti, G., Pradhan, U., Helmi (Eds.), Redefining Diversity and Dynamics of Natural Resources Management in Asia. Elsevier, Amsterdam. (Chapter 12), this volume.

Cochard, R., Ranamukharachchi, S.L., Shivakoti, G., Shipin, O., Edwards, P.J., Seeland, K.T., 2008. The 2004 tsunami in Aceh and Southern Thailand: a review on coastal ecosystems, wave hazards and vulnerability. Perspect. Plant Ecol. Evol. Syst. 10, 3–40.

Currie, D.J., 1991. Energy and large-scale patterns of animal and plant species richness. Am. Nat. 137, 27–49.

Das, S., Vincent, J.R., 2009. Mangroves protected villages and reduced death toll during Indian super cyclone. Proc. Natl. Acad. Sci. U. S. A. 106, 7357–7360.

De Koninck, R., Bernard, S., Girard, M., 2012. Aceh's forests as an asset for reconstruction? In: Daly, P., Feener, R.M., Reid, A. (Eds.), From the Ground up. Perspectives on Post-Tsunami and Post-Conflict Aceh. Institute of Southeast Asian Studies, Singapore, pp. 156–179.

Diemont, W.H., Nabuurs, G.J., Rieley, J.O., Rijksen, H.D., 1997. Climate change and management of tropical peatlands as a carbon reservoir. In: Rieley, J.O., Page, S.E. (Eds.), Tropical Peatlands. Samara Publishing Limited, Cardigan, UK, pp. 363–368.

Dudgeon, D., 2000. Riverine biodiversity in Asia: a challenge for conservation biology. Hydrobiologia 418, 1–13.

Elliott, S., Brimacombe, J., 1987. The medicinal plants of Gunung Leuser National Park, Indonesia. J. Ethnopharmacol. 19, 285–317.

Fadli, N., Campbell, S.J., Ferguson, K., Keyse, J., Rudi, E., Riedel, A., Baird, A.H., 2012. The role of habitat creation in coral reef conservation: a case study from Aceh, Indonesia. Oryx 46, 501–507.

Foster, R., Hagan, A., Perera, N., Gunawan, C.A., Silaban, I., Yaha, Y., Manuputty, Y., Hazam, I., Hodgson, G., 2006. Tsunami and Earthquake Damage to Coral Reefs of Aceh, Indonesia. Reef Check Foundation, Pacific Palisades, CA. http://www.reefcheck.org/PDFs/reefcheck_aceh_jan2006_web.pdf.

Fourqurean, J.W., Duarte, C.M., Kennedy, H., Marbà, N., Holmer, M., Mateo, M.A., Apostolaki, E.T., Kendrick, G.A., Krause-Jensen, D., McGlathery, K.J., Serrano, O., 2012. Seagrass ecosystems as a globally significant carbon stock. Nat. Geosci. 5, 505–509.

Gaillard, J.C., Clavé, E., Kelman, I., 2008. Wave of peace? Tsunami disaster diplomacy in Aceh, Indonesia. Geoforum 39, 511–526.

Gaston, K.J., 2000. Global patterns in biodiversity. Nature 405, 220–227.

Giesen, W., 2004. Causes of peatswamp forest degradation in Berbak NP, Indonesia, and recommendations for restoration. Arcadis Euroconsult. International Agricultural Centre (IAC), Wageningen University, Netherlands. http://www.waterfoodecosystems.nl/docs/AirHitamLaut/Rehabilitation.PDF.

Gopala, A., Hadian, O., Sunarto, Sitompul, A., Williams, A., Leimgruber, P., Chambliss, S.E., Gunaryadi, D., 2011. *Elephas maximus ssp. sumatranus*. The IUCN Red List of Threatened Species. Version 2014.2. www.iucnredlist.org (accessed October 2014).

Gössling, S., 1999. Ecotourism: a means to safeguard biodiversity and ecosystem functions? Ecol. Econ. 29, 303–320.

Greenpeace, 2014. Sumatra: going up in smoke. New evidence shows peat and forest protection is key to stopping the haze wave. http://www.greenpeace.org/international/Global/international/briefings/forests/2013/Peat-Forest%20Fires_Briefer_May28-2014.pdf (accessed October 2014).

Griffin, C., Ellis, D., Beavis, S., Zoleta-Nantes, D., 2013. Coastal resources, livelihoods and the 2004 Indian Ocean tsunami in Aceh, Indonesia. Ocean Coast. Manag. 71, 176–186.

Guardian Environment Network, 2014. Fires in Indonesia at highest levels since 2013 haze emergency. http://www.theguardian.com/environment/2014/mar/14/fires-indonesia-highest-levels-2012-haze-emergency (accessed August 2014).

Gurevitch, J., Scheiner, S.M., Fox, G.A., 2006. The Ecology of Plants. Sinauer Associates, Inc., Massachusetts.

Hadi, A., Inubushi, K., Furukawa, Y., Purnomo, E., Rasmadi, M., Tsurata, H., 2005. Greenhouse gas emissions from tropical peatlands of Kalimantan, Indonesia. Nutr. Cycl. Agroecosyst. 71, 73–80.

Hahn-Schilling, B., 1994. Struktur, sukzessionale Entwicklung und Bewirtschaftung selektiv genutzter Moorwälder in Malaysia. Göttinger Beiträge zur Land- und Forstwirtschaft in den Tropen und Subtropen, Heft 94. Dissertation, Georg-August-Universität, Göttingen.

Hance, J., 2015. Officials: Sumatran rhino is extinct in the wild in Sabah. http://news.mongabay.com/2015/0423-hance-sumatran-rhino-sabah-extinct.html (accessed May 2015).

Hardus, M.E., Lameira, A.R., Menken, S.B.J., Wich, S.A., 2012. Effects of logging on orangutan behavior. Biol. Conserv. 146, 177–187.

Harvenist, E., 2014. From 'production' forests to protected forests, groups work to save Sumatran orangutan habitat. But will it be enough? Mongabay. http://news.mongabay.com/2014/0916-gfrn-harfenist-sumatra-orang.html (accessed November 2014).

Hedges, S., Tyson, M.J., Sitompul, A.F., Kinnaird, M.F., Gunaryadi, D., Aslan, 2005. Distribution, status, and conservation needs of Asian elephants (*Elephas maximus*) in Lampung Province, Sumatra, Indonesia. Biol. Conserv. 124, 35–48.

Herdiana, Y., Kartawijaya, T., Ardiwijaya, R.L., Setiawan, F., Prasetia, R., Pardede, S.T., Campbell, S.J., 2008. Ecological survey on coral reefs of Simeulue and Banyak Islands, Aceh 2007. Wildlife Conservation Society, Indonesia Marine Program, Bogor, Indonesia.

Hoeksema, B.W., Cleary, D.F.R., 2004. The sudden death of a coral reef. Science 303, 1293.

Houghton, J.T., Ding, Y.D.J.G., Griggs, D.J., Noguer, M., van der Linden, P.J., Dai, X., Maskell, K., Johnson, C.A., 2001. Climate Change 2001: The Scientific Basis, vol. 881. Cambridge University Press, Cambridge.

Inubushi, K., Furukawa, Y., Hadi, A., Pur nomo, E., Tsurata, H., 2003. Seasonal changes of CO_2, CH_4 and N_2O fluxes in relation to land-use change in tropical peatlands located in coastal area of South Kalimantan. Chemosphere 52, 603–608.

Jakarta Post, 2013. Aceh earthquake death toll reaches 35. http://www.thejakartapost.com/news/2013/07/05/aceh-earthquake-death-toll-reaches-35.html (accessed September 2014).

Jandl, R., Schüler, S., Schindlbacher, A., Tomiczek, C., 2013. Forests, carbon pool, and timber production. In: Lal, R., Lorenz, K., Hüttl, R.F., Schneider, B.U., von Braun, J. (Eds.), Ecosystem Services and Carbon Sequestration in the Biosphere. Springer, Heidelberg, pp. 101–130.

Kelman, I., 2012. Disaster Diplomacy. How Disasters Affect Peace and Conflict. Routledge, London.

Koh, L.P., Wich, S.A., 2012. Dawn of drone ecology: low-cost autonomous aerial vehicles for conservation. Trop. Conserv. Sci. 5, 121–132.

Kumar, P. (Ed.), 2010. The Economics of Ecosystems and Biodiversity: Ecological and Economic Foundations. Earthscan, London.

Kuniyasu, M., 2002. Environments and people of Sumatran peat swamp forest II: distribution of villages and interactions between people and forests. Southeast Asian Stud. 40, 87–108.

Lambert, F.R., Collar, N.J., 2002. The future for Sundaic lowland forest birds: long-term effects of commercial logging and fragmentation. Forktail 18, 127–146.

Laso Bayas, J.C., Marohn, C., Dercon, G., Dewi, S., Piepho, H.P., Joshi, L., van Noordwijk, M., Cadisch, G., 2011. Influence of coastal vegetation on the 2004 tsunami wave impact in west Aceh. Proc. Natl. Acad. Sci. U. S. A. 108, 18612–18617.

Laumonier, Y., 1997. The Vegetation and Physiography of Sumatra. Kluwer Academic, Publishers, Dordrecht, Netherlands.

Laurance, W.F., 1997. Biomass collapse in Amazonian forest fragments. Science 278, 1117–1118.

III. LEARNING FROM THE FIELD CASES/ISSUES

Laurans, Y., Pascal, N., Binet, T., Brander, L., Clua, E., David, G., Rojat, D., Seidl, A., 2013. Economic valuation of ecosystem services from coral reefs in the South Pacific: taking stock of recent experience. J. Environ. Manag. 116, 135–144.

Leader-Williams, N., 2013. Fate-riding on their horns – and genes? Oryx 47, 311–312.

Levang, P., Sitorus, S., Gaveau, D., Sunderland, T., 2012. Landless farmers, sly opportunists, and manipulated voters: the squatters of the Bukit Barisan Selatan National Park (Indonesia). Conserv. Soc. 10, 243–255.

Limin, S.H., Putir, P.E., 2000. The massive exploitation of peat swamp forest potentially has not successfully increased the local people's prosperity in Central Kalimantan. In: Iwakuma, T., Inoue, T., Kohyama, T., Osaki, M., Simbolon, H., Tachibana, H., Takahashi, H., Tanaka, N., Yabe, K. (Eds.), Proceedings of the International Symposium on: Tropical Peatlands. Bogor, Indonesia, 22–24 November 1999. Sapporo Editors, Japan, pp. 491–498.

Linkie, M., Wibisono, H.T., Martyr, D.J., Sunarto, S., 2008. *Panthera tigris ssp. sumatrae*. The IUCN Red List of Threatened Species. Version 2014.2. www.iucnredlist.org (accessed September 2014).

Marchal, V., Hill, C., 2009. Primate crop-raiding: a study of local perceptions in four villages in North Sumatra, Indonesia. Primate Conserv. 24, 107–116.

Margono, B.A., Turubanova, S., Zhuravleva, I., Potapov, P., Tyukavina, A., Baccini, A., Goetz, S., Hansen, M.C., 2012. Mapping and monitoring deforestation and forest degradation in Sumatra (Indonesia) using Landsat time series data sets from 1990 to 2010. Environ. Res. Lett. 7, 034010. http://dx.doi.org/10.1088/1748-9326/7/3/034010.

Mbaiwa, J.E., 2011. Hotel companies, poverty and sustainable tourism in the Okavango Delta, Botswana. World J. Entrep. Manag. Sustain. Dev. 7, 47–58.

Mbaiwa, J.E., Stronza, A., Kreuter, U., 2011. From collaboration to conservation: insights from the Okavango Delta, Botswana. Soc. Nat. Resour. 24, 400–411.

McAdoo, B., Dengler, L.D., Prasetya, G., Titov, V., 2006. Smong: how an oral history saved thousands on Indonesia's Simeulue Island during the December 2004 and March 2005 tsunamis. Earthquake Spectra 22 (S3), 661–669.

McArthur, R.H., McArthur, J.W., 1961. On bird species diversity. Ecology 42, 594–598.

McCarthy, J.F., 2005. Between adat and state: institutional arrangements on Sumatra's forest frontier. Hum. Ecol. 33, 57–82.

McCarthy, J.F., 2006. The Fourth Circle: A Political Ecology of Sumatra's Rainforest Frontier. Stanford University Press, Stanford, CA.

Meijaard, E., 2014a. 10 years following tsunami, Aceh aims to create its own, new, and totally preventable disaster. Mongabay. http://news.mongabay.com/2014/1211-meijaard-rps-acehs-pending-disaster.html (accessed December 2014).

Meijaard, E., 2014b. Indonesia's silent wildlife killer: hunting. http://news.mongabay.com/2014/1226-rsp-meijaard-indonesia-hunting.html (accessed December 2014).

Mohibbi, A.A., Cochard, R., 2014. Residents' resource uses and nature conservation in Band-e-Amir National Park, Afghanistan. Environ. Dev. 11, 141–161.

Muchlisin, Z.A., Siti Azizah, M.N., 2010. Diversity and distribution of freshwater fishes in Aceh water, Northern-Sumatra, Indonesia. Int. J. Zool. Res. 6, 166–183.

Nalbant, S.S., Steacy, S., Sieh, K., Natawidjaja, D., McCloskey, J., 2005. Earthquake risk on the Sunda trench. Nature 435, 756–757.

Natawidjaja, D.H., Sieh, K., Chlieh, M., Galetzka, J., Suwargadi, B.W., Cheng, H., Edwards, R.L., Avouac, J., Ward, S.N., 2006. Source parameters of the great Sumatran megathrust earthquakes of 1797 and 1833 inferred from coral microatolls. J. Geophys. Res. 111, B06403.

Nater, A., Arora, N., Greminger, M.P., van Schaik, C.P., Singleton, I., Wich, S.A., Fredriksson, G., Perwitasari-Farajallah, D., Pamungkas, J., Krützen, M., 2013. Marked population structure and recent migration in the critically endangered Sumatran orangutan (*Pongo abelii*). J. Hered. 104 (1), 2–13.

Ng, J., Nemora, 2007. Tiger Trade Revisited in Sumatra, Indonesia. TRAFFIC Southeast Asia, Petaling Jaya, Malaysia.

Nichol, J., 1997. Bioclimatic impacts of the 1994 smoke haze event in Southeast Asia. Atmos. Environ. 31, 1209–1219.

Nijman, V., 2010. An overview of international wildlife trade from Southeast Asia. Biodivers. Conserv. 19, 1101–1114.

Ogden, F.L., Crouch, T.D., Stallard, R.F., Hall, J.S., 2013. Effect of land cover and use on dry season river runoff, runoff efficiency, and peak storm runoff in the seasonal tropics of central Panama. Water Resour. Res. 49, 8443–8462.

Ormsby, A., Mannle, K., 2006. Ecotourism benefits and the role of local guides at Masoala National Park, Madagascar. J. Sustain. Tour. 14, 271–287.

Page, S.E., Rieley, J.O., Shotyk, Ø.W., Weiss, D., 1999. Interdependence of peat and vegetation in a tropical peat swamp forest. In: Newberry, D.M., Clutton-Brock, T.H., Prance, G.T. (Eds.), Changes and Disturbance in Tropical Rainforest in Southeast Asia. Imperial College Press, London, pp. 161–173.

Page, S.E., Siegert, F., Rieley, J.O., Boehm, H.V., Jaya, A., Limin, S., 2002. The amount of carbon released from peat and forest fires in Indonesia during 1997. Nature 420, 61–65.

Page, S.E., Wüst, R.A.J., Weiss, D., Rieley, J.O., Shotyk, W., Limin, S., 2004. A record of late Pleistocene and Holocene carbon accumulation and climate change from an equatorial peat bog (Kalimantan, Indonesia): Implications for past, present and future carbon dynamics. J. Quat. Sci. 19, 625–635.

Palis, U.T., 2000. Livelihood role of inland floodplain ecosystem for local community related to fisheries commodity. In: Iwakuma, T., Inoue, T., Kohyama, T., Osaki, M., Simbolon, H., Tachibana, H., Takahashi, H., Tanaka, N., Yabe, K. (Eds.), Proceedings of the International Symposium on: Tropical Peatlands. Bogor, Indonesia, 22–24 November 1999. Sapporo Editors, Japan, pp. 503–514.

Perdani, Y., 2013. Aceh performs worst in forestry protection: UNDP. The Jakarta Post. http://www.thejakartapost.com/news/2013/04/30/aceh-performs-worst-forestry-protection-undp.html (accessed September 2014).

Pet-Soede, C., Cesar, H.S.J., Pet, J.S., 1999. An economic analysis of blast fishing on Indonesian coral reefs. Environ. Conserv. 26, 83–93.

Phelps, N.A., Bunnell, T., Miller, M.A., 2011. Post-disaster economic development in Aceh: neoliberalization and other economic-geographical imaginaries. Geoforum 42, 418–426.

Phillips, V.D., 1998. Peatswamp ecology and sustainable development in Borneo. Biodivers. Conserv. 7, 651–671.

Prasodjo, E., Mukarwoto, 1997. Development of peat mining in Indonesia within a context of ecologically sustainable principles. In: Rieley, J.O., Page, S.E. (Eds.), Tropical Peatlands. Samara Publishing Limited, Cardigan, UK, pp. 281–287.

Purwaningsih, Yusuf, R., 2000. Vegetation analysis of Suaq Balimbing peat swamp forest, Gunung Leuser National Park, South Aceh. In: Inoue, T., Kohyama, T., Osaki, M., Simbolon, H., Tachibana, H., Takahashi, H., Tanaka, N., Yabe, K. (Eds.), Proceedings of the International Symposium on: Tropical Peatlands. Bogor, Indonesia, 22–24 November 1999. Sapporo Editors, Japan, pp. 275–282.

Ramakrishna, S., 2005. Conservation and sustainable use of peat swamp forests by local communities in South East Asia. Suo 56, 27–38.

Reliefweb, 2014a. More than 94,000 evacuated from floods in North Aceh. http://reliefweb.int/report/indonesia/more-94000-evacuated-floods-north-aceh (accessed December 2014).

Reliefweb, 2014b. Over 15,000 people flee homes after floods in Aceh, Indonesia. http://reliefweb.int/report/indonesia/over-15000-people-flee-homes-after-floods-aceh-indonesia (accessed December 2014).

Reliefweb, 2014c. Indonesia air drops aid, sends in teams after flooding. http://reliefweb.int/report/indonesia/indonesia-air-drops-aid-sends-teams-after-flooding (accessed December 2014).

Ricklefs, M.C., 2001. A History of Modern Indonesia Since c.1200. Stanford University Press, Stanford, USA.

Rieley, J.O., Page, S.E., Shepherd, P.A., 1997. Tropical bog forests of South East Asia. In: Stoneman, L.P.R.E., Ingram, H.A.P. (Eds.), Conserving Peatlands. CAB International, Wallingford, UK, pp. 35–41.

Rijal, S., Cochard, R., 2015. Invasion of *Mimosa pigra* on the cultivated Mekong River floodplains near Kratie, Cambodia: farmers' coping strategies, perceptions, and outlooks. Reg. Environ. Change. http://dx.doi.org/10.1007/s10113-015-0776-3.

Rijksen, H.D., Diemont, W.H., Griffith, M., 1997. The Singkil swamp: the kidneys of the Leuser Ecosystem in Aceh, Sumatra, Indonesia. In: Rieley, J.O., Page, S.E. (Eds.), Tropical Peatlands. Samara Publishing Limited, Cardigan, UK, pp. 355–362.

Rini, C., 2012. Road construction in Aceh fragment wildlife habitat. Endoftheicons.org. http://endoftheicons.org/2012/07/31/road-construction-in-aceh-fragment-wildlife-habitat/ (accessed September 2014).

Rood, E., Ganie, A.A., Nijman, V., 2010. Using presence-only modelling to predict Asian elephant habitat use in a tropical forest landscape: implications for conservation. Divers. Distrib. 16, 975–984.

Sambas, E.N., 2003. Possible long-term effects of the haze within the Leuser Ecosystem. Institut Pertanian Bogor. http://tumoutou.net/702_07134/edy_n_sambas.htm (February 2007).

Sarkis, S., Sarkis, S., van Beukering, P.J.H., McKenzie, E., Brander, L., Hess, S., Bervoets, T., Looijenstijn-van der Putten, L., Roelfsema, M., 2013. Total economic value of bermuda's coral reefs: a summary. In: Sheppard, C.R.C. (Ed.), Coral Reefs of the United Kingdom Overseas Territories. Coral Reefs of the World, vol. 4. Springer, Heidelberg, pp. 201–211.

Settle, W.H., Ariawan, H., Astuti, E.T., Cahyana, W., Hakim, A.L., Hindayana, D., Lestari, A.S., 1996. Managing tropical rice pests through conservation of generalist natural enemies and alternative prey. Ecology 77, 1975–1988.

III. LEARNING FROM THE FIELD CASES/ISSUES

Shah, R., Cardozo, M.L., 2014. Education and social change in post-conflict and post-disaster Aceh, Indonesia. Int. J. Educ. Dev. 38, 2–12.

Shepherd, C.R., Magnus, N., 2004. Nowhere to Hide: The Trade in Sumatran Tiger. TRAFFIC Southeast Asia, Petaling Jaya, Malaysia.

Siapno, J.A., 2009. Living through terror: everyday resilience in East Timor and Aceh. Soc. Identities 15, 43–64.

Sidle, R.C., Ziegler, A.D., Negishi, J.N., Nik, A.R., Siew, R., Turkelboom, F., 2006. Erosion processes in steep terrain – truths, myths, and uncertainties related to forest management in Southeast Asia. For. Ecol. Manag. 224, 199–225.

Singleton, I., van Schaik, 2000. Orangutan hone range size and its determinants in a Sumatran swamp forest. Int. J. Primatol. 22, 877–911.

Singleton, I., Wich, S.A., Griffiths, M., 2008. Pongo abelii. The IUCN Red List of Threatened Species. Version 2014.2. www.iucnredlist.org (accessed October 2014).

Smulders, M.J.M., van 't Westende, W.P.C., Diway, B., Esselink, G.D., van der Meer, P.J., Koopman, W.J.M., 2008. Development of microsatellite markers in *Gonystylus bancanus* (Ramin) useful for tracing and tracking of wood of this protected species. Mol. Ecol. Resour. 8, 168–171.

Soejarto, D.D., Fong, H.H.S., Tan, G.T., Zhang, H.J., Ma, C.Y., Franzblau, S.G., Gyllenhaal, C., Riley, M.C., Kadushin, M.R., Pezzuto, J.M., Xuan, L.T., Hiep, N.T., Hung, N.T., Vu, B.M., Loc, P.K., Dac, L.X., Binh, L.T., Chien, N.Q., Hai, N.V., Bich, T.Q., Cuong, N.M., Southavong, B., Sydara, K., Bouamanivong, S., Ly, H.M., Thuy, T.V., Rose, W.C., Dietzman, G.R., 2005. Ethnobotany/ethnopharmacology and mass bioprospecting: issues on intellectual property and benefit-sharing. J. Ethnopharmacol. 100, 15–22.

Sorensen, K.W., 1993. Indonesian peat swamp forests, and their role as a carbon sink. Chemosphere 27, 1065–1082.

Stem, C.J., Lassoie, J.P., Lee, D.R., Deshler, D.D., Schelhas, J.W., 2003. Community participation in ecotourism benefits: the link to conservation practices and perspectives. Soc. Nat. Resour. 16, 387–413.

Sunarto, Kelly M.J., Klenzendorf, S., Vaughan, M.R., Zulfahmi, Hutajulu M.B., Parakkasi, K., 2013. Threatened predator on the equator: multi-point abundance estimates of the tiger *Panthera tigris* in central Sumatra. Oryx 47, 211–220.

Tacconi, L., 2003. Fires in Indonesia: causes, costs and policy implications. CIFOR Occasional Paper No. 38. CIFOR, Bogor, Indonesia.

Takakai, F., Morishita, T., Hashidoko, Y., Darung, U., Kuramochi, K., Dohong, S., Limin, S.H., Hatano, R., 2006. Effects of agricultural land-use change and forest fire on N_2O emission from tropical peatlands, Central Kalimantan, Indonesia. Soil Sci. Plant Nutr. 52, 662–674.

Tata, H.L., van Noordwijk, M., Ruysschaert, D., Mulia, R., Rahayu, S., Mulyoutami, E., Widayati, A., Ekadinata, A., Zen, R., Darsoyo, A., Oktaviani, R., Dewi, S., 2014. Will funding to reduce emissions from deforestation and (forest) degradation (REDD+) stop conversion of peat swamps to oil palm in orangutan habitat in Tripa in Aceh, Indonesia? Mitig. Adapt. Strateg. Glob. Chang. 19, 693–713.

Thampanya, U., Vermaat, J.E., Sinsakul, S., Panapitukkul, N., 2006. Coastal erosion and mangrove progradation of Southern Thailand. Estuar. Coast. Shelf Sci. 68, 75–85.

Thirgood, S., Mlingwa, C., Gereta, E., Runyoro, V., Malpas, R., Laurenson, K., Borner, M., 2008. Who pays for conservation? Current and future financing scenarios for the Serengeti ecosystem. In: Sinclair, A.R.E., Packer, C., Mduma, S.A.R., Fryxell, J.M. (Eds.), Serengeti III: Human Impacts on Ecosystem Dynamics. University of Chicago Press, Chicago, pp. 443–470.

Tomascik, T., Mah, A.J., Nontji, A., Moosa, M.K., 1997. The Ecology of the Indonesian Seas. Periplus Editions (HK) Ltd., Singapore.

Tribun Network, 2014. Gajah Sumatera hanya tersisa 460 ekor die Aceh. http://jogja.tribunnews.com/2014/08/19/gajah-sumatera-hanya-tersisa-460-ekor-di-aceh/ (accessed October 2014).

Tsumura, Y., Kado, T., Yoshida, K., Abe, H., Ohtani, M., Taguchi, Y., Fukue, N., Tani, N., Ueno, S., Yoshimura, K., Kamiya, K., Harada, K., Takeuchi, Y., Diway, B., Finkeldey, R., Na'iem, M., Indrioko, S., Ng, K.K.S., Muhammad, N., Lee, S.L., 2011. Molecular database for classifying *Shorea* species (Dipterocarpaceae) and techniques for checking the legitimacy of timber and wood products. J. Plant Resour. 124, 35–48.

UNEP, 2005. After the Tsunami: Rapid Environmental Assessment. United Nations Environment Programme, Nairobi. http://www.unep.org/tsunami/reports/TSUNAMI_report_complete.pdf.

UNEP, 2007. The last stand of the orangutan. State of emergency: illegal logging, fire and palm oil in Indonesia's National Parks. UNEP Rapid Response Assessment. Prepared by C. Nellemann, L. Miles, B.P. Kaltenborn, M. Virtue and H. Ahlenius. UNEP, Nairobi. http://www.grida.no/_documents/orangutan/full_orangutanreport.pdf.

van Beukering, P.J.H., Cesar, H.S.J., Janssen, M.A., 2003. Economic valuation of the Leuser National Park on Sumatra, Indonesia. Ecol. Econ. 44, 43–62.

van Beukering, P.J.H., Grogan, K., Hansfort, S.L., Seager, D., 2009. An economic valuation of Aceh's forests. The road towards sustainable development. Report Number R-09/14. Institute for Environmental Studies, VU University, Amsterdam.

van Schaik, C.P., Rao, M., 1997. The behavioural ecology of Sumatran orangutans in logged and unlogged forest. Trop. Biodivers. 4, 173–185.

van Schaik, C.P., Monk, K.A., Robertson, J.M.Y., 2001. Dramatic decline in orang-utan numbers in the Leuser Ecosystem, Northern Sumatra. Oryx 35, 14–25.

van Strien, N.J., Manullang, B., Sectionov, Isnan, W., Khan, M.K.M., Sumardja, E., Ellis, S., Han, K.H., Boeadi, Payne, J., Bradley, M.E., 2008. *Dicerorhinus sumatrensis*. IUCN Red List of Threatened Species. Version 2014.2. http://www.iucnredlist.org/details/6553/0 (accessed November 2014).

ver Berkmoes, R., Atkinson, B., Brash, C., Butler, S., Noble, J., Skolnick, A., Stewart, I., Stiles, P., 2013. The Lonely Planet Guide to Indonesia. Lonely Planet Publications, Singapore.

Veron, G., Gaubert, P., Franklin, N., Jennings, A.P., Grassman, L.I., 2006. A reassessment of the distribution and taxonomy of the endangered otter civet *Cynogale bennettii* (Carnivora: Viverridae) of South-east Asia. Oryx 40, 42–48.

Whitten, T., Damanik, S.J., Anwar, J., Hisyam, N., 1997. The Ecology of Sumatra. Periplus Editions (HK) Ltd., Singapore.

Wibisono, I.T.C., Suryadiputra, I.N.N., 2006. Study of lessons learned from mangrove/coastal ecosystem restoration efforts in Aceh since the tsunami. Wetlands International Indonesia Programme, Bogor, Indonesia. .

Wibisono, H.T., Linkie, M., Guillera-Arroita, G., Smith, J.A., Sunarto, Pusparini, Asriadi, W., Baroto, P., Brickle, N., Dinata, Y., Gemita, E., Gunaryadi, D., Haidir, I.A., Herwansyah, Karina I., Kiswayadi, D., Kristiantono, D., Kurniawan, H., Lahoz-Monfort, J.J., Leader-Williams, N., Maddox, T., Martyr, D.J., Maryati, Nugroho A., Parakkasi, K., Priatna, D., Ramadiyanta, E., Ramono, W.S., Reddy, G.V., Rood, E.J.J., Saputra, D.Y., Sarimudi, A., Salampessy, A., Septayuda, E., Suhartono, T., Sumantri, A., Susilo, Tanjung I., Tarmizi, Yulianto K., Yunus, M., Zulfahmi, 2011. Population status of a cryptic top predator: an island-wide assessment of tigers in Sumatran rainforests. PLoS One 6. http://dx.doi.org/10.1371/journal.pone.0025931.

Wich, S.A., Singleton, I., Utami-Atmoko, S.S., Geurts, M.L., Rijksen, H.D., van Schaik, C.P., 2003. The status of the Sumatran orang-utan *Pongo abelii*: an update. Oryx 37 (1), 49–54.

Wickramasinghe, S., Borin, M., Kotagama, S.W., Cochard, R., Anceno, A.J., Shipin, O.V., 2009. Multi-functional pollution mitigation in a rehabilitated mangrove conservation area. Ecol. Eng. 35, 898–907.

Wikipedia, 2014. Aceh. http://en.wikipedia.org/wiki/Aceh; Indonesia. http://en.wikipedia.org/wiki/Indonesia (accessed September 2014).

Wösten, J.H.M., Ismail, A.B., van Wijk, A.L.M., 1997. Peat subsidence and its practical implications: a case study in Malaysia. Geoderma 78, 25–36.

Wösten, J.H.M., van den Berg, J., van Eijk, P., Gevers, G.J.M., Giesen, W.B.J.T., Hooijer, A., Idris, A., Leeman, P.H., Rais, D.S., Siderius, C., Silvius, M.J., Suryadiputra, N., Wibisono, I.T., 2006. Interrelationships between hydrology and ecology in fire degraded tropical peat swamps. Water Resour. Dev. 22, 157–174.

Yamada, I., 1997. Tropical Rain Forests of Southeast Asia: A Forest Ecologist's View. University of Hawai'i Press, Honolulu.

Yonariza, Webb, E.L., 2007. Rural household participation in illegal timber felling in a protected area of West Sumatra, Indonesia. Environ. Conserv. 34, 73–82.

Zafir, A.W.A., Payne, J., Mohamed, A., Lau, F.C., Sharma, D.S.K., Alfred, R., Williams, A.C., Nathan, S., Ramono, W.S., Clements, G.R., 2011. Now or never: what will it take to save the Sumatran rhinoceros *Dicerorhinus sumatrensis* from extinction? Oryx 45, 225–233.

Targeting Deforestation Through Local Forest Governance in Indonesia and Vietnam

Ngo, T.D.*, Mahdi†

*Hue University of Agriculture and Forestry, Hue City, Vietnam †Andalas University, Padang, Indonesia

14.1 INTRODUCTION

Community-based forest management (CBFM) is recognized as an effective solution to find a balance between forest conservation and livelihood improvement in localities. Both Indonesia and Vietnam initiated CBFM during the 1990s due to similar contexts of rapid deforestation, high rate of poverty, and urgent demand for local participation in forest management. Given the difference in geographical, socioeconomic and sociopolitical settings, the progress of CBFM setting and operation in each country was different. In Indonesia, the government has adopted traditionally favored large-scale, capital-intensive industries that had monopolized the forest economy, resulting in serious ecological and economic problems. This management style has caused serious illegal logging and forest conversion activities that destroyed 70% of country forest areas during last 50 years (Rukmantara, 2006). In Vietnam, the government has managed forests through its state forest enterprises (SFEs) since the country's reunification in 1975. During the period 1943–1993, Vietnam lost about 5 million hectares of forests, which reduced forest cover from 43% to 27.8% (de Jong et al., 2006). Both countries have adopted centrally managed forest management systems where all decision-making power comes from the state. This system excludes local people from participating in the decision-making process, abandoning them from access to forest rights, and eliminating them from their forest-dependent livelihood practices. Consequently, it has caused serious impacts on ecological and economic issues at global, national, and local levels.

This chapter identifies some common elements that influenced local forest governance through CBFM and their implications to ensure the access for the poor to natural resources for improving their livelihoods and guaranteeing the sustainability of the forest ecosystem service.

14.2 TRENDS OF DEFORESTATION AND FOREST DEGRADATION IN INDONESIA AND VIETNAM

Deforestation in Indonesia has attracted worldwide attention due to its high rate. The Food and Agriculture Organization of United Nations reported that the rate of the world's forest cover change during 1990–2000 was −0.22% and during 2000–2005 was −0.18% annually. During 1990–2000, the worlds' forest cover declined about 8,868,000 ha and during 2000–2005 it declined 7,317,000 ha annually. The highest rate of decline was found in Indonesia: −1.61% during 1990–2000 and −1.91% during 2000–2005 annually (FAO, 2006). The last report shows the deforestation rate in Indonesia is continuing to increase despite a high-level pledge to combat deforestation and a nationwide moratorium on new logging and plantation concessions (Murdiyarso et al., 2011). For the period of 2000–2012, Indonesia lost 6 million hectares of primary forest. This situation put Indonesia in the first row on deforestation rate, even higher than Brazil. Primary forest lost in Sumatra island contributed the most, 2.86 million hectares, following by Kalimantan, Sulawesi, and Papua for the same period. More than half of the deforestation occurred in lowland forests. Wetlands forest loss increased at a faster rate, accounting for 2.6 million hectares or 43% of loss overall within the same period (Margono et al., 2014). For the two decades, 1990–2010, Sumatra island alone lost 7.54 million hectares of primary forest (Margono et al., 2012). Beside deforestation, Indonesia is also facing the problem of forest degradation. Forest degradation does not represent a change in land use and the outcome is by definition still a forestland cover, but the forest is destroyed. Forest degradation has mainly occurred in state-owned forest. Sumatra island has degraded its primary forest by 2.31 million hectares (Margono et al., 2012). The Ministry of Forestry reported that for the period of 2000–2010 Indonesian primary forest was degraded by 50,000 ha a year.

Forest cover decline involves agents, both institutional and environmental, that influence agents themselves (Contreras-Hermosilla, 2000; Sunderlin et al., 2001). The agents could be an individual, government institution, estate firm, or others. The factors for deforestation and forest degradation have been widely reported. The main driver is demand for forest-related production in the international market and domestic infrastructure development (Angelsen and Kaimowitz, 1999; Kaimowitz and Angelsen, 1998). A study by Wheeler et al. (2013) pointed out that palm oil future prices, saw-log prices, global palm oil production, global saw-log production, exchange rate, and mobile phone coverage were the variables that significantly correlate with forest clearing in Indonesia. The draft national Reducing Emissions from Deforestation and Forest Degradation (REDD+) strategy indicates that the main causes of deforestation and forest degradation are weak spatial planning, problems with tenure, ineffective forest management, weak governance, and weak law enforcement (REDD-Monitor.org).

Poverty is the main reason for participation of rural households in illegal logging in the absence of secure and sustainable alternative livelihoods (Yonariza and Webb, 2007). Weak governance during social and political turmoil during 1997–2000 led to a high rate of forest degradation within protected and conservation forests. The breakdown of the "new order" Suharto regime and unclear responsibility among regions and government agencies in managing forest provided space for people to illegally take timber from the state owned forest (Mahdi et al., 2009).

In Vietnam, deforestation occurred seriously during the years 1943–1993. In this period, the core mandate of SFEs was to harvest timber for export to cover the national budget shortage after the war. Forest cover declined from 43% (14.3 million hectares of forests) to 27.8% (9.2 million hectares) over this 50-year period (de Jong et al., 2006; Meyfroidt and Lambin, 2008). By the 1980s, of Vietnam's 33 million hectares of total land area, 19 million hectares had been legally classified as state forestland. Publicly managed companies (SFEs) held over 4 million hectares, and conducted logging operations on 150,000 ha each year, rapidly exploiting them for commercial timber production (Poffenberger, 1998). Among the major causes of deforestation in this period were (1) timber logging, (2) resettlement of people from delta and coastal areas to the central highland for new economic zones, and (3) destruction by war and Agent Orange. To address deforestation, the government of Vietnam has launched huge plantation programs such as Program 327 (1992–1997) and Program 661 (1998–2010) with the objective of adding 5 million hectares of forests by the end of 2010. As a result, forest cover increased to 39.1% in 2009 and 41% in 2013 (MARD, 2014). The increase of forest cover was due to plantation forest (Sunderlin and Huynh, 2005; de Jong et al., 2006; Meyfroidt and Lambin, 2008) while the natural forest was still decreased and degraded. A study by de Jong et al. (2006) showed that poor-quality natural forests, with a forest stock of less than $80 \, m^3/ha$, occupied up to 80% of the total forest area at that time.

In both countries, central planning and management in the forestry sector led to the high rate of deforestation and exacerbated poverty due to preventing local people from having access to forests through transmigration in Indonesia or resettlement programs in Vietnam. Enacting the Basic Forestry Law of 1967 in Indonesia resulted in deposing about 100 million people of their land rights, with rights formally transferred to the Forestry Department (USAID, 2012). Similarly, approximately 4 million people were resettled mostly into the Da River and, after 1975, the central highlands under the New Economic Development Zone policy of the 1960s and 1970s in Vietnam (Poffenberger, 1998). Specifically, both governments have relied too much on state forestry departments and enterprises in timber logging, management planning, and collaborating with private sectors to pursue economic benefits. Under the context of sustainable development set by Earth Summit Rio 1992, both countries had to revise their policies on forest management, which incorporated local participation to harmonize socioeconomic and environmental aspects in the country development. These contexts helped to facilitate the CBFM setting in both countries.

14.3 HISTORY AND CONCEPTS OF CBFM IN INDONESIA

In Indonesia, CBFM was first initiated during the 1990s. Following-up this initiative, in 2007 the government of Indonesia (GoI) issued regulations to allow individuals and communities to manage the forests near to where they were living. Currently, there are four types of CBFM as follows:

- Group or cooperative rights, under a regime known as Community Forestry (*Hutan Kemasyaratan, HKm*).
- Cooperative and individual rights in timber production, under a regime known as People's Timber Plantations (*Hutan Tanaman Rakyat, HTR*).

- The delegation of forest management rights to village administrations within the framework of Village Forests (*Hutan Desa, HD*).
- Company-community partnerships in which communities may gain access to forest resources based on an agreement with holders of business licenses or concessions (*Kemitraan*).

Among them, two major types of community forest tenure were community-based forests (*Hutan Kemasyarakatan-HKm*) and village forests (*Hutan Desa*). Community-based forests allow groups of farmers with 35-year contracts to manage selected production or protection forests and rights to harvest forest products. Village forests enable village-based institutions to obtain a 35-year lease to manage and protect state forestlands (USAID, 2012). Although Indonesia has many customary tenure systems operating at varying levels of functionality, the centralized government has strongly resisted efforts to implement legislation that would recognize customary ownership claims to forest resources. So far, the total area under either type of arrangement is very small. The contracting arrangements in particular are relatively cumbersome and successful contracts usually involve nongovernmental organizations (NGOs) or research organizations.

According to the CBFM plan for the period 2009–2014, about 2.5 million hectares of forests (out of 132 million hectares of forests or 1.9%) would be allocated to CBFM with 35-year renewable permits. However, according to the Partnership for Governance Reform (known locally as *Kemitraan*), only 326,000 ha (13% of the target) had been allocated for CBFM by the end of 2013. The main reason was due to the bureaucratic procedures that required permits to pass through 29 levels and took about 3 months to 3 years (Satriastanti, 2014). Weak coordination between the central and local governments was another obstacle that delayed the process of allocating community forest or village forest permits.

14.4 CBFM IN VIETNAM

In Vietnam, local communities have been in different positions in the forest management system during historical development of the country. Before the colonized period (1954), local communities actively participated in forest management led by the village patriarch council and maintained good forest cover via sustainable uses (Tran, 2004). This traditional forest management system proved its sustainability in terms of meeting local livelihood demand and maintaining ecological services (Sikor and Apel, 1998). After reunification of the country in 1975, the state nationalized forest tenure and prioritized timber production for national economic development (Poffenberger, 1998; Sam and Trung, 2001). During this period, local communities were excluded from forest management despite their long tradition of forest uses and conservation. Crisis in both economic development and environmental management has led to the *Doi moi* (renovation) in 1986 in which the government called for participation of multistakeholders in developing a market economy. In the forestry sector, local participation in forest management was officially encouraged in a forest allocation program expressed by a series of government forest policies (eg, Decree 01/1995/ND-CP, Decree 02/1994/ND-CP, and Decree 163/1999/ND-CP). The local communities, however, were only recognized as "legal" forest users until the issuance of the Land Law (2003), the Forest Protection and

Development Law (2004), and the Civil Law (2005). These laws, however, emphasized local communities' duties rather than the benefits they could gain from forest management; and local communities continued managing allocated forest with very limited rights compared with other stakeholders (Sikor and Tan, 2007; Clement and Amezaga, 2008).

Even though there were still some disputes on concepts and management structure of CBFM in Vietnam, most of experts and policy makers agreed that forest areas under CBFM in Vietnam could be classified in three different models (Nguyen et al., 2006).

14.4.1 Traditional Forests

Forest areas were claimed and managed by local communities for generations. These areas are often located in remote areas where local communities are ethnic people and traditional regulations are still strong in forest use and management. These forests play important roles in a community's living such as watershed protection, graveyards, NTFP-produced forest, and grazing land. Forests in this category are mainly located in northern Vietnam where the majority of ethnic people are living. In the Forest Development and Protection Law (2004), these forests are officially recognized and planned to allocate their traditional management to local communities.

14.4.2 Allocated Forests

Forest areas that were allocated to local communities by the government were based on legal documents such as Decree 02/1994/ND-CP, Decree 163/1999/ND-CP, Decree 181/2003/ND-CP, or by particular projects. Until 2004, there were 18 provinces that allocated forests to local communities as pilot programs (Nguyen et al., 2006). In the course of allocating land and forests to organizations, households, individuals, and based on real conditions, some local authorities (province and district) have allocated land and forests to the community for management and long-term forestry use. The community has become a forest manager. Also, in this group, there are forests that had been previously allocated to cooperatives, which no longer exist; they are being managed by local communities. Under this category, the Community Forest Management Pilot Program was implemented in the period of 2007–2009 in 40 communes nationwide, which try to establish a good model of CBFM in terms of allocation procedures, monitoring, benefit sharing, and long-term secure tenure for local communities. Forest areas under this category are managed well, although there are still shortcomings in allocation procedures and unclear benefit sharing mechanisms.

14.4.3 Contracted Forests

Contracted forests are forest areas that were contracted with local communities for protection, natural regeneration, and plantation. These forests belong to state forest enterprises and are subcontracted to local people for protection or plantation following Decree 01/1995/ND-CP, Decision 661/1998/QD-TTg, and Decree 181/2003/ND-CP. These forests were not owned by communities, but the communities were entitled to share benefits from the forests, depending on the time, labor, and funds such communities have invested in the course of their management, protection, and development of the forests. This was a joint management model between local communities and state organizations. Because the amount of payment for protection was very low (about $3/ha/year), most of the forests in this category are not very well protected. As proposed in Forest Allocation Proposal 2007–2010 (signed by Minister of MARD on Sep. 20, 2007), forest under a contracted group will be allocated to local communities for long-term management.

In both countries, the government only provides "forest use rights" to local communities within a period of time (35 years in Indonesia and 50 years in Vietnam). From the discussion above, the community-based forest (HKm) in Indonesia is similar to the "allocated forest" in Vietnam, with harvesting rights in both production and protection forests. However, in Vietnam the harvesting rights were very limited, especially timber. The village forest (HD) is similar to the "contracted forest" in Vietnam, which provided limited rights over the forests through community management.

14.5 ASSESSMENT OF CBFM CAPACITY TOWARDS FOREST CONSERVATION AND LIVELIHOOD IMPROVEMENT

In overall assessment, we analyzed local forest governance structure based on several theories developed recently by researchers and practitioners. The first one was a Program on Forests (PROFOR) toolkit (Kishor and Rosenbaum, 2012) that explained that a good model of forest governance should include three major components: policy framework, planning and decision-making process, and implementation-enforcement-compliance. These components are analyzed by six principles: accountability, effectiveness, efficiency, fairness/equity, participation, and transparency. The added value of this toolkit is to generate a standard method of overall assessment on the sustainability of forest governance regardless of who is the operator (state, private, or communities). Using this toolkit can help to identify why some policies worked in this context but not in the other ones. In the long-term, results of assessment by this toolkit can contribute to the process of designing a rationale policy that could reduce the deforestation and increase local capacity in dealing with climate change impacts. The second one was a critical review of facilitating and enabling conditions for sustainable governance of resources (Agrawal, 2001). This review showed elements of a self-sustaining system of resource governance that can be grouped into four main categories: resource system characteristics, group characteristics, institutional arrangements, and external environment. Besides, relationships between resource system and group as well as resource system and institutional arrangements were also discussed. To come up with our approach, we selected five common and relevant elements that can be used to analyze the current situation of CBFM in both Indonesia and Vietnam. The five elements are secure forest tenure, sound business practices, committed government supports, engaged local participation, and integrated global initiatives.

14.5.1 Secure Forest Tenure for Local Communities and Households

By forest tenure, we mean all aspects of bundles of property rights (Schlager and Ostrom, 1992). Bundles of property rights expressed in five levels from less to most power: access, withdrawal, management, exclusion, and alienation. Clearly defined property rights was the most important condition for a sustainable system of local forest governance. This helps to secure forest tenure and provide people with strong incentives for long-term planning, executing, and mobilizing resources to effectively manage their forests. Without land (forest) being secure, all activities and efforts would be temporary, and that may cause resources to be exhausted in a short time. However, the security of forestland tenure may have different influences on local behavior depending on forest status, conditional support from external forces, and local people's awareness about tenure.

In Vietnam, land belongs to the entire people and the government is the only representative owner (Land Law 2003). Therefore, there is no absolute "ownership" meaning when discussing forest tenure in Vietnam. In reality, the government allocates only "use rights" (usufructs) to local people for 50 years of use (forestland) or 30 years of use (agricultural land). In any case, with the user rights or long-term tenure, local people would tend to invest more resources and time on their forestland due to their long-term tenure over the allocated forestland areas. Ngo et al. (2014) compared the management of forests in relation to five levels of property rights in the Thua Thien Hue province of Vietnam, and reported that households that have individual rights in forest management tend to manage the forest more sustainably in comparison with groups and villages types of forest management. An individual household's forest management has the strongest power in terms of rights. It has access, withdrawal, management, and exclusion as well as alienation rights, while the members of both groups and villages forest management have weaker property rights, and tend to harvest short-term benefits from their respective plots of forest.

In Indonesia, forest tenure has a strong relationship with legalizing community property rights by state law. However, forest tenure security requires more, as it results from an interplay between state and/or community normative systems, actual practices, and actors' perceptions. As pointed out by Safitri (2010), the degree of forest tenure would be determined by three elements: the rights' robustness, proper duration, and strong legal protection. Unfortunately, Indonesian national legislation had not been able to achieve satisfactory results on those three elements. The bundles of rights in forest areas continued to be limited; communities were not allowed to hold any ownership rights in these areas. The legal basis of the Ministry of Forestry to physically control all land in forest areas is also unclear. In addition, legal protection of community rights was weak (Safitri, 2010).

14.5.2 Appropriate Business Practices

The concept of business practices was employed from community-based enterprises or community forest enterprises (CFEs). CFEs are businesses based on collective ownership or secured resource access that serve multiple functions and multiple goals (Antinori and Bray, 2005). More simply stated, Peredo and Chrisman (2006) define a community-based enterprise as a community engaging corporately as an entrepreneur and enterprise toward the collective good. The underlying theory of the approach is that by linking a viable community enterprise to the biodiversity or ecosystem of an area—and thereby generating sufficient livelihoods for community stakeholders—the stakeholders are enabled and motivated to counteract the threats to the resources (Salafsky et al., 2001). Our research used case studies from different projects operated in both countries to illustrate how business practices can motivate local villagers in protecting their forest. Also, the concept of "appropriate business practices" implied long-term business strategy and nature-friendly measures that applied are by local people. A sound business practice based on the forest resource can be defined if it is compatible with livelihood strategy (Measham and Lumbasi, 2013) and the potential diversified products can be ensured for the long-term sustainability (Stoian and Roda, 2006). So the conversion of the forest to plant economic exotic tress (as *Acacia mangium*) was not a sound practice because it destroyed the provision of environmental service provided by natural forest. Hunting wild-life was a similar case of unsound business practice because it challenges the biodiversity value of the wildlife population.

In Vietnam, business practices were a set of activities that local people apply to generate incomes for short- and medium-term benefits while investing in long-term benefits derived from their resources. By the short and medium terms, we mean that local people could have income daily or monthly from their activities based on the forest resources. Examples are palm leave collection, rattan harvest, or wild fruit collection (*Scaphium macropodum*). The long-term benefit can be either timber or environmental services such as water resources or carbon sequestration. Other examples of business activities include nontimber forest product (NTFP) cultivation, ecotourism, wildlife hunting, or nursery garden. Other models of sound business practices in study areas such as rattan plantation, nursery garden of native tree species, beekeeping, bamboo plantation (Nam Dong district), and ecotourism (Phong My commune). Most of these business models were funded either through the state program (Decision 147/2007) or by nongovernmental funding through projects (eg, International Cocoa Organization (ICCO), Extension and Training Support Project (ETSP), PROFOR, Consultative and Research Center on Natural Resources Management (CORENARM)). Some villages didn't receive any support and they continued their NTFP harvest in protected areas (Pa Hy village in A Luoi district). In general, business practices were more common at groups and household levels than at the village level. The reason could be the nature of business activities that require some knowledge of finance management such as investment, cost and benefit management, and simple business model without linking to a large-scale market.

In Indonesia, business practices in CBFM were limited due to slow progress in allocating forest tenure to local communities. However, the potential for applying business practices—especially for timber plantation—in CBFM in Indonesia is very huge given the massive gap between timber supply and industrial processing capacity of the country. As discussed in Macqueen (2012), forest-owning families and communities could be key players in potentially tackling this issue in Indonesia. For example, in Java alone, between 1 and 1.5 million hectares of private woodlots managed by local farmers and communities already produce up to 8 million cubic meters of timber to industrial processing units—worth US$360 million a year in income to local farmers. Prospects for expanding this contribution are huge. There were other projects that supported business practices under the CBFM. During 2008–2010, the Ford Foundation supported the establishment of one of the country's first village forests in Sulawesi. With support from the project, villagers from three communities in Bantaeng district attained Village Forest Management Licenses in November 2010, securing tenure over local forestlands for 35 years. Those resources will be managed by local village enterprise bodies (referred to as Badan Usaha Milik Desa (BUMDes)) under Village Forest Management Regulations. On average, the sale of coffee grown in the village forest augments a family's income by 50%. Another case study was the CBFM-certified villages in Wonogiri district, Central Java. There were two villages, Sumberejo and Selopuro, which were entitled to community-based forest certification under the Indonesian Ecolabelling Institute (LEI) system. Right after certification was granted in 2004, timber and wood materials produced in the village were sold at a 15% to 30% markup (Takahashi, 2008).

14.5.3 Engaged Local Participation in Decision Making of Forest Allocation and Management

The engagement process includes the early steps to the ending step of forest allocation and management. We explored how local people were consulted in the forest allocation process including forest types, areas, location, and in preparing 5-year local forest management plan. Local participation also included aspects of decision-making power on the rights (and duties) over the allocated forests and integration of traditional rules into new/state regulations in both formal and informal ways. It was also important to identify whether those rules/ norms were working or nonworking in reality (Thomson and Freudenberger, 1997). The idea of bringing local participation into the framework was to emphasize how local people feel about their ownership over the decision-making process, which could further facilitate their ability to make future decisions in forest management.

In Vietnam, during the process of forest allocation, local people participated in different stages, and levels of participation varied. Local people participated in the following activities:

- Preparing application forms for receiving the forest (based on guidelines of the district forest protection unit).
- Taking a forest inventory to collect data with provincial technicians. In most cases, local villagers just help field work and showing boundary rather than attending in technical process.
- Building a 5-year forest management plan in some sections (majority was done by the district forest protection unit).
- Dividing into groups to conduct a forest patrol weekly or monthly.
- Attending some trainings on technical and management issues.

A general concern was the lack of a presurvey on people readiness in receiving forests for long-term management. Many villages entered the forest allocation without technical and management skills and even knowing little about their rights over the forest allocated. As a result, many villages could not manage their forests and returned the task to the local government after 2–5 years of receipt.

With support from international funding and through NGOs, forest allocation was carried out in a more participatory way and thus local people participated more actively. The results were a forest allocation map that clearly defined the forest in different zones of management (plantation, conservation, regeneration). This approach has greatly facilitated local people in managing the forest after the project ends. Local participation was more effective in group management due to their homogeneity and commitment rather than the villages. For example, local groups were encouraged to initiate their business activities. Then the project was funded for most feasible business activities and helped them to create a village fund for long-term management. These pilots were only available in international funding projects.

14.5.4 Committed Government Supports in Law Enforcement and Establishment of Local Institutions

This element came from the fact that local people were not yet fully aware about forest management practices and their capacity might not be strong enough in law enforcement. Based on our working experience, law enforcement was the most challenging part that local people often met after forest allocation. In theory, the government supports should lead to create "incentives" for local people to actively participate in resource management, protection, and utilization. The questions for exploring were then what kinds of incentives that the government could provide to local communities for sustainable resource management. We also focused on types of government supports such as a particular program (eg, buffer-zone development in VN) or a policy (eg, subsidies for planting native tree species) to analyze for an incentive structure for local forest management.

The government supported in-forest allocation through land-use planning, forest management planning, and forest inventory. These components were essential during and after the forest allocation process. Besides, other supports came from integrated programs such as buffer-zone development (agriculture and forestry sectors), sustainable forest development and conservation (funded by the World Bank), or projects on protected area conservation (Carbon and Biodiversity (CarBi) project).

During the period 2000–2007, about 5300 ha of forests were allocated without inventory data (Program 430). This caused a lack of clarity in the benefit sharing mechanism because there was no baseline data to prove forest increment in the future. Program 430 has addressed these shortcomings during the period of 2010–2014.

In most of the forest allocation supported from state budgets, there were no supports for income generation activities after allocation. This created a big challenge for local people who had no benefits from protecting degraded forests without any payment or compensation. It was a fact that many villages were reluctant to continue protecting forests and returned allocated areas back to the local government. In recent years, the government tried to incorporate Payment for Environmental Services (PES) into community forest management. However, very limited amounts of forest areas were located within the watershed, which was a condition for receiving payment from the PES fund.

A more critical issue was that most of the local forest owners (villages, groups, or households) were not strongly supported by local government in terms of law enforcement. In a workshop on local forest management in the postallocation period, several farmers reported that they didn't receive any local government's support in preventing illegal encroachment that happened in their allocated forests. Together with low or zero payment for forest protection, this action could create a disincentive for local people to continue to protect their own forest. In a worse situation, local people could let their forest be destructed to start planting other commercial crops such as rubber or *Acacia* spp.

Indonesia forest governance and policy are ambiguous between forest protection as well as conservation and forest resources extraction. On the one hand most policy was taken mostly for economic development purposes. Good sound policy on paper is not in line with implementation on the ground level. Wood and forest-related products is the source of state revenue until the present (Wardojo and Masripatin, 2002). This policy encouraged entrepreneurs and government agencies to exploit forest and their resources. Investors, both domestic and international, were invited to develop forest product industries. For this purpose, some forest concessions were issued and given to some enterprises both private and state-owned enterprises. In addition, agricultural development has also been done extensively. Plantation estates were opened in Sumatra, Kalimantan, Sulawesi, and other islands. This was also attracted by handsome demands from both domestic and international sources. Economic development encouraged the exploitation of natural resources. Forest resources were one of the important sources of income. It was in third position after oil and mineral and garments and textiles product. GoI received Rp. 2.8 trillion in revenue from the forestry sector and increased this to Rp. 4.2 trillion in 2013 (Ministry of Finance, 2014). Furthermore, economic development policy encourages business communities to expand agricultural land, especially for oil palm plantation. There are recorded 423 forest concessions operating in Indonesia that cover more than 13 million hectares of forest (APHI, 2014). Oil palm plantation grown by 4.14% annually for 2012–2014 (Directorate General of Estate, 2015). Peatland fire has still happened regularly for the last 5 years due to land clearing for plantation.

On the other hand, GoI also has adopted natural forest protection principles. Fighting between these two groups of interest results in a lack of policy implementation at the ground level (Mahdi et al., 2014). GoI adopted the natural forest protection principles in 1990 by enacting Law No. 5 in 1990, Conservation of Living Resources, and their Ecosystems Act. The management of natural resources conservation in a terrestrial area is grouped into six different types: strict natural reserve (*cagar alam*), wildlife sanctuary (*suaka margasatwa*), nature recreational park (*taman wisata alam*), game hunting park (*taman buru*), national park (*taman nasional*), and grand forest park (*taman hutan raya*). In addition, Indonesia ratified the United Nations Convention on Biodiversity in 1994. This convention supports Law No. 5 in 1990. Based on this law, Indonesia has 368 units of conservation in a terrestrial area that covered more than 17 million hectares of forestland throughout the country by the end of the 1990s. In addition, Indonesia has also introduced forest rehabilitation policies since 1967. Some policies and programs were launched to rehabilitate destructed forest by forest planting. Total forest plantation in Indonesia from 1997 to 2004 was 645,376.6 ha (MoF, 2005). However, forest cover declining over the same period is much higher that forest plantation, almost 14 million hectares, that mentions clearly that the forest policies at that time was ambiguous.

Some scholars argued the factors for this ambiguous policy. Structural problems and conflicts of interest between government agencies (Siscawati, 1999), and lack of law enforcement and corruption (Barr et al., 2010; Smith et al., 2003) also lead to lack of policy implementation.

Political change at the end of 1990s forced GoI to introduce new Forestry Law No. 41/1999. Article 66 of the law stipulates that forest management should be transferred to the local government. It means that the law encourages the adoption of a decentralized management model. The Ministry of Forestry has no direct line to local government for forest service. Instead, the forest service of the district government is likely more independent in decision making. Furthermore, the new law provides a greater chance for the participation of all stakeholders. The district government was given the authority to manage the forest within its territory, while the provincial government was given the authority over transdistrict boundaries. The central government was relegated to the role of national planning and providing guidance for forest management (Mahdi et al., 2013).

There are structural problems related to the governance of forests. By implementation of Law 1967, there were conflicts between government agencies and local communities over forest management. Some local communities, de facto, only acknowledge local customary rule in forest management instead of national law. The conflict produced forest destruction due to lack of local communities participation. In addition, conflict between government agencies arose on the issue of forest conversion to other purposes versus protection as natural forest. In the context of economic development, the agriculture sector is supposed to expand agricultural land to generate income in an effort to improve people's living standard.

Lack of law enforcement in forest management is acknowledged by the Ministry of Forestry in its report document to stakeholders in 2003 (MoF, 2003). Some problems emerge due to it. Illegal logging is the main issue that destroyed the Indonesian forest. The total loss because of this problem was estimated at about US$3.38 billion a year. The illegal practice was exacerbated by a high level of corruption practiced during the new order era. Presently, corruption in forest management is still rampant.

GoI issued a permit moratorium in 2011 to slow down deforestation. However, the policy was not effectively implemented at the ground level mainly due to unresolved conflict of interest among stakeholders. Although the central government banned license issuance within certain state-owned forest areas, the deforestation within the area is continuing. The forest was converted to be palm oil plantation both by private companies and by local communities (Mahdi et al., 2014).

14.5.5 Integrated Global Initiatives in CBFM Planning and Management

The integrated global initiative in CBFM planning and management might help to reduce transaction costs as well as increase financial sustainability of local forest management. Such initiatives include PES or REDD+, which help to create sustainable financing mechanisms for local forest management. The recent involvement of Vietnam into the Forest Law Enforcement, Governance, and Trade (FLEGT) process was also discussed in the context of global initiatives that contribute to large-scale forest management in Vietnam.

Recently, there were some global initiatives that help contribute to sustainable local forest management. The contribution could be technical aspects (ie, forest inventory, silvicultural practices), strengthen local participation (eg, Free, Prior, and Informed Consent (FPIC) rule), or a better benefit sharing mechanism (eg, PES). The following programs are active in Vietnam and have strong impacts on the local forest management.

REDD+ required many conditions from developing countries to participate in a "payment mechanism" such as Monitoring, Reporting and Verification (MRV) and FPIC that call for strong participation from local people. Payment from carbon credit did not happen, but the process of setting conditions for that payment is being set up through two phases of the REDD+ program and one project of the Forest Carbon Partnership Facility (FCPF).

The PES program started in 2010 and strongly redistributed benefits to local forest owners. However, a condition for this payment was the forest areas must be located within a watershed where environmental services were produced. Therefore, not all allocated forest areas could benefit from this program. In addition, the amount of payment depended largely on the total budget regenerated from that watershed. Some areas received very low payment (estimated at 8–10 USD/ha/year).

FLEGT is another initiative that was being negotiated in 2015. This initiative, together with a voluntary partnership agreement (VPA) could potentially have negative impacts on forest-dependent communities who do not have land tenure over forest resources. However, the preparation process for FLEGT/VPA could accelerate the provision of land tenure during forest allocation.

Because of an ambiguous policy between forest protection and extraction, Indonesian commitments on emission reduction, which were submitted in the Cancun Agreements in 2010 (COP16), is questioned (Mahdi et al., 2014). Recent political changes at the national level have diminished REDD+ Agency, which was established in respond to LoI with the Norway Government. REDD+ scheme implementation in Indonesia is now becoming uncertain.

The policy in the near future is likely to focus more on economic development and resources extraction. Although the frequency of peatland forest fires was reduced in the last year, agricultural land expansion and issuance of forest concessions are still continuing. Indonesia still needs agricultural land expansion for food production to meet food demand from a growing population and high prices of palm oil in the international market (Greenpeace Southeast Asia—Indonesia, 2012). The situation is becoming worse as opportunity cost to comply with the pledges on REDD+ is much higher than for BAU. Irawan et al. (2013) reported that the opportunity cost of an oil palm plantation on mineral soil is much higher than compensation from the REDD+ scheme, except for oil palm plantation on peatland. The business lobby has not tried to get a concession license and even bribed the officials (Dermawan et al., 2011).

14.6 CONCLUSION

Allocation of forest should facilitate both access to and control over the forest with full rights and suitable local knowledge. There is a need to have better benefit sharing policies on local forest management, especially during the first several years after forest allocation.

Livelihood should be an important part that integrated with forest allocation. Beside forest protection, local people need to have sufficient land for agroforestry production so that they do not clear forest for agriculture cultivation. Supporting off-farm jobs showed potential benefits to local people as case studies on local weaving and bee keeping recorded.

The local governments (province, district, and commune) must strongly commit to support local forest management, at least in terms of law enforcement. Besides, local people expressed their request in other types of hands-on training on forest management, conflict resolution, and conservation of traditional practices. To reduce conflicts among local forest users, it is important to create a dialogue where local communities, district forest rangers, and district/commune authorities can sit together to discuss "supporting local communities on legal aspects of protecting forest." This issue is really important when local people start to earn benefits from their efforts to protect and develop forest resources. Otherwise, they will lose their interests in village forest protection and management and the forest will become "open access," which is subject to deforestation quickly.

With the new funding scheme from PES, those forest areas in watershed and subject to PES need to be allocated to villages that have set up a village fund for development. The village fund can receive future payment from carbon trading (REDD+) and thus contribute to sustainable forest management.

References

Agrawal, A., 2001. Common property institutions and sustainable governance of resources. World Dev. 29 (10), 1649–1672.

Angelsen, A., Kaimowitz, D., 1999. Rethinking the causes of deforestation: lessons from economic models. World Bank Res. Obs. 14 (1), 73–98.

Antinori, C., Bray, D.B., 2005. Community forest enterprises as entrepreneurial firms: economic and institutional perspectives from Mexico. World Dev. 33 (9), 1529–1543.

APHI (Asosiasi Pengusaha Hutan Indonesia), 2014. Daftar Anggota APHI. http://www.rimbawan.com/anggota/daftar-anggota (retrieved 06.06.2015).

Barr, C., Dermawan, A., Purnomo, H., Komarudin, H., 2010. Financial Governance and Indonesia's Reforestation Fund During the Soeharto and Post-Soeharto Periods, 1989–2009: A Political Economic Analysis of Lessons for REDD+. Center for International Forestry Research (CIFOR), Bogor, Indonesia.

Clement, F., Amezaga, J.M., 2008. Linking reforestation policies with land use change in northern Vietnam: why local factors matter. Geoforum 39 (2008), 265–277.

Contreras-Hermosilla, A., 2000. The Underlying Causes of Forest Decline. Centre for International Forest Research (CIFOR), Bogor, Indonesia (Occasional Paper No. 30).

Dermawan, A., Petkova, E., Sinaga, A., Muhajir, M., Indriatmoko, Y., 2011. Preventing the Risks of Corruption in REDD+ in Indonesia. Center for International Forestry Research (CIFOR), Bogor, Indonesia.

Directorate General of Estate Crops of Ministry of Agriculture, 2015. Indonesian Palm Oil Statistics 2014. Badan Pusat Statistik, Jakarta. https://www.bps.go.id/website/pdf_publikasi/Statistik-Kelapa-sawit-Indonesia-2014.pdf (retrieved 06.06.2015).

Food and Agriculture Organization of United Nations (FAO), 2006. Global Forest Resources Assessment 2005. Available online: http://www.fao.org/forestry/foris/webview/forestry2/index.jsp?siteId=101&sitetreeId=16807&langId=1&geoId=0 (retrieved 15 April 2006).

Irawan, S., Tacconi, L., Ring, I., 2013. Stakeholders' incentives for land-use change and REDD+: the case of Indonesia. Ecol. Econ. 87, 75–83.

de Jong, W., Do, D.S., Trieu, V.H., 2006. Forest Rehabilitation in Vietnam: Histories, Realities and Future. CIFOR, Bogor, Indonesia.

Kaimowitz, D., Angelsen, A., 1998. Economic Models of Tropical Deforestation: A Review. Center for International Forestry Research (CIFOR), Bogor, Indonesia.

Kishor, N., Rosenbaum, K., 2012. Assessing and Monitoring Forest Governance: A User's Guide to a Diagnostic Tool. Program on Forests (PROFOR), Washington DC.

Macqueen, D., 2012. "Needed: Local Farmers and Communities to Plant Trees in Indonesia" – International Institute for Environment and Development. http://www.iied.org/needed-local-farmers-communities-plant-trees-indonesia (accessed 25.06.2015).

Mahdi, Shivakoti, G., Schmidt-Vogt, D., 2009. Livelihood change and livelihood sustainability in the uplands of Lembang subwatershed, West Sumatra, Indonesia, in a changing natural resource management context. Environ. Manag. 43 (1), 84–99.

Mahdi, Arbain, A., Senatung, M., Helmi, 2013. Developing organizational structure of Kesatuan Pengelolaan Hutan (KPH) Limapuluh Kota District, West Sumatra, Indonesia, for sustainable forest management. Paper Presented at 14th Global Conference of the International Association for the Study of the Commons. Retrieved from http://hdl.handle.net/10535/8928.

Mahdi, Shivakoti, G.P., Yonariza, 2014. Assessing Indonesian commitments and progress on emission reduction from forestry sector. Paper Presented at the Proceedings of International Conference on Forests, Soil and Rural Livelihoods in a Changing Climate. Kathmandu University, Dhulikhel, Nepal, pp. 92–106.

Margono, B.A., Turubanova, S., Zhuravleva, I., Potapov, P., Tyukavina, A., Baccini, A., et al., 2012. Mapping and monitoring deforestation and forest degradation in Sumatra (Indonesia) using Landsat time series data sets from 1990 to 2010. Environ. Res. Lett. 7 (3), 16.

Margono, B.A., Potapov, P.V., Turubanova, S., Stolle, F., Hansen, M.C., 2014. Primary forest cover loss in Indonesia over 2000–2012. Nat. Clim. Change 4, 730–735.

Measham, T.G., Lumbasi, J.A., 2013. Success factors for community-based natural resource management (CBNRM): lessons from Kenya and Australia. Environ. Manag. 52 (3), 649–659.

Meyfroidt, P., Lambin, E.F., 2008. The causes of reforestation in Vietnam. Land Use Pol. 25, 182–197.

Ministry of Agriculture and Rural Development (MARD), 2014. Decision No. 3322/QD-BNN-TCLN Dated on 28/7/2014 on the National Forest Cover of 2013. MARD, Hanoi, Vietnam.

Ministry of Finance of Republic of Indonesia, 2014. Data Pokok APBN 2007–2013. Kementerian Keuangan Republik Indonesia, Jakarta. http://www.anggaran.depkeu.go.id/dja/acontent/Data%20Pokok%20APBN%202013.pdf (retrieved 06.06.2015).

MoF (Ministry of Forestry of Republic of Indonesia), 2003. Statement on Progress towards implementing sustainable forest management at the twelfth CGI meeting. Report to stakeholders: current condition of forestry development. Denpasar. Indonesia January 2003.

MoF (Ministry of Forestry of Republic of Indonesia), 2005. Forestry Statistics of Indonesia 2005. Ministry of Forestry of Republic of Indonesia, Jakarta.

Murdiyarso, D., Dewi, S., Lawrence, D., Seymour, F., 2011. Indonesia's Forest Moratorium: A Stepping Stone to Better Forest Governance? Center for International Forestry Research (CIFOR), Bogor, Indonesia (Working Paper 76).

Ngo, T.D., Tran, N.T., Nguyen, V.H., 2014. Evaluating local forest governance in the context of climate change in a central province of Vietnam. SEARCA report.

Nguyen, H.Q., Pham, X.P., Vu, L., 2006. Handbook for Forestry Sector in Vietnam: Community Forestry. Forest Support Sector Partnership (FSSP), MARD, Hanoi, Vietnam.

Peredo, A.M., Chrisman, J.J., 2006. Toward a theory of community-based enterprise. Acad. Manag. Rev. 31 (2), 309–328.

Poffenberger, M. (Ed.), 1998. Stewards of Vietnam's Upland Forests. Research Network Report Number 10, Asian Forestry Network, pp. 118–138.

Rukmantara, A., 2006. Govt. told to Share Forest Management with Communities. http://www.thejakartapost.com/news/2006/06/20/govt-told-share-forest-management-communities.html (accessed 16.06.2015).

Safitri, M.A., 2010. Forest Tenure in Indonesia: The Socio-Legal Challenges of Securing Communities' Rights. Unpublished Doctoral thesis. Leiden University, Leiden.

Salafsky, N., Cauley, H., Balachander, G., Cordes, B., Parks, J., Margoluis, C., Margoluis, R., 2001. A systematic test of an enterprise strategy for community-based biodiversity conservation. Conserv. Biol. 15 (6), 1585–1595.

Sam, D.D., Trung, L.Q., 2001. Forest policy trends in Vietnam. In: Policy Trend Report 2001. Institute for Global Environmental Strategies, Japan, pp. 69–73.

Satriastanti, F.E., 2014. In Indonesia, Community Forest Management Choked by Bureaucracy. http://www.trust.org/item/20140228143109-gmzhv/ (accessed 16.06.2015).

Schlager, E., Ostrom, E., 1992. Property-right regimes and natural resources: a conceptual analysis. Land Econ. 68 (3), 249–262.

Sikor, T., Apel, U., 1998. The Possibilities for Community Forestry in Vietnam, vol. 1. Asian Forestry Network, California, USA (Working Paper).

Sikor, T., Tan, N.Q., 2007. Why may forest devolution not benefit the rural poor? Forest entitlements in Vietnam's central highlands. World Dev. 35 (11), 2010–2025.

Siscawati, M., 1999. Forest policy reform in Indonesia: has addressed the underlying causes of deforestation and forest degradation? In: Paper Presented at the 3rd IGES International Workshop on Forest Conservation Strategies for The Asia and Pacific Region, Tokyo, Japan, pp. 177–186.

Smith, J., Obidzinski, K., Subarudi, Suramenggala, I., 2003. Illegal logging, collusive corruption and fragmented governments in Kalimantan, Indonesia. Int. For. Rev. 5 (3), 293–302.

Stoian, D., Rodas, A., 2006. Community forest enterprise development in Guatemala: a case study of Cooperativa Carmelita RL. Community-based Forest Enterprises in Tropical Forest Countries: Scoping Study.

Sunderlin, W.D., Angelsen, A., Resosudarmo, D.P., Dermawan, A., Rianto, E., 2001. Economic crisis, small farmer well being, and forest cover change in Indonesia. World Dev. 29 (5), 767–782.

Sunderlin, W.D., Huynh, T.B., 2005. Poverty Alleviation and Forests in Vietnam. CIFOR, Bogor, Indonesia.

Takahashi, S., 2008. Challenges for local communities and livelihoods to seek sustainable forest management in Indonesia. J. Environ. Dev. 17 (2), 192–211.

Thomson, J.T., Freudenberger, K.S., 1997. Crafting Institutional Arrangements for Community Forestry. Community Forestry Field Manual 7. FAO, Rome.

Tran, N.T., 2004. Forest Use Pattern and Forest Dependency in Nam Dong District, Thua Thien Hue Province, Vietnam. M.Sc. Thesis, Asian Institute of Technology, Thailand.

III. LEARNING FROM THE FIELD CASES/ISSUES

USAID, 2012. Devolution of Property Rights and Sustainable Forest Management Volume 1: A Review of Policies and Programs in 16 Developing Countries. Property Rights and Resource Governance Project (PRRGP). Tetra Tech ARD, Vermont, USA.

Wardojo, W., Masripatin, N., 2002. Trends in Indonesian Forest Policy. Policy Trend Report, pp. 11–21.

Wheeler, D., Hammer, D., Kraft, R., Dasgupta, S., Blankespoor, B., 2013. Economic dynamics and forest clearing: a spatial econometric analysis for Indonesia. Ecol. Econ. 85, 85–96.

Yonariza, Webb, E.L., 2007. Rural household participation in illegal timber felling in a protected area of West Sumatra, Indonesia. Environ. Conserv. 34 (1), 73–82.

LOOKING FORWARD

Prospect of Sustainable Peatland Agriculture for Supporting Food Security and Mitigating Green House Gas Emission in Central Kalimantan, Indonesia

A. Surahman[*,†], *G. Shivakoti*[‡,§], *P. Soni*[*]

[*]Asian Institute of Technology, Klong Luang, Pathumthani, Thailand [†]Indonesian Agency for Agricultural Research and Development (IAARD), Jakarta Selatan, Indonesia [‡]The University of Tokyo, Tokyo, Japan [§]Asian Institute of Technology, Bangkok, Thailand

15.1 INTRODUCTION

Recently, agricultural production has become important both in terms of climate change and food security. An exponentially increasing world population leads to increased demand for food. According to a United Nation Report (2013), the current world population of 7.2 billion will reach 9.6 billion by 2050 and to feed those people the Food and Agriculture Organization of the United Nations (FAO) estimates that cereal production should be higher than the 2006 level by 2050 (FAO, 2009). On the other hand, increasing agriculture production to meet the food demand will also have a negative impact on global climate due to high energy input and emissions from agricultural practices. Smith et al. (2007) cited in Tubiello et al. (2012) reported that agriculture contributes 10–12% of total greenhouse gas (GHG) emissions.

Indonesia, which is the fourth most populous country of the world, also faces the challenge of increasing food production for food security. Indonesia's population growth has reached an alarming number; in 2010 the population of Indonesia reached 237.56 million people with a 1.49% per year growth rate (BPS, 2011). This means every year there would be three million more of the recent population. Another challenge is to provide food, as much as approximately 34 million tons, on 139.15 kg per capita per year (BPS, 2011). It is increasingly

http://dx.doi.org/10.1016/B978-0-12-805454-3.00015-3

difficult to rely on Java as a consistent national food supplier, while the rate of demand for rice outside Java is also increasing. The limited agricultural potential of Java due to land conversion suggests that the food supply in the future can no longer rely on Java. Extensive irrigated lowland converted to nonagricultural use has reached 110,000 ha per year. A number of issues concerning food security also require solutions, including increasing the production of agricultural commodities such as soybeans, maize, and rice toward self-sufficiency, then sustainable self-sufficiency and eventually achieving a surplus of 10 million tons of rice.

Strategic steps that need to be taken are to give greater attention to the use of marginal lands outside Java Island, and strictly controlling the rate of agricultural land conversion in Java. Marginal land such as peatland and upland has the potential to be used for producing food even though the productivity is low. Using marginal areas for producing food has consequences in terms of technology innovation to improve land productivity and to reduce environmental degradation.

One of the marginal lands with great potential to be developed as agricultural land is peatland. BBSDLP (2008) reported the Indonesian peatland area is spread across the islands of Sumatra, Kalimantan, and Papua, where about 33% of them are fit for agricultural uses. Furthermore, BBSDLP reported that Indonesia's peat area is currently 14.9 million hectares (Mulyani et al., 2012). Peatlands are an important terrestrial carbon pool. Peatland stores about 39% of the world soil carbon even though peatlands' area is only 3% of the Earth's surface, but peatlands are highly vulnerable and a major source of carbon emissions; therefore, agricultural peatland use has become the main interest in the climate protection debate in terms of GHG emissions. Climate-friendly peatlands agriculture management is the strategy for utilization peatlands for agricultural purposes and also requires a policy to mitigate GHG emissions. Economically, peatlands play an important role because of their potential to be developed into rice farming. Peatlands can be used for food crops on shallow peat (<100 cm). A primary consideration is that the shallow peat has relatively higher fertility and lower environmental risk than does deep peat (BBSDLP, 2008). However, peatlands also have a variety of constraints and problems in the area of GHG emissions.

The Mega Rice Project (MRP) in Central Kalimantan several years ago has a sour record in the development of peatland agriculture. The government argued that a million hectares of peatland clearance was intended to rescue national food self-sufficiency. Now the peatland in the former MRP become degraded. Flood in the rainy season and drought in the dry season will occur due to decreased capability of water discharge (Noor and Sarwani, 2004).

Peatlands also become targets for the development of large-scale oil palm plantations. Utilization of peatlands for palm oil plantations is promising and brings equitable benefits. There are 348 oil palm plantation companies with permission for exploitation of peat. Utilization of peatlands for palm oil plantations is growing rapidly in Riau Province and is followed by other areas such as Jambi, Sumatra South, West Kalimantan, and Central Kalimantan. The other plantation commodity that can be developed in the peatlands is rubber. The utilization of peatlands for rubber plantations has been carried out by communities in Central Kalimantan. Rubber plantations also have significant socioeconomic importance to Indonesian society, such as a foreign exchange source, providing jobs for residents, and a source of income. But, development of large-scale oil palm plantations in peatlands needs serious attention to minimize negative impact on the environment and economic losses with respect to the rate of subsidence of the peat.

Sustainable peatlands management aims to optimize the function of peatlands to support the improvement of the welfare of farmers and efforts to reduce GHG emissions without compromising the rights of future generations to meet their needs (Gandasasmita and Barus, 2012). In the management of tropical peatlands, forests will continue to face a conflict between the concept of preservation and development with the concept of land for crop production. Indeed, recognizing that their use is not by choice but because of the insistence of a need, then what is needed is how to balance (reconcile) the two concepts. One of the main ideas is to utilize only a relatively small portion of the expanse of peat dome, which are located at the edges, to the cultivation of plants by leaving largely intact its center to perform its natural function. Then, regulation of land use and the selection of crops are the important aspects of principles of sustainability in reclamation and land management. Therefore, the huge amount of tropical peatlands located in Indonesia, and implications for the development of regional and global interests, requires a policy for its use that is based on universal wisdom.

This chapter reviews the prospects of sustainably reclaiming peatlands for agricultural purposes to support food security and also addresses to a certain extent the GHG emissions from peatlands with an emphasis on strategy for sustainable development in terms of supporting food security and mitigating GHG emissions.

15.2 BRIEF OVERVIEW OF STUDY METHODOLOGY

This chapter uses a dynamic modeling approach, which is developed based on the system of dynamic modeling methodology. Dynamic modeling is an abstraction or simplification of a complex system while representing the real condition. Based on this model, scenario simulation is done with logical assumptions (Sterman, 2002).

Six steps were followed for developing a dynamic system of sustainable peatlands agriculture. First, understanding the complexity of causal relationship in peatlands management, which consists of population, food demand/consumption, food supply/production, land-use change, and GHG emissions. Second, constructing those causal relationships in a causal loop diagram. Third, setting basic assumptions. In terms of CO_2 emissions, the government of Indonesia (GoI) has a target to reduce GHGs by 26% from the GHG prediction in 2020. Fourth, formulating the model based on the basic assumptions and causal loop diagram. Fifth, validating the model by sensitivity analysis. And sixth, simulating the model with a different policy scenario. The case of degraded peatlands in the former MRP in West Kalimantan is used as a sample site for simulation.

The simulation model was built using the dynamics modeling software STELLA v.10. STELLA is a stock and flow-based software where the practitioner can create and run simulations over time. In this chapter the scenario was projected into 2025.The data used in this study are mainly secondary, which are available from various sources.

15.3 RESULT AND DISCUSSION

Economically, peatland plays an important role to be developed as a food crop area. However, utilization of peatland for agriculture is still being debated among experts. On the one hand, environmentalists expect peat can be used for conservation but on the other hand, the need of agricultural land for food security is urgently required. Peatlands are not only valued for their ecosystem services (water quality and storage, biodiversity, carbon, etc.) but they also have been valued to fulfill many human needs including food, energy, construction material, livestock bedding, and health (Clarke and Jack, 2010). Therefore, wise use of peatlands is essential to ensure that the remaining peatland area still has an important role for environmental conditions for both the present and next generations (Joosten and Clarke, 2002).

15.3.1 Status of Peatlands Agriculture in Indonesia

Koorders is recorded as the first person to discover peat in Indonesia in 1895, through his observations of coastal swamp forest in East Sumatra (Noor and Sarwani, 2004). This discovery has broken the previous opinion that the peat only formed due to the cold climate (temperate), which is limited by altitude (Notohadiprawiro, 1997 in Noor and Sarwani, 2004). However, peatlands had been exploited for a long time before this discovery. Agricultural practices in these peatlands reportedly are also due to the introduction of Chinese farmers in the 13th century when they invaded Borneo to trade (Noor and Jumberi, 2007). Furthermore, repairing of the irrigation canal pushes land use in this region gradually changing from swamp to the paddy field and now becomes rice producing region.

Peatlands agriculture in Indonesia has developed rapidly since the 1970s, along with government planning to open the tidal marshland area of 5.25 million ha for supporting the transmigration program and increasing national rice production. But not all peatlands that were opened and occupied by transmigration are working well, some have been abandoned by the farmers and the displaced migrants because those locations were not suitable for agriculture (Noor, 2001). Peatlands are classified as marginal and fragile land, which are characterized by being flammable in the dry season, and easily decline on the surface (subsidence). Therefore utilization of peatland should be done carefully and cautiously. Planning should be based on the results of the in-depth study of the peat soil characteristics and local environmental impact (Agus and Subiksa, 2008).

Utilization of peatlands will be more extensive in the future because of the limited arable land for agricultural development and the growing need for food commodities, as well as the conversion of productive agricultural land. On the other hand, prospects and opportunities for peatlands use for agriculture with proper management are promising. But because of the high diversity of peatland and difference of peatlands' character in different locations, it is necessary to do a land suitability evaluation before utilization of peatland for agriculture. Land resources data and information such as soil, climate, and other biophysical environment issues is needed for doing the land suitability evaluation. Based on Presidential Decree 32/1990, peatland with <3 m depth can be used for agricultural practices but peatland with >3 m depth will be used for conservation area. According to the direction of the Ministry of Agriculture (BBSDLP, 2008), shallow peat (<100 cm thickness) is recommended for food crops such as rice, corn, soybeans, cassava, and many other vegetables and perennial crops that can be planted in the 2–3 m depth of peatland (Sabiham, 2008). The reasons are shallow peat is more fertile and offers less environmental risk than depth peat.

15.3.2 Status of Peatlands in the Former MRP

Peatlands clearance through the MRP in Central Kalimantan and by Presidential Decree 82/1995 is the compensation of agriculture land conversion in Java. MRP is located between the River Sebangau, Kahayan, Kapuas Murung, Kapuas, and Barito in Kapuas District (Suriadikarta, 2009). This project was directed to convert the peat swamp forests (wetland) into rice fields to maintain and continue the national rice self-sufficiency that had been achieved by Indonesia in 1984. But, the MRP project implementation without prior environmental impact assessment (EIA) has a negative impact on the environment, physical, biological, and social. This is caused by inadequate planning and design (Suriadikarta, 2009). Construction of irrigation channels in the former MRP also caused an estimated 400,000 ha of wet tropical forest to be converted into open land and triggered peat fires.

The former MRP area is dominated by plains and river estuaries peatlands. Peat with a depth of more than 0.5 m covers about 920,000 ha, of which about 450,000 ha has a depth of more than 3 m. The remaining area of 532,000 ha are mainly composed of mineral soil. Most river and canal banks found in the traditional settlement are suitable for agricultural irrigation and water management practices based on the experience of local agriculture.

The current state of the land is mostly made up of forests, shrubs, damaged forests, agricultural land (including plantations), and forest and natural forest burning bush, healthy and damaged forest partially covering 550,000 ha or 38% of the total area, while badly damaged forest covers up to 14%, shrubs 37%, and 12% of agricultural land. There are currently about 400,000 ha of peatlands, more than 1 m are without the protection of forests, and most of these areas need to be replanted as part of the treatment rehabilitation of peatlands and forests. In addition, there are about 130,000 ha of peat shallow (0.5–1 m) without forest protection that could also be targeted for reforestation, although it is likely that the majority of the area will be utilized by local communities for agriculture.

15.3.3 Future Scenario

Based on the various challenges and biophysical and socioeconomic conditions in the former location of MRP, there are three scenarios in the model simulation for sustainable agriculture. The first scenario is focused on the reforestation of the former MRP areas. The second scenario is focused on the utilization of the former MRP areas for agricultural practices. The third scenario is focused on a compromise between reforestation and utilizing of peatland for agriculture.

In the first scenario, GHG emissions can be reduced more than the target. All of the former MRP areas were reforested. This is a good achievement from the environment side but there is no direct economic benefit for the farmers in the area. There is no improvement in farmer income and farmer welfare. In the second scenario, when the all former MRP areas were utilized for agricultural practices, direct social and economic benefits can be perceived by farmers in these areas. But this region will remain as a source of global carbon emissions. In the third scenario, there is a compromise between reforestation and peatland agriculture. The peatland agriculture area will increase due to improvement in the soil, and water management then will contribute to the increasing crop yield for rice and other plantation crops. Reforestation and fire management will reduce the GHG emissions. Farmer household income will increase due to an increase in crop and plantation yield. This scenario will bring balance and sustainability in the development of peatland management. The socioeconomic condition will grow, and GHG emissions will decrease. For details of the land-use change area, a map overlay between degraded peatland and the peat depth map with the administration map should be done. From the result, the location for crops and plantations can be determined.

15.3.3.1 Prospective Peatland Agriculture for Mitigation of GHG Emissions in Indonesia

Peatlands conversion to agricultural practices will increase decomposition rate and decrease peat soil stabilization. Emissions from drained peatlands used for agriculture are an important source of GHG emissions. Land degradation and deforestation also contribute to increases in GHG. The GHG emissions are a result of peat oxidation resulting from drainage, peatland fires, and loss of aboveground biomass from illegal deforestation and degradation.

Drainage of peat introduces oxygen into the surface, which promotes decomposition. The result is that organic matter in the peat will be lost through oxidation and the land surface. Drainage and deforestation create degraded peatland that can become highly susceptible to fires. Peat fires were also caused by land burn in the land preparation by plantation companies and small holder farmers.

In the first scenario, GHG emissions can be reduced more than the target. All of the former MRP areas were reforested. This is a good achievement from the environment side but there is no direct economic benefit for the farmers in the area. There is no improvement in farmer income and farmer welfare. When all former MRP areas are utilized for agricultural practices, the socioeconomic condition of the farmer will increase but the GHG emissions will increase also. In the third scenario, there is a balance and sustainability for peatland management because GHG emissions will be reduced and farmer welfare will increase. The GHG emissions condition from the three scenarios can be shown in Fig. 15.1.

The simulation model, which allocated 70% from 530,000 ha of degraded peatlands in the MRP to reforestation, will need 10 years to complete. The remaining uses are as a food crop area (20% or 106,000 ha) and 10% (53,000 ha) for plantation. The CO_2 emissions from business as usual (BAU) means there are no actions done, with sustainable peatlands management. In terms of reducing GHG for a food crop farming system, research for mitigating GHG in the peatland has been already done by the Indonesian Agency for Agricultural Research and Development (IAARD) under the Indonesia Climate Change Trust Fund (ICCTF) program. According to Setyanto et al. (2014), recent results show that using ameliorant can reduce 15% to 35% of CO_2 and CH_4 emissions. Among the ameliorant material, soil material and animal manure are the most consistent ameliorant with decreasing rates of 24.1% and 25.1%, respectively. The first ameliorant application can be done in between the 5th and 7th weeks.

Maintaining the water level is important in the plantation in the reducing CO_2 emissions. Processing GHG emissions from peatlands will be faster after drainage. It is associated with changes in ground water levels, which leads to changes in the environment from anaerobic into aerobic conditions. The aerobic conditions will increase the activity of soil microorganisms and result in increasing the release of CO_2 to the atmosphere (GHG emission) (Dariah et al., 2011).

15.3.3.2 Prospective Peatlands Agriculture for Supporting Food Security

Without solving the food problem, such a country cannot succeed in achieving rapid economic growth. The objectives of development are difficult to realize without having to meet

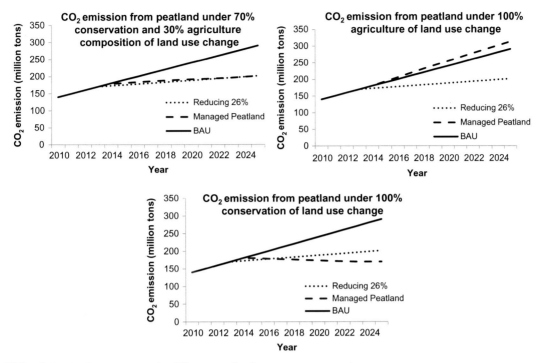

FIG. 15.1 GHG emissions in the different peatland management scenarios.

the community need of food. Therefore, the availability of food for fulfilling the community need is absolutely important for achieving national security, especially food security.

Based on the simulation model, it can be shown that from all of the degraded peatlands only 20% (106,000 ha) will be used as a food crop area, which consists of rice, maize, and soybeans area. Most of the area will be used as rice area (86,000 ha) and the remaining area for maize (15,000 ha) and soybeans (5000 ha) (Fig. 15.2).

From those food crop areas, the peatlands will be a source of food production in Central Kalimantan province, especially for rice production. Rice production in the area is 291,113 tons, far above the rice consumption in the area, which is 61,887 tons and 77% from the total rice consumption in the province (377,120 tons). Maize and soybean production were also above the consumption in the area. Maize and soybean production are 52,624 and 6592 tons, respectively, and the consumption was only 29,634 and 4370 tons, respectively.

Utilization of peatland should be considered with the characteristics of peatland. Based on the Ministry of Agricultural decree, peatlands that can be used for crop cultivation are shallow peat with less than 100 cm thickness. The reason is this peat soil is more fertile and has low environment risk. Rice, maize, soybeans, and vegetables are suitable to be cultivated in this area.

There are technological innovations available for supporting utilization of peatland for agriculture such as water management and soil management. Water management application should be considered with the peat soil characterization. Construction of a microirrigation canal 10–50 cm in depth is needed for cultivation of food crops in peatlands; for example, a

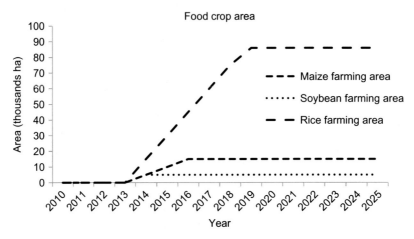

FIG. 15.2   Development of a food crop area of degraded peatlands.

rice crop needs a 10–30 cm irrigation canal. This canal is used to maintain water level in the soil for soil respiration development and leaching organic acid in the soil. Peat soil is acidic; therefore, pH needs to increase by amelioration to improve media of the crop root system. Lime, mineral soil, animal manure, and ash are used as ameliorant (Agus and Subiksa, 2008).

The most important thing in peatlands agriculture is land preparation. Recently, slash and burn practices are still done by farmers. Slash and burn practices have negative impacts on peatland management in terms of peatland fires, peatland destruction, and increasing GHG emissions. A peatland fire is more dangerous than other forest fires because a layer of peat burning will generate heavy smoke due to the imperfect combustion that takes place.

Oil palm plantations and rubber plantations are the most profitable plantations that can be used to improve farmer welfare in the area and also support the national export production. Rubber production in the area was only 10% of the total rubber production of the Central Kalimantan province (491,950 tons). This area will produce 58,855 tons of oil palm. This is about 5% of the total oil palm production in this province. The minimum area used for the plantation is due to the Presidential Instruction, which only allowed an area of 10.000 ha for plantations.

The Primary Cooperative Credit for Members' scheme (*Koperasi Kredit Primer untuk Anggota* (KKPA)) can be used as a model for partnership between a small holder and a private company in the rubber and palm oil plantation. The KKPA schemes rely on a contract signed between a company, small holders grouped in cooperatives, and banks, under the supervision of the government (Feintrenie et al., 2010). According to Feintrenie et al. (2010), a small holder will hand over the land to the nucleus (the company), which will be responsible for plants, management, and harvesting the crops. The farmers are paid a percentage of the harvest revenue after deduction of the initial investment for land preparation and management cost for maintenance cost. Local governments have to facilitate the discussions between the partners and land titling. Banks will keep land titles as collateral, and the company is responsible for collecting the repayments from the farmers. A 70/30 contract is common in the KKPA, which means the farmer will receive 30% from the net profit and 70% will go to the company. But farmers also will get wages as they are workers in the company.

TABLE 15.1   Farmer Income From Various Farming Systems

	Farming System				
	Oil Palm	Rubber	Rice	Maize	Soybean
Farmer income (US$)	2339	2137	1124	1157	956

1 US$ = IDR 11,000.

The other aspect of food security is accessibility to food by people. One of the factors in the accessibility to the food is income. Farmer income from the various farming systems in the area are shown in Table 15.1.

Table 15.1 shows that plantation farming system farmers will get more income than food crop farming system farmers. This is because the plantation produce export commodity has a high price in international trade. The combination of plantation and food crops is a better option to improve farmer welfare as well as maintain household food security. The farmer will have a 1 ha area for food crop and a 1 ha area for plantation and the farmer will get wages from the plantation plus the net benefit sharing. But overall, the farming system income is above the poverty line, which is US$267 per year according to BPS of Central Kalimantan (2011). Therefore, all the farmers in the area will not be categorized as poor farmers.

15.3.3.3 Strategic Pathways toward Sustainable Peatland Agriculture in Indonesia

From the scenario above, peatlands will become more important in the future because of the limitation of the lowland for increasing the area for food crops to fulfill food demand. However, it should be recognized that the use of peatlands has environmental risks, because peat is very susceptible to degradation. Peatland degradation can occur when land management is not done properly, so that the rate of decomposition is too big and causes substantial GHG emissions. Eliminating GHG emissions in peatlands agriculture is impossible, because the process of decomposition is a natural process that is also needed for the supply of nutrients to the plants. The concept of sustainable peatland management should be done by increasing the maximum productivity and reducing the level of GHG emissions. Therefore the strategy for sustainable peatlands agriculture should focus on improving peatland productivity and reducing GHG from peatlands agriculture and also improving farmer welfare.

(i) Strategy for improving peatlands productivity
- Water management

    Proper drainage systems are urgently needed in peatland agriculture, both for food crops or plantations. Improper drainage systems will accelerate land degradation. Water management should be adjusted with the characteristics of the local peat and commodities that will be developed. Landscaping activities include setting up a network of drainage channels, land leveling, manufacturing "*surjan*," mounds, and intensive shallow drainage. The dimensions and density of the drainage developed should be fitted to the commodity whether for food crops, vegetables, plantations, or plantation forests. Leveling of land is essential for food crops and vegetables. The function of drainage channels is to remove excess water, creating unsaturated conditions for plant roots respiration, and removing most organic acids (Agus and Subiksa, 2008).

- Amelioration

    Peatlands are very acidic due to the high levels of organic acids from the weathering of organic matter. Most of these organic acids, phenolic acid groups in particular, are toxic and inhibit the development of plant roots and plant growth. Amelioration is needed to overcome the problem of acid soil reaction and the presence of toxic organic acids; therefore, it will improve plant rooting. Lime, mineral soil, manure, and ash can be given as ameliorant material to raise the soil pH. But unlike mineral soil, peat soil pH should be increased up to 5 because peat does not have the potential toxic for Al (Subiksa et al., 2011). A high increase in the peatland pH will increase the peat decomposition rate.

- Fertilizer

    Fertilization is necessary because peatland is characterized by very poor peat soil minerals and nutrients that plants need. The type of fertilizer that is needed is a complete fertilizer containing mainly N, P, K, Ca, Mg and microelements Cu, Zn, and B. Fertilization should be done gradually with lower doses because low sorption power, peat soil nutrient fertilizer is easily leached. The use of slow-release fertilizer such as phosphate rock and *"Pugam"* (innovation in peat fertilizer) is better than the SP-36, because it would be more efficient, low-cost, and can increase soil pH (Subiksa et al., 2011). In *"Pugam"* the main nutrient, P, is also classified as a slow-release fertilizer that can improve nutrient uptake, reducing nutrient leaching P, and significantly improving plant growth compared to SP-36, Peat soil is also known for micronutrient deficiency because it is bounded by organic matter (Subiksa et al., 2011). Micronutrient deficiencies can cause sterile male flowers causing a vacuum in rice, corncobs on blank, or empty pods in peanuts. "Pugam" as ameliorant and fertilizer also contains the necessary microelements of plants, so that additional micronutrient fertilization is not needed anymore.

(ii) Strategy for reducing GHG emissions from peatlands agriculture

- Controlling groundwater level

    Peatlands have high hydraulic conductivity, either vertically or horizontally. Therefore, a drainage channel is crucial in controlling water table conditions. Key control groundwater level is setting drainage channel dimensions, particularly depth, and set up the floodgates. Lowering the water table is very necessary to keep the rooting medium conditions remaining in aerobic conditions. But a high decrease in the water table will damage the peat. Therefore, the groundwater level should be controlled so that plant roots get enough oxygen, but still moist peat to avoid large emissions and peat dries. Control of the water table by adjusting the water level in the drainage channel is one of the mitigations of CO_2 emissions. Therefore, the groundwater level should be maintained at a level that is safe for plants, and minimal emissions are a very effective mitigation of peatlands degradation. Another technique is paludicultures that can help stop peat oxidation and simultaneously provide sustainable harvests from peatlands. Paludicultures may have a double role to play in climate change mitigation; they avoid GHG emissions (by preventing peatlands from being drained or by rewetting drained peatlands) and the biomass produced may replace fossil raw materials and fossil fuels.

Maintaining the water level is important in the plantation in the reducing CO_2 emissions. According to Agus and Subiksa (2008) commodities that require shallow drainage (20 cm) such as rubber plantation, sago, and rice, can reduce the amount of emissions compared to commodities that require deep drainage (50–80 cm) like palm oil. Processing GHG emissions from peatlands will be faster after draining. This is associated with changes in groundwater levels, which leads to changes in the environment from anaerobic into aerobic conditions. Aerobic conditions will increase the activity of soil microorganisms and resulted in releasing CO_2 to the atmosphere (Dariah et al., 2011). Therefore maintaining the water content in the peatlands should be done to prevent a peat fire, also reducing the decomposition process in the soil.

- Amelioration

    GHG emissions are the result of the decomposition process of the peat into carbon compounds in short chains. The decomposition process can be further suppressed by the complexation process of changing simple organic compounds into complex compounds. Complexation can be done by adding ameliorant-rich ingredients to polyvalent cations. The complex compounds formed are resistant to decomposition so that carbon emissions can be reduced. Giving ameliorant peat 5–10 tons per hectare is expected to reduce cumulative CO_2 emissions as much as 15.5% ± 5.5% compared to not giving ameliorant. When the amelioration programs incorporate with the application of fertilizer to replace the traditional burning techniques, emissions are expected to be reduced by 19% ± 7%. These results can still be improved by using low-emission "*Pugam.*" According to Subiksa et al. (2011), fertilization with "*Pugam*" is able to reduce GHG emissions by 47% and increase biomass production.

- Nonburning tillage

    The most massive carbon emissions occur because of either deliberate or accidental peat fires. Land preparation by a burning system will cause the loss of carbon stocks, subsidence, and ultimately lead to endless layers of peat. From peat burned during planting (twice per year), carbon emission is around 110.1 ton CO_2 per hectare per year (Subiksa et al., 2011). Therefore information about land preparation without burning should be disseminated to the farmer. One of the methods is land clearing using a bioharvester, but this equipment is too expensive for the farmer. So, amelioration and fertilization efforts are needed to optimize plant growth and the burning habit should be avoided. Efforts to prevent fires can reduce CO_2 emissions by 0.284 Gt, or 25% of the projected emissions in 2025.

(iii) Strategy for improving farmer welfare

    Beside food availability through increasing food production, food accessibility by the farmer is also important factor in food security. Improving food accessibility can be done by improving farmer income as well as famer welfare. The strategy for improving farmer welfare in peatlands management in the food crop farming system can be done by increasing farmer adoption of innovative technology. Innovative technologies for the food crop farming system in the peatlands were already available in research centers in the universities or the Ministry of Agriculture. Dissemination of information about innovative technologies should be done more intensively by extension workers.

IV. LOOKING FORWARD

As a plantation partnership can be used for improved farmer income, farmer bargaining in the plantation partnership should be improved. A 70/30 sharing profit was not relevant with increasing land prices, therefore government should encourage revising the contract to a 60/40 split in the KKPA model partnership between the farmer and the private company. This means 40% of the net profit will be received by the farmer and 60% will go to the private company as a nucleus.

Because of the promising profit that can be received by the farmer, more attention should be paid by government to land titling so the farmer has a guarantee in the land rights.

15.4 CONCLUSION

The limited agricultural potential of Java due to land conversion means the food supply in the future can no longer rely on Java. Extensive irrigated lowland converted to non-agricultural use reached 110,000 ha per year. Therefore, rice production is not expected to be able to meet national food requirements. A number of issues concerning food security also require finding a solution, including the increasing number of agricultural commodities toward self-sufficiency and sustainable self-sufficiency and a surplus of 10 million tons. Another thing that needs to be done is to give greater attention to the use of marginal lands outside the Java Island, and strictly control the rate of agricultural land conversion in Java. One marginal land area with great potential to be developed as agricultural land is peatlands.

Degraded peatlands caused by MRP should be reforested and rehabilitated to get more benefit to the farmer in terms of food security and also reducing GHG emissions as a target of the GoI. Food crop area and plantation area should be set up in the best composition to optimize the benefit and maximize reducing GHG emissions. The best simulation model comprises the proportion area for reforestation and agriculture in the degraded peatland. In the best simulation model, reducing GHG emissions target can be achieved and farmer income will increase greater than the poverty line in the location. For detailed mapping of the land-use change, the next step should be overlaying the degraded peatland and peatland depth map with the administration map.

Utilization of peatlands for producing food has greater consequences. Unproductive land and GHG emissions are the issues that should be overcome in peatlands agriculture. Various strategies for sustainable peatlands agriculture can be done such as improving peatlands productivity by water management, amelioration, and fertilization and also reducing GHG emissions by controlling the water table via agricultural practices, amelioration, and preventing burning in land preparation. Farmer income and farmer welfare are also important in sustainable peatland agriculture. It will also improve farmer accessibility to get their food needs met. Implementation of those strategies will improve the role of peatlands for supporting food security and also reducing GHG emissions in Indonesia.

References

Agus, F., Subiksa, I.G.M., 2008. Peatland: Potency for Agriculture and Environment Aspect. Soil Research Institute, Indonesian Agency for Agriculture Research and Development-World Agroforestry Centre, Bogor.

BBSDLP, 2008. Final report 2008 on consortium research and development on climate change impact for agriculture. Indonesian Center for Agriculture Land Resources Research and Development.

BPS, 2011. Statistic of Indonesia 2010. Central Bureau of Statistic of Indonesia, Jakarta, Indonesia.

BPS of Central Kalimantan, 2011. Statistic of Central Kalimantan, Indonesia. Bureau of Statistic of Central Kalimantan, Palangkaraya, Central Kalimantan.

Clarke, D., Jack, R., 2010. Strategy for responsible peatland management on the international peat society annual report 2010.

Dariah, A., Erni, S., Fahmuddin, A., 2011. Carbon storage and CO_2 emission in the peatlands. In: Nuraida, N.L., et al. (Eds.), Sustainable Management of Peatland. National Soil Research Institute, Center for Agriculture Land Resources Research and Development, Bogor, Indonesia.

FAO, 2009. How to Feed the World in 2050. Food and Agriculture Organization of the United Nation, Rome, Italy.

Feintrenie, L., Chong, W.K., Levang, P., 2010. Why do farmers prefer oil palm? Lessons learnt from Bungo district, Indonesia. Small-scale For. 9, 379–396. http://dx.doi.org/10.1007/s11842-010-9122-2.

Gandasasmita, K., Barus, B., 2012. National Strategy on Sustainable Peatland Management in Indonesia. In: Paper presented on "National Strategy on Sustainable Peatland Management in Indonesia Seminar", Bogor, 12 October 2012.

Joosten, H., Clarke, D., 2002. Wise Use of Mires and Peatlands. International Mire Conservation Group and International Peat Society, Finland.

Mulyani, A., Susanti, E., Dariah, A., Maswar, Wahyunto, Agus, F., 2012. Basic Data of Peat Soil Characteristic in Indonesia. Indonesian Center for Agriculture Land Resources Research and Development, Bogor, Indonesia.

Noor, M., 2001. Peatland Agriculture: Potency and Problems. Kanisius, Yogyakarta.

Noor, M., Jumberi, 2007. Peatland utilization and management. Farmer knowledge and experience. Proceeding of National Seminat on Assessment Institute of Agriculture Technology. December 2007, Jambi.

Noor, M., Sarwani, M., 2004. Indonesian peatland agriculture; past, present and future. Paper on the Wetland International Seminar on "Climate Change, Forests and Peatlands in Indonesia". Wetland International. http://www.peat-portal.net/view_file.cfm?fileid=296.

Sabiham, S., 2008. Suitability and sustainability aspects of peatland utilization for agriculture. Proceeding of National Seminar and Dialogue on Agriculture Land Resources, Bogor 10–20 November 2008. Center for Agriculture Land Resources Research and Development, Bogor, Indonesia.

Setyanto, P., Titi, S., Terry, A.A., Ali, P., Anggri, H., Sri, W., Wihardjaka, A., 2014. Green house gas emission from peatland exploitation and ameliorant application: synthesis from five research location. In: Wiharjaka, et al. (Eds.), Proceeding of National Seminar "Sustainable Management of Degraded Peatland for Mitigation of GHG Emission and Increasing Economic Value Added", Jakarta, 18–19 Agustus 2004.

Sterman, J.D., 2002. All models are wrong: reflections on becoming a systems scientist. On line publication http://jsterman.scripts.mit.edu/docs/Sterman-2002-AllModelsAreWrong.pdf (accessed 05.01.2014).

Subiksa, I.G.M., Wiwik, H., Fahmuddin, A., 2011. Sustainable peatland management. In: Nuraida, N.L., et al. (Eds.), Sustainable Management of Peatland. National Soil Research Institute, Center for Agriculture Land Resources Research and Development, Bogor, Indonesia.

Suriadikarta, D.A., 2009. Lesson learn from the failure of Mega million Rice Project to sustainable peatland management. Agric. Innov. Dev. 2 (4), 229–242.

Tubiello, N., Mirella, S., Simone, R., Alessandro, F., 2012. Analysis of global emissions, carbon intensity and efficiency of food production. Energia, Ambiente e Innovazione 4-5/2012. http://www.enea.it/it/pubblicazioni/pdf-eai/luglio-ottobre-2012/prima-parte/studi-research-analysis-emissions-food-production.

United Nation, 2013. World population projected to reach 9.6 billion by 2050. United Nation Report 2013, New York, US. https://www.un.org/en/development/desa/news/population/un-report-world-population-projected-to-reach-9-6-billion-by-2050.htm.

Decentralization of Forest Management, Local Institutional Capacity, and Its Effect on Access of Local People to Forest Resources: The Case of West Sumatra, Indonesia

Mahdi[*], *G. Shivakoti*[†,‡], *M. Inoue*[†]

[*]Andalas University, Padang, Indonesia [†]The University of Tokyo, Tokyo, Japan
[‡]Asian Institute of Technology, Bangkok, Thailand

16.1 INTRODUCTION

Decentralization of natural resources management is believed to be a better way to accommodate diverse interests of people at the local level in the process of making resource management more sustainable. Decentralized policy guides policy makers in deciding the appropriate measures for decentralizing natural resources management in most developing countries (Agrawal and Gupta, 2005; Andersson et al., 2006; Larson, 2002; Pacheco, 2006; Ribot, 2002). The level and scope for decentralization are different for different places and times depending upon the political, social, and developmental characteristics. Several scholars have found that decentralization is rendered by giving the property right and authority over natural resources management to local communities (Agrawal and Gupta, 2005; Agrawal and Ostrom, 1999; Enters and Anderson, 1999; Fisher, 1999; Lindsay, 1999). However, decentralization may not work in the absence of clear specific institutions that include mechanisms of accountability, oversight, and resource transfer (Andersson et al., 2004; Ostrom, 1990). Strong and dynamic local institutions are the prerequisite for successful implementation and good performance of decentralization (Ostrom, 1990; Uphoff, 1986).

The Indonesian government launched decentralization at the local level by enacting Law no. 22/1999 and then revised with Law no. 32/2002. The law was followed with new regulations regarding forest that emphasize decentralization of forest management. It recognizes the role of local institutions in forest management (Mahdi et al., 2009; Yonariza and Shivakoti, 2008). West Sumatra province of Indonesia responded to these laws, government and forest management decentralization, by transfering management rights to the lowest level of the local organization, the *nagari*, by introducing Provincial Regulation no. 10/2000. *Nagari* is now the formal institution for the lowest level of government after it was restored by replacing the *Desa* system of lowest level of uniform administration policy implemented throughout the country during the new order regime (Benda-Beckmann and Benda-Beckmann, 2001). The Solok district of West Sumatra issued District Regulation no. 4/2001 regarding the local government that mandates transfering natural resources management to *nagari*.

Nagari is the lowest-level political unit of the *Minangkabau* ethnic group who practices the matrilineal system. Nagari is composed of several neighboring hamlets that represent a clan (*suku*). A clan has several lineages (*kaum*) (Mahdi et al., 2009). Each nagari has its own rules and laws because nagaris are independent institutions. The nagari has a democratic, autonomous, and informal structure, with the clan and hamlet leaders placed on top (Naim, 1984). In the West Sumatra province of Indonesia, nagari had been the village organization of local government during colonial and postcolonial times up until 1983, when it was replaced with a system based on smaller administrative villages, *desa*. Nagari had been the informal institution since 1983 until the enacting of Provincial Regulation no. 10/2000 (Benda-Beckmann and Benda-Beckmann, 2001).

In this chapter, we assess the capacity of the nagari in regulating the sustainable use of resources and resolving conflict in natural resources management. Changes in local institutions for natural resources management directly affects the changing local livelihood (Batterbury and Fernando, 2006; Mahdi et al., 2009; Sunderlin et al., 2005; Tacconi, 2007). While forest resources are the main sources of livelihood for the people who are living in and around the forest, especially for the poor (Sunderlin et al., 2005; Yonariza and Webb, 2007), changing access to forest resources has a significant effect on poor household livelihood. We continue assessing local institution capacity with its effects on local livelihood.

16.2 DECENTRALIZATION, LOCAL INSTITUTION, AND LIVELIHOOD

Decentralization of natural resources management takes place in two ways. First is to devolve property rights over natural resources to local communities. Second is to hand over the formal powers of government to its own subunits. Both ways of decentralization claim that outcomes will be more efficient, flexible, equitable, accountable, and participatory (Andersson et al., 2004). In some publications, decentralization is rendered by giving the property rights and authority over natural resources management to local communities (Agrawal and Gupta, 2005; Agrawal and Ostrom, 1999; Enters and Anderson, 1999; Fisher, 1999; Lindsay, 1999). However, decentralization may not work in the absence of institutions that specify mechanisms of accountability, oversight, and resource transfer (Andersson et al., 2004; Ostrom, 1990). Strong and dynamic local institutions are the prerequisite for successful implementation and good performance of decentralization (Ostrom, 1990; Uphoff, 1986). To develop strong and dynamic local institutions takes time and effort (Lam, 2001).

Local communities are the main actors in decentralization of natural resources management. They can play a key role in shaping outcomes and will be the beneficiaries of positive outcomes as well as the victim of negative ones (Agrawal and Gupta, 2005; Agrawal and Ostrom, 1999). Local institutions, those whose policies, institutions, and processes are found in a specific geographical area and are more likely to directly affect the households living there (Messer and Townsley, 2003), play important roles in decentralized natural resources management. It is can be a bridge between decentralization and resources users. Linkages between local institutions and livelihood is taken to mean any way in which an institution influences or affects a livelihood strategy undertaken by a particular group or individual, or, vice versa, any way in which a livelihood strategy influences or affects an institution. By controlling access to assets, local institutions affect different livelihood assets or capital that people use for their livelihoods. Furthermore, an institution may change the context or interact with other institutions in which people live in a way that affect their vulnerability.

Decentralization may affect people at a local level individually or collectively to pursue their own interests regarding resources utilization (Holling et al., 2002). Individually, resource users will try to get benefits for their own livelihood security and improvement. Collectively, they will find new agreement among themselves and with other stakeholders. Therefore, decentralization will lead to changes in local communities' livelihood and local institutions, and it will uncover latent conflicts among local people and between local communities and other stakeholders.

Decentralization also affects the livelihood of local people (Dupar and Badenoch, 2002). Change of rules at the macro level has an influence on how local people manage resources (Lam, 2001). Theoretically, the connection between decentralization and change in local livelihood is initially assessed through the change of local institutions. Decentralization should ideally include the handing over of rights and authority to manage resources to local communities, often through the agency of local government (Andersson et al., 2006; Larson, 2002; Sunderlin, 2006).

16.3 RESEARCH METHOD

16.3.1 Rural Rapid Appraisal and Household Survey

Rapid rural appraisal (RRA) and household survey were carried out at three nagaris within the Lembang subwatershed, Nagari Selayo Tanang Bukik Sileh, Nagari Koto Laweh, Nagari, and Dilam. RRA was carried out to identify local institutions and their dynamism. In the course of RRA, in-depth interviews and focus group discussions (FGDs) were carried out. In-depth interviews were carried out with key informants, such as local leaders and forest- and water-related government officials who served as nodal points for FGD. The FGDs were organized at *nagari* and district levels. Household survey was carried out by clustering households into three groups based on their recent income: low, middle, and high. In each group, household samples were taken randomly. Table 16.1 shows the household sample characteristics by area and income groups.

TABLE 16.1   Household Samples Characteristics

No.	Household Characteristics	Income Group			Total
		Low	Middle	High	
1	Household number (N)	94	41	25	160
2	Average age of the head of household (year)	45.85	45.73	49.40	46.38
3	Average monthly household income per capita (Rupiah)[a]	149,438.86	349,378.05	994,963.47	332,786.50
4	Average household size	5.27	3.88	3.40	4.62
5	Average years of formal education	6.45	7.32	6.60	6.69

[a] During the study period, 1 US dollar was equal to 9000 Indonesian Rupiah.

16.3.2 Data Analysis

Both qualitative and quantitative analyses were done to achieve the objectives. Qualitative analysis was done to identify local institutions for forest management, and to study their changes during the last 10 years, their interaction with other institutions, and the problems they have faced. The RRA was carried out in three steps: preparatory activities; field activities; and validation forum. Preparatory and field activities were carried out to collect data and information. The researcher then organized the data and information, and brought them to validation forum. The validation fora were organized for three purposes: first to clarify data and information that were gathered in RRA field activities; second to confirm differences, conflicts, and overlapping of rules and organizational tasks on forest and water management in the nagari; and third to analyze data and information and to make conclusions in a participative way. The same techniques were employed to learn about problems of resources management among stakeholders at the district level.

Statistical analyses including t-test and one-way ANOVA test were carried out to examine the significance of differences in access to resources at two points in time: before and after decentralization (1996 and 2006). A one-way ANOVA test was done to examine differences in resources access among the three income groups. Access to resources was measured by developing the index, comprising combined indices of variables; that is, access to forest and land as shown in Table 16.2.

TABLE 16.2   The Variables for Indexing

No.	Variable	Index
1	Access to forest resources	The percentage of family income from forest product gathering both timber and Nontimber Forest Product (NTFP)
2	Access to land	Security of land ownership. Private land is given 1 Lineage land 0.5 and Rent/sharecropper 0.2 Quality of land. Paddy field is given 1, dry land 0.5, and other 0.2

16.4 STUDY SITE OVERVIEW

Lembang subwatershed is part of Sumani watershed, the most important watershed in the central part of West Sumatra. It is situated in the southern part of the watershed, and is under Solok District administration. In the upland part of the study site lies Talang Mountain, the most active volcanic mountain in Sumatra. The lowland consists of plains area and is the center of West Sumatra's rice production. Its altitude ranges from 400 to 1700 m above sea level with 7768 mm average rainfall with a range of 34–212 rainy days in a year. The lowest rainfall is usually recorded in July, while November to February receive the higher amount of rainfall. Temperature ranges between 12.5°C and 24.60°C. The population density in the study area is the highest compared to other subwatersheds: 329 persons per square kilometer. The population has grown at about 1.3% annually during the last 5 years (BPS, 2005).

Almost all of the people belong to the *Minangkabau* ethnic group. This ethnic group has the matrilineal lineage system, which involves four characteristics (Kato, 1982). First, the descent and descent groups are organized according to the female line; thus, all children belong to the mother's clan. Second, a lineage possesses communal properties such as agricultural land, ancestral treasures, and miscellaneous *adat* titles. Third, the residential pattern is uxorilocal; that is, the husband resides in the house of his wife's kin after marriage. And fourth, authority within a lineage is in the hands of the *mamak rumah* (mother's eldest brother), and not in the hands of the father.

This is an intricate and unique sociopolitical system. Most sociopolitical units in West Sumatra are formed as a result of interactions among families and clans. *Mamak rumahs* of the same lineage choose a leader whom they call *Mamak Barih*. Then, from among the *Mamak Barih* within a clan, *Datuk* is elected. Many clans, then, form hamlets and many hamlets form *nagari*. A *nagari* is led by a *Wali Nagari*. A *nagari* has a democratic, autonomous, and informal structure dominated by the clan and hamlet leaders (Naim, 1984). Because of independent institutions, each *nagari* has its own rules and laws, which can be different from others.

16.5 DECENTRALIZATION PROCESS IN INDONESIA AND RESTORATION OF THE NAGARI

The Indonesian government decided to decentralize its political and governmental administration after the economic crisis of 1998 that led to social and political turmoil. Decentralization of the regional government was implemented in 2000 based on Law no. 22/1999 for regional government autonomy (replacing Law no. 5/1974). Bahl (2001, cited in Esden, 2002) described the decentralization in Indonesia as "one of the largest decentralization programs that has been seen, it was done quickly, and there still is not a detailed transition plan."

In 2004, the government further changed the law at the regional level with regional government Law no. 32/2004. The main highlight of this law is the specification of resources, revenue, and role-sharing mechanisms among central, provincial, and district governments (Cahyat, 2005). Previously, Law no. 22/1999 gave wider role for district government and less power for provincial government both in governmental administration and natural resources management while the Law no. 32/2004 has assigned balanced role among central, provincial, and local government (Cahyat, 2005). The new law gives a greater opportunity for local government to get a higher income share from natural resources management, provided the local government takes over the management task. With respect to the lowest level of governmental administration, the new law gives more power to the head of village. Law no. 32/2004 ended the controlling function of the village legislative bodies (*badan perwakilan desa*).

In addition, the central government formulated forestry Law no. 41/1999 to replace Law no. 5/1967, and launched Law no. 7/2004 regarding water resources to replace Law no. 11/1974. The new forest law authorizes the partial transfer of forest authority to the local government and the recognition of local customary ownership and management systems. The new water resources law gives clearer instructions to farmers concerning their duties and responsibilities to finance, operate, and maintain tertiary irrigation canals under government or local government assistance. The two laws give more attention to watershed resources conservation and protection.

During field research, some changes in resources management at the local level were identified. Governmental decentralization programs responded by returning to *nagari* as the main local institution. Consequently, local people in West Sumatra were given a chance to apply their customary laws in governance and resources management of natural resources. In forest management, for instance, *nagari* recently made the rules concerning the harvesting of forest products with the *nagari's* government performing evaluation and monitoring. However, in three *nagaris* of the upland research site, small forest plots owned by *nagari* were found to suffer from lack of attention.

Changing the national law has also affected the *nagari's* legislative body (*badan perwakilan nagari*). The body cannot control the wali *nagari* (the head of *nagari*), but it can work collaboratively in drafting *nagari* regulations. *Nagari's* natural resources management, actually, cannot be intervened, because the ownership in *adat* law is in the hands of a board of the *nagaris* clan leader (*kerapatan adat nagari* (KAN)).

16.5.1 Local Property Rights

The land tenure system of West Sumatra still follows traditional (*adat*) principles (Balzer et al., 1987; Yonariza, 1996). Complex land rights and ownership existed in a society based on matrilineal cultural background. There are four main type of land status (Balzer et al., 1987; Kato, 1982; Yonariza, 1996). First, *nagari's* land (*tanah ulayat nagari*) refers to land that is controlled by *nagari*. All members of *nagari* have a right to access. Second, clan land (*tanah ulayat suku*), which can only be accessed by members of the clan and only they can withdraw benefit from the land. Third, lineage land (*tanah ulayat kaum*) refers to land that is collectively controlled by the lineage. In many publications, clan and lineage land are also called "high inherited land" (*tanah pusako tinggi*) because these lands are inherited through more than two generations from the ancestors. Fourth, private or individual land refers to the land owned by individuals or groups privately. Private land can be procured in two ways; first is to buy the land by an individual and second is to inherit from parents. The latter is also called "low inherited land" (*tanah pusako rendah*). Our analysis will focus on the *nagari's* land or *tanah ulayat nagari*, because land management change has mostly affected this kind of land ownership.

The utilization of *nagari's* land is decided collaboratively among *nagari's* clan leaders in the forum that is called the board of *nagari's* clan leader (*kerapatan adat nagari*/KAN). *Nagari* people can either utilize or cultivate a plot of land if no one else does. KAN can give the utilization right of the land to someone else, if he or she has not cultivated the land for 4 or 6 years (each *nagari* has a different time range). The land utilization right is like open access for the *nagari's* people. When someone cultivates a plot of land, the utilization right of the land is in his/her hands and can be inherited by his/her children as well as his/her nephews/nieces. Because of population growth, most of *nagari's* land is utilized and land use tends to get more intensive. Because of such pressure, forest cover in the entire area of the Lembang subwatershed has been reduced significantly.

There are three categories of forestland ownership and utilization rights in the studied *nagaris*. First is *nagari's* forest where utilization rights have not yet changed and that is still covered by forest. It is a tiny part of the area and most of it is infertile land in the high slopes. Some of the land was left by the people and regreened by the government during the last 30 years by planting *Pinus merkusii*. Second is the *nagari's* land that is either covered by forest, or cultivated for agriculture, or consists of degraded land. Most of *nagari's* land is in this category. Third is clans'/lineage's land that is covered by forest or cultivated for agriculture, or consists of degraded land.

Table 16.3 shows land-use change in the Lembang subwatershed since 1890. In 2004, forest cover was only 2.98% compared to 24.25% in 1890, 8.07% in 1976, and 7.74% in 1993. Land for settlement has increased rapidly due to high population growth. The highest cumulative change of settlement at a record 565% was from 1890 to 1976. This trend was due to high in-migration to this zone because of its fertile soil. Volcanic soil is the main soil type here and is very suitable for horticulture in upland and for paddy cultivation in lowland.

The high intensity of shifting cultivation during the same period led to land degradation. The percentage of degraded land (that occupied by *imperata cylindrica*) is the highest among other land use, ever since the Dutch colonial era. Degraded land occupied 42.88% of land in 1890 and increased to 48.78% in 1976. It reduced to 41.83% in 1993 and increased again to 46.65% in 2004.

16.5.2 Forest Management Within the Nagari

The main idea of return to *nagari* is to empower *adat* law by implementing it through formal regulation within *nagari* both for governance and management of natural resources. Based on *adat* law, natural resources owned by *nagari* were employed in a sustainable way for *nagari* sustenance.

To effectively implement *adat* law within the *nagari*, there are two challenges of forest resources management. On one hand, *adat* leaders tend to retain their role and right over forest resources and land after return to *nagari*. They like to readminister the *nagari's* land and impose their own regulation upon it. On the other hand, most of the land utilization right is either in the hands of *nagari's* people or some even in the hands of outsiders. Based on the *adat* rule, KAN cannot take over the *nagari's* land if people continue to live and use it for longer than 4–6 years (different *nagari* have different ranges) even though they have not cultivated it for productive purposes. KAN only can take it for public purposes such as for public infrastructure construction, public open space, and so on, if it provides compensation to these people.

TABLE 16.3   Lembang Subwatershed Land Use Change 1890, 1976, 1993, and 2004

Land Use	1890		1976		% Cumm. Change	1993		% Cumm. Change	2004		% Cumm. Change
	Ha	%	Ha	%		Ha	%		Ha	%	
Forest	4142.67	24.25	1379.29	8.07	−66.71	1322.36	7.74	−68.08	509.71	2.98	−87.70
Shifting cultivation	1628.30	9.53	661.47	3.87	−59.38	5666.33	33.16	248.00	940.94	5.51	−42.21
Bush	2851.06	16.69	1488.34	8.71	47.79	443.90	2.60	−84.43	3158.78	18.49	10.79
Settlement	713.91	4.18	4750.92	27.81	565.48	1983.30	11.61	177.80	3907.74	22.87	447.37
Paddy field	423.40	2.48	472.34	2.76	11.56	522.57	3.06	23.42	597.96	3.50	41.23
Degraded	7326.84	42.88	8333.82	48.78	13.84	7147.72	41.83	−2.44	7971.06	46.65	8.79
Total	17,086.18	100.00	17,086.18	100.00		17,086.18	100.00		17,086.19	100.00	

Source: Istijono, B., 2006. Konservasi Daerah Aliran Sungai dan Pendapatan Petani: Studi Tentang Integrasi Pegelolaan Daerah Aliran Sungai. Studi Kasus DAS Sumani Kabupaten Solo/Kota Solok, Sumatera Barat. Unpublished Dissertation, Universitas Andalas, Padang.

This situation has two implications for forest management within the *nagari*. First, administering forestland causes conflict either with neighbors or with *nagari's* people. Conflicts with neighboring *nagari* are linked with two conditions: the absence of definitive borders among *nagaris* when return to *nagari* was implemented, and the recent increase of the real value of land and forest resources. Traditionally, each *nagari* has an oral story/legend about *nagari's* border. Each *nagari* has different versions of such stories so that in the field the claimed areas are overlapping. Second, most of the land is degraded and difficult to rehabilitate. Even though people do not cultivate it, they still keep their right by planting something to signal that the land is still utilized and should not be taken.

16.5.3 Conflict Over Forest Resources Management

The earlier period of decentralization implementation in Indonesia faced some conflicts and our study area also experienced it. Table 16.4 recapitulates the conflicts within the Lembang subwatershed due to changes in the rules of government administration related to natural resources management. There were three conflicts recorded on issues related to forest and land resources management, and three others in water and irrigation management. Conflicts have been recorded among communities, between communities and *nagari*, between *nagari*, *nagari* and local government, and between *nagaris* and other resource user groups. Until recently, the conflicts are not yet resolved completely because of the absence of conflict resolution channels, in particular conflicts involving more than one *nagaris*.

There are two main roots of conflict: increasing resources scarcity and unclear rules of the game with respect to resources utilization. Conflicts are caused by population growth on the one hand and resources depletion on the other. Population has increased sharply and, therefore, the subwatershed has the highest population density in the Sumani watershed. Rapid land-use changes produce big portions of degraded land. Because of depletion of upland forest, high fluctuation in water availability and quality is a common phenomenon in the entire subwatershed. Istijono (2006) reported that the subwatershed has a high degree of soil erosion as well.

All these issues are not yet tackled because of weak and in some cases due to the absence of effective local institutions. When central authoritarian power declines and new institutions

TABLE 16.4   Conflicts on Forest and Water Management Within the Lembang Subwatershed

No.	Type of Conflicts	Root of Conflict	Stakeholders Involved
1	*Forest and land resources*		
	• Land titling	• Unclear ownership	• Among communities
	• Forest resources exploitation	• Unclear ownership	• Among communities
	• *Nagari's* border	• Local customary law allows people from other *nagaris* to obtain forest land in their neighboring *nagari* • Ancestor's agreement on *nagari's* border are not formally documented	• Between *nagaris*

are not yet established, conflict is the consequence. However, during the FGD at the district level, where all stakeholders participated, a wish to establish a new forum for conflicts and problems solving was expressed. Local government officials have initiated facilitating this aspiration.

16.6 HOUSEHOLD ACCESS TO FOREST RESOURCES

16.6.1 Household Participation in Forest Management

Table 16.5 recapitulates the distribution of households by their expressed personal interest in the involvement of managing forest in the upland area. In general, most households participated in forest management to get a short-term job from the forest protection and land rehabilitation project, followed by preventing a natural disaster, maintaining social relationships, and making water more available in the future. Although, forest resources contribute to household income, fewer and fewer locals expect to get more benefit from forest resources. Different interests are found among the different income groups. The high-income group participated in forest management within their *nagari* mostly to prevent natural disaster, while the low-income group would like to get short-term economic support from the government project for rehabilitation and protection of forest and land.

During the last 5 years, the government has launched forest and land rehabilitation programs within this area, by involving local communities in replanting degraded forestland. Their motivation to participate was to get the economic incentive from the project. This is the main reason for participating in forest management for more than one-third of the respondents. To prevent natural disaster is the second reason why locals participated in forest management (23.91%). When the government of *nagari* asks them to work voluntarily (*gotong-royong*) to replant degraded forest and land, they participate because they want to maintain social relationships. This reason is in the third priority for 18.48% of the respondents.

TABLE 16.5   The Distribution of Household by Their Interest to Participate in Forest Management Within the *Nagari*

No.	Vested Interest to Participate in Forest Management	Low Income	Middle Income	High Income	Total
1	To get job forest protection and rehabilitation project	41.67	26.67	21.43	33.70
2	To make water more available in the future	14.58	13.33	14.29	14.13
3	To get more benefit from forest resources	10.42	13.33	–	9.78
4	To maintain social relationship	20.83	13.33	21.43	18.48
5	To prevent natural disaster	12.50	33.33	42.86	23.91
	Total	100.00	100.00	100.00	100.00

When we look at income groups, there are different interests in different groups. For locals in the low-income group, their top most interest is to get short-term job from forest protection and rehabilitation project (41.67%) and maintaining social relationship and making water more available in the future are in the second and third reasons respectively. The richest group, on the other hand, participate in forest management within the *nagari* mostly because the want to prevent natural disaster (42.86%). It means the low-income group needs alternative sources of income to secure their livelihood.

16.6.2 Access to Forest Resources and Land

Although it is difficult to claim that the changes of resource access in the research site are due to decentralization alone, decentralization, followed by institutional changes both at the national and local levels, seem to be the main factor that propels resources access change. The dynamics of local people's access to natural resources before and after decentralization is shown in Table 16.6. The table shows differences in access to resource indexes in 1996 and 2006 in upland as well as the differences in access among different income groups. Generally, upland people experienced a huge change in resources access during the last 10 years. The differences in 1996 and 2006 of resources access indexes in the upland area are statistically significant.

There was no significant change in forest resource access in the upland in general among different income groups; the people from the middle-income group obtained more forest resources from upland in 1996, 15% of their income came from it, while the low- and high-income groups got 9% each. These differences were statistically significant. However, the people from low- and high-income groups have taken a bigger portion of resources during the last 10 years than the middle group. In 2006, forest extraction contributed 15% of income of the people from the low-income group and 13% of income of the people from the high-income group, while the middle-income people have gotten less than before. Briefly, we can conclude that the people from the low-income group got higher access to forest in the upland area during the last 10 years.

The distribution of land access index shows a significant difference among income group in upland, where the people from the middle-income group have received the highest benefit.

TABLE 16.6 The Average of Resources Access Indexes in 1996 and 2006, Its Change and Statistical Test

Household Access to Resource	Year/Change	Income Group				One-way ANOVA	t-Test
		Low	Middle	High	Total		
Forest resources	1996	0.09	0.15	0.09	0.10	3.0006[a]	1.9484
	2006	0.15	0.11	0.13	0.14	0.9989	
	Change	0.06	−0.05	0.04	0.03	4.2053[a]	
Land	1996	0.40	0.29	0.45	0.38	1.8753[a]	7.9330[a]
	2006	0.59	0.66	0.73	0.63	4.5440[a]	
	Change	0.19	0.37	0.28	0.25	3.0810[a]	

a Significant at 95% confidence.

The variation of access among income group is also substantial, with the middle-income groups enjoying high and increasing access. The low- and high-income groups accessed 0.40 and 0.38 of index, respectively, while the middle-income group accessed 0.29 in 1996. Then, in 2006, the middle-income group access index increased to 0.66, the highest change among other income groups. In the same period, people from low- and high-income access index increased to be 0.59 and 0.73, respectively. In summary, access to land resources increased significantly and the middle-income group has received a greater benefit. In summary, access of upland people to forest resources has increased significantly during the last 10 years. The middle-income and high-income groups received more benefit from the change in access to land resources.

Changes in local institutions due to decentralization of forest management lead to changing access of local people to forest resources for their main livelihood. Due to a weak local institution, the elites, high- and middle-income groups, within the *nagaris* occupied more land for their livelihood enhancement. Meanwhile, the low-income groups extracted more forest resources for surviving against economic shock and less jobs during the last 10 years.

These findings provide two important implications for future forest management within the context of Indonesian decentralization. First, income disparities among groups will increase. With less forest resources, the low-income group is facing difficulties to exploit more forest resources to support their livelihood. On the other hand, the high- and middle-income groups are enjoying more land for their investment in intensive agriculture for higher livelihood security. Second, pressure on forest resources is becoming higher especially from the low-income group. It is becoming worse because the capacity of the local institution is still weak to formulate and implement regulations regarding resources use.

16.7 CONCLUSION

Decentralization has been the main political and administrative decision in most developing countries during the last three decades. In many countries, decentralization in forest resources management means transfer of part of the authority and the right of management to people at the local level. Most attention has been paid to the effect of such decisions on management. A wealth of publications concerning the pros and cons of decentralization have been produced. However, less attention has been paid to the capacity of existing local institutions to receive it.

Decentralization, which has been implemented during the last 8 years in Indonesia, and which is continuing to be implemented, still causes uncertainty to people at the local level. On the one hand, local institutions have changed the institutional arrangement in forest management, but confusion and conflicts between central government and local customary rule still exist. On the other hand, conflicts over resources extraction have become more rampant because of the absence of a prior agreement among stakeholders. Coupled with high tension from growing population and economic development, this situation has led to accelerated resources depletion. In conclusion, local institutions are still too weak to receive management rights transfer at this time. Some adjustment and empowerment are needed to improve the capability of the local institution to handle forest management effectively. Individually, people at the local level respond to the new governance of resources by pursuing their own interests.

At the local level within the Lembang subwatershed, a decentralization policy leads to changes in the local institutions that shift access of local people to forest resources and land. Both changes eventually produce conflict in resources utilization, disparities in income distribution due to different access among different income groups, and increasing pressures of local people on forest resources.

16.8 POLICY IMPLICATION

The lessons that could be extracted from this study that might be the basis for policy implications include:

1. Similar to the study of Lam (2001) in Taiwan, this study also found that strong and dynamic local institutions need time and effort to evolve over time. Although decentralization has been implemented during last 8 years, the new local institutions have still not been able to tackle their collective actions, especially actions among two or more *nagaris*.
2. The facilitation from upper-level institutions to establish and empower local institutions is needed to pursue an effective decentralization implementation process in the future. *Nagaris* need to be empowered in strengthening and implementing their own regulations for forest resources use and access.
3. Creating alternative income sources, especially for the low-income group, is a way to reduce pressure on forest resources.
4. Rising conflict among local communities, provincial, and district government is the main problem to develop collective action in forest management within the study area. Assisting a local institution in adopting alternative dispute resolution is a way to encourage them to find agreement and develop an institution for sustainable forest management.

Acknowledgments

This study was funded by a grant from the Ford Foundation—Jakarta Office made to the Andalas University and Asian Institute of Technology, which is duly acknowledged. We are grateful to local residents of the Lembang subwatershed for their participation in interviews and surveys. Special thanks are also extended to the head of the *nagaris* for their help. The earlier version of the paper was the poster note presented at the international conference on Poverty Reduction and Forests: Tenure, Market, and Policy Reforms in Bangkok, Thailand, Sep. 3–7, 2007.

References

Agrawal, A., Gupta, K., 2005. Decentralization and participation: the governance of common pool resources in Nepal's Terai. World Dev. 33 (7), 1101–1114.

Agrawal, A., Ostrom, E., 1999. Collective action, property rights, and devolution of forest and protected area management. In: Paper Presented in International Conference: Collective Action, Property Rights and Devolution of Natural Resource Management: Exchange of Knowledge and Implications for Policy. Puerto Azul, The Philippines. 21–25 June.

Andersson, K.P., Gibson, C.C., Lehoucq, F., 2004. The Politics of Decentralized Natural Resource Governance [Electronic Version]. *PSOnline*, pp. 421–426. Retrieved May 18, 2006 from www.apsanet.org.

Andersson, K.P., Gibson, C.C., Lehoucq, F., 2006. Municipal politics and forest governance: comparative analysis of decentralization in Bolivia and Guatemala. World Dev. 34 (3), 576–595.

Balzer, G., Deipenbrock, N., Ecker, R., Eisenbeis, M., Focken, V., Gihr, K., 1987. Shifting Cultivation in West Pasaman, Sumatra. Centre for advanced training in agriculture development. Institute of socio-economics of agriculture development, Technical University of Berlin, Berlin.

Batterbury, S.P.J., Fernando, J.L., 2006. Rescaling governance and the impacts of political and environmental decentralization: an introduction. World Dev. 34 (11), 1851–1863.

Benda-Beckmann, F.v., Benda-Beckmann, K.v., 2001. Recreating the Nagari: Decentralisation in West Sumatra. Max Planck Institute for Social Anthropology, Halle (Working Paper No. 31).

BPS, 2005. Sumatera Barat Dalam Angka 2004 (West Sumatra in Figures 2004). Badan Pusat Statistik Sumatera Barat (BPS), Padang.

Cahyat, A., 2005. Perubahan Perundangan Desentralisasi Apa Yang Berubah? Bagaimana Dampaknya Pada Upaya Penanggulangan Kemiskinan? Dan Apa Yang Perlu Dilakukan? (No. 22). Centre for International Forest Research (CIFOR), Bogor, Indonesia.

Dupar, M., Badenoch, N., 2002. Environment, Livelihoods, and Local Institutions Decentralization in Mainland Southeast Asia. World Resources Institute, Washington, DC.

Enters, T., Anderson, J., 1999. Rethinking the decentralization and devolution of biodiversity conservation. Unasylva 50 (199), 6–11.

Esden, B.P., 2002. INDONESIA: Rising Above Challenges. In: Brillantes, A.B., Cuachon, N. (Eds.), Decentralization and Power Shift: An Imperative for Good Governance. A Sourcebook on Decentralization Experiences in Asia, Vol. 1. Asian Resource Center for Decentralization, University of the Philippines, Diliman, Quezon City, pp. 115–126.

Fisher, R.J., 1999. Devolution and decentralization of forest management in Asia and the Pacific. Unasylva 50 (199), 3–6.

Holling, C.S., Gunderson, L.H., Ludwig, D., 2002. In a quest of a theory of adaptive change. In: Gunderson, L.H., Holling, C.S. (Eds.), Panarchy: Understanding Transformations in Human and Natural Systems. Island Press, Washington.

Istijono, B., 2006. Konservasi Daerah Aliran Sungai dan Pendapatan Petani: Studi Tentang Integrasi Pegelolaan Daerah Aliran Sungai. Studi Kasus DAS Sumani Kabupaten Solo/Kota Solok, Sumatera Barat. Unpublished Dissertation, Universitas Andalas, Padang.

Kato, T., 1982. Matriliny and Migration: Evolving Minangkabau Traditions in Indonesia. Cornell University Press, Ithaca.

Lam, W.F., 2001. Coping with change: a study of local irrigation institutions in Taiwan. World Dev. 29 (9), 1569–1592.

Larson, A.M., 2002. Natural resources and decentralization in Nicaragua: are local government up to the job? World Dev. 30 (1), 17–31.

Lindsay, J.M., 1999. Creating a legal framework for community-based management: principles and dilemmas. Unasylva 50 (199), 28–34.

Mahdi, Shivakoti, G., Schmidt-Vogt, D., 2009. Livelihood change and livelihood sustainability in the uplands of Lembang subwatershed, West Sumatra, Indonesia, in a changing natural resource management context. Environ. Manag. 43 (1), 84–99.

Messer, N., Townsley, P., 2003. Local Institutions and Livelihoods: Guidelines for Analysis. Rural Development Division of Food and Agriculture Organization of the United Nations, Rome.

Naim, M., 1984. Merantau: Pola Migrasi Suku Minangkabau (Merantau: Migration Pattern Among Minangkabau Ethnic). Gajah Mada University Press, Yogyakarta.

Ostrom, E., 1990. Governing the Commons: The Evolution of Institutions for Collective Action. Cambridge University Press, Cambridge.

Pacheco, D., 2006. Opening Common-Property Forests to Timber Production: Bolivia's Community Forestry Policies, Indigenous Timber User Groups' Performance and Local Perceptions of Forests' Livelihoods. In: Paper Presented in IASCP 2006 Conference, Bali, Indonesia, June.

Ribot, J.C., 2002. Democratic Decentralization of Natural Resources: Institutionalizing Popular Participation. World Resources Institute (WRI), Washington, DC.

Sunderlin, W.D., 2006. Poverty alleviation through community forestry in Cambodia, Laos, and Vietnam: an assessment of the potential. Forest Policy Econ. 8, 386–396.

Sunderlin, W.D., Angelsen, A., Belcher, B., Burgers, P., Nasi, R., Santoso, L., Wunder, S., 2005. Livelihoods, forests, and conservation in developing countries: an overview. World Dev. 33 (9), 1383–1402.

Tacconi, L., 2007. Decentralization, forests and livelihoods: theory and narrative. Glob. Environ. Chang. 17 (3–4), 338–348.

Uphoff, N., 1986. Local Institution Development: An Analytical Sourcebook With Cases. Kumarian Press, West Hartford, CT.

Yonariza, 1996. Agricultural Transformation and Land Tenure Systems: A Study of a Shifting Cultivation Community in East Rao Pasaman District, West Sumatra, Indonesia. Ateneo de Manila University, Manila.

Yonariza, Shivakoti, G.P., 2008. Decentralization policy and revitalization of local institutions for protected area co-management in West Sumatra, Indonesia. In: Webb, E.L., Shivakoti, G.P. (Eds.), Decentralization, Forests and Rural Communities: Policy Outcomes in South and Southeast Asia. Sage, New Delhi, pp. 128–149.

Yonariza, Webb, E.L., 2007. Rural household participation in illegal timber felling in a protected area of West Sumatra, Indonesia. Environ. Conserv. 34 (1), 73–82.

Can Uplanders and Lowlanders Share Land and Water Services? (A Case Study in Central Java Indonesia)

I. Andriyani, D. Jourdain[†,‡], G. Shivakoti[§,∥], B. Lidon[‡], B. Kartiwa[¶]*

*Jember University, Jawa Timur, Indonesia †Asian Institute of Technology, Klong Luang, Pathumthani, Thailand ‡CIRAD UMR G-EAU, Montpellier, France §The University of Tokyo, Tokyo, Japan ¶IAHRI, Indonesian Agriculture Research Institute, Bogor, Indonesia ∥Asian Institute of Technology, Bangkok, Thailand

17.1 INTRODUCTION

Food sovereignty is one of the government of Indonesia's most important programs, which requires full support from the best natural resource condition as well as technology and regulations. On the other hand, natural resource degradation increases rapidly and threatens the food sovereignty program. Deforestation to agricultural and the settlement causes natural resources degradation. In Indonesia forests are one of the natural resources that affect human life, so that by authority of its 1945 Constitution, the government is both owner and manager of the forest with the aim of improving people's welfare. The deforestation rate in Indonesia in 2003–2006 reached 1.17 million hectares per year (Rusli, 2008). The 1950–2000 deforestation was 40% of the total forest area, and the fastest rate of deforestation started in 1998 (Arnold, 2008).

According to Minister of Forestry Zulkifli Hasan (Ministry cabinet for the period 2009–2014), the total forest area in Indonesia is 180 million hectares, of which 21% of forests have been destroyed and 25% is deforested and in bad condition due to farmers' activities and industry exploitation. Deforestation is caused not only by population explosion but also by the political situation in Indonesia. The collapse of the new order under the Suharto regime in 1998 led to the pillage of forest (deforestation). Moreover, decentralized governance through regional autonomy makes deforestation worse. Decentralization changes of authority and responsibilities in forest management leads to overexploitation of forest.

The second cause is the development of agroindustries. Conversion of forest to plantation areas (such as coffee, rubber, sugarcane, and oil palm) had destroyed more than 7 million hectares of forest up to the end of 1997 (Angelsen, 1995, 1999; van Noordwijk, 2002).

Another cause is transmigration. The aim of transmigration is the distribution of population, improvement of the economy, and development of other regions outside of Java Island. Forest clearing carried out to provide housing and agricultural land for the citizens of transmigration is causing deforestation (Arnold, 2008).

The transformation of forests into agricultural areas both for food crops and plantation crops have many negative impacts on natural resources as well as the hydrology function (Angelsen, 1995, 1999; Gaveau et al., 2009; van Noordwijk et al., 2004). The negative impacts of deforestation, especially in mountainous areas, can be on-site or off-site. On-site impacts include landslides and loss of soil fertility due to erosion. Off-site impacts include sedimentation and water shortage in the dry season. Moreover, deforestation creates many problems for water resources services and increases conflicts among water users.

Java Island, Indonesia, only has 7% of the total area of Indonesia but is home to 65% of the Indonesian population, and it faces water problems. Total area for forest decreases continually while agricultural and housing areas increase. This induces problems on-site and off-site: land degradation, increased sedimentation and pollution, erosion, floods, water shortages, and so on. In the dry season when water shortages occur, conflicts between water users escalate and can threaten the sustainability of both business and agricultural production.

Agricultural activities not only produce main products but also create pollution, such as sediments, pesticides, animal manures, fertilizers, and other sources of inorganic and organic matter into receiving waters. For example, the agriculture system in Indonesia mostly is irrigated for the lower area and rainfed for the upper area (Hermiyanto et al., 2004).

Agricultural activities in the upper area are greatly influenced by agricultural performance in the lower area (van Noordwijk, 2004). So it could be stated that agriculture in the upper area could be constraints for the agricultural activities in the lower area. Kurnia (2003) states that the biophysical constraints for land use are associated with inappropriate land use and environmental management as well as soil characteristics. For examples, the area with high rainfall (ie, 1.500–3.000 mm per year), located on the slope area, is susceptible to erosion, which can cause loss of fertile surface soil layers.

Intensification of agricultural activity has led to land degradation, so the opposite of this activity can reduce the erosion that may occur. Activities of deintensification include replanting of open land or replacement of plants that are less protected with cover crops that reduce erosion and sedimentation effects (Vanacker et al., 2005).

Intensive agriculture on a sloping area causes increased erosion and sedimentation rate (Bakker et al., 2005). This is caused by the nature of the land itself that is susceptible to erosion and intensification in this case is the intensification of agricultural inputs (such as fertilizer, improved seed, water uses, and pest eradication) rather than the intensification of land cover. Erosion will cause the displacement layer of soil in the form of sedimentation into the river, which in turn would reduce the river's hydrological functions (Bakker et al., 2005; Esen and Uslu, 2008; Van Rompaey et al., 2007).

The sustainability of natural resources and economic development require a feasible method of shared management of natural resources among all stakeholders both upstream and downstream of the watershed. The Payment for Environmental Services (PES) method seems promising to implement in a watershed that has conflicts with both upstream and downstream environment users.

We selected three subwatersheds in Central Java where conflicts among water users both upstream and downstream occur. In the upstream, we analyzed types and activities of water users using cluster analysis and predicted the impact of their activities on water supply. Moreover we identified farmers' constraints and analyzed the relationship of farmers' characteristics and activities and how these influence erosion yields and constraints. We found water user conflicts in the downstream caused by water storage. Deforestation and farmers' activities in the upstream cause land degradation (high erosion yields 212 ton/ha/year) and reduce water supply for the downstream area.

17.2 STUDY AREA

As a big rice producer in Indonesia, Central Java faces some land and water degradations. Land use and land cover changes (LULCC) also occurred in Central Java, more specifically in the Samba, Soka, and Pusur watersheds, which are located at the crossing of two districts named Boyolali District and Klaten District. The upper watershed (above 400 m a.s.l.) is in Boyolali and the lower part (below 400 m a.s.l.) is in Klaten. It is on the foothills of the Merapi volcano (Fig. 17.1).

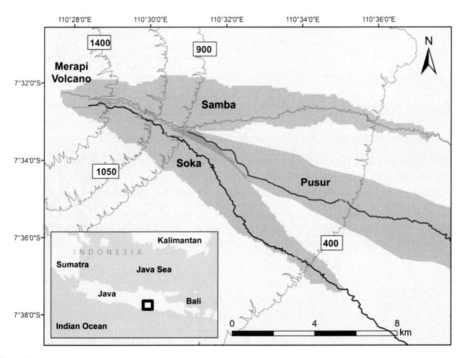

FIG. 17.1 Study area.

The Samba (2492 km²), Soka (2317 km²), and Pusur (5781 km²) watersheds are located at 7°31'50"S–7°37'29"S and 110°27'35"E–110°37'30"E. Those three subwatersheds are part of the Bengawan Solo watershed, the biggest watershed on Java Island. The altitude of the research area varies between 400 and 1400 m a.s.l.

The average minimum temperature is about 20°C and the average maximum temperature is 30°C. The relative humidity is varied from 50% to 90% during the day. The average rainfall is around 1458 mm/year, maximum rainfall occurs between Nov. and Jan. and could reach around 450 mm/day, whereas minimum rainfall usually occurs between Jul. and Aug. with only around 12 mm/day.

In the upper watershed, the forested area has declined steadily as a result of increasing agricultural activities. The majority of the land is situated on steep slopes. Nevertheless, intensive seasonal crops using quality seeds, chemical fertilizers, and pesticides, are cultivated on the fragile soil. Moreover, soil and water conservation technologies are seldom used. These activities cause land degradation by erosion and increase the occurrence of landslides.

On the other hand, lower-catchment areas face different problems, as the important agricultural activity in the lower area of the Pusur watershed is irrigated rice farming called *sawah*. Klaten is a major rice producer area of Central Java. Nowadays conflicts among water users in the district of Klaten are more and more frequent. These conflicts occur because of decreasing water availability. Water discharge of the Pusur River decreased by 33% between 1994 and 2005 and sedimentation also increased in the river and irrigation channels (Kartiwa, 2006).

17.3 METHODOLOGY

17.3.1 Upstream

A survey and an interview 180 respondents was conducted in the upstream. Data collected included structural data, on-farm data, off-farm data, cropping system, and plots details. Analysis used the following:

1. Cluster analysis using the factor analysis for mixed data (FAMD) method to identify farm types, characteristics, and activities,
2. Erosion yield in plot level calculated using the Revised Universal Soil Loss Equation (RUSLE) method,
3. Farmers face labor and water constraints. We identified he constraints using labor balance analysis and survey of water resource availability.

17.3.2 Downstream

Water user conflicts were identified using survey, interview, and focus group discussion (FGD). We identified three water users (a) farmers, (b) bottled water company as representative of the industry sector, and (c) the local water company (*Perusahaan Daerah Air Minum*) (PDAM) as representative of the urban water user sector.

17.4 RESULTS

17.4.1 Water User in the Upstream

Uplanders are rainfed agricultural farms and small-scale livestock. The majority of the land is situated on steep slopes. Main crops in the upstream are vegetables and maize in the rainy season and tobacco in the dry season. Grass fodder and roses grow all year long. Nevertheless, intensive seasonal crops (with the use of quality seeds, chemical fertilizers, and pesticides) are cultivated on those fragile soils while soil and water conservation technologies are seldom used.

Farm-type characteristics based on structural data were identified using cluster analysis. Five important factors influencing farm types were farm cultivating aspect, off-farm activities, income, labor force, and plots location. On the other hand, erosion yields for each farm type were identified using the RUSLE method, which calculated erosion yield at the plot level. Erosion yields for each farm type are presented in Table 17.1.

The average erosion yield is high compared to other studies using the same method (Verbist et al., 2010; Valentin et al., 2008; Bruijnzeel, 2004; Kusumandari et al., 2013). Moreover, this result is very high compared with the tolerable soil loss of 13.5 ton/ha/year defined by Arsyad (2000). The important factors that influenced erosion yield in the study area were plot location (on the steep, slope, or flat) and cropping system. Tobacco cultivating has the potential to produce high erosion yield because farmers need to (1) avoid planting trees because the farmer indicated that the tree's shade is reducing tobacco quality; (2) clear weeds and leave the bare soil to reduce tobacco diseases for crops; and (3) make soil mounds following the highest slope to reduce waterlogging. The combination of those three practices increase water run-off speed and lead to the high erosion.

We found five types of farmer typology and their impacts on erosion. The first type is "OFF-farm" (OFF). Farmer's income in this group is mainly from off-farm activities rather than on-farm. Farmers manage a small area for agriculture and their plot is mainly in the low-slope area. They plant less tobacco. In this sense OFF farmers' cultivating system contributes the lowest erosion yield. The second type is farmers who are tagged as "Diversified Agricultural Production" (DivAg). Farmers in this group are the poorest. The majority of farmers are old and manage a small plot and get small revenue from their efforts as well as lower erosion yield. The third type is farmers who are tagged as "Land and Capital Shortage" (LCS). Farmers in this group are poor and young and depend on tobacco production, They manage a small plot area in the steep and slope area. Their practices create the third-highest erosion yield. The fourth type is farmers who are tagged as "Grass and Milking Cow" (GMilk). A farmer's income in this

TABLE 17.1   Erosion Yields for Each Farm Type

Index	Off	DivAg	LCS	GMilk	UpAg	Average
Erosion index (ton/ha/year)	17[e]	70[d]	100[c]	162[b]	531[a]	212
Standard deviation	87.2	43.3	65.2	145.4	563	363.5

[a]*very high erosion yield*
[b]*high erosion yield*
[c]*medium erosion yield*
[d]*low erosion yield*
[e]*very low erosion yield*

group is mainly from tobacco as well as livestock. On the other hand, GMilk's farmers create the second-highest erosion yield. Farmers who have plots in the slope and are cultivating tobacco create high erosion yield. The last type is farmers who are tagged as "Uplanders Agricultural System" (UpAg). Farmers in this group are the richest. Their main income is from tobacco that is planted in a large plot in the steep and slope area and they produce the highest erosion yield.

The high erosion yields (212 ton/ha/year) in the study area influenced natural resources both on-site and out-site. Out-site impacts such as high sedimentation in the river body will reduce river capacity to convey water. In the irrigation network, sedimentation will reduce irrigation system performance. In the general sedimentation both in the river body and the irrigation network, in downstream will reduce river capacity to collect and convey water, which reduces the water supply.

Water resources for household and livestock upstream are from springs. However, to get water from the springs is difficult and costly because in the 1000 m a.s.l. discharge of the springs is low and it is not possible to distribute water to houses using pipes. In some areas where discharge is lower than 1000 m a.s.l., some springs have big discharge and farmers are able to distribute the water to houses using pipes. Moreover, farmers can extract water from a deep well and sell the water to their neighbor. Some rich farmers even can buy water from downstream. However, due to high operation cost water prices upstream are expensive. However, the area with an altitude of 900–400 m a.s.l. is known as the springs belt. This area is an agroforestry system and supplier of water for both upstream and downstream.

As irrigation is not feasible for hydrogeological reasons, the cropping system calendars are constrained by the weather conditions, mainly rainfall, but farmers also need water for household consumption and raising their livestock, and their main sources of water are springs. In these areas, the lack of water can be a constraint for increasing the number of livestock raised.

17.4.2 Water User in Downstream

The majority of water users downstream use water from the Pusur watershed. They are farmers in the Pusur irrigated area, local municipal water company (PDAM) Klaten and Surakarta, which provides water for the urban area, bottled water companies, other industries, and tourism.

17.4.2.1 Farmers

The Pusur irrigation system covers 4000 ha irrigated area downstream of the Pusur watershed. Water resources for the Pusur irrigation system are from the Pusur River, Cokro spring (the biggest spring), Sigedang spring, and groundwater. The cropping system in the Pusur irrigated area recommend by the government are rice, secondary crops (maize, ground nut, mung bean, soybean), industrial crops (tobacco and sugarcane), and fishery. The cropping calendar is rice-rice-rice/industrial crops for rainy-rainy-dry season, respectively. But farmers prefer to plant rice-rice-rice all year long because rice is more profitable than maize and industrial crops. In the past (before 1994) water was sufficient for a rice-rice-rice cropping system. But the water discharge has been decreasing steadily.

17.4.2.2 Industries

Bottled water industries are the biggest water users in the industries sector downstream. A company named Danone Aqua is the biggest bottled water company in Indonesia that operates in the Pusur watershed. Water resources for Aqua Danone are from artesian wells in the Sigedang spring, which has a discharge rate of 60 L/s. According to a memorandum of understanding (MoU) between Aqua Danone and the government of Klaten District, Aqua Danone has the right to use 23 L/s of Sigedang artesian only, which is extracted using a deep well in the Sigedang area while the rest of the water is flowing to an irrigation system. Bottled water industries operate in the Pusur watershed are the biggest bottled water industries in Indonesia owned by international industries, which get very high benefit from water exploration compared to the farmers.

Besides Aqua Danone in the study area there are several private water bottled companies that use groundwater for their production system.

17.4.2.3 Municipal Water Company (PDAM)

PDAM is a government company that provides water for the urban area mainly outside of the Pusur watershed area named the Surakarta District. PDAM has operated since Dutch colonization. The water resource for PDAM is Cokro spring, which is the biggest spring in the Pusur watershed. Water requirements for the urban area increases because of increasing population. In this sense, PDAM uses more water to fulfill the demand. However, nowadays the discharge of Cokro spring is decreasing. The urban area outside of the Pusur watershed gets its water supply with no worries about a water shortage, contrary to the situation of farmers.

17.5 DISCUSSION

The survey found that 20% of 162 springs in the Pusur watershed are dead. Moreover, along of the Pusur River and irrigation channel, sedimentation is high. It caused water discharge in the main gate of the Pusur irrigation system to decrease significantly (33%) and it is no longer sufficient for all farmers. Problems arise in the Pusur watershed because farmers face water shortages in the dry season. Farmers complain to the government and blame other parties such as bottled water companies and other industries as well as PDAM. According to the farmers, those industries overuse water and cause water supplies for agriculture to decrease. Bottled water industries that extract groundwater are blamed for causing the decrease of the Sigedang springs. On the other hand, PDAM is blamed for causing the decreasing supply of water from the Cokro springs for agriculture. Nowadays, conflicts among water users have become larger and more serious. Farmers protest with anarchy to the government as well as other water users regarding threats for the sustainability of water resources.

On the other hand, water demand for agriculture increases because the government tries to boost the food security program. Farmers try to increase productivity of the rice crop by planting rice three times per year. As a result, farmers need more water; however, water is insufficient, especially during the dry season. To get water some farmers in the middle and downstream of the Pusur irrigation system violate the law and perform vandalism to irrigation, resulting in 47% of irrigation facilities in the middle of the network and 95% downstream of the irrigation network being damaged. Some farmers pump water from the irrigation channel or from their plots. Their activities lead to a decrease in the water table of villagers surface wells and an increase in problems between the agricultural sector and rural household water users. In this sense, implementation of Law No. 7 (2004) on Water Resources and Government Regulation No. 20 (2006) on Irrigation failed.

According Law No. 7 (2004) on Water Resources and Government Regulation No. 20 (2006) on Irrigation, operation and maintenance of an irrigation system on tertiary levels are under authorization of the farmer's water user association. To be able to manage the operation and maintenance (OnM) of an irrigation system, farmers must have skill and knowledge about OnM. However, farmers and farmer's water user associations still lack skill and knowledge about OnM. Moreover, farmers' awareness of how to use water in an irrigation system effectively and efficiently is low. In this sense, farmers need capacity building to improve their skill and knowledge. Government provides assistance for a capacity building program for farmers, but the assistance of OnM activities by government was not sustainable because it was a projects-based program.

To reduce thievery of water and violation of irrigation facilities government, in collaboration with universities, introduced a program to change the cropping system in the Pusur irrigated area from paddy-paddy-paddy to paddy-paddy-maize. Implementing this cropping system change, the water supply from irrigation will be sufficient for the entire year. However, this program was not interesting to farmers because maize offers less revenue compared to rice.

Water users' low awareness about integrated irrigation networks and water resource management can enlarge the conflict. This can be evidenced by a lack of awareness about sharing water sources services; for example, farmers upstream of the irrigation channel insist on growing paddy-paddy-paddy all year and, as a result, they open the water gate for their plots and use water as much as they need. This leads to insufficient water for farmers' plots in the middle area and downstream. Because of water shortages in the middle area and downstream, some farmers steal the water by pumping it directly from the primary channel irrigation of rice fields, and perform vandalism against the irrigation facilities.

17.6 CONCLUSION

LULCC of a forest to an agricultural system changes the hydrology patterns in the watershed and creates conflicts among water users. Moreover, farmers upstream of the watershed create erosion due to a lack of conservation practices in their agricultural system. As off-site impacts of uplanders activities, some springs are dead and sedimentation in the river and irrigation system is high. Those lead to water shortages for farmers downstream and create some problems among water users.

Conservation of natural resources in the Pusur watershed is absolutely necessary. Deforestation will threaten the sustainability of agriculture (including water availability) and other economic activities. Conservation of natural resources requires cooperation of all stakeholders, including government, communities, universities, and industries. While conservation of natural resources is a complex problem associated with various interests and objectives of the parties involved. For Pusur watershed there are not yet available methods or tools that can be used to achieve the conservation of natural resources.

The problems of water shortages during the dry season are experienced by farmers downstream of Pusur irrigation areas caused by (1) reduced discharge of water from the Pusur River; (2) increased sedimentation in rivers and irrigation infrastructure; (3) extensive damage to infrastructure, especially intake gate water upstream of the irrigation system; and (4) increased extraction of water by public or private companies.

Problems (1) and (2) probably are correlated with LULCC in the upper area; problem (3) is correlated with irrigation network management; and problem (4) is influenced by social and economic aspects. These situations are causing conflict between farmers and other actors. To tackle this problem, one of two alternative solutions can be used: decreasing water demand or increasing water supply. Decreasing water demand can be obtained by changing cropping patterns, and/or by improving water management (better infrastructures and institutions).

Further study for integrated natural resources management is necessary for sustainability of natural resources services. The PES concept seems promising for sustainability of natural resources services with the assumption that:

(1) Inducing LULCC in the upper catchment to increase water supply and decrease erosion may reduce sedimentation for the lower area.
(2) Reducing sedimentation in the lower area is costly.
(3) Changing the cropping system both upstream and downstream is providing good benefits for all parties.
(4) Educating all water users (both upstream and downstream) that integrated water resources management is necessary.

References

Angelsen, A., 1995. Shifting cultivation and "deforestation": a study from Indonesia. World Dev. 23, 1713–1729.

Angelsen, A., 1999. Agricultural expansion and deforestation: modelling the impact of population, market forces and property rights. J. Dev. Econ. 58, 185–218.

Arnold, L.L., 2008. Deforestation in decentralised Indonesia: what's law got to do with it? Law Environ. Dev. J. 4 (2), 75.

Arsyad, S., 2000. Konservasi Tanah dan Air (Soil and Water Conservation). Bogor Institute of Agricultural IPB Press, Bogor, Indonesia.

Bakker, M.M., Govers, G., Kosmas, C., Vanacker, V., Oost, K.v., Rounsevell, M., 2005. Soil erosion as a driver of land-use change. Agric. Ecosyst. Environ. 105, 467–481.

Bruijnzeel, L.A., 2004. Hydrological functions of tropical forests: not seeing the soil for the trees? Agric. Ecosyst. Environ. 104, 185–228.

Esen, E., Uslu, O., 2008. Assessment of the effects of agricultural practices on non-point source pollution for a coastal watershed: a case study Nif Watershed, Turkey. Ocean Coast. Manag. 51, 601–611.

Gaveau, D.L.A., Linkie, M., Suyadi, Levang, P., Leader-Williams, N., 2009. Three decades of deforestation in southwest Sumatra: effects of coffee prices, law enforcement and rural poverty. Biol. Conserv. 142, 597–605.

Hermiyanto, B., Zoebisch, M.A., Singh, G., Agus, F., 2004. Soil quality of three small watersheds in Central Java, Indonesia. Int. J. Agr. Rural Dev. 4, 28–33.

Kartiwa, B., 2006. Identification and Analysis of Potential Water Resources to Support Water Management Strategies in Klaten District. In Indonesia: Identifikasi Dan Analisis Potensi Sumberdaya Air Mendukung Strategi Pengelolaan Air Di Kabupaten Klaten. Research Report Year 2006 Indonesian Agroclimat and Hydrology Research Institute, Agirculture Departement RI.

Kurnia, U., Sudirman, I.J., Soelaeman, Y., 2006. Effect of Land Use Change on River Discharge and Flooding in Downstream of Kaligarang Watershed. Proceedings of the National Seminar on Wetland Versatility. Research and Development Centre for Soil and Agro-climate, Agricultural Department Indonesia.

Kusumandari, A., Widiyatno, Marsono, D., Sabarnurdin, S., Gunawan, T., Nugroho, P., 2013. Vegetation Clustering in Relation to Erosion Control of Ngrancah Sub Watershed, Java, Indonesia. Procedia Environ. Sci. 17, 205–210.

Rusli, Y., 2008. Statisitik Kehutanan Indonesia (Forestry Statistics of Indonesia) 2007. D.O. Forestry, Jakarta, Indonesia, Ministry of Forestry.

Valentin, C., Agus, F., Alamban, R., Boosaner, C., Bricquet, C.P., Chaplot, V., de Guzman, T., de Rouw, A., Janeau, L.J., Orange, D., Phachomphonh, K., Do-Duy, P., Podwojewski, P., Ribolzi, O., Silvera, N., Subagyono, K., Thiébaux, J.P., Tran Duc, T., Vadari, T., 2008. Runoff and sediment losses from 27 upland catchments in Southeast Asia: Impact of rapid land use changes and conservation practices. Agric. Ecosyst. Environ. 128 (4), 225–238.

Van Noordwijk, M., 2002. Scaling trade-offs between crop productivity, carbon stocks and biodiversity in shifting cultivation landscape mosaics: the FALLOW model. Ecol. Model. 149, 113–126.

Van Noordwijk, M., 2004. The Role of Agroforestry to Maintaining Hydrological Functions of Watersheds (DAS) in Indonesia: Peran Agroforestry dalam Mempertahankan Fungsi Hidrologi Daerah Aliran Sungai (DAS). Agrivita 26 no 1.

Van Noordwijk, M., Poulsen, J.G., Ericksen, P.J., 2004. Quantifying off-site effects of land use change: filters, flows and fallacies. Agric. Ecosyst. Environ. 104, 19–34.

Van Rompaey, A., Krasa, J., Dostal, T., 2007. Modelling the impact of land cover changes in the Czech Republic on sediment delivery. Land Use Policy 24, 576–583.

Vanacker, V., Molina, A., Govers, G., Poesen, J., Dercon, G., Deckers, S., 2005. River channel response to short-term human-induced change in landscape connectivity in Andean ecosystems. Geomorphology 72, 340–353.

Verbist, B., Poesen, J., van Noordwijk, M., Widianto, Suprayoga, D., Fahmuddin, A., Deckers, J., 2010. CETANA 80 (1), 34–46.

The Role of Information Provision on Public GAP Standard Adoption: The Case of Rice Farmers in the Central Plains of Thailand

D. Jourdain*, *S. Srisopaporn*[†], *S. Perret**, *G. Shivakoti*[†,‡]

*CIRAD, UMR G-EAU, Montpellier, France [†]Asian Institute of Technology, Bangkok, Thailand [‡]The University of Tokyo, Tokyo, Japan

18.1 INTRODUCTION

The development and adoption of new agricultural technologies is still heavily researched because it is central to agricultural growth and, in developing countries, contributes to poverty reduction efforts (Mendola, 2007). Many researches have studied the adoption of efficiency-increasing technologies such as improved seeds (Smale et al., 1995; Ouma and De Groote, 2011), fertilizers (Alene et al., 2008), or mechanization (Larkin et al., 2008), and recently more complex and systemic changes within farming systems, such as conservation agriculture (see Knowler and Bradshaw, 2007, for a review of the literature).

While efficiency-enhancing technologies are still high on the agenda to maintain farmers' competitiveness in globalized markets, food safety and production sustainability criteria are becoming increasingly important for consumers (Othman, 2007) and are reshaping the technology push toward farmers. Good agricultural practices (GAPs) are now increasingly promoted by the public and private sectors. Government agencies' public goals are to ensure that food is produced in a safe and sustainable manner, and eventually to sustain incomes in the agricultural sector. In addition, agricultural products value chains are increasingly dominated by large firms that want to impose their production and traceability standards.

In many cases, GAP standards are developed for niche markets such as high-value horticultural crops or fruits. They are then managed by the private sector and usually require farmers' investment to participate (Kariuki et al., 2012; Subervie and Vagneron, 2013). In contrast, public GAP programs are also promoted by governments in Southeast Asia. These public standards are covering a broader set of crops including less value-adding crops such as rice (Premier and Ledger, 2006). Introducing GAP for rice is important from a food safety point of view because it is the main staple food in Asia. Besides, rice is also the main source of livelihood for farmers and, in the case of Myanmar, Vietnam, and Thailand, rice is also an important export revenue earner.

The main objective of this chapter is to investigate the influence of farm-level factors on farmers' participation in a certification program that requests the alteration of their farming practices and the abidance to a set of rules. The case of the Rice Q-GAP program in the Central Plains of Thailand is analyzed. As this program was intensively promoted to farmers by government agencies, we will also investigate the influence of the Q-GAP training sessions and information provision on adoption. Finally, we will analyze whether Q-GAP participants have different farming practices as compared to nonparticipants. This study differs from previous studies (Amekawa, 2009; Kersting and Wollni, 2012; Schreinemachers et al., 2012) in at least two ways. First, we are looking at a pure public standard, promoted mainly by government agencies, that does not require monetary investment from farmers. Therefore, we are expecting that farmer's barriers to adoption are not conditioned by wealth and investment factors but that other factors will be related to adoption or rejection. Second, the role of information will be considered specifically, as it has been considered of chief importance by a number of studies (Feder and Slade, 1984; Hussain et al., 1994). Information delivered to farmers is hypothesized to be largely influencing the first adoption. In fact, the final decision of a farmer to adopt a new technology depends on his ability to acquire, decipher, and process the information related to farming practices and the innovation itself (Stoneman and David, 1986). On the other hand, a farmer's choice to adopt innovation affects his decision to gather technical information from various sources. A farmer who is interested in applying new farming practices will likely search for relevant information either through his information network or by actively seeking new information sources. Certain elements of context have influenced the methodology used: Government agencies made an intensive promotion of the Rice Q-GAP standard via information and training sessions given to farmers. However, information and training provided by government might have been delivered to farmers in a nonrandom way: agencies have probably tried to engage farmers who were most likely to adopt Q-GAP, and farmers attending the training and information sessions were possibly the most interested in food safety and sustainability issues (for additional and probably unobservable reasons). As a consequence, there is a great possibility that the influence of training sessions may be endogenous. This requires attention from a methodological viewpoint and has been considered.

The chapter proceeds as follows. The next section provides background on Q-GAP in Thailand. Second, the modeling framework is presented. Third, the study areas and data collection process are described. Fourth, factors influencing Rice Q-GAP participation are identified and discussed, followed by a comparison of cultivation practices. The final section summarizes the key results of the chapter.

18.2 RICE Q-GAP IN THAILAND

Several Southeast Asian countries have recently introduced public standards of GAP such as the IndoGAP (Indonesia), VietGAP (Vietnam), PhilGAP (Philippines), SALM (Malaysia), and Q-GAP where Q stands for Quality (Thailand). They all aim at increasing the supply of safe and high-quality food by promoting more sustainable crop production practices, but they were also set up in response to pressures from the export market (Premier and Ledger, 2006; Schreinemachers et al., 2012). The Association of Southeast Asian Nations (ASEAN) economic community will require all agricultural produce in the member countries to meet the same standards; Q-GAP is part of a larger set of national policies to standardize agricultural production and improve farms' competitiveness. ASEAN countries have agreed to harmonize national GAP standards and to create a new ASEAN-GAP. This regionally recognized GAP standard will be more comprehensive than the Thailand Q-GAP standard with additional areas including, among others, planting materials, soil and substrates, biodiversity, and workers' welfare. However, at this stage ASEAN-GAP is limited to fruits and horticultural products.

A Q-GAP program specific for rice was initiated in 2004 to promote sustainable practices and improve the quality of rice produced. Rice Q-GAP is fully managed by the government, from setting standards to training, auditing, and the issuing of certificates (Sarsud, 2007; Schreinemachers et al., 2012). The production standards are set by the National Bureau of Agricultural Commodity and Food Standards, and the program implementation is managed by two departments of the Ministry of Agriculture and Cooperatives: the Department of Agricultural Extension (DoAE), which has overall responsibility for the program, and the Rice Department, which is in charge of the farm auditing and issuing of GAP certificates. The program's aim is to certify that Q-GAP rice is produced according to the best-known practices for (a) farm-level hygienic conditions, (b) management of agricultural equipment and tools, (c) management of input factors, (d) production control and practices, and (e) bookkeeping and document control (Table 18.1). More specifically, Rice Q-GAP is also promoting cost-saving technical solutions, production of quality rice (including variety purity), and prevention of chemical residues from rice farming operations. In 2011, the Office of Rice Product Development of the Rice Department has summarized these recommendations in a booklet entitled "Rice Production and Cost Saving" to be used during GAP-related training. Chemical fertilizers and pesticides use are not forbidden. However, Q-GAP features lists of authorized and banned pesticides, as well as indicative spraying calendars (with a special focus on preharvest intervals). In the same way, quantity and number of applications of fertilizers are detailed under various rice cropping systems. The Q-GAP label is promoted by the Rice Department and can be identified on rice sold to consumers. As such, one of the expectations of the program is to provide market opportunities (price mark-up or higher sales volumes) to participating farmers.

To participate in the Rice Q-GAP program, farmers' rice plots must be registered; then, on these plots, they have to follow a set of practices listed in the detailed Q-GAP guidelines (Table 18.1). Later, the agriculture extension services at the provincial office or a local Rice Department office will send an officer for auditing. The results of the audit are submitted to the committee under the Rice Department or Rice Research Center for evaluation and the farmer is eventually given a Q-GAP certificate. The certificate is issued free of charge every

TABLE 18.1　Criteria Points for GAP Rice Thailand

Criteria Points[a]	Method of Verification
A: Water is not from sources at risk from hazardous substances	Inspect or submit water test if area is at risk
B: Plot location is not at risk to hazardous substances	Inspect or submit soil test if area is at risk
C: Nationally banned chemicals may not be used or for exports of rice; importing countries' banned chemicals may not be used	Inspect chemical storage facility Record information on chemical substances used Randomly test rice produce in case of doubts
D: Preharvest off-type rice management	Review bookkeeping for seed and test off-type rice; Inspect plot
F: Preharvest pest and weed management	Inspect rice produce affected from pests Review bookkeeping for pest observations and prevention methods; Inspect plot
G: Harvest time and rice quality according to national standards	Review bookkeeping for production management
H: Tool used in harvesting, container, and approach to harvesting to prevent off-type rice and cleanliness of threshing machine or combine harvester	Inspect equipment, packaging, and harvesting technique Review bookkeeping records Inspect postharvest rice
I: Humidity of rice and its reduction management	Review bookkeeping Inspect facility; interview
J: Cleanliness and well-maintained transportation, storage facility, product collection, as well as off-type rice management	Inspect facility, equipment, packaging, procedure, and methods of transportation
K: Availability of bookkeeping and information records	Review bookkeeping

[a] Based on Department of Rice GAP-06 Manual for Auditors for GAP Rice.

three years for each plot registered so participants are required to record their practices (eg, application date, dose, and input used) and are subjected to an annual audit, evaluation, and review by the Rice Department.

　　Thai government agencies have been quite successful in fostering the participation to the Rice Q-GAP certification program. As of 2012, the Rice Q-GAP program had already been promoted in 72 of the 76 provinces of Thailand, and approximately 40,000 rice farmers had registered[1] (Ms. Viriyangkura, Rice Department, personal communication, 2012). This is particularly true in the two main rice-producing areas. The northeastern part of the country, known as *Isaan* was targeted first by the agencies. In Isaan, rice production is mainly lowland rainfed (ie, produced on rice bundled areas without real control of water flows); Isaan farmers are mainly cultivating low-yielding quality rice fragrant varieties in a relatively extensive way. By contrast, the Central Plains area benefits from better soil and irrigation; two to three rice crops are produced per year, using high-yielding varieties and much higher doses of

[1] Although this should be compared to the number of rice farming household in Thailand estimated to be around 3.5 million.

fertilizers and pesticides (Thanawong et al., 2014). Rice Q-GAP has also been introduced in the Central Plains of Thailand, but at a later stage. Although rice farmers of the Central Plains are also concerned with securing markets and price, they are more intensive rice farmers producing rice mostly for the food processing industry (and not for direct household consumption) and are less likely to trade-off yields for quality and food safety standards.

18.3 DATA AND METHODS

18.3.1 Empirical Model

In Thailand, Rice Q-GAP adoption is closely related to the work of the DoAE who organized information and training sessions in 72 out of the 76 provinces of the country. Farmers who participated received a comprehensive set of information about the Q-GAP program, its objectives, and the benefits they can expect from participation, but also the duties involved (bookkeeping, ban on chemical products, promotion of practices, etc.). At the end of these sessions, the farmers are supposed to be able to make an informed decision about their participation. Therefore, the roles of information and training need to be carefully analyzed.

Unfortunately, two selection biases may occur. First, DoAE has an ambitious plan in terms of farm enrollment and is likely to have targeted farmers with greater potential for adoption (nonrandom, purposive distribution of the information). Second, farmers accepting to attend the training sessions are also likely to be the ones that are more receptive to new technologies for unobservable reasons such as "dynamism" of the head of the household (self-selection). In short, farmers who received primary information and support on Q-GAP certification were probably those who had more chance to adopt the standard than randomly selected farmers. As a result, the factors that influence training participation might also influence Q-GAP adoption. To control these potential biases, we modeled two separate, but potentially connected, decisions: participation in training sessions and Rice Q-GAP registration. Two common approaches could be used: the first disregards the binary nature of the adoption and training variables and uses linear instrumental variables (IVs) model (Angrist, 2001); the second uses estimates of a recursive bivariate probit model (RBPM), which assumes that the adoption and training variables are each determined by latent limited dependent models with jointly normal error terms (eg. Greene, 1998; Bhattacharya et al., 2006). From both a theoretical and empirical point of view, simultaneous likelihood estimation methods are superior to two-stage IV procedures (Wooldridge, 2002; Chiburis et al., 2012). Therefore, we used a RBPM formulation that allows for two separate probit models with correlated error terms. The specification of the RBPM is best described with the use of two latent (not observed) variables associated with Q-GAP participation (*qgap*), and rice training participation (*train*); it is given by

$$\begin{cases} qgap^* = x'_1 \beta_1 + \gamma.train + u_1 \\ \quad train^* = x'_2 \beta_2 + u_2 \end{cases} \tag{1}$$

where x_1 is a vector of K_1 covariates explaining Q-GAP adoption, and x_2 is a vector of K_2 covariates explaining farmers participating to the training. The error terms $(u_1; u_2)$ have a bivariate normal distribution with zero mean, unit variance, and correlation ρ.

The variables *qgap* and *train* can be interpreted as the utility the farmers are deriving from the two participation variables. The observed decisions to participate (as binary variables) result from positive utilities derived from participating (as compared to nonparticipation):

$$\begin{cases} qgap = 1 \text{ if } qgap^* > 0 \text{ otherwise } qgap = 0 \\ train = 1 \text{ if } train^* > 0 \text{ otherwise } train = 0 \end{cases} \tag{2}$$

The correlation coefficient between the disturbances measures the effect of *train* on *qgap* after the influence of the endogenous variable *train* is accounted for in the first equation (Greene, 2007). The key null hypothesis of the model is that the disturbances u_1 and u_2 are not correlated ($H_0 \rho = 0$). If the null hypothesis is not rejected, the two equations could have been estimated with separate probit models (Monfardini and Radice, 2008).

RBPM can be estimated by the full information maximum likelihood (FIML) method as the endogenous nature of train in the first equation can be ignored in formulating the likelihood function (Greene, 2007). The log-likelihood function to be maximized can be summarized as

$$\log L = \sum_{i=1}^{n} \log \left(\Phi_2 \left(\omega_{i1}, \omega_{i2}, \rho^*_i \right) \right) \tag{3}$$

where *i* are sampled farmers, $j = 1; 2$ are indices corresponding to *qgap* and *train*, $w_{ij} = (2y_{ij} - 1)$ $x'_{ij}\beta_{ij}$ for $j = 1; 2$, $\rho^*_i = (2y_{i1} - 1)(2y_{i2} - 1).\rho$; and Φ_2 is the bivariate normal cumulative distribution function.

Because RBPM is nonlinear, the impact of an independent variable on predicted probabilities is not constant. Therefore, simply reporting parameter estimates and their significance is not helpful for understanding the effects of independent variables. Besides, analysis is further complicated by the interactions between the two equations.

First, we are reporting the changes in the expectation $E(qgap \mid train, x_1, x_2)$ for a marginal change in explanatory variables (also interpreted as the change in the probability that $qgap = 1$). The marginal effects can be decomposed into the sum of a direct and of an indirect effect depending on which equation(s) the variable is included in (Greene, 1998). The direct effect measures the direct influence of the variable on the probability of *qgap* via the first equation. The indirect effect is present for variables entering the train equation; that is, influencing the probability that *train* equals one. This effect is transmitted back to *qgap* probability (see Eq. 4).

$$E(qgap \mid x_1, x_2) = E(qgap \mid train = 1, x_1, x_2).\text{Prob}(train = 1) \\ + E(qgap \mid train = 0, x_1, x_2).\text{Prob}(train = 0) \tag{4}$$

The average marginal effects (AME) presented in this chapter were calculated using the formula presented in Greene (1998) and extended in Baslevent and El-Hamidi (2009) for continuous, binary, and endogenous variables. An exception was made for the evaluation of the marginal effect of the binary variable groups present in the two equations that were considered as a continuous variable to be able to compute the direct and indirect effects as suggested in Baslevent and El-Hamidi (2009). Finally, the marginal effect of the endogenous binary variable *train* is defined as

$$E(qgap \mid train = 1) - E(qgap \mid train = 0) = \Phi\left(x'_1\beta_1 + \gamma\right) - \Phi\left(x'_1\beta_1\right) \tag{5}$$

Reported standard errors of the marginal effects were calculated using a nonparametric bootstrap method with 900 replications.

18.3.2 Variables Potentially Influencing Q-GAP Adoption and Training Outcomes

We expect the two decisions to be driven by possibly different variables. Drawing upon existing literature on agricultural technology adoption (Feder et al., 1985; Adesina et al., 2000; Doss, 2003), and adoption of standards (Okello and Swinton, 2007; Kersting and Wollni, 2012), hypotheses about the expected influences of the independent variables are made (Table 18.2). Factors can be divided into three broad categories: (a) farm and household livelihood assets (age, experience, education, land available, and land-to-labor ratio); (b) behavioral variable to capture connectedness of the household (information sources, membership in farmer groups, networks of farms having adopted Q-GAP, participation in agricultural trainings, assistance by governmental and nongovernmental organizations); and (c) anticipations at the time of the decision adoption (expected price increase due to rice quality mark-up, increased sales due to quality, production cost reduction, etc.).

TABLE 18.2   Independent Variables Used for the Models and Hypotheses

Variable (Code)	Variable (Definition)	Q-GAP	Training
Educ	No. years in public education	+	
Exp	Farming experience (years)	+/−	
Groups	Groups member (yes/no)	++	++
govCont	Freq. of government contact	+	+
Training	Training attendance (binary)	+	
otherChan	Other information channels	+	
kAdopt	No. of known Q-GAP adopters	+	
Off Farm	Off-farm labor opportunities	−	
Area	Total area grown with rice (ha)	−	++
labha	Labor-land ratio (pers./ha)	++	
Own	Pure renters		
	Mixed land ownership		
	Pure land owners	+	
ePrice	Expected price increase (yes/no)	+	
eCost	Expected cost reduction (yes/no)	+	
eSale	Expected easy sales (yes/no)	+	
eYield	Expected yield increase (yes/no)	+	
pImpact	Perceived negative impact on environment	+	+

Previous studies have found a positive relation between adoption of production standards and household head education level. Although the public Q-GAP is probably less knowledge intensive than private Global-GAP, record keeping requirements and grasping sustainability concepts is probably easier done by more educated farmers. It is difficult to anticipate the possible effects of experience on Q-GAP adoption. On one hand, experience is highly correlated with age (which was not considered here precisely because of high collinearity).

As younger farmers are likely to be more innovative and interested in new production modes, a negative relationship with age is possible. On the other hand, longer farming experience provides farmers with a better practical understanding of the intricacies of managing cropping systems. As such, a positive effect of experience on adoption is also possible.

Social capital and the diversity of information channels are likely to have a positive effect on adoption. Farm group membership is a sign of connectedness with other farmers, which should enhance the possibilities to have access to information and to share experiences during the learning process. As such, a positive effect is likely. In addition, farmer groups are often targeted by extension officers for this particular reason, but also because of possible gains of time (scale effect). The number of contacts with extension services and other government offices refers to the same idea that the contacts already established may be a sign of active behavior, and conversely increase the likeliness of being targeted for the rice training. Finally, farmer-to-farmer learning, and nongovernmental channels of information have long been seen as important factors of diffusion of farming innovations (Foster and Rosenzweig, 1995). We included one dummy variable to account for other types of information channels (otherChan) and another variable to account for possible farmer-to-farmer diffusion (kAdopt).

Land and labor constraints are also known to play an important role in the adoption of technologies (Feder et al., 1985). Record keeping and the practices promoted are more time-consuming. As such, labor constraints should play an important role in the ability to adopt Q-GAP. The number of family labor per cultivated area (labha) should have a positive relation with adoption. Conversely, off-farm opportunities are likely to reduce time available for farming and to be negatively correlated to adoption.

We are not expecting a strong and direct relationship between landownership and Q-GAP adoption. However, we are making the hypothesis that renters would be less concerned by pollution issues as they can possibly move to other land in case of trouble. In the same way, pure renters may be afraid of losing some of their productivity gains through increased subsequent rents. Therefore, we anticipate a mild positive correlation between ownership and adoption. As no registration fees are collected, we have considered that household wealth is unlikely to be related to Q-GAP adoption. Besides, reliable information about farmers' wealth is difficult to obtain. So we have not considered any wealth variable as a possible explanatory variable.

Finally, we have included the expectations farmers were having before adopting the Q-GAP (based on recall). Expected cost reduction, yield increase, price increase (for labeled rice products), or access to markets should all be related to positive adoption. Among the perceptions, we also asked about the perceptions farmers were having about their own impact on the environment. This last variable was included in the training equation.

18.3.3 Study Area and Data Collection

We selected the province of Ayutthaya, in Central Thailand, for farm-level data collection because it was one of the first provinces where Q-GAP was introduced in the Central Plains. The survey was conducted in 13 of the 16 districts of the province. This province, located 80 km north of the Thai capital, is having good access to input and output markets. Rice cropping is irrigated and intensive. Two to three rice crops, mainly high-yielding but low-quality rice varieties, are produced per year by farmers in that province.

In collaboration with the Rice Research Center of Ayutthaya and the Rice Department, a comprehensive list of farmers being promoted with Q-GAP in Ayutthaya Province between 2007 and 2012 was retrieved. Out of a total of 692 rice farmers being promoted with Q-GAP, 200 farmers were randomly selected and interviewed in December 2012. An additional 50 farmers who did not get introduction training were also interviewed. The structured questionnaire used included both closed and open-ended questions. Key environmental, social, and economic characteristics along with rice production practices were recorded. After careful checking of questionnaires answers and detection of outliers, 244 cases out of the initial 250 were considered valid for the statistical analysis.

18.4 RESULTS

18.4.1 Descriptive Statistics of the Sampled Farmers

Around 72% of our sample did register at least one rice plot since 2006. A slightly lower proportion participated in the Rice Q-GAP training. However, these two populations are not strictly overlapping because the chi-square test does not show significant differences with regard to rice training between adopters and nonadopters (Table 18.3)

Data on human and social capital (education, experience, membership in groups or associations) do not exhibit special patterns when compared with the general farming population of the area (Table 18.4). Nearly half of the sampled population have more than 25 years of experience as farm managers, which is symptomatic of the aging of the agricultural population in Thailand.

On another note, one-third of the sampled population belongs to a farmer group (being either a cooperative or an association). There is an important diversity in terms of rice cultivated areas and family farm labor per area (0.37 person/ha on average for a standard deviation of 0.35). This are also representatives of the dual farming situations in Thailand's Central Plains where family farms purely dedicated to rice cultivation are coexisting with farms relying on external labor for most of their farm operations ("phone farmers" as they are sometimes

TABLE 18.3   Relationship Between Training and Q-GAP Adoption

Variable		Sample	No Q-GAP	Q-GAP	Sig.
Rice training	No	31.56	37.14	29.48	0.312
	Yes	68.44	62.86	70.52	

Column percentages are presented; Chi-square to test the hypothesis of different distributions.

TABLE 18.4   General Characteristics of the Sampled Farmers

Variable	Units/Coding	Mean or %
Educ	No. years	6.39 (3.25)
Exp	Low (≤10)	18.4%
	Medium (>10; ≤30)	36.1%
	Large (>30)	45.5%
Groups	No	68.8%
	Yes	31.1%
govCont	No	48.4%
	Yes	51.6%
Area	Rice area (ha)	7.0 (4.8)
Labha	(man/ha)	0.37 (0.36)
offFarm	No	50.8%
	Yes	49.2%
Own	Pure renters	46.3%
	Mixed	17.2%
	Pure owners	36.5%
pImpact	No	55.3%
	Yes	44.7%

Standard errors (in parentheses) added to mean values for continuous variables; percentage of the sample population for nominal variables.
Educ: No. years in public education; *exp*: Farming experience; *groups*: Groups member; *govCont*: Frequency of government contacts; *area*: Total area grown with rice; *labha*: Labor-land ratio; *offFarm*: Off-farm labor opportunities; *own*: Land ownership; *pImpact*: Perceived negative impact on environment.

labeled). In the latter case, farm managers do not always reside in the village near the plots but in larger cities nearby. In the same way, land renting is very important in the study area and the sample, with 47% of the farmers interviewed being pure renters, and an additional 36% are both owners and renters, leaving only 17% of farmers being pure landowners. These figures differ only slightly from the ones of the agricultural population of Ayutthaya province where 20% of farmers are pure landowners and 50% are both owners and renters. Besides, of all the cultivated land in Ayutthaya province, 28% is owned land, 64% is rented land, and 3.5% is free land (Office of Agriculture Economics, 2010). Finally, around 45% of the sampled population did not view their farming activities as potentially harming the environment.

18.4.2 First Time Adoption

When testing the null hypothesis $H_0 := 0$, the Wald test statistics of 2.59 (Prob($>\chi^2$) = 0.107), and the LR statistic of 0.437 (Prob($>\chi^2$) = 0.51) do not allow us to reject the null hypothesis of

TABLE 18.5   Maximum Likelihood Estimates of Separate and Recursive Probit Models

Variables	Sep. Probits Coefficients (Robust SD)	RBPM Coefficients (Robust SD)
(intercept)	−2.35 (0.55)***	−1.53 (0.61)**
Education (years)	0.09 (0.04)**	0.08 (0.03)**
Experience—medium	0.42 (0.36)	0.38 (0.31)
Experience—high	0.39 (0.36)	0.34 (0.31)
Groups	0.38 (0.29)	0.64 (0.26)**
Gov. contacts—frequent	1.36 (0.27)***	1.21 (0.30)***
Know GAP farmers—high	0.22 (0.31)	0.19 (0.26)
GAP other channels—yes	1.59 (0.39)***	1.43 (0.37)***
Labor per ha	1.85 (0.53)***	1.75 (0.48)***
O-farm	0.34 (0.21)	0.29 (0.19)
Full ownership	0.14 (0.25)	0.15 (0.22)
Exp. cost reductions—yes	1.18 (0.24)***	1.01 (0.30)***
Rice training—yes	−0.26 (0.22)	−1.23 (0.39)***
Log likelihood	−75.54	
Rice training		
Labor per ha	0.34 (0.17)*	0.35 (0.17)**
Perception impact—yes	0.27 (0.14)*	0.27 (0.14)*
Groups—yes	0.92 (0.19)***	0.93 (0.19)***
		0.74 (0.43)
Log likelihood	−140.6	−215.9
N	244	244

Wald test of $\rho = 0$; $\chi^2(1) = 2.59$; Prob $> \chi^2 = 0.107$.
Likelihood ratio test of $\rho = 0$; $\chi^2(1) = 0.59$; Prob $> \chi^2 = 0.441$.
Standard errors in parentheses; *$p < 0.1$, **$p < 0.05$, ***$p < 0.01$.

no correlation. However, since the Wald test is close to a 10% threshold, and that outright rejection would be more convincing than inability to reject, estimation results for the individual probit models and for the RBPM are presented in Table 18.5. The two formulations are giving similar results in terms of signs of the relationship between adoption and potential explanatory variables. However, the RBPM formulation is showing higher significance of the relationship with the variables groups and training. Variables included in both equations have the same signs in the two equations, meaning the potentially indirect effects are reinforcing effects on adoption.

The AME of each explanatory variable on adoption and training are presented in Table 18.6. These AME give more meaningful information as they can be interpreted in terms of impact

TABLE 18.6　Average Marginal Effects of the Dependent Variables on Q-GAP Adoption

Variable	Direct	Indirect	Total	St. Err.	Sig.[a]
QGAP equation					
Education (years)	0.016		0.02	0.008	0.047
Experience—medium	0.074		0.07	0.069	0.286
Experience—high	0.068		0.07	0.066	0.307
Groups—yes	0.067	0.41	0.48	0.091	0.000
Gov. contacts—frequent	0.290		0.29	0.061	0.000
GAP other channels—yes	0.254		0.25	0.050	0.000
Know GAP farmers—high	0.039		0.04	0.060	0.518
Labor per ha	0.327	0.15	0.48	0.121	0.000
O-farm—yes	0.060		0.06	0.041	0.145
Full ownership—yes	0.025		0.03	0.046	0.587
Exp. cost reductions—yes	0.245		0.25	0.053	0.000
Perception impact—yes			−0.004	0.005	0.38
Rice training—yes	−0.046		−0.05	0.002	0.000
Training equation					
Labor per ha	0.111		0.11	0.060	0.063
Groups—yes	0.278		0.28	0.046	0.000
Perception impact—yes	0.089		0.09	0.046	0.050

[a] p *value of the Wald test.*

on the probability of adoption, and also integrate the potentially indirect effects captured in the recursive system of equation. Results presented in Table 18.6 were calculated with the hypothesis of no correlations between the two errors. The variables with the strongest positive impact on Q-GAP adoption are related to farm labor available (*labha*), farmers' affiliations in farmer groups (*groups*), and farmers connections to sources of information (with an equal strength of the effects of extension contacts Govcont and other sources of information GAPcha).

Education, measured by the number of schooling years of the household head, has a positive statistically significant relationship. This result extends to standard adoption the findings of literature on agricultural technologies adoption that the longer the farmers' schooling experience, the higher the tendency to adopt new technologies (Feder et al., 1985; Chouichom and Yamao, 2010; Liu et al., 2011). Farmers who have been through school are probably more equipped to understand the reason behind Q-GAP efforts and can follow the instructions of the program.

Besides, as the registration also requires participant to record their practices, more educated farmers are probably less impressed by this administrative exercise. However, the magnitude of the relationship is relatively limited (around 2% increase in adoption for an additional schooling year).

In the same way, farmers' experience has a positive (but not significant) relationship with Q-GAP adoption. This is in line with Knowler and Bradshaw (2007) who did not find consistent and clear impacts of experience on adoption of conservation agriculture across the studies they reviewed. If we retain the positive correlation, this indicates that experienced farmers evaluate more positively the potential of the Q-GAP program than inexperienced ones. It concurs with Chouichom and Yamao (2010) who showed that longer experience in farming and more years of education were related to conversion to organic rice farming in Surin Province in Thailand. However, this link is tenuous in our case. Farmers' participations in associations, cooperatives, and groups have positive and highly significant effects on both training and Q-GAP adoption. This confirms results found for adoption of conservation agriculture (Adesina et al., 2000). In our case, the positive effect for Q-GAP adoption is only significant under the RBPM formulation, probably as a result of the endogeneity of the rice-training variable. Common unobservable variables, such as dynamism and dedication to rice agriculture, is likely to explain both training attendance and Q-GAP adoption. Besides, AME results are showing that most of the effect of the variable group on Q-GAP adoption is indirect (via the training variable) reinforcing the possibility of a selection bias.

Frequent contacts with government and extension officers have a positive and statistically significant impact on Q-GAP adoption. This is consistent with the technology adoption literature. Feder et al. (1985), summarizing a large spectrum of adoption studies, concluded that education and extension services contacts improves farmers' ability to adjust to changes. Similarly, Moser and Barrett (2006) found that learning from extension agents influenced the decision to adopt low-input rice production methods. More recently and in a reverse relationship, low rate of adoption of sustainable agriculture in China was linked to inadequate agricultural extension efforts (Liu et al., 2011). A dual relationship may be at work: (a) more contacts are improving farmers' skills as extension officers are transmitting knowledge, but on the other hand, the farmers that maintain close contact with extension offices are probably more dedicated to agriculture. Other channels of information (variable *GAPChannel*) have also some positive and significant impact on Q-GAP adoption. Other channels in this case included family members, friends, village chief, community leader, experience with GAP for vegetable crops, and local soil doctors (ie, trained volunteers providing soil recommendation services to other farmers in the community).

It was expected that farmers having observed many neighbors adopting Q-GAP would be more likely to adopt: it is very common that community members decide to follow similar management patterns as each may not want to be left out (social cohesion). For example, the social cohesion factor was found to be one of the key variables of local community adoption of conservation agriculture in Laos (Lestrelin et al., 2012). However, contrary to our expectations, the number of neighbors known to be adopting Q-GAP did not have a significant relationship with adoption. Several hypotheses can be made about this counterintuitive result. First, social cohesion might be low in the agricultural zone we chose. Central plains are now cultivated by relatively larger farms and a substantial number of farms are managed by farmers that do not live permanently in the area and/or are passing orders to contracted labor.

As a result, the farm-to-farm transmission is likely to be slower than expected. A second and more worrying interpretation for the Q-GAP program would be that farmers that did observe earlier adopters were not really convinced that it would fit their needs and constraints. Under such an assumption, farmer-to-farmer connections might not be efficient in spreading the program.

Labor availability is often affecting farmers' adoption decisions, especially for smallholders (White et al., 2005; Lee, 2005). Farmers adopting Q-GAP have to dedicate more time for rice cultivation. First, it requires recording all activities conducted on the farm and encourages some practices that are likely to substitute time-saving but potentially polluting practices with more knowledge and time-intensive practices. For example, using less pesticides requires more pest monitoring of the rice fields. Not surprisingly the variable *labha* (ie, the amount of family labor available for rice farming per ha) has a positive and highly significant effect on both adoption and training attendance. Contrary to the variable group, the influence of *labha* is mainly a direct effect, as the time constraints are more likely to be important once Q-GAP has been adopted. Contrary to our expectations, off-farm opportunities have a positive effect on adoption. However, this relation is not significant and cannot really be commented on.

Not all perception variables could be included in the analysis because of collinearity issues: for example, farmers who anticipated some cost-reduction potential before adopting Q-GAP were also anticipating better market access for their certified products. Among the different perceptions elicited during the interviews, we retained only the anticipations farmers had about the cost-reduction potential of Q-GAP. The relationship between expected cost reduction and adoption was both strong (25% probability increase) and significant, meaning farmers who adopted were really convinced that adopting Q-GAP would reduce their expenditures, possibly through a more rational use of chemical inputs. A nontested hypothesis here is that participation in Q-GAP could be associated by farmers with dedicated external advice leading to their more efficient use of inputs. Although not included in the model, adopters' expectations were probably high in terms of access to new markets, and price mark-up (because these variables are positively correlated to the variable on cost-reduction perceptions). We found a positive relationship between farmers attending training sessions and their perceived negative impact on the environment, but it is difficult to decide on the "direction" of the relationship. However, farmers' perception of negative impact on the environment did not translate into a significant effect on adoption.

The coefficient for training is giving unexpected results as we were expecting that farmers having attended a dedicated presentation about Q-GAP would be more likely to adopt. In fact, both individual probit and RBPM models are showing a negative relationship between training attendance and Q-GAP adoption. This should not be immediately interpreted as a sign of poorly conducted training (although this cannot be ruled out). An equivalently possible interpretation is that farmers are forming some positive expectations from the different contacts they had (justifying their training attendance) but are actually disappointed once they understand clearly the costs and benefits associated with Q-GAP adoption. On the one hand, farmers may be expecting higher "costs" in terms of labor requirements, which may prevent larger landholders from adopting (identified by the variable *labha*). On the other hand, some farmers may not be convinced about the potential benefits presented to them; as the government agencies are not responsible for the marketing of the Q-GAP rice, they can only suggest that the rice produced under Q-GAP will be more attractive, but cannot guarantee it. In the same way, farmers may not be confident in the capacity of the new practices in reducing production costs (for example, by using less pesticides or different types of fertilizers). In other words, farmers may well be interested by the general concept of Q-GAP but may ultimately make a rational decision related to labor issues as we showed earlier. Finally, one should also note that the negative impact is relatively small (−5% for farmers attending the training).

18.4.3 Field Practices

We found significant differences in cultivation practices between adopters and nonadopters (Table 18.7), but on a very limited set of practices.

The number of fertilizer applications is not significantly different between adopters and nonadopters. Furthermore, quantities of nitrogen applied are not significantly different. In both cases, farmers are complying with the recommended range of 60–90 kg of nitrogen per ha and per season. However, average fertilizer costs are slightly and significantly higher for nonadopters. Similar nutrient doses at higher prices for the nonadopters indicate a qualitative change in fertilizer use. For example, this may be the result of farmers applying organic fertilizer or compost that are made on the farm instead of purchased chemical fertilizers. However, further and more careful investigation should be conducted to take into account the differential costs of making and applying those alternative types of fertilizers instead of purchasing outside ready-made fertilizers.

TABLE 18.7 Comparisons of Cultivation Practices and Costs Between Adopters and Nonadopters of Rice Q-GAP

Variable	Adopters, Mean (St. Dev.)	Nonadopters, Mean (St. Dev.)	Sig.[a]
Season 1 fertilizer (no.)	2.25 (0.5)	2.17 (0.4)	0.12
Season 2 fertilizer (no.)	2.27 (0.5)	2.20 (0.4)	0.21
Season 1 nitrogen (kg/ha)	80.73 (45.1)	89.00 (46.4)	0.12
Season 2 nitrogen (kg/ha)	81.63 (46.2)	89.27 (46.1)	0.17
Average fertilizer costs[b]	911.59 (568.7)	1087 (516.4)	0.003
Season 1 herbicide (no.)	1.91 (0.5)	1.95 (0.5)	0.58
Season 2 herbicide (no.)	1.92 (0.5)	1.97 (0.5)	0.54
Average herbicides costs[b]	165.37 (119.7)	181.27 (124.3)	0.597
Season 1 pesticides (no.)	3.35 (2.2)	4.32 (2.0)	0.0012
Season 2 pesticides (no.)	3.28 (2.2)	4.00 (1.8)	0.004
Average pesticide costs[b]	314.93 (387.2)	371.22 (319.4)	0.28

[a] p value of the Kruskal-Wallis rank sum test.
[b] All costs expressed in THB/ha/season.

The number of application of herbicides is not significantly different between adopters and nonadopters. As the herbicides costs are also not significantly different, the use of herbicides does not seem to be different between adopters and nonadopters.

By contrast, Q-GAP adopters apply pesticides (mostly insecticides) less often than nonadopters, with a significant mean difference of 1 application less per season. Yet, pesticides costs are not significantly different between adopters and nonadopters. One possible explanation is that Q-GAP adopters are on average using more expensive pesticides. In particular, one hypothesis that would need to be further researched is that Q-GAP adopters stopped using banned pesticides, usually less expensive but less harmful for human health and the environment than authorized products. This would require a comprehensive list of the chemicals that were used by farmers (as in Schreinemachers et al., 2012). As our survey was based on recalls from farmers, we could not make sure we had a complete and accurate list for all surveyed farmers.

Overall, farm practices are mainly differing in terms of number of pesticide applications, and probably in terms of pesticides that are used. Adopters spray pesticides less often but seem to use higher-quality pesticides. This indicates production techniques that are less harmful for the environment and human health (if we assume the use of more expensive pesticides than of cheap unauthorized pesticides). This finding contrasts with the results of Schreinemachers et al. (2012) and Amekawa (2013), who found a lack of Q-GAP standard compliance among fruit and vegetable farmers in Thailand. Reducing pesticide use arguably is less difficult to implement when producing rice than when producing vegetables, as Integrated Pest Management technologies for rice had been intensively researched and promoted well before the Rice Q-GAP program (Praneetvatakul and Waibel, 2006). However, as it was not the main aim of the project, we have not looked at the detailed use of pesticides. For example, we haven't looked at issues such as the timing of the pesticides used and the compliance with preharvest intervals. This should be explored further, as it could have an important influence of the quality of the rice produced.

18.5 CONCLUSION AND POLICY IMPLICATIONS

We estimated the potential effects of farming household characteristics and information channels on the adoption of public Q-GAP by rice farmers of the Central Plains of Thailand. We used FIML to estimate a RBPM to control for direct effects of farm and household characteristics on adoption as well as the indirect effects through endogenous participation to training organized for the promotion of Q-GAP.

Our result showed that information channels used by farmers and membership in farmers' groups are having an important effect on the probability of Q-GAP adoption. However, the model results suggest some indirect effects: farmers connected to groups are more likely to participate in the training and information sessions, either through self-selection or nonrandom assignment of invitations by the organizers. We also found a small but significant negative relationship between training and Q-GAP adoption. Our interpretation is that the different information channels used by farmers have very positive impacts on the adoption of Q-GAP. However, regardless of the amount of information that can be channeled, farmers still have to make an informed decision based on their assets and expectations about the costs and the benefits of adopting.

As such, the training should probably be credited for giving farmers an honest representation of the additional costs and benefits. As a result, it has probably discouraged (given the requirements) a number of farmers that were initially interested in improving their practices. Besides, some farmers have probably not been convinced by the positive arguments put forward during the training that included the ability to reduce their production costs, or improved market access. We also found that labor constraints had an important influence on adoption. Higher availability of labor per cultivated rice area increased the probability of adoption. Again, labor factors seem to have influenced rice training participation and as such also have an indirect influence on Q-GAP adoption. Other traditional factors influencing adoption such as education had some influence on the probability of adoption but to a much lesser scale than labor-related factors. Finally, the farmer-to-farmer diffusion does not seem to function well, at least in that region.

Cultivation practices were found to be different between adopters and nonadopters of Q-GAP, and potentially less harmful for farmers and the environment. However, without a baseline survey, we are unable to say if these differences are the consequences of Q-GAP adoption or the result of a selection process (eg, only farmers with more sustainable practices registered for the program). We initially asked farmers to recall their practices, but the initial interviews showed us that recall was inducing very long interviews that farmers were not ready to give, and in many cases resulted in very unreliable estimates of the practices. So we had to drop this procedure. More sophisticated statistical procedures such as propensity score analysis (Blackman and Naranjo, 2012; Ruben and Zuniga, 2011) would be required to confirm our first intuition that changes in practices are a consequence of Q-GAP adoption. More generally, additional research is needed to analyze the dynamic of adoption and the Q-GAP real impact. A similar survey could be conducted on a regular basis to observe adoption and potential disadoption dynamics (Barham et al., 2004). Panel data would also allow us to identify the real impact of the project over time.

From a policy perspective we found that the training provided by government agencies to be important and useful tools for sound decision making by farmers. More informed farmer's decision is likely to conduce to higher rate of success and lower disadoption rate over time. Again, this could be confirmed by observing the adoption/disadoption patterns over time.

We also found that Q-GAP adopters have slightly reduced their production costs. However, these cost reductions are small and it may be difficult to keep farmers' long-term interest in the program if no specific markets (paying higher prices) for the Q-GAP produced rice emerge. Qualitative data collected during the survey show that many adopters are finding that their initial expectations of increased prices or access to new markets have not materialized. The Q-GAP program is, by design, not interfering with rice prices and the lack of specific demand for Q-GAP labeled products by millers and intermediate stakeholders of the value chain is a weak point of the program that needs to be addressed. In addition, since the early 1990s, the successive Thai governments have been guaranteeing or supporting rice prices paid to farmers through different complex mechanisms. These transfers to rice producers, mainly through price mechanisms, have been made without requiring any specific changes in production practices. Therefore, the signals sent to farmers by the different government agencies are inconsistent: on one side, price mark-ups are given to farmers without any requirements in terms of agricultural practices; on the other side, farmers are encouraged to change toward more sustainable farming practices without being given any particular incentives to do so.

Overall, Thai farmers are asked to produce a rice that is safe to consume and produced in a way that is less harmful for the environment. However, from their point of view, the incentives for achieving those "societal goals" are unclear. First, adoption is not resulting in important cost reductions and nonadopters are in many cases anticipating higher production costs. Second, adopters are not rewarded by higher prices or higher sales volumes when they achieve these societal goals. Therefore, the long-term success of the Q-GAP program will depend upon the government's ability to develop new incentive mechanisms that bring into line the interests of society with those of rice producers. Given the structure of the rice value chain, several such mechanisms could be envisioned. A first mechanism would be to condition the current support to rice farmers to the compliance of Q-GAP and to make sure that government support is proportional to the areas where these GAP are followed and not channeled through price mechanisms (Latacz-Lohmann and Hodge, 2003). A sophistication of this first mechanism would be to rethink the support not in terms of agricultural practices but in terms of outcomes sought out by society (Burton and Schwarz, 2013; Hasund, 2013). This mechanism can be seen as targeted income subsidies that would be similar to the European agroenvironmental policy (Latacz-Lohmann and Hodge, 2003). More market-oriented solutions could be designed. A specific market for Q-GAP produced rice could be developed, through a mechanism of Q-GAP labeling and price differentiation. Under this second mechanism, the rice seller should be able to prove that 100% of the physical product sold under this label originates from a certified sustainable rice farm. As such, it requires that the physical sustainable product from a certified rice farm is kept separate from conventional rice throughout the entire supply chain. This may be difficult to establish given the number of stakeholders between farmers and consumers but would create a specific market for Q-GAP produced rice. Finally, a Book and Claim (B&C) mechanism, an innovative market approach inspired from incentive mechanisms developed for other commodity chains like oil palm (eg, van Duijn, 2013) could be developed. The main idea of B&C is to create a new market for certificates of sustainable rice production separated from the rice market. Certified producers sell certificates on a market to end users who choose to support (and claim they do support) specific areas or volumes of certified rice. In any of these settings, it would be necessary to anticipate the farmers' willingness to participate in such programs (eg, Schulz et al., 2014; Jaeck and Lifran, 2014).

References

Adesina, A.A., Mbila, D., Nkamleu, G.B., Endamana, D., 2000. Econometric analysis of the determinants of adoption of alley farming by farmers in the forest zone of southwest Cameroon. Agric. Ecosyst. Environ. 80 (3), 255–265.

Alene, A.D., Manyong, V., Omanya, G., Mignouna, H., Bokanga, M., Odhi-ambo, G., 2008. Smallholder market participation under transactions costs: maize supply and fertilizer demand in Kenya. Food Policy 33 (4), 318–328.

Amekawa, Y., 2009. Reflections on the growing influence of good agricultural practices in the global South. J. Agric. Environ. Ethic. 22 (6), 531–557.

Amekawa, Y., 2013. Can a public GAP approach ensure safety and fairness? A comparative study of Q-GAP in Thailand. J. Peasant Stud. 40 (1), 189–217.

Angrist, J.D., 2001. Estimation of limited dependent variable models with dummy endogenous regressors: simple strategies for empirical practice. J. Bus. Econ. Stat. 19 (1), 2–16.

Barham, B.L., Foltz, J.D., Jackson-Smith, D., Moon, S., 2004. The dynamics of agricultural biotechnology adoption: Lessons from series rBST use in Wisconsin, 1994–2001. Am. J. Agr. Econ. 86 (1), 61–72.

Baslevent, C., El-Hamidi, F., 2009. Preferences for early retirement among older government employees in Egypt. Econ. Bull. 29 (2), 554–565.

Bhattacharya, J., Goldman, D., McCaffrey, D., 2006. Estimating probit models with self-selected treatments. Stat. Med. 25 (3), 389–413.

Blackman, A., Naranjo, M.A., 2012. Does eco-certification have environmental benefits? Organic coffee in Costa Rica. Ecol. Econ. 83, 60–68.

Burton, R.J., Schwarz, G., 2013. Result-oriented agri-environmental schemes in Europe and their potential for promoting behavioural change. Land Use Policy 30 (1), 628–641.

Chiburis, R.C., Das, J., Lokshin, M., 2012. A practical comparison of the bivariate probit and linear IV estimators. Econ. Lett. 117 (3), 762–766.

Chouichom, S., Yamao, M., 2010. Comparing opinions and attitudes of organic and non-organic farmers towards organic rice farming system in north-eastern Thailand. J. Org. Syst. 5 (1), 25–35.

Doss, C.R., 2003. Understanding farm-level technology adoption: lessons learned from CIMMYT's micro surveys in Eastern Africa. No. 03-07 in Economics Program – Working Paper. CIMMYT.

Feder, G., Slade, R., 1984. The acquisition of information and the adoption of new technology. Am. J. Agric. Econ. 66 (3), 312–320.

Feder, G., Just, R.E., Zilberman, D., 1985. Adoption of agricultural innovations in developing countries: a survey. Econ. Dev. Cult. Change 33 (25), 255–298.

Foster, A., Rosenzweig, M., 1995. Learning by doing and learning from others: human capital and technical change in agriculture. J. Polit. Econ. 103 (6), 1176–1209.

Greene, W.H., 1998. Gender economics courses in liberal arts colleges: further results. J. Econ. Educ. 29 (4), 291–300.

Greene, W.H., 2007. Econometric Analysis, sixth ed. Prentice Hall, London.

Hasund, K.P., 2013. Indicator-based agri-environmental payments: a payment-by-result model for public goods with a Swedish application. Land Use Policy 30 (1), 223–233.

Hussain, S.S., Byerlee, D., Heisey, P.W., 1994. Impacts of the training and visit extension system on farmers' knowledge and adoption of technology: evidence from Pakistan. Agric. Econ. 10 (1), 39–47.

Jaeck, M., Lifran, R., 2014. Farmers' preferences for production practices: a choice experiment study in the Rhone river delta. J. Agric. Econ. 65 (1), 112–130.

Kariuki, I.M., Loy, J.-P., Herzfeld, T., 2012. Farmgate private standards and price premium: evidence from the GlobalGAP scheme in Kenya's French beans marketing. Agribusiness 28 (1), 42–53.

Kersting, S., Wollni, M., 2012. New institutional arrangements and standard adoption: evidence from small-scale fruit and vegetable farmers in Thailand. Food Policy 37 (4), 452–462.

Knowler, D., Bradshaw, B., 2007. Farmers' adoption of conservation agriculture: a review and synthesis of recent research. Food Policy 32 (1), 25–48.

Larkin, S.L., Larson, J.A., Paxton, K.W., English, B.C., Marra, M.C., Reeves, J.M., 2008. A binary logit estimation of factors affecting adoption of GPS guidance systems by cotton producers. J. Agric. Appl. Econ. 40 (1), 345–355.

Latacz-Lohmann, U., Hodge, I., 2003. European agri-environmental policy for the 21st century. Aust. J. Agric. Resour. Econ. 47 (1), 123–139.

Lee, D.R., 2005. Agricultural sustainability and technology adoption: issues and policies for developing countries. Am. J. Agric. Econ. 87 (5), 1325–1334.

Lestrelin, G., Quoc, H., Jullien, F., Rattanatray, B., Khamxaykhay, C., Tivet, F., 2012. Conservation agriculture in Laos: diffusion and determinants for adoption of direct seeding mulch-based cropping systems in smallholder agriculture. Renew. Agric. Food Syst. 27 (1), 81–92.

Liu, M., Wu, L., Gao, Y., Wang, Y., 2011. Farmers' adoption of sustainable agricultural technologies: a case study in Shandong Province, China. J. Food. Agric. Environ. 9 (2), 623–628.

Mendola, M., 2007. Agricultural technology adoption and poverty reduction: a propensity-score matching analysis for rural Bangladesh. Food Policy 32 (3), 372–393.

Monfardini, C., Radice, R., 2008. Testing exogeneity in the bivariate probit model: a Monte Carlo study. Oxford B. Econ. Stat. 70 (2), 271–282.

Moser, C.M., Barrett, C.B., 2006. The complex dynamics of smallholder technology adoption: the case of SRI in Madagascar. Agric. Econ. 35 (3), 373–388.

Office of Agriculture Economics, 2010. Socio-economic characteristics of farm household and farm labour at provincial level for the 2010/2011 production year. http://www.oae.go.th.

Okello, J.J., Swinton, S.M., 2007. Comparison of a small- and a large-scale farm producing for export. Appl. Econ. Persp. Pol. 29 (2), 269–285.

Othman, N.M., 2007. Food safety in Southeast Asia: challenges facing the region. Asian J. Agric. Dev. 4 (2), 83–92.

Ouma, J., De Groote, H., 2011. Determinants of improved maize seed and fertilizer adoption in Kenya. J. Dev. Agric. Econ. 3 (11), 529–536.

Praneetvatakul, S., Waibel, H., 2006. Impact assessment of farmer field school using a multi period panel data model. In: 26th Conference of the International Association of Agricultural Economists, Gold Coast, pp. 12–18.

Premier, R., Ledger, S., 2006. Good agricultural practices in Australia and Southeast Asia. HortTechnology 16 (4), 552–555.

Ruben, R., Zuniga, G., 2011. How standards compete: comparative impact of coffee certification schemes in Northern Nicaragua. Supply Chain Manag. 16 (2), 98–109.

Sarsud, V., 2007. Challenges and opportunities arising from private standards on food safety and environment for exporters of fresh fruit and vegetables in Asia: Experiences of Malaysia, Thailand and Vietnam. Ch. National Experiences: Thailand, pp. 53–69.

Schreinemachers, P., Schad, I., Tipraqsa, P., Williams, P.M., Neef, A., Ri-wthong, S., Sangchan, W., Grovermann, C., 2012. Can public GAP standards reduce agricultural pesticide use? The case of fruit and vegetable farming in northern Thailand. Agric. Hum. Values 29 (4), 519–529.

Schulz, N., Breustedt, G., Latacz-Lohmann, U., 2014. Assessing farmers' willingness to accept "greening": insights from a discrete choice experiment in Germany. J. Agric. Econ. 65 (1), 26–48.

Smale, M., Heisey, P.W., Leathers, H.D., 1995. Maize of the ancestors and modern varieties: the microeconomics of high-yielding variety adoption in Malawi. Econ. Dev. Cult. Change 43 (2), 351–368.

Stoneman, P.L., David, P.A., 1986. Adoption subsidies vs. information provision as instruments of technology policy. Econ. J. 96 (380a), 142–150.

Subervie, J., Vagneron, I., 2013. A drop of water in the Indian Ocean? The impact of GlobalGAP certification on lychee farmers in Madagascar. World Dev. 50, 57–73.

Thanawong, K., Perret, S., Basset-Mens, C., 2014. Eco-efficiency of paddy rice production in Northeastern Thailand: a comparison of rain-fed and irrigated cropping systems. J. Clean. Prod. 73 (2014), 204–217.

van Duijn, G., 2013. Traceability of the palm oil supply chain. Lipid Technol. 25 (1), 15–18.

White, D.S., Labarta, R.A., Leguia, E.J., 2005. Technology adoption by resource-poor farmers: considering the implications of peak-season labor costs. Agric. Syst. 85 (2), 183–201.

Wooldridge, J.M., 2002. Econometric Analysis of Cross Section and Panel Data. The MIT Press, Cambridge.

A Multiple Case Study on Analyzing Policy and Their Practice Linkages: Implications to REDD+

S. Sharma*, G. Shivakoti[†,‡]

*WWF-Nepal, Kathmandu, Nepal [†]The University of Tokyo, Tokyo, Japan
[‡]Asian Institute of Technology, Bangkok, Thailand

19.1 INTRODUCTION

A series of progressions on annexation and state formations on policies has classified forest under the term "public good" that relatively offers poorer management privileges to forest users in Southeast Asia. In Thailand, Indonesia, and Vietnam, state forest ranges between 32% and 51.4% of the nation's complete land zone (Kartodihardjo, 2002; FAO, 2009). Choices on usages and management of these forests are determined by the central government and then passed to regional, provincial, and then local agencies (Sharma et al., 2015; Yonariza et al., 2015; Mahdi et al., 2015). Especially, these forests are set aside for protection and production whereby the use rights and management responsibilities are governed by the central administration.

In the process, the central administration could not efficiently manage these forests; deforestation was high, further intensified by land-use changes for industrial logging. The major reasons for failure to manage forest are clarified as a lack of efficient governance and discontent over hierarchal authority outright by corruption transparency and accountability (Sharma et al., 2015; Yonariza et al., 2015; Mahdi et al., 2015). For above 200 million people located in the land categorized as public forest (Fisher et al., 1997), this process of forest governance has not only impacted their livelihoods but crushed overall forest ecosystems. Realizing the restrictions put forward by this centralized module of forest management, countries in Southeast Asia are devolving powers to local administration, exclusively for small forest patches. Vietnam's Land Law of 1993, Thailand's Tambon Administrative Act of 1994, and the Regional Autonomy Law of 1999 in Indonesia are some of the initial provisions foreseeing forest management at the local level (Sharma et al., 2015; Yonariza et al., 2015; Mahdi et al., 2015). This is in fact recognition to rural communities of their roles in forest management and monitoring, assumed by their closeness to the forest and their habitual observations and decisions on forest management that are often mediated by informal operational rules.

Nevertheless, deforestation is still alarmingly high. Thailand, Vietnam, and Indonesia have approximately 28.4%, 39.7%, and 52.1% of forest area, respectively (FAO, 2005; FAO, 2010). Thailand drops approximately 115,100 ha of forest extent yearly with an annual deforestation rate of 0.72% (FAO, 2005). Not long after the Tambon Administrative Act, Thailand's forest decreased by 44.9% to 0.40% per annum throughout 2000 to 2005. Between 1990 and 2005, Thailand lost 9.1% of its forest, or around 1,445,000 ha. Likewise, from 1990 to 2005, Vietnam expanded 38.1% of its forest, or about 3,568,000 ha (FAO, 2005). However, Vietnam lost 299,000 ha of its primary forest throughout that period. Conversely, deforestation rates of primary forest cover had been reduced 77.9% by the 1990s (FAO, 2005). In addition, calculating total habitat transformations for 1990 to 2005, Thailand lost 14.3% of forest habitat while Vietnam gained 48.8% of forest habitat. Indonesia lost about 120.3% of forest between 1990 and 2010 (Butler, 2014). This drift of managing resources with local arrangement is not analogous across Southeast Asia, with substantial dissimilarity among all on bundles of rights and extent of rights given.

In some ways, polices put forward to reduce deforestation and forest degradation were not efficiently achieving their objectives. The literature shows that deforestation is linked to a country's economy; shaped by the political economy coupled with the institutional landscape while overall affecting the forestry sector (Indrarto et al., 2012). The failure of other sectorial polices to comply with the decentralization process may envision such results, too. Also, the actual implementation of policy issued by the government may be significantly low; creating a huge gap between policy and practice (Yonariza et al., 2015; Soriaga and Mahanty, 2008). Although reassuring outlines may exist, operation gaps are common because management resource and capacity limitations and confrontation exist in certain sections of government (Soriaga and Mahanty, 2008).

Also meaningfully imperative is that the governance structure be in line with the policy provisions. The concept of good governance has been common jargon in forestry since the 1980s. Even if policy had a broader goal of devolving powers to the general public, just as vital are the questions on: Who made the decisions and did the decisions incorporate the needs and aspirations of the forest-dependent community, both at the national and subnational level? A good policy implementation in the forestry sector should consider issues such as decisions being fair and transparent, rights being appreciated, rules being implemented equitably, and overall decisions being based on creating a favorable condition for both the forest and the people in common and not in individual importance (Larson and Petkova, 2011).

Reducing Emissions from Deforestation and Forest Degradation (REDD+) is one of the most discussed topics in the international forestry arena. Together with discussion, huge investments are made in the international climate change forum to prepare countries themselves for future REDD+. Thailand, Indonesia, and Vietnam are such countries, whereby preparation includes formulating strategies to combat deforestation and forest degradation. At this stage, acknowledging the current governance status will aid in recognizing areas that need improvements.

In this current backdrop, whereby the polices are in fact not enough to reduce deforestation rates, the governments of Thailand, Indonesia, and Vietnam are planning to implement REDD+ without acknowledging the fact that forest management is a socioecological interaction that involves institutions, political pressures, and users actions that have come up through experiences and methodologies.

Thailand, Vietnam, and Indonesia are preparing themselves for REDD+ with numerous pilot studies and fund mobilization. This is the right time for countries to adjust policies and strategies inclined to REDD+ and also new goals and ideologies would be forwarded. REDD+ requires a good benefit sharing mechanism often dominated by effective governance. Prior to adoption, it is vital to acknowledge the existing legal provisions on which the countries are planning REDD+ and the relative efficiency of existing policies to future emission reduction program implementation; in terms of REDD+ benefit sharing and creating scope for better livelihood through forest restoration. At question is whether REDD+ generates more prospects than livelihood menaces for forest users and managers.

Evidence of the effectiveness of existing policies for local practices require acknowledged treasuring if policy changes local practices. For policies relevant in a national context may not be suited locally; simultaneously, more policy-practice gaps are foreseen. Either way, a national context may be least favorable for REDD+ in terms of unclear benefit sharing and governance criteria; locals may have managed resources putting forward best frameworks for REDD. At both circumstances, new modifications could be made suited to REDD+ while promoting strong extensions and delivery of services in the national arena. Hence, this chapter focuses on analyzing policy existences that are suited for REDD+ in Thailand, Indonesia, and Vietnam; while also analyzing actual practices undertaken locally. With these change-overs and encounters in observance, we emphasize ensuing inquiries: What polices favorable for successful REDD+ implementation exist? How are local communities applying these forest polices implemented by the government? Based on a review of REDD+, pilot cases policy briefs of Thailand, Indonesia, and Vietnam in terms of forest governance are analysed, and the links between policy practice gaps of the forest-dependent communities are identified.

19.2 METHODS

This study was undertaken through two steps. A framework on existing policy provisions were developed in the first step while the second step is an application to a former constructed framework on policy provisions from case studies.

19.2.1 Framework Identifying Existing Policy Provisions That Has Implications for REDD+

The work began with evaluating policy provisions through desk study; mostly favorable to REDD+ governance and benefit sharing in a framework developed by Program on Forests. This has been used intensively to identify and analyze forest governance; mostly endorsed by the European Union, the Food and Agriculture Organization of the United Nations (FAO), the World Bank, the Swedish International Development Agency (SIDA), and lately UN-REDD (Kishor and Rosenbaum, 2012). It has indicators, 13 components and subcomponents and 3 pillars, out of which only a few indicators were considered for the study (Table 19.1).

These indicators are represented in scaled units. After comprehensive examination of prevailing nations' legal documents, the authors have ranked each indicator on a scale of 0 to 3, in accord with the subsequent characteristics: 0, no considerations; 1, policy makers are aware of it and being discussed; 2, agreed in principle; 3, rules exist in laws. This aids in comparing different governance levels to evaluate differently. A series of key informants from

TABLE 19.1   Framework for Analyzing Policies with Implications to REDD+

Pillars	Components	Indicators
Policy, legal, institutional and regulatory frameworks	Forest related polices and laws	1. Policy existence 2. Clarity of these polices 3. Extent to supporting adaptive management 4. Consistency with international commitments
	Land tenure, ownership, and use	1. Extent to recognizing carbon rights 2. Conflict resolution mechanism
	Institutional frameworks	1. Extent to which forest mandates/national agencies are mutually supportive 2. Adequacy, predictability and stability of forest agency budgets and organizational resources 3. Availability of information, technology tools
	Financial incentives and benefit sharing	1. Equity in forest resources, rights and rents 2. Openness and competitiveness of procedures, such as auctions for forest resources allocation 3. Mechanisms for internalization of social and environmental externalities
Planning and Decision-making processes	Stakeholder participation	1. Participation in planning and decision making
	Transparency and Accountability	1. Transparency to public access to information, allocation of timber
Implementation, enforcement and compliance	Forest Law implementation	1. Extent, appropriateness and effectiveness of enforcement agencies 2. Land tenure documentation and administration

government and universities engaged in the forestry sector, particularly at the REDD+ initiation level, also took part as primary sources. In each country, roughly 10–15 key informants were questioned. As REDD+ is a new, multidisciplinary, and specializes in subjects only a few informants could comprehend the detail is perceptions. The scales, as mentioned as above, were also discussed with these informants.

19.2.2 Case Studies

Our efforts were next followed by examining the status of policy implications using a multiple case study method. We studied five pilot schemes for REDD+ to synthesize findings on breaches between what's written on policy and how people are implementing it. Cases were selected to signify the broadest probable series of REDD+ schemes, tenure, and states to exploit prospective lessons from undergoing forest governance in a diverse context. Therefore, the cases were selected on the basis of

a. From geographic regions, likely to target REDD+ for on-going conflicts and high deforestations
b. Representing diverse tenure arrangements and management
c. Characterizing a variety of financial plans (project funded, government funded).

TABLE 19.2   Individualities of Case Studies

Case Study	Geographical Coverage	Tenure Arrangement	Scheme
Sor Por Kor and Inpang Networks	Northeast Thailand	State owned; management rights	Funded by research project
Community forestry	Northern Thailand	State owned; demanding management and exclusion rights	Royal Department of Forest (RFD)
Krang Krachan National Park	South Thailand	National park	Department of National Park, Wildlife and Plant Conservation (DNPWPC)
Bukit Panjang Farmer's Group	Indonesia	Common property	Forest rehabilitation fund
PES program in Lam Dong province	Vietnam	Common property; private	

An additional feature was the value and accessibility of data. Each case study was individually reviewed by researchers from particular countries.

Five of the selected cases are projects from Southeast Asia (Table 19.2). As most of the researches have noted, tenure rights are usually overlapping and not always clear in terms of entree, usage, and management over a particular boundary.

Communities from these sites were visited and information regarding policy effects in their forest conditions and livelihood improvements were gathered through focus group discussions and compared with existing norms in policies. The information gathered was further triangulated in a provincial level workshop including donors, community-based organizations (CBOs), and governmental agencies. Queries arose in the field and a provincial workshop forum was discussed further in regional workshops.

19.3 EXISTING FORESTRY POLICIES IN THAILAND, INDONESIA, AND VIETNAM IN RELATION TO REDD+

19.3.1 Policy, Institutional, and Regulatory Legal Framework

Fig. 19.1 summarizes the first component of the first pillar "policy, legal, institutional, and regulatory frameworks," governance attributes that are prerequisites for successful REDD+ implementation in Vietnam, Indonesia, and Thailand as discussed in Table 19.1.

Fig. 19.1 shows the diverse and assorted status in three countries. Most of the countries have relatively good forest policies and associated laws, however, Vietnam has better policies and the institutional arrangements needed for successful REDD+ implementation.

According to Sharma et al. (2015) "Formal legislation governing the forest sectors in Thailand includes: (1) The Forest Act of 1941, (2) The National Parks Act of 1961, (3) The National Reserved Forests Act of 1964, (4) The Commercial Forest Plantation Act of 1992, (5) The Forest Plantation Act of 1992, (6) The Community Forest Bill of 2007."

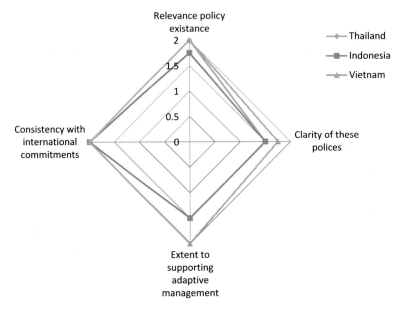

FIG. 19.1   Existing policies and laws: clarity and supporting international commitments and adaptive management.

The key official legislation prevailing in the forest segments in Vietnam comprises: (1) Law of Forest Protection and Development 2004; (2) Decision 178/2001/QD-TTg on the sharing of benefits between local households/communities participating in forest protection and management with different forest types: production, protection, and special use forest; and (3) National Degree No. 99 and Decision 380 on Payment for Environmental Services.

Finally, the key official legislation prevailing in the forest segments in Indonesia comprises: (1) Law on Development of Industrial Plantation Forest (Hutan Tanaan Industri-HTI); (2) Forestry Law of 1967; (3) PP No. 21/1970 on forest use rights; (4) Land Use by consensus in 1984; Tebang Pilih dan Tanam Jalur-1998 on plantation crops; (5) regional autonomy policy in 1999; and (6) Reforestation Fund in 2002 (Indrarto et al., 2012).

There are policies and related cabinet decisions or their amendments. These laws overlay and add difficult dimensions to implementation. For example, per Thailand's National Reserved Forest Act 1964, public discussion and consent is voluntary during reserve forest declaration; whereas, the Indonesian Constitution provides a mandate for locals to participate in decision making in forest management. There has been no consideration of local consent as policy makers tend to follow former law, while locals are protesting the judgments with respect to the constitution.

Other examples from Vietnam: the Law of Environmental Protection, Land Law, and Law of Biodiversity Conservation. Although all these laws have the objective to protect forests, they use an inconsistent definition of terms that has led to complexity. Similarly, the Basic Forestry Law in Indonesia tries to address major drivers of deforestation, while the economic policy is promoting activities for economic development without environmental considerations. These provisions are creating confusion and conflicts due to diverse interpretations.

REDD+ application with a key foundation on the legal background makes the status quo further disordered. On the other hand, among the three countries, Vietnam has relatively fewer ambiguities in terms of policy understanding.

Initially, forest policy clarified that 50%, 60%, and 40% of total land area needs to be forested in Thailand, Indonesia, and Vietnam, respectively, while other policies were framed to provision the earlier policy, explicitly laid out by the government for diminishing natural disasters while remaining by the cabinet. In Thailand and Indonesia, the community is not once given the facts, nor is consensus taken. Some are ad hoc with no adaptive learning elaborated. Quite the opposite, REDD+ cannot be undertaken through these policies.

Thailand, Indonesia, and Vietnam are cosigners to "the Convention on International Trade in Endangered Species of Wild Fauna and Flora, the Convention on World Heritage, the Ramsar Convention, the Convention on Climate Change (UNFCCC), the Convention on Biological Diversity (CBD), and the International Tropical Timber Agreement (ITTA)" (Nalampoon, 2003; Indrarto et al., 2012; Sharma et al., 2015). Utmost rules and laws forced these three countries to dedicate to these agreements. Particularly in northeast Thailand and Indonesia, individuals existing in the conservation zone are abandoned to declare heritage sites (Hares, 2009; Indrarto et al., 2012).

19.3.2 Legal Framework to Support Land Tenure, Rights, and Ownership

As shown in Fig. 19.2, Vietnam has relatively good provisions for addressing forest-dependent communities residing in the vicinity of protected areas. The state owns all forests in Vietnam, which are then allocated to households and organizations for short-term or long-term benefits. This is a relative advantage of Vietnam over Indonesia and Thailand even though there are individual issues within Vietnam. For instance, many studies show that local stakeholders have no ownership and the only rights given are for use and management.

Thailand's constitution confers on communities the right to defend their civilizations and to the management, conservation, and usage of natural environment resources and biodiversity in a sustainable manner. On the other hand, there are state-run forests in Thailand. Issues are debatable on how this complexity works for communities to manage.

The challenges and ambiguities of indigenous rights and tenure arrangements in state forests remain baffling in Indonesia. The country has various provisions at the policy level but,

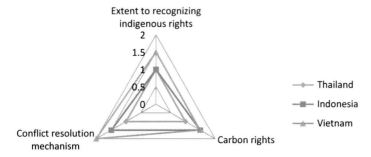

FIG. 19.2   Land tenure, ownership, and use.

according to Indrarto et al. (2012), these are neglected in reality. Issues of concern, such as carbon tenure, are mentioned in several Indonesian government legal documents that emphasize carbon sequestration as one of the forms of ecosystem services. It is noticeable that those benefitting from carbon capture have the same stake as those using environmental services. In the abovementioned descriptions on state forest, prominent mention of carbon tenure is restricted as a right to enterprise and does not guarantee ownership. However, the product of carbon sequestration may be sold as other forest product.

Conflicts between government and communities are arising in Indonesia and Thailand, especially at protected zones due to contradictory dual goals of government and community as preservation and utilization. Examples include struggles among the administration and industrial logging concessionaires, or clashes flanked among government and individuals pretentious by estate plantation (evicted/affected by plantations). This has adverse effects on forests, because to fulfil their basic needs, local affected communities break rules and are engaged in haphazard harvesting. Conflict is also ascending from industrial concessions for high-value timber logging and profitable estate plantations.

There aren't any conflict resolution devices in policy documents. Conversely, conflict against national parks is usually against the people because the national park has strong legal support and does not offer a local-based conflict resolution process. Most of the cases are piled up in the court as resource-poor farmers cannot pay high court and legal fees. Conflicts are also rampant in Vietnam, which turn into physical damages. Lack of an effective conflict resolution mechanism has failed to settle land conflicts locally. There have been no policy talks about the conflict resolution mechanism.

19.3.3 Institutional Frameworks

There are overlaps in administrative policies and functions. The Royal Forest Department (RFD) and the Department of National Parks, Wildlife and Plant Conservation (DNWPC), both administrative agencies under the Ministry of Natural Resources and the Environment (MNRE), which has the major turf. RFD is responsible for management, logging, forest harvesting, and promoting public participation, while DNWPC basically manages wild flora and fauna inside the national park. In addition, watershed conservation is also associated with DNWPC. During administrative reformation, the process of REDD+ implementation went by to DNWPC. Community forestry with a suitable benefit sharing mechanism is managed under RFD. This has created confusion and fueled conflicts. As one of the REDD+ pilot sites, Krang Kachen National Park has been chosen with no consideration of public involvement, rules, and benefit-sharing ideologies. At most, the community residing within the national park are not involved in the process (Fig. 19.3).

There is overlap between the Ministry of Agriculture and Rural Development (MARD) and the Ministry of Natural Resources and Environment (MONRE) in Vietnam. MARD is responsible for forest development and management, while MONRE especially focuses on land and environmental issues. Mostly, confusion and overlapping rules often characterize differently in the implementation arena at the local, provincial, and district levels. In Indonesia, decentralization and authority devolution to district and regional government increases deforestation through the number of forestry industrial permits granted without considering the socioecological consequences. On the background of conflicting ideologies among economy

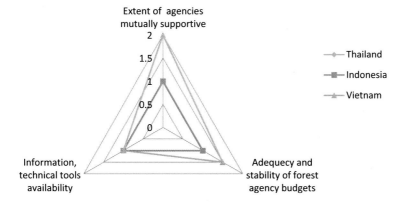

FIG. 19.3   Institutional frameworks.

and emission reduction, REDD+ is likely to be operative if it reimburses management and entities for "opportunity cost" suffered due to "disappointment" of economic inducements that REDD+ substitutes. Regional government is likely to device REDD+ if it is relatively advantageous economically.

Investment in climate change comes from countless sources, mainly through bilateral sources such as the U.S. Country Study Program and World Bank National CDM Strategy Study. A part of this investment is also contributed by national budgets. Nevertheless, monetary practical applications for technology handover on environmentally feasible appliances are relatively less in these three countries. All rules and information are available but essentially not outside country capitols. Forest laws, rules, and related court documents are typically circulated in difficult languages not easily understood by local communities.

19.3.4 Financial Incentives and Benefit Sharing

Thailand's Forest Act includes a teak and timber proclamation and the royalty involved, as forest in Thailand belongs to the state (Sharma et al., 2015). The law has no provision for equitable distribution of this money even if the harvested timber is from community forestry. The National Park Act confers on government the power to demark any area into national park with or without local consensus (mostly done without consensus). Though the cabinet backed this act by clarifying communities to be settled elsewhere, in practice this is often not done. Communities are forcefully moved without incentives and as a result communities are compelled to undertake haphazard destruction to manage settlement in forest areas (Sharma et al., 2015). Though clear mechanisms on timber harvesting and sale procedures is mentioned in the policy, other products have no such procedures. As such, Thailand has no competitiveness and openness in forest product marketing, harvesting, and sale.

In Vietnam, the Law of Forest Protection and Development mentioned in Point b, Clause 3, Article 24—forest allocation regulates that the government allocate natural production forest and get usage payments from all forest enterprises. However, most of the remaining natural forests are average or poor; there is almost no income from forest within the long investment time frame (25–30 years). Resource poor households that depended on these unproductive

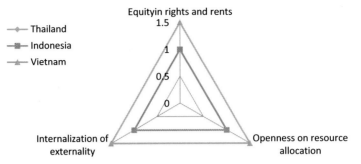

FIG. 19.4   Financial incentives and benefit sharing.

forests may face hard times balancing their livelihood priorities with government usage payments. However, as shown in Fig. 19.4, Vietnam has relatively higher equity in terms of rights and rents, mostly due to forest land allocation.

Equity is not taken into consideration in Indonesian forestry as much of the attention is toward attaining a national target of palm plantations and the national economy. The participation in usage, management, and decision making of customary participation is constrained and to date there are no unpretentious, cheap, user-friendly measures to permit customary immigration to accomplish forest areas.

Vietnam and Indonesia have unclear criteria to distinguish between the products and by-products from the forest. The ratio of benefit sharing confirms an allocation procedure should be based on forest status at the time of allocation. Conversely, these policies do not mention anything about conducting forest inventories to determine forest conditions both before and after allocations. In conditions whereby local authorities tend to conduct baseline inventories, the process is expensive and no financial support from the central state is rendered. Owing to the fact that corruption is relatively lower in Vietnam, openness on resource allocation can be considered relatively high.

There are several policy document guidelines in Degree 99/2010/NĐ-CP of the Prime Minister on Payment for Forest environmental services in Vietnam, including Circular of the Minister of Agriculture and Rural Development on the principle and methodology for determining the watershed area to serve for the payment of environmental services. The decision of the prime minister explicates on service, amount of payment, payment method, and benefit-sharing mechanism. These provisions are not available in Thailand and Indonesian policy; hence, internalization of externality could be considered higher in Vietnam as shown in Fig. 19.4.

19.3.5 Stakeholder Participation

A community forest bill is under contemplation and reflection in the parliament in Thailand. The aim is to allot management rights to local communities to manage the resources according to their needs. Because the forestry sector contributes a major amount of the nation's gross domestic product (GDP), the slow process of according community rights reflects an unwillingness to devolve power to communities and undermine stakeholder participation. Similarly, Indonesia's major revenue is drawn from estate plantations, and this has led to the ambiguity as made by Phelps et al. (2010) as the country may recentralize the forest with less priority to stakeholder participation.

There was no public consultation process for promulgating forest-related laws in Vietnam, Indonesia, and Thailand. However, since the introduction of Doi Moi, the government of Vietnam managed the forest through a forest allocation process; this is believed to have greater community participation in planning and implementing. A number of people who had previously settled and practiced agriculture within the areas that became reserve areas were categorized as illegal forest encroachers in Indonesia. These people were forced to relocate from those protected areas. Sometime those now landless people will become criminals if they continue living in the forests and can suffer heavily with imprisonment or even their lives.

19.3.6 Transparency and Accountability

Even though the Forest Act of Thailand clearly talks about marketing and distribution of natural resources, market transparency is limited in the log trade; forest wood dimension measurement practices are beyond the effective control of sellers and therefore provide opportunities for misuse.

The present statistical reporting in the forestry sector in Vietnam is far better than Thailand and Indonesia but is far from satisfactory. Information is a powerful tool in managing the sector, but it has been neglected in the past. There is an average strategy for information management. The purpose of data collection is, however, unclear to middle-level staff, which tends to perceive it as an administrative burden rather than a management tool. The entire forest statistical system needs to be reviewed.

In Indonesia, the processes for issuing permits and distributing revenues are corrupted and they contravene rules and legislation not only in issuing permits but also during supervision. Exacerbating corruption is the lack of clear penalties for abuse of authority. The lack of transparent permit issuance has also facilitated the process of corruption. For instance, prior to issuing permits, parties are required to undergo environmental impact assessment. But most of the time either this is not properly done or it is just copied from another region.

19.3.7 Implementation, Enforcement, and Compliance

Usually financial and technical support from the government is lacking, which has compelled poor farmers to invest themselves in protecting and managing forests. High taxes are imposed on users and they have no loan provision, coupled with strict harvesting and transportation rules. This is likely to reduce the willingness of local forest dwellers to participate and invest in natural forest management. Ironically, no government document has revealed improvements in the forest sector through the collected tax amounts; usage of the forest resource tax is inconsistent. This may lead to the cynical situation that REDD+ will be successful in Vietnam, Thailand, and Indonesia.

In Vietnam, Circular No. 38/2007/TT-BNN, dated 4/25/2007, and issued by MARD instructing the process and procedure of forest allocation, forest lease for organization, households, individual and local communities, and when doing the allocation of natural forest, it is required to determine forest stock. However, for the normal allocation process at the household level, forest allocation area is determined along with generalizing the forest into type II, III, IV or rich, average, poor, restoration forest; there are no quantitative measurements on the amount and quality of timber in allocated forest. Thus, there are no baseline data to determine the added growth value of the forest. In reality, most ethnic minority households have yet to take full advantage of these policies. According to Tebtebba (2010), this is a difficult to solve social problem that poses challenges in the face of REDD+ implementation in a background that land allocation in Vietnam is based on the individual's financial capacity to invest. Especially with ethnic minorities, the policy has often excluded them from receiving a larger share of land allocation (Dinh, 2005). Similar conditions may be found in Thailand and Indonesia.

19.4 GOVERNANCE COMPARISON FOR REDD+ IMPLICATIONS

REDD+ is one of the most discussed topics in the international forestry arena. Together with discussion, huge investments are made in the international climate change forum to prepare countries for future REDD+. Thailand, Indonesia, and Vietnam are such countries, whereby preparation includes formulating strategies to combat deforestation and forest degradation. At this stage, acknowledging the current governance status will aid in recognizing areas that need improvements.

Fig. 19.5 summarizes and compares governance attributes that are prerequisites for successful REDD+ implementation in Vietnam, Indonesia, and Thailand as discussed in Table 19.1. Fig. 19.5 also shows the diverse and assorted governance status in these three countries. Most of the countries have relatively good forest policies and associated laws; however, Vietnam has relatively better policy and the institutional arrangement needed for successful REDD+ implementation.

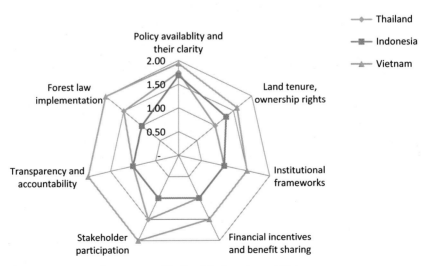

FIG. 19.5   Governance comparisons among Thailand, Indonesia, and Vietnam.

All three countries are rated moderately on a 0–3 scale with none of the attributes scoring 3. All three countries have relevant polices that through some modification may aid in successful REDD+ implementations. However, countries need to focus on land tenure and ownership arrangements. Unclear tenure arrangements have been identified as major causes for deforestation. Accordingly, ameliorating tenure is deliberated as a significant measure to manage deforestation. Explicatory tenure is well understood as a technique of endorsing reasonable the carrying out of REDD+. By instructive tenure it will be inflexible for administrations or influential actors to gain the welfares of REDD+. Strong and confident tenure may defend deprived forest tenants and local groups of people from marginalization or even being thrown out of forest areas and deliver better leverage to the country's REDD+ progress to demonstrate how indispensable REDD+ is as it raises the worth of forests.

A benefit-sharing mechanism for REDD+ could be considered as a common understanding among related stakeholders in terms of circulation of benefits. Such benefit-distribution mechanism should be effective enough to provide equitable recompense to communities for their interventions to avoid emissions. We see from Fig. 19.5 that Vietnam has relatively better institutional frameworks through their experience with provisioning ecosystem services (PES) programs, which is achieved through treating the public equitably. Vietnam also has a well-defined provision through PES that contemplates 10% distribution at the state, local, and district stages, for organization and execution rate. "The remainder is distributed based on a set formula known as the K-factor" (Minang et al., 2014). The K-factor indicates the ability of natural forest to provide environmental services often considered by forest status, origin, type, and degrees of management difficulties (Minang et al., 2014). For countries like Indonesia and Thailand, Fig. 19.5 prominently shows the need for orientation programs at the community and national levels for involving community in the decision-making route. The promising functionality of miscellaneous mechanisms that deeds good governance by cultivating objectives and usefulness as well as authority and transparency together is required.

19.5 POLICY PRACTICE GAPS

Governance emphasizes decision making and status of governance in these three countries, as clarified above. But specific interest is in the enquiry on how these policies and positions are communicated, translated, and supported into actions and programs (Dovers, 2005). To investigate this intricacy it will be crucial to make a distinction between policy systems and the implementation mechanism through which the policies are established and carried out. This is one of the challenges that pervade policy procedures with local standpoints, achieving a level of acknowledgment for prevailing community inventiveness and understanding in carrying out indefinable privileges and accountabilities that REDD+ polices articulate.

Government has formulated forest polices and implemented them in their own ways. but local stakeholders are unaware of it. This reflects a huge gap between policy and practice. One way to see the gap between policy and practice is by comparing the guideline with what was carried out. On the basis of the above-stated policy details, we have tried to list the observation practices (Table 19.3) in the case studies as listed in Table 19.1.

TABLE 19.3 Policy Practice in Case Study

Indicators	Sor Por Kor and Inpang	Community Forest of Thailand	Krang Krachan National Park	Bukit Panjang Farmer's Group (Mutoliba et al., 2012)	PES in Lam Dong
Clarity on these polices	No statistics exists on policies and documents. But, data exist on uncompleted schemes undertaken locally				
Adaptive management	No scope; though inhabitants grip adaptive forest management rendering to prevailing social values	No scope	Farmers are allowed to plant local crops that meets company's deeds	PES been piloted with considerations to adaptive management	
Consistent with international commitment	Disclosing to international obligations part of this provision been assembled within national park	Conceivable concluded declaring conservation expanse	Yes, but local needs considered	Yes, local needs undertaken	
Extent to indigenous right	Have use and management rights with no sale rights	Probable to follow carbon rights if community bill validated	No scope	If volunteer carbon company does not renew contracts, farmers not allowed to use existing plants on their lands	Capacity building and orientation on social inclusion and indigenous rights undertaken but sustainability not envisioned
Conflict resolution mechanism	Minor conflicts managed locally. Encounters demanding government's consideration are baffling	No scope	No such considerations undertaken. Most of conflicts requiring governments' attention are unsolved		
Adequacy of budgets and resources	Sustainability issues	Consistent budgets	Sustainability is major issue after project phases out		

Agencies supportive	No harmonization among agencies	Not supportive;	Supportive; then again no thought on indigenous communities	Not supportive	MARD, Ministry of Finance, Ministry of Justice and other departments have good coordination
Available of information	Slow and less information flows, least availability in the rural areas	Indigenous residing within are beyond reach	Strong communication among institutions and stakeholders	Number of progress reviews and financial reviews considered	
Equity in rights and rents	Unbiased locally	No scope	Demands land ownership and dispute free lands	Poverty alleviation is a primary importance; focused on poor	
Competitiveness	No scope at policy level but unbiased locally		Does not consider such provisions	Payment based on performance	
Internalization of externality	No considerations		Yes, because associated ecosystem challenges incorporated	Yes	
Participation	Participation in operational planning no connexion in constitutional and policy planning	No scope	Participating in selecting what plants to crops, cropping and preserving the plants, training, and monitoring	Good participation from resource poor households	
Transparency	Local users are highly transparent on resource allocation and operational planning	No transparency mechanisms	Due to strong communications, transparency believed to be enhanced		

IV. LOOKING FORWARD

19.5.1 The Gap

Often what happens is, knowledge gained through research is over and over again not considered by knowledge practitioners in policy formations. Similarly, implicit familiarity from the local field seldom influences the investigators or those responsible for making decisions. Even more effective intermediates are required between policy and practice. This is prominent in Fig. 19.1, clearly showing diverse and distorted status in terms of existing forest polices in policy existence, legal clarity, and legalized support to adaptive management at the national level. However, through examinations from case studies in Table 19.3, it clearly shows gaps between policy provisions at the national level and those practiced at the local level. Nevertheless, those practiced at the local level appear more favorable toward REDD+ implications. Local people can put together a diverse range of understanding of a problem, undertake it through either rejection or improvement, and display perfect examples of evidence-informed decision making, which is one of the characteristics of REDD+ governance.

In terms of comparisons among these five case studies, the least policy application gaps are seen especially in the national park areas. This is because activities undertaken under the national park umbrella are replicas of national polices that do not consider the associated socioecological conditions prevailing. But in terms of actual REDD+ implementation, national parks cannot be suitable sites for piloting. However, it fails against criteria with good governance. There are a number of people residing inside the national park and their legitimate right to stay and have a successful livelihood is restricted. Though governments have identified national parks as the potential piloting site for REDD+ implementation, the real issues are not taken under consideration.

Indonesian, Vietnamese, and Thai forest-dependent groups are proficient at carrying out best practices that require bringing up in the planning process. Currently, more than 2 million indigenous people are residing in the vicinity of protected areas and their livelihood relies on these areas. Additionally, another 20 to 25 million individuals are believed to be living nearby national forest reserves using forest harvests both for consumption and sale (Witchawutipong, 2005). Livelihoods of these communities are snatched for conservation's sake. Continuous incidences of forest area encroachments in these three countries, together with politics over the endorsement of a community forest bill in Thailand, has raised questions about the feasibility of REDD+. The scheme of REDD+ necessitates communities, indigenous groups, and the government to work together.

PES lessons from Vietnam show full ownership is not required for PES programs, but indicators and criteria on which those participants were chosen may have important determinants. Some levels of benefits were given to the participants in addition to their regular income, while a fund for infrastructure development at the community level was created. But the payments made were lower than the opportunity costs rendered by participants during the scheme and consequently may obstruct sustainability of these schemes. The same implies with Indonesia. These conditions may have a negative effect on overall REDD+ implementation.

19.6 CONCLUSION

Forest polices are determined by the central government and then passed on to local governments for implementation; thereby neglecting the needs and problems of individuals. The supportive text is grounded in which REDD+ undertaken is filled with inconsistencies and anomalies. There are numerous rules, polices, and amendments that overlays and generates a state of misperception and makes circumstances confused. Overlying and intricacies can upsurge high transaction costs during the process of law enforcements while simultaneously necessitating additional examination and evidence. Furthermore, policies are impromptu with the least scope for an intricate adaptive learning method. There are often disagreements among forestry communal officials. Instead of formulating better management for conflict resolutions they linger in courts, whereby the resource poor opt out being unable to pay legal charges. This state is more intensified by coinciding of managerial tasks and unreliable government improvement programs. No decent governance structures exist in the forestry region that are noticeable from end to end by lack of equity, fairness, and public participation in forest resource distribution, especially in Thailand and Indonesia. Among the three countries—Thailand, Indonesia, and Vietnam—Vietnam's national forest legislation is more inclined toward good governance. In this regard, major changes in the national legislation of Thailand and Indonesia are required; especially at recognizing indigenous and carbon rights, formulating a better conflict resolution mechanism, adjusting clear national mandates, and providing better delivery of services while at the same time promoting locals in the decision-making process.

Government has formulated forest polices and implemented them in their own ways, but local stakeholders are unaware of it. This reflects a huge gap between policy and practice. Nevertheless, those practiced at the local level show more favorable attributes toward REDD+ implications. Local people have put together a diverse range of understanding of a problem and undertaken it through either rejection or improvement and displayed perfect examples of evidence-informed decision making, which is one of the characteristics of REDD+ governance. Vietnam can undertake REDD+ with some point of tenure arrangement as the benefit-sharing mechanisms and associated attributes have previously been successfully piloted. The national policies are equally contingent toward REDD+, while local communities are simultaneously managing resources accordingly. The fact that least policy practice gap observed, Vietnam can opt for minor changes in policies by slotting in field learning.

In Thailand, REDD+ mechanisms cannot be implemented successfully with the available provisions in the policy, especially through identifying national parks as one of the pilot areas for REDD+. In a way, rather than focusing on formulating new policies, an attempt should be made to monitor the effectiveness of existing policies and then capture essential local learning; smallholders are contributing to sustainable forest management. Serious workout is required in Indonesia; with the country's major focus on the economy, REDD+ cannot be successful. It will further marginalize the forest-dependent communities. These communities are helping to sequester carbon through plantations, restoring degraded land, and providing a wonderful opportunity for the REDD+ initiative.

Acknowledgment

This study is part of the project "Bridging Policy Practice Gap in the Effective Implementation of REDD+ Programs in SE Asia: Collaborative Learning among Indonesia, Thailand, and Vietnam," with financial and technical support from the Toyota Foundation from November 2013 to November 2014.

References

Butler, R.A., 2014. Despite moratorium, Indonesia now has world's highest deforestation rate. Mongabay.com. Retrieved online from http://news.mongabay.com/2014/0629-indonesia-highest-deforestation-rate.html.

Dinh, D.T., 2005. Forestry, poverty reduction and rural livelihoods in vietnam. Ministry of Agriculture and Rural Development, Hanoi, Vietnam.

Dovers, S., 2005. Environment and Sustainability Policy: Creation, Implementation, Evaluation. Federation Press, Sydney.

FAO, 2005. Global Forest Resource Assessment 2005: Progress towards sustainable forest management. Food and Agricultural Organization of the United Nations, Rome.

FAO., 2009. Vietnam Forestry Outlook Study. Food and Agricultural Organization of the United Nations. Regional Office for Asia and the Pacific. Retrieved from http://www.fao.org/docrep/014/am254e/am254e00.pdf.

FAO, 2010. Global Forest Resources Assessment 2010. Food and Agriculture Organization of the United Nations, Rome.

Fisher, R.J., Srimongkontip, S., Veer C., 1997. People and Forests in Asia and the Pacific: Situation and Prospects. Asia-Pacific Forestry Sector Outlook Study. Forestry Policy and Planning Division Rome and Regional Office for Asia and the Pacific, Bangkok. Retrieved from http://www.fao.org/docrep/w7732e/w7732e00.htm# Contents.

Hares, M., 2009. Forest conflict in Thailand: northern minorities in Focus. Environ. Manage. 43, 381–395.

Indrarto, G.B., Murharjanti, P., Khatarina, J., Pulungan, I., Ivalerina, J., Prana, N.N., Resodudarmo, I.A.P., Muhrrom, E., 2012. The context of REDD+ in Indonesia: drivers agents, and institutions. Center for International Forestry Research, Bogor, Indonesia (online). http://www.cifor.org/online-library/browse/view-publication/publication/3876.html.

Kartodihardjo, H., 2002. The Condition and Degradation of Indonesia's Forest. Retrieved from http://globetrotter.berkeley.edu/GreenGovernance/papers/Indonesiabckgrd.pdf.

Kishor, N., Rosenbaum, K., 2012. Assessing and Monitoring Forest Governance: A User's Guide to a Diagnostic Tool. Program on Forests (PROFOR), Washington, DC.

Larson, A.M., Petkova, E., 2011. An introduction to forest governance, people and REDD+ in Latin America: obstacles and opportunities. Forest 2011 (2), 86–111.

Mahdi, Shivakoti, G.P., Yonariza, 2015. Assessing Indonesian commitments and progress on emission reduction from forestry sector. In: Bajracharya, R.M., Sitaula, B.K., Sharma, S., Shrestha, H.L. (Eds.), Proceedings of the International Conference on Forests, Soils and Rural Livelihoods in a Changing Climate, Kathmandu, Nepal, p. 368.

Minang, P.A., Noordwijk, M.V., Duguma, L.A., Alemagi, Do, T.H., Benard, F., Agung, P., Robiglio, V., Catacutan, D., Suvanto, S., Armas, A., Aquad, C.S., Feudjio, M., Galudra, G., Maryani, R., White, D., Widayati, A., Kahurani, E., Namirembe, S., Leimona, B., 2014. REDD+ Readiness progress across countries: time for reconsideration. Clim. Pol. 14 (6).

Mutoliba, A., Yonariza, Mahdi, Hanung, I., 2012. Community based forest rehabilitation through incentive mechanisms (case study in nagari paninggahan solo district and nagari amang agam district west sumatra). In: International Conference on Development Management and Public Policy, Padang, Indonesia. 20–21 November 2014.

Nalampoon, A., 2003. National forest policy review, Thailand. In: Durst, P.B. (Ed.), An Overview of Forest Policies in Asia. FAO, Bangkok, pp. 293–311. Available online: ftp://ftp.fao.org/docrep/fao/005/AC921E/AC921E11.pdf.

Phelps, J., Webb, E.L., Agrawal, A., 2010. Does REDD+ threaten to recentralize forest governance. Science 328 (5976), 312–313. http://dx.doi.org/10.1126/science.118777.

Sharma, S., Shivakoti, G.P., Sakayarote, K., 2015. Analyzing Thailand forest policy practice gaps with emission reduction in retrospect. In: Bajracharya, R.M., Sitaula, B.K., Sharma, S., Shrestha, H.L. (Eds.), Proceedings of the International Conference on Forests, Soils and Rural Livelihoods in a Changing Climate, Kathmandu, Nepal, p. 368.

Soriaga, R., Mahanty, S., 2008. Strengthening local forest governance lessons on the policy-practice linkage from two programs to support community forestry in Asia. Int. J. Soc. For. 1 (2), 96–122. Retrieved from https://www.academia.edu/550834/Strengthening_Local_Forest_Governance_Lessons_on_the_PolicyPractice_Linkage_from_Two_Programs_to_Support_Community_Forestry_in_Asia.

Tebtebba, 2010. Indigenous people, forest and REDD+. State of Forest Forward, Tebtebba, Baguio City, Philippines.

Witchawutipong, J., 2005. Thailand Community Forestry. Royal Forest Department, Bangkok.

Yonariza, Shivakoti, G.P., Mahdi, 2015. The gap between policy and practice in Indonesia forest rehabilitation. In: Bajracharya, R.M., Sitaula, B.K., Sharma, S., Shrestha, H.L. (Eds.), Proceedings of the International Conference on Forests, Soils and Rural Livelihoods in a Changing Climate, Kathmandu, Nepal, p. 368.

CONCLUDING SECTION

Managing Dynamic Natural Resources in 21st Century in Asia

R. Ullah*, G. Shivakoti[†,§], Helmi[‡]

*The University of Agriculture, Peshawar, Pakistan †The University of Tokyo, Tokyo, Japan
‡Andalas University, Padang, Indonesia §Asian Institute of Technology, Bangkok, Thailand

20.1 INTRODUCTION

This chapter presents a summary of some key issues related to natural resources and the integrated natural resources management practices in Asia discussed in earlier chapters. The chapter starts with a summary of issues concerning natural resources in Asia, including deforestation and forest degradation, ecosystem degradation and biodiversity losses, conflicts, unsustainable use of natural resources, and issues related to rights, social security, and legal complexity. The integrated natural resources management practices in Asia, including land use (peatland agriculture in Indonesia and quality-good agricultural practices (Q-GAP) in Thailand), collaborative governance of forest resources, polycentric water governance, Reducing Emissions from Deforestation and Forest Degradation (REDD+) implementation, and community-based natural resources management are discussed in the second section of the chapter. Recommendations based on the findings of the studies discussed in the above chapters are provided at the end.

20.2 ISSUES CONCERNING NATURAL RESOURCES IN ASIA

20.2.1 Deforestation and Forest Degradation

The decentralization and authority devolution to district and regional governments in Indonesia increased deforestation through the number of forestry industrial permits granted without stopping to consider the socioecological consequences. The annual rate of deforestation and forest degradation in Indonesia increased from 700,000 ha in the mid-1980s to approximately 2.4 million in the beginning of 1999 (see, for example, Chapter 11). Among the major causes of deforestation in East Kalimantan, one of the most forest-rich provinces in

Indonesia, are excess utilization, illegal logging, and conversion of forest area for plantations (particularly for oil palm plantations) and coal mining industries. Owing to large-scale commercial logging and agroindustrial development, vast areas of Sumatra, particularly in the lowland forest in South and Central Sumatra, have also been deforested during the last decades. Forest conversions for agricultural activities both for food crops and plantation crops have adverse impacts on natural resources as well as for hydrological functions including landslides, loss of fertility due to erosion (on-site impacts), and sedimentation and water shortages in the dry season (off-site impacts). The water shortages ultimately lead to conflicts among water users. The implementation of decentralization and regional autonomy that was hoped would promote effective resource governance by empowering the local community in decision making and policy formulation and to create product sustainability has generally failed to produce the desired outcomes. Lack of transparency, less participation, low accountability, and weak coordination made the decentralized mode of resource management far from good governance principles. Moreover, local governments were not able to optimize their authority for increasing community welfare; their policies had even degraded the remaining forests. Financial policies for stimulating regional incomes to support local economic development have also aggravated forest and land governance. Under such a situation a good approach is to implement collaborative forest governance (CFG) where government and nongovernment institutions (including local communities) are believed to be key for successful resource management.

20.3 ECOSYSTEM DEGRADATION AND BIODIVERSITY LOSSES

Biodiversity is an important aspect in maintaining the functions of ecosystems. Each species is expected to play a certain and distinct function of greater or of minor importance within species communities. Biodiversity is also seen as a pool of genetic and molecular resources that can potentially be tapped for economic benefit by industries such as pharmaceutical firms; for agricultural breeding; as a genetic source of domestic, zoo, and pet animals or plants of botanical gardens; or as a source of scientific wonder and technical inspiration. Species extinctions often lead to changes in ecosystems with sometimes dramatic effects.

Aceh Province in Sumatra, Indonesia, is one of the biodiversity-rich provinces harboring vast tropical lowland, upland, and mountain forest of which a significant proportion is under formal protection within Leuser National Park and within the extended Leuser Ecosystem Conservation Area, as well as in other minor protected areas (see, for example, Chapter 13). The province also harbors the most extensive and intact rain forest ecosystem in Southeast Asia and many near-pristine coastal ecosystems. These forests represent one of the last and most important refuges for highly endangered Southeast Asian megafauna, such as the Sumatran rhinoceros, tiger, elephant, and orangutan, as well as other rare, endangered, and endemic animal and plant species. The coasts of Aceh Province harbor many natural treasures, such as species-rich coral reefs, sea grass beds, mangroves, and beach forests. The coral reefs of Weh Island on the northwestern tip of Sumatra are world-famous for scuba diving, whereby whale sharks and manta rays are occasionally sighted in the surrounding waters. The Banyak Islands archipelago in the south, with its extensive sea grass beds, harbor some of the last habitats and breeding sites of dugongs (sea cows) and sea turtles around Sumatra.

Extensive tropical peat swamp forests are found in the southern lowlands of Aceh Province. These ecosystems are not only rich in unique fauna and flora; they also represent formidable carbon stores and may therefore play a significant role for the mitigation of global warming.

With Aceh's low population density (81 persons per km^2) it was expected the resources would be sufficient to support a healthy and prospering population if the resources are efficiently managed and the living forests, with all their unique and rare wildlife representing an ever more precious asset, must not be squandered carelessly. However, this does not seem certain. Many resources are increasingly exploited in often unnecessarily destructive ways. The forests of Aceh are increasingly under pressure from encroachment through illegal logging, mining, and wildlife poaching.

Among the most threatened species are the Sumatran rhinoceroses, tigers, elephants, and orangutans that are classified by the International Union for Conservation of Nature (IUCN) as critically endangered (see Chapter 13). The Sumatran rhinoceroses are threatened by habitat loss, but the main threat is posed from poaching, particularly for their horns which are highly priced in Chinese medicine. Sumatran tigers, like the Sumatran rhinoceroses, occur from lowland forests up to mountain forests and are also under threat of extinction. The population of the Sumatran elephant has declined by at least 80% over the last 75 years and it was classified as a "critically endangered" species by the IUCN in 2011. The Sumatran orangutan has become rare in many parts of its former range and is now listed as "critically endangered" by the IUCN. There is a need for significant conservation efforts focusing on these species to avoid their entire extinction. Clear legislative frameworks for forest protection (in the optimal case sufficiently supported and understood by local communities) would be fundamental for effective and longer-term sustainable conservation.

20.4 CONFLICTS

Owing to the contradictory dual goals of government and community for preservation and utilization of protected zones in Indonesia and Thailand, conflicts between governments and communities are arising. Estate plantations effect/enlarge the struggles among the administration and the industrial logging concessionaries or conflicts among government and individuals that adversely affect the forest as the basic needs of local communities induce them to break rules and engage in haphazard harvesting of the natural resources. Conflicts are also arising from the industrial concessions for high-value timber logging and profitable estate plantations. The lack of an effective conflict resolution mechanism and the lack of policy interventions to settle land conflicts locally resulted in widespread conflicts in Vietnam that erupted into physical damages (see, for example, Chapter 19).

There is serious competition and conflict over land allocated for large-scale, commercially oriented concessions and upland peasant farming. There are also serious issues of inequality within upland peasant societies for those who are engaged in multiple opportunities provided by large-scale agricultural and forestry concessions. The manifestations of this inequality are clear: land conflicts between peasants and the state over large-scale forestry concessions, plantations, and other allocations of state land for forest conservation and protection, mining, and oil palm plantations; a high rate of land degradation; and a considerable proportion of households living below the poverty line. At small-scale, local levels, pressure on the resources has increased and fueled conflicts over natural resources as a means of production for the market as well as a means for subsistence (see Chapter 6). The political and economic value of natural resource property rights has also increased interethnic tensions and struggles.

Members of various ethnic groups are involved in struggles over natural resources as a consequence of spontaneous and planned transmigration. Although the interethnic conflicts in Indonesia cannot be reduced to economic conflicts, and economic conflicts cannot be reduced to conflicts over natural resources only, it is nevertheless clear that such conflicts do play an important role, and may continue to do so in the future. Conflicts between local communities and state agencies and with private investors who derive their exploitation rights from government concessions and licenses have also increased. Moreover, the allocation of profits from natural resources exploitation put the relationship between central and regional government under immense stress. Such struggles for control over natural resources are increasingly connected with struggles for self-determination or political autonomy.

The hydrological pattern in the watershed is affected by the land use and land cover changes (LULCC) of forest to agricultural system creating conflicts among water users. Moreover, farmers at the upstream of the watershed create erosion due to lack of conservation practices on their agricultural system (see, for example, Chapter 17). As off-site impacts of uplanders activities, some springs are dead and sedimentation in the river and irrigation system is high. Those lead to water shortages for farmers downstream and create conflict between farmers and other actors.

20.5 UNSUSTAINABLE USE OF NATURAL RESOURCES

The overexploitation and unsustainable use of natural resources in Indonesia has changed the quantity, quality, and distribution of the natural capital. Deforestation is not always followed by reforestation. Many agricultural areas have been subject to overapplication of chemical fertilizers. In many areas the industrial plants and mining operations are increasingly polluting land and water resources. The stress upon natural resources has also been exacerbated by the improvement of infrastructure and extraction techniques and expanding markets that create additional opportunities for extraction by other than concession holders and local populations. In various ways, the quality of land, water, and forests, and soil fertility is threatened, the renewability of resources is no longer guaranteed, and depletion of nonrenewable resources is approaching at great speed. As a result, the resource base, both for economic development and for social security purposes, is contracting and severely weakening and the negative economic and ecological externalities are passed on to future generations.

20.6 RIGHTS, SOCIAL SECURITY, AND LEGAL COMPLEXITY

20.6.1 Land Rights

According to Agrarian Law no. 5/1960, the Indonesian National Land Agency define and classify land as public (state) land or private land and allocate land for large-scale plantations including rubber, coffee, and tobacco. On the other hand, the Ministry of Forestry (MoF) claims that the prevailing forestry legislation (Law no. 5/1967 and Law no. 41/1999) classifies two-thirds of Indonesia's total land area as state forest areas, and that this land is, therefore, under the ministry's jurisdiction (see, for example, Chapter 9). There is a conflict of land tenure arrangements implemented under these two sets of laws resulting in the two concerned agencies being locked in competition over the control of vast areas of land. Almost 50 million Indonesian peasants live in and around these state forests and cultivate land under conditions of unclear and insecure land tenure. Access of local communities to land and natural resources is restricted through regulations that do not allow the cutting of trees, hunting, cultivating land, or house construction. In several cases, a local community's access to land and natural resources has been completely terminated by leasing rights over the land and resources to private sector companies within production forests, or by classifying the area as protected forests.

The MoF initiated collective community forestry stewardship program, *Hutan Kemasyarakatan* (HKm), under the state forest areas to provide upland farmers with limited access to land and resources and a village forest stewardship program, *Hutan Desa* (HD), since 2007. However, land tenure security under both of the programs does not permit residential settlements in these areas and they are exclusively designed for timber-based farming or forest protection and do not include food-based agriculture and indigenous mixed farming. The indigenous cultural community forests are an alternative tenure option for communal forest ownership; however, they involve complex administrative requirements that are difficult for the intended beneficiaries to fulfill.

After a long struggle between the forest agency, land agency, private sector, local governments, and peasant movements, upland peasant communities have succeeded in obtaining individual land rights within converted forest areas under the public land redistribution policy. The MoF allowed the implementation of land reform through small-scale state forest land redistribution to tillers in densely populated areas in Java and Sumatra to minimize food insecurity and reduce pressure for land demands by landless peasants. However, due to the effect of open competition and surplus appropriation by other classes, significant numbers of landless peasant households appeared a few years after the land redistribution, which reflects the unequal land distribution, and those who controlled more land under informal tenure before the land redistribution received more formal individual land title. The distribution of land that was expected to address the problems of tenure security, productive farm investment, access to credit, and so on, has only proved valid for some of the peasant households. The main beneficiaries of the program were the landlord class and the absentee landowners from the cities. The lower socioeconomic class, the near landless, and some of the middle-class peasant households were dispossessed and became landless only a few years after the land was redistributed.

20.6.2 Water Rights

Despite the presence of a formal water rights framework in the Philippines, courtesy of Spanish and American colonial law, conflicts over water allocation still exist and the framework is of little help in defining water allocations. This indicates that the issue is not simply the formal presence of water rights, but instead the extent to which such resource rights become an effective part of water allocation and dispute resolution. Simply establishing a legal framework, or issuing formal rights, may have little impact, unless the rights become a meaningful part of how claims to water are disputed and defended (see Chapter 4).

20.6.3 Social Security and Legal Complexity

The concept of neediness and social security are closely related to rights to natural resources. The general distribution and availability of natural resources and other forms of wealth in a community are important for social security; however, the actual mix of sources of wealth each individual has is equally important for the overall situation of social security. Because of resource competition the social security conditions pose serious problems at all levels of social organization. Resource competition may even take place between different organizational mechanisms with a social security function. Similar competition for resource allocation also occurs within villages and families. Most social relationships through which goods and services for social security are transferred are multifunctional. State social security policies and programs are also generally multifunctional and depend on multifunctional relationships. Education and migration are two important strategies of social security. Migration shapes the relationships of social security in complex ways. Families sacrifice considerable amounts of resources to make migration possible, often causing considerable hardship at first, softened only by the expectation of a greater security for the future.

The coexistence of government law, *adat* law, and religious law, and also new forms of local legal regulations along with international law and conventions that have introduced human rights issues, regulate environmental issues, access and exploitation rights, and with respect to the legal status and political and economic rights of local communities and indigenous people have come to play an increasingly important role. The complex constellations of law, high extent of legal insecurity, and legal pluralism have aggravated the problems. The legal status and area of action of these subsystems have always been dynamic and contested and their role for socioeconomic organization has changed with the economic and political power of their proponents (see Chapter 6).

20.7 INTEGRATED NATURAL RESOURCE MANAGEMENT

20.7.1 Ostrom's Models for Effective Management of Common Pool Resources in Asia

Key issues of inquiry in managing common pool resources (CPRs) include the formation and operation of institutions, how they change over time, and their influence on behavior in society. Individuals involved in CPR management are strongly motivated in solving common problems to enhance their own productivity as they gain a major part of their economic return from CPRs. The cooperative governance of CPRs, therefore, could be more effective in formulating and enforcing rules than governance by either a government agency or a corporation. There is a need for institutions that can weave linkages across hydrological and administrative boundaries, facilitating the sharing of information and collective action, without mandating a single form. Exchange of ideas within epistemic communities may be more influential than formal structures, and create shared ideas and contacts that then facilitate working together to address shared concerns.

The Institutional Analysis and Development framework and design principles developed by Elinor Ostrom have become the most widely used research framework for robust performance and management of CPRs. The design principles have wider applicability in analyzing governance of natural resources in Asia. Findings indicate that the design principles are applicable in irrigation management in Nepal and Thailand; however, the level of applicability of each design principle varies from rarely applicable to fully applicable depending on the local context and other factors across the two countries. These design principles can help analyze the robustness of irrigation, forestry, fisheries, and coastal resources; however, it is advisable to consider (i) rephrasing the definition of local condition, including social and economic context, and (ii) adding an enforcement mechanism to ensure rule compliance. This will help to provide wider applicability of these design principles in specific conditions.

The Ostrom's model of polycentric governance is based heavily on their experience in California where institutional circumstances enabled and facilitated self-organization through mechanisms that are generally missing or weak in Asia. These mechanisms include legal authority for specialized governmental units such as irrigation districts, courts as forums for resolving disputes and giving legal force to agreements that bred new institutions, and financial mechanisms that enabled substantial investments by irrigators' organizations and other water management authorities. Polycentric governance is possible in Asia, but it takes different forms and may develop in diverse ways through institutional improvisation, including adaptive comanagement in irrigation and weaving network governance and practical problem-solving in river basins. They seem to be heavily shaped by subsidy politics and by the goals and instruments of broader national policies toward rural areas and environmental protection. Nevertheless, there is scope for farmers, village residents, and other stakeholders to influence the discourse and direction of change, including the development of polycentric water governance through which they may have a measure of autonomy, and opportunities to work together with others in pursuing their aims.

The social-ecological system (SES) is another framework proposed by Elinor Ostrom that, with some modifications, applies to a well-defined domain of CPR management systems where resource users (actors) harvest resources from the resource system and in turn provide maintenance of the resource system under the rules set out by the governance system in the context of related ecological systems and broader sociopolitical-socioeconomic settings. In Ostrom's initial work on the SES framework, the processes of extraction and maintenance were recognized as the most important forms of interaction and outcomes and were located in the center of this framework. However, it is not clear how CPR theory can be applied to large-scale resource systems as most of the previous studies have provided conflicting opinions on the use of CPR in large-scale resource systems and urged the need for reinterpretation of the meaning of CPR theories.

20.7.2 Collaborative Governance of Forest Resources

Although 70% of the authority has been handed over to local government in Indonesia under regional autonomy, forestry performance has not improved. Extensive development of the oil palm and coal mining industries in particular and other land uses in general make the forestry sector suffer from wider disturbances. The main factors affecting the infectivity of forest governance under the regional autonomy are (1) it is the lowest level of authority, which led to a wide span of control for a better policy and administrative implementation; (2) forestry has been made the least priority by the local government and even in many districts it has been merged with its competitors (crop estate and mining sectors); and (3) lack of sufficient capacity, time, space, and financial supports of the local forestry governed services. In addition, the indirect cause of ineffective forest governance under regional autonomy is the more attractive offers for nonforestry investments, particularly oil palm and coal mining for local income generation. This not only attracted local governments but also the local communities who find it difficult to absolutely depend on forests and forest products for their livelihoods. These complex problems of forest management under limited human resource capital and insufficient financial capital cannot be solved solely by the local government without any support from all related actors, particularly the existing forest concession holders and local communities in collaborative ways.

CFG is organized through collaboration among various stakeholders who have a range of interests in the use and management of local resources. Participation of government and nongovernment parties with a common purpose of sustaining forest functions and its benefits should be followed by authority sharing, responsibilities, and indeed benefits based on the roles of each parties. At the policy level, appropriate arrangements should be assigned to allow a sufficient degree of local autonomy in the use and management of local forestlands and the resources therein. While at the field implementation level it is imperative to identify relevant stakeholders who must get involved in the collaboration, particularly the local communities whose livelihood depends on the forests and in most cases also have traditional rights over the lands. For appropriate CFG arrangements, clarity and understanding of customary land tenure and the local communities' rights over the forestlands are crucial.

According to Indonesian policy the whole forested areas will be totally divided into Forest Management Units (FMUs) or in Indonesian *Kesatuan Pemangkuan Hutan* (KPH) under three categories based on the most dominant area, such as KPHP (production FMU); KPHL (protection FMU); and KPHK (conservation FMU). That understanding means it is possible to have different functions under one category, for instance under KPHP it is possible to have protection and/or conservation. Moreover, different large- and small-scale utilization and use rights schemes including timber concessionaires, mining industries, and others, may also be found in one KPH (including CBFM). An area without use rights can be directly managed by KPH independently or under partnership management with the local communities. The main functional tasks of FMUs include performing all elements of efficient forest management (production, protection, conservation, rehabilitation, etc.), implementation of national and regional forest polices, conducting managerial aspects (planning, organizing, actuating, controlling, etc.) and monitoring and evaluating forest management. The advantages of KPHs for promoting good forest cogovernance in Indonesia include (1) resolution of resource conflicts, an important element of collaborative management, through intensive contacts and communications and building mutual understanding with other stakeholders; (2) debureaucratization for achieving low-cost economy (KPHs located on-site and equipped with professional foresters guarantee better forest management and more optimum services for forest users); and (3) socioeconomic facilitation of local institutions. Giving authority to

KPHs as on-site management units assures direct and indirect solutions to problems related to profit sharing of large-scale forest enterprises (see, for example, Chapter 11).

In South Sumatra, the local government has developed a multistakeholder approach, based on the establishment of a multistakeholder forum (MSF) that serves as an umbrella organization for various cross-institutions stakeholders, including government, nongovernmental organizations (NGOs), local community organizations, private companies, and academia. The main objective of a MSF is to provide a foundation for learning and joint action. The MSF actively contributed in the reduction of fire risk and its impacts through increased coordination and consultation among MSF members, proposing policies and helping to initiate the establishment of new functions for selected district agencies (see Chapter 10).

20.7.3 Polycentric Water Governance

The potential for standardized models, such as water users' associations (WUAs), irrigation management transfer, and river basin organizations, to lead dramatic reforms seems limited in Asia. These have been tried in most parts of Southeast Asia but there seems little reason to expect big breakthroughs from trying more of the same reforms and policies. The historical pathways through which water resources development has occurred, and in particular the strong role of centralized hydraulic bureaucracies, has entrenched particular patterns of interests that are unlikely to change easily or quickly, especially if change is primarily driven by international funding agencies and not supported by irrigation agencies. The main challenge related to water governance in Asia is to find workable solutions that will be more achievable if they include a broader range of alternatives that are politically feasible and can use a variety of mechanisms. Water management needs not only knowledge from engineering, but also from public health, economics, political science, and other academic disciplines. The need for integration has justified efforts to reform policies and develop suitably comprehensive plans and institutional arrangements.

Concepts of polycentric water governance can contribute to research and policy in Asia; however, it needs to consider some important institutional aspects such as the role of village governments, national intervention in basin management, the strength of hydraulic bureaucracies, and subsidy politics. Though there may be functional equivalents or pathways, mechanisms enabling polycentric governance are either absent or inefficient in Asia. These mechanisms include special districts, court-authorized agreements, water rights, and bonds. Moreover, without the ability to make major capital investments and hire the services of engineers and professionals, the scope for greater involvement of farmers in polycentric governance is limited. There is also a need to move from merely orientation to achieving the social missions toward a social entrepreneurship orientation as institutions only concerned with social mission without strategy to support the achievement of the social mission would not be sustainable.

Polycentric water government in Asia faces many challenges. In important ways governance is already at least partially polycentric, with a variety of stakeholders and organizations involved, linked at different scales and in different sectors. The flows of information through larger epistemic communities frame the discourse and define problems and acceptable solutions with patterns of network governance. In contrast to assumption of top-down hierarchical command and control, the concept of polycentric governance figures out the potential for alternative ways of organization. Some space for change still exists; however, dramatic shifts are not easily achieved as the historic pathways have entrenched a very large role for national hydraulic bureaucracies.

20.7.4 REDD+ Implementation

REDD+ is one of the most discussed topics in the international forestry arena, and huge investments are directed toward an international climate change forum to prepare countries for future REDD+. In Southeast Asia, Thailand, Indonesia, and Vietnam are countries planning to implement REDD+, whereby the preparation includes formulating strategies to overcome deforestation and forest degradation issues with numerous pilot studies and fund mobilization. REDD+ requires an effective benefit-sharing mechanism, often dominated by good governance. It is crucial to recognize the existing legal provisions on which the countries are planning REDD+ and their relative efficiency of existing policies to future emission reduction program implementation. Acknowledging the current governance status will aid in recognizing areas that need improvements.

The constitution of Thailand, on one hand, gives communities the right to defend their civilizations and to manage, conserve, and use natural environmental resources and biodiversity in a sustainable way. On the other hand, forests are managed by the state in Thailand. This creates debatable and complicated issues on how this complexity works for communities to manage. In Indonesia the challenges and ambiguities of indigenous rights and tenure arrangements in the state forest remain unsolved. Though the country has various provisions at a policy level, these provisions are neglected in reality, creating a policy and practice gap.

Vietnam has relatively better policy, an institutional arrangement essential for successful REDD+ implementation, and relatively good provisions for addressing forest-dependent communities residing in the vicinity of protected areas. The relatively better institutional framework in Vietnam is attributed to their experience with the Payment for Environmental Services (PES) program, which is achieved through treating the public equitably. All forest areas in Vietnam are state-owned and are then allocated to households and organization for short-term and long-term benefits. Though Vietnam has a relative advantage over Thailand and Indonesia in forest governance, there are individual issues within Vietnam; for example, local stakeholders have no ownership and they only have the rights to use and manage.

All three countries have relevant polices that may aid in successful REDD+ implementations through some modification. Unclear tenure arrangements have been identified as major causes for deforestation; therefore, countries need to focus on land tenure and ownership arrangements. Effective and confident tenure may defend deprived forest tenants and local communities from marginalization or even from being thrown out of forest areas, and provide better leverage in the country's REDD+ progressions as REDD+ raises the worth of forests. With strong tenure it will be inflexible for administrations or other influential actors to gain the welfare of REDD+.

Provided the conflicting ideologies between the economy and emission reduction are resolved, REDD+ is likely to be functioning provided that it reimburses management and entities for the opportunity costs suffered due to disappointment of economic inducements that REDD+ substitutes. Regional government is likely to devise REDD+ if it is relatively advantageous economically. The benefit-sharing mechanism for REDD+ should be effective and could be considered as a common understanding among related stakeholders in terms of circulation of benefits and it should provide equitable recompense to communities for their interventions in emission reduction. The scheme of REDD+ requires communities, indigenous groups, and the government to work together; therefore, there is a need for orientation programs at the community and national levels for involving the community in the decision-making route in Indonesia and Thailand (see, for example, Chapter 19).

20.7.5 Community-Based Natural Resource Management

Societal attempts to handle environmental management challenges rely on the power of a particular way of social organization. The community has emerged as a distinct way of human organization in environmental management in a variety of forms from place-based groups to interest-based coalitions. Owing to the more recent challenges of climate change and poverty, expectations from communities are rising to new levels and across new dimensions. Community-based solutions to environmental issues rest on a number of expected socioenvironmental benefits: ensuring effective environmental stewardship, creating collective action to address social and environmental issues together, absorbing risks and creating safety nets for the vulnerable members, strengthening human capacity to adapt to climate risks, and providing economic benefits to the poor through social enterprises. However, the potential of the community to drive environmental management is often left unfulfilled. Development agencies, state organizations, and even the dominant market players use community-based management as a discursive weapon to legitimize other strategic actions. Moreover, communities are undergoing rapid changes in form, functionality, and boundaries of geography and interests due to technological advancement, social and cultural dynamics, rapid urbanization, exposure to an increasing array of risks, and changes to how localized communities are networked. These dynamics are, however, ignored by current environmental and resource management policy approaches and rely on the simplistic, place-based, largely homogeneous and localized view of community.

20.7.5.1 Community-Based Forest Resources Management

In Asia development annexation and state formation on policies has classified forest as a public good and allow poorer management privileges to forest users. The usage and management choices of these forests are usually determined by the central government and then passed to regional, provincial, and then local agencies. In Indonesia, a number of people who had previously settled and practiced agriculture within the areas that became reserve areas were categorized as illegal forest encroachers and were forced to relocate from those protected areas.

In response to the restrictions put forward by the centralized module of forest management, countries in Asia are devolving powers to local administration, particularly for small forest patches. Vietnam's Land Law of 1993, Thailand's Tambon Administrative Act of 1994, and the Regional Autonomy Law of 1999 in Indonesia are some of the initial provisions foreseeing forest management at the local level.

A set of community forestry initiatives have been designed and implemented in Indonesia over the past 15 years, allowing local community control of forest management, and safeguarding the nation's diminishing primary forests. However, these initiatives have had mixed results in terms of contributing to sustained economic development. Despite a degree of tension between forest agencies and forest-dependent people over the control of forests, local approaches to forest management are increasingly regarded as a promising way to manage forests. The active involvement of farmer forest groups (FFGs) appears to be a key component of effective CBFM in Indonesia. The village-based groups of farmers (smallholders) are brought together by government forest agencies to facilitate reforestation programs. These FFGs are growing in confidence to initiate and expand their interactive (two-way) networks to include government agencies, NGOs, market brokers, and regional timber processors. Local community's interest in forestry are guided by the interests of group members who, in turn, seek information and partnerships with a diverse and expanded network. In Vietnam, since the introduction of Doi Moi, forests are managed through a forest allocation process that is expected to have greater community participation in planning and implementation. A community forest bill to allot management rights to local communities to manage resources according to their needs is under contemplation and reflection in the parliament in Thailand. However, due to the major contribution of the forestry sector to gross domestic product, the process on community rights reflects the unwillingness of the government to devolve power to communities and undermine stakeholder participation.

20.7.5.2 Community-Based Water Resources Management

Historic and current trends maintain and even strengthen the role of village authorities in irrigation and water management in Asia. Access to project benefits is often sufficient to create the appearance of hydraulically based WUAs. However, after project completion these tend to disappear and may represent fascia under which communities continue to manage irrigation in old ways or mix components of old and new institutions.

The local community has created irrigation institutions at the Karya Mandiri Irrigation System (KMIS) following the principles proposed by Elinor Ostrom. These institutions empowered the local community to adapt to the pressures and changes, make necessary investments, perform various management functions, and made the system continue to exist as a self-organizing irrigation system serving the farmers. The stakeholders at KMIS have developed social entrepreneurship principles, in addition to the eight design principles of the Ostrom framework, consisting of two main aspects: (i) provision of irrigation services to earn money out from their activities and to distribute the benefits through cofinancing of the infrastructure's development and (ii) mechanisms to guarantee stable revenue from irrigation services provision, irrigation water availability at the farm level, development of a planting schedule, assistance in land preparation, provision of agricultural inputs, and agriculture technology transfer (see, for example, Chapter 8).

In Indonesia village water masters, *ulu-ulu*, had been a standard part of village government in Java since the time of colonial administration. The formal position of *ulu-ulu* was eliminated by the reforms to village administration in 1979; however. villages usually continued the practice and often put *ulu-ulu* in the more general position of head of economic and development activities, *kaurekbang*, or continued to elect and supervise the work of *ulu-ulu* informally (see, for example, Chapter 4). The WUAs along hydraulic lines are set up by the development projects in Indonesia. In most parts of Java, irrigation systems are developed following village boundaries, reducing the need for a cross-cutting pattern of governance. Since the mid-1970s, Thai governments routed funds through subdistricts for development of rural infrastructure, which progressed from council of village head to Tambon Administrative Organization (TAO) with substantial offices, staff, and budgets. These funds are utilized for repairing and improving weirs, water tanks, and other small-scale water resources infrastructure, strengthened by the availability of equipment and political causes. This increased the opportunities for villages and subdistricts to control more funds and have more influence over the decision about construction and use of water resources. In Vietnam, the agricultural cooperatives managed irrigation during the period of collectivized agriculture. The irrigation teams not only controlled canals but also water deliveries and water levels in individual fields. Even after decollectivization, cooperatives continued to play a significant role in many areas where gravity irrigation and drainage were important. Corporate villages historically played a significant autonomous role in local governance in northern and central Vietnam, where village elites had a leading role in managing their internal affairs. Cooperatives functioned under the guidance of commune government, and satisfactory performance of leadership roles in cooperatives was a path to becoming a commune leader. These cooperatives were formally restructured in the 1990s to be more democratic and accountable, and they continue to be active in irrigation and to work closely with commune governments. The presence of these cooperatives and the strength and interest of commune and cooperative leaders, lack of established institutional arrangements for intercommune coordination, and the interest of district governments in retaining their roles usually hamper the efforts to develop higher WUA along the hydraulic lines of canals or transfer irrigation management at wider scales.

20.7.6 Land-Use Change

Under the current trend of land demand in Yogyakarta, Indonesia, future projections for future land use indicates that the wet agricultural land will be reduced to less than the required area in 2030 due to land conversion (see, for example, Chapter 5). The driving factors of land-use change, particularly agricultural land conversion to nonagricultural land uses are numerous, including distance to city, distance to road, population density, elevation, terrain slope, irrigation availability, land tenure, and land suitability for rice cultivation. Besides these factors there are some other factors that affect decision making; these factors include revenue from farming activity, socioeconomic status of household, access to land-related regulation, sustainability of household farming, perception about farmland protection, and land tenure. Farmland loss due to land conversion affects rice production, food security, and food self-sufficiency in the region. Owing to a higher demand for services and jobs, land conversion is almost unavoidable. Future land-use projections for 2030 urge policies with the objective to preserve productive farmland and urban expansion in less important farmland. In this way the land-use change can be controlled to maintain the productive function of farmland and the nonagricultural function of other land uses.

20.7.6.1 Peatland Agriculture in Indonesia

Agricultural production has gained importance recently because of climate change and food security. The exponential increase in the world's population leads to an increased demand for food. Although productivity of marginal lands such as peatland and upland is low, they can still be utilized for producing food to cater to the food needs of an increasing population. To support a transmigration program and increasing national rice production, the government of Indonesia supported the development of peatlands agriculture on the islands of Sumatra, Kalimantan, and Papua. These peatlands are capable of producing food crops including rice, corn, soybeans, cassava, and other horticultural crops on shallow peat having relatively higher fertility and lower environmental risk compared to deep peat.

Utilizing peatlands for the production of food crops has greater environmental consequences. Infertile land and greenhouse gas (GHG) emissions are issues requiring attention. There are various strategies for sustainable peatland agriculture, including improvement of peatlands productivity by water management, amelioration and fertilization while GHG emissions in the peatland agriculture can be reduced by controlling the water table in the agricultural practices, and amelioration and preventing burning in land preparation. Implementation of such strategies will greatly improve the role of peatlands for supporting food security and in reducing GHG in Indonesia. Farmers' income and welfare are also important factors in enhancing their accessibility to food and sustainable peatland agriculture. The plantation farming system yields more income than the food crop farming system as the export commodity produced during plantation has a high price in international markets. The combination between plantation and food crop is a better option to improve farmer welfare as well as to maintain household food security. Degraded peatlands caused by the Mega Rice Project (MRP) in Central Kalimantan should be reforested and rehabilitated to get intended benefits to the local farmers in terms of food security and also reducing GHG emissions as a target of the government of Indonesia. Sustainable peatlands management aims to optimize the function of peatlands to support an improvement of the welfare of farmers and efforts to reduce GHG emissions without compromising the rights of future generations to meet their needs. The food crop area and the plantation area should be set up in appropriate proportion to optimize the benefit and maximize the reduction in GHG emission (see, for example, Chapter 15).

20.7.6.2 *Good Agricultural Practices: Q-GAP Thailand*

Throughout Southeast Asia, government agencies are promoting good agricultural practices (GAP) standards for improving food safety, avoiding potential nontariff barriers to exports, and ultimately securing farmers' health and revenues. Several Asian countries have recently introduced public standards of GAP such as IndoGAP (Indonesia), VietGAP (Vietnam), PhilGAP (Philippines), SALM (Malaysia), and Q-GAP (Thailand). They all aim at increasing the supply of safe and high-quality food by promoting more sustainable crop production practices, but they were also set up in response to pressures from the export market (see Chapter 18).

Rice being an important staple food crop and a source of livelihood for farmers in Asia and, in the cases of Myanmar, Vietnam, and Thailand, rice is also an important export earner. Introducing GAP for rice is important from a food safety point of view. Thai government agencies have been quite successful in fostering the participation in the Rice Q-GAP certification program. As of 2012, the Rice Q-GAP program had already been promoted in 72 of the 76 provinces of Thailand, and approximately 40,000 rice farmers had registered. Information channels used by farmers and membership in farmer's groups are having an important effect on the probability of Q-GAP adoption.

20.8 RECOMMENDATIONS

1. Governments in Asia should guarantee land allocation for all peasant households with maximum land under the control of peasant household units. This is important to limit accumulation by rich peasant households. A special effort is needed to emphasize women and youth as they are more innovative and will continue farming the land in the future.
2. Under the new act (Act no. 23/2014) on regional governance, almost all authority in the forestry and mining sector has been withdrawn from district governments and moved over to provincial government, causing the district government to be the least interested party in supporting any program of the central government. Therefore, anticipated actions have to be taken immediately to prevent wasting hard efforts and energy, and incurring expensive costs and losing significant progress in these programs.
3. Benefit-sharing mechanisms and associated attributes have previously been successfully piloted in Vietnam. There is a need for some point of tenure arrangement to implement REDD+ in Vietnam. However, in Thailand attempts should be made to monitor the effectiveness of existing policies, rather than focusing on new policies formulation, and then capture essential local learning. In Indonesia, the forest communities are helping to reduce carbon emissions through plantations and restoring degraded land and making a conducive environment for REDD+ implementation. However, keeping in view that the country's major focus is on the economy and economic development, REDD+ cannot be successfully implemented and requires some serious workout.
4. The land demand for nonagricultural use and the higher farmland conversion rate are threatening and raising issues of food availability. Combined land-use options should be in place to control land-use change and to manage the remaining farmland. Promoting farmer households' understanding of how to avoid converting their farmland and reduce nonagricultural land demands is important. Strategies for developing building that require less land and zoning for residential or industrial areas will be beneficial to suppress competition with agricultural land use.
5. Incentives to empower farmers in providing higher revenue from farming activities might be more effective to discourage further land conversion. These incentives may include subsidized farming inputs, better financial capital access, guaranteed income from farming activity, high-yielding farming technology, and reduced land tax. Providing these incentives will effectively sustain farming activity maintenance of farmland. Moreover, increasing land productivity to produce sufficient food within the reduced farmland area requires high-yielding varieties and related farming technologies.

Index

Note: Page numbers followed by *f* indicate figures *t* indicate tables, and *np* indicate footnotes.

Printed in the United States
By Bookmasters